DIGITAL HOLOGRAPHY AND DIGITAL IMAGE PROCESSING:
Principles, Methods, Algorithms

DIGITAL HOLOGRAPHY AND DIGITAL IMAGE PROCESSING:
Principles, Methods, Algorithms

by

Leonid Yaroslavsky
Tel Aviv University, Israel

SPRINGER SCIENCE+BUSINESS MEDIA, LLC

Library of Congress Cataloging-in-Publication

CIP info or:

Title: Digital Holography and Digital Image Processing: Principles, Methods, Algorithms
Author (s): Leonid Yaroslavsky
DOI 10.1007/978-1-4757-4988-5

Originally published by Kluwer Academic Publishers in 2004
MyCopy version of the original edition 2004

Printed on acid-free paper.
www.springer.com/mycopy

TABLE OF CONTENTS

Chapter 1

INTRODUCTION

1.1 DIGITAL HOLOGRAPHY AND EVOLUTION OF IMAGING TECHNIQUES

The history of science is, to a considerable degree, the history of invention, development and perfecting imaging methods and devices. Modern science began with the invention and application of optical telescope and microscope in the beginning of 17-th century by Galileo Galilei (1564-1642), Anthony Leeuwenhoek (1632-1723) and Robert Hook (1635-1703).

The next decisive stage was invention of photography in the first half of the 19-th century. Photographic plates were the major means in discoveries of X-rays by Wilhelm Conrad Roentgen (1845-1923, The Nobel Prize Laureate, 1901) and radio activity by Antoine Henri Becquerel (1852-1908, The Nobel Prize Laureate, 1903) at the end of 19-th century. These discoveries, in their turn, almost immediately gave birth to new imaging techniques: to X-ray imaging and radiography.

X-rays were discovered by W.C. Roentgen in experiments with cathode rays. Cathode rays were discovered by Julius Plücker (1801-1868) in 1859 who used vacuum tubes invented in 1855 by German inventor Heinrich Geissler (1815-1879). These tubes, as modified by Sir William Crookes (1832-1919) led eventually to the discovery of the electron and finally brought about the development of electronic television and electron microscopy in 30-th - 40-th of 20-th century. Designer of the first electron microscope E. Ruska (1906-1988) was awarded The Nobel Prize in Physics for 1986. This award was shared with G. Binning and H. Rohrer, who were awarded for their design of the scanning tunneling microscope.

The discovery of diffraction of X-rays by Max von Laue (1889-1960, the Nobel Prize Laureate, 1914) in the beginning of 20-th century marked the advent of a new imaging technique, imaging in transform domain. Although Von Laue's motivation was not creating a new imaging technique but rather proving the wave nature of X-rays, lauegrams had very soon become the main imaging tool in crystallography. Lauegrams are not conventional images for visual observation. However, using lauegrams, one could numerically reconstruct the spatial structure of atoms in crystals. This shoot gave its crop in less then half a century. One of the most remarkable scientific achievements of 20-th century is based on X-ray crystallography. It is discovery by J. Watson and F. Crick of spiral structure of DNA (the Nobel Prize, 1953). And about the same time a whole bunch of such new transform domain imaging methods had appeared: holography, synthetic aperture radar, coded aperture imaging, tomography. Two of these inventions were awarded the Nobel Prize: D. Gabor for "his invention and developments of the holographic method" (the Nobel Prize in Physics, 1971) and A. M. Cormack and G. N. Hounsfield for "the development of computer assisted tomography" (the Nobel Prize in Physiology and Medicine, 1979).

Denis Gabor invented holography in 1948 ([1]). This is what D. Gabor wrote in his Nobel Lecture ([2]) about the development of holography: " Around 1955 holography went into a long hibernation. The revival came suddenly and explosively in 1963, with the publication of first successful laser hologram by Emmett N. Leith and Juris Upatnieks of the University of Michigan, An Arbor ([3]). Their success was due not only to the laser, but to the long theoretical preparation of Emmett Leith (in the field of the "side looking radar") which started in 1955... Another important development in holography [happened] in 1962, just before the "holography explosion". Soviet physicist Yu. N. Denisyuk published an important paper ([4]) in which he combined holography with the ingenious method of photography in natural colors, for which Gabriel Lippman received the Nobel Prize in 1908".

Denis Gabor received his Nobel Prize in 1971. The same year, a paper "Digital holography" was published in Proceedings of IEEE by T. Huang ([5]). This paper marked the next step in the development of holography, the use of digital computers for reconstructing, generating and simulating wave fields, and reviewed pioneer accomplishments in this field. These accomplishments prompted a burst of research and publications in early and mid 70-th. At that time, most of the main ideas of digital holography were suggested and tested ([6-16]). Numerous potential applications of digital holography such as fabricating computer generated diffractive optical elements and spatial filters for optical information processing, 3-D holographic displays and holographic television, holographic computer vision stimulated a great enthusiasm among researchers.

However, limited speed and memory capacity of computers available at that time, absence of electronic means and media for sensing and recording optical holograms hampered implementation of these potentials. In 1980-th digital holography went into a sort of hibernation similarly to what happened to holography in 1950-th - 1960-th. With an advent, in the end of 1990-th, of the new generation of high speed microprocessors, high resolution electronic optical sensors and liquid crystal displays, of a technology for fabricating micro lens and mirror arrays digital holography is getting a new wind. Digital holography tasks that required hours and days of computer time in 1970-th can now be solved in almost "real" time for tiny fractions of seconds. Optical holograms can now be directly sensed by high resolution photo electronic sensors and fed into computers in "real" time with no need for any wet photo-chemical processing. Micro lens and mirror arrays promise a breakthrough in the means for recording computer generated holograms and creating holographic displays. Recent flow of publications in digital holographic metrology and microscopy indicate revival of digital holography from the hibernation[§].

The development of optical holography, one of the most remarkable inventions of the XX-th century, was driven by clear understanding of information nature of optics and holography ([17-19]). The information nature of optics and holography is especially distinctly seen in digital holography. Wave field recorded in the form of a hologram in optical, radio frequency or acoustic holography, in digital holography is represented by a digital signal that carries the wave field information deprived of its physical casing. With digital holography and with incorporating digital computers into optical information systems, information optics has reached its maturity.

The most substantial advantage of digital computers as compared with analog electronic and optical information processing devices is that no hardware modifications are necessary to solving different tasks. With the same hardware, one can build an arbitrary problem solver by simply selecting or designing an appropriate code for the computer. This feature makes digital computers also an ideal vehicle for processing optical signals adaptively since, with the help of computers, they can adapt rapidly and easily to varying signals, tasks and end user requirements. In addition, acquiring and processing quantitative data contained in optical signals, and connecting optical systems to other informational systems and networks is most natural when data are represented and handled in a digital form. In the same way as monies are the general equivalent in economics, digital signals are the general equivalent in information handling. Thanks to its universal

[§] Curiously enough that the initial meaning of digital holography is getting forgotten and, in many recent publications, digital holography is associated only with numerical reconstruction of holograms .

nature, the digital signal is an ideal means for integrating different informational systems.

This is not a coincidence that digital holography appeared in the end of 60-th, the same period of time to which digital image processing can be dated back to. In the same way, in a certain sense, as inventing by Ch. Towns, G. Basov and A. Prokhorov lasers in mid 1950-th (the Nobel Prize in Physics, 1964) stimulated development of holography, two events stimulated digital holography and digital image processing: beginning of industrial production of computers in 1960-th and introducing Fast Fourier Transform algorithm made by J. W. Cooley and J. M. Tukey in 1965 ([21]).

Digital holography and digital image processing are twins. They share common origin, common theoretical base, common methods and algorithms. This is the purpose of the present book to describe these common principles, methods and algorithms for senior-level undergraduate and graduate students, researchers and engineers in optics, photonics, opto-electronics and electronic engineering.

1.2 CONTENTS OF THIS BOOK

The theoretical base for digital holography and image processing is signal theory. Ch. 2 introduces basic concepts of the signal theory and mathematical models that are used for describing optical signals and imaging systems. The most important models are integral transforms. All major integral transforms, their properties and interrelations are reviewed: convolution integral; Fourier transform and such its derivatives as Cosine, Hartley, Hankel, Mellinn transforms; Fresnel transform; Hilbert transform; Radon and Abel transforms; wavelet transforms; sliding window transforms. In the last section of the chapter, stochastic signal transformations and corresponding statistical models are introduced.

One of the most fundamental problems of digital holography and, more generally, of integrating digital computers and analog optics is that of adequate representation of optical signal and transformations in digital computers. Solving this problem requires consistent account for computer-to-optics interface and for computational complexity issues. These problems are treated in Chs. 3 , for signals, and in Ch.4, for signal transforms.

In Ch. 3, general signal digitization principles and concepts of signal discretization and element-wise quantization as a practical way to implement the digitization are introduced. Signal discretization is treated as signal expansion over a set of discretization basis functions and classes of shift, scale and combined shift-scale basis function are introduced to describe traditional signal sampling and other image discretization techniques implemented in coded aperture imaging, synthetic aperture imaging, computer and MRI tomography. Naturally, image sampling theory, including 1-D and 2-D sampling theorems and analysis of sampling artifacts, is treated at the greatest length. Treatment of element-wise quantization is mostly focused on the compander - expander optimal non-uniform quantization method that is less covered in the literature then Max-Lloyd numerical optimization of the quantization. A separate section is devoted to peculiarities of signal quantization in digital holography. The chapter concludes with a review of image compression methods. As this subject is very well covered in the literature, only basic principles of data compression and a classification and brief review of the methods are offered.

Ch. 4 provides a comprehensive exposure of discrete representation of signal transforms in computers. In digital holography, computers should be considered as an integral part of optical systems. This requires observing mutual correspondence principle between discrete signal transforms in computers and analog transforms in optical system that is formulated in Sect. 4.1. On the base of this principle, discrete representations of the convolution integral, Fourier and Fresnel integral transforms are developed in Sects. 4.2 - 4.4. Conventional Discrete Fourier and Discrete Fresnel Transforms are represented here in a more general form that takes into consideration that image or hologram sampling and reconstruction devices can be placed in optical setups with arbitrary shifts with respect to their optical axes. This requires introducing into discrete Fourier and Fresnel transforms arbitrary shift parameters. The presence of these parameters provides to modified in this way transforms some new useful features such as flexibility of image resampling using Discrete Fourier Transform (DFT) and allows to treat numerous modifications of DFT and, primarily, Discrete Cosine Transform (DCT), in the most natural way as special cases of DFT for different shift parameters and different type of signal symmetry.

A critical issue in digital holography and image processing is the computational complexity of the processing. The availability of fast computation algorithm determine feasibility of their practical applications. These issues are treated in Chs. 5 and 6.

Ch. 5 describes efficient computational algorithms for image and hologram filtering in signal domain. Described are separable, recursive, parallel and cascade implementations of digital filters; new implementation of signal convolution in DCT domain that is practically free of boundary effects of the traditional cyclic convolution with DFT; recursive filtering in sliding window in the domain of DCT and other transforms; combined algorithms of 1-D and 2-D DFT and DCT for real valued signals.

Ch. 6 exposes basic principles of the so called fast transform of which Fast Fourier Transform is the most known special case. The fast transforms are treated here in the most general matrix form that allows to formalize their design and provides a very compact and general formulas for the fast algorithms. From the practical point of view, pruned algorithms and quantized DFT described in this chapter may represent the most interest for the reader.

Many phenomena in holography and, generally, in imaging are treated as random, stochastic. In addition, one of the task of digital holography is statistical simulation of holographic and imaging processes. Hence, the next chapter, Ch. 7, deals with statistical methods and algorithms. It covers basically all involved issues, from measuring signal statistical characteristics for different statistical models of signals to building digital statistical models and to generating arrays of pseudo-random numbers with prescribed statistical characteristics. Practical applications are represented in this chapter by methods for generating correlated phase masks and diffusers and by demonstration of statistical simulation for studying speckle noise phenomena in imaging systems that use coherent radiation.

The primary task of processing images and holograms after they are digitized and put into computer is usually correcting distortions in imaging and holographic systems that prevent images from being perfect for the end user. We call this task sensor signal perfection. In image processing, the term image restoration is commonly accepted to designate this task. Image reconstruction in holography and tomography may also be regarded a similar task. Methods for sensor signal perfection, image restoration and image reconstruction are discussed in Ch. 8. The methods assume a canonical imaging system model, described in Sect. 8.1, that consist of a cascade of units that perform signal linear transformation, nonlinear point-wise transformation and stochastic transformation such as adding to the signal signal independent noise. As a theoretical bench mark, the design of optimal linear filters is used. The filters correct signal distortions in the linear filtering units of the system model such as image blur, and suppress additive signal independent noise in such a way as to minimize root mean square restoration error measured over either a set of images in an image data base or image ensemble or even over a particular individual image. In the latter case, filters are adaptive as their parameters depend on the particular image to which they are applied. In view of the implementation issues, the filters are designed and work in the domain of orthogonal transforms that can be computed with fast transform algorithms. The design of such filters is discussed in Sect. 8.2. In Sect.8.3, this approach is extended to local adaptive filters that work in transform domain of a window sliding over the image and, in each window position, produce filtered value of the window central pixel. Illustrative examples are presented that demonstrate edge preserving noise suppression capability of local adaptive filters. In particular, it is shown that such filters can be efficiently used for filtering signal dependent noise, such as speckle noise. Sect. 8.4 extends this approach further to the design of optimal linear filters for multi component images such as color, multi spectral or multi modality images.

In real application, it very frequently happens . Very that noise randomly replaces signal values that in this case are completely lost. In such situation, the model of impulse noise is used. Filtering impulse noise, in general, requires nonlinear filtering. Efficient and fast impulse noise filtering algorithms that use simple linear filter and point wise threshold detector of pixels distorted by impulse noise are described in Sect. 8.5. Other filters for filtering impulse noise are described in Ch. 12.

All above mentioned filtering methods are aimed at correcting signal distortions attributed to linear filtering and stochastic transformation units of the imaging system model. Sect. 8.6 deals with processing methods for correcting image and hologram gray scale distortions in the point-wise nonlinear transformation unit.

In a broad sense, image restoration can be treated as applying to the distorted signal a transformation inverse to that that caused the distortions provided that the inverse transformation is appropriately modified to allow for random interferences and other distortions that are always present in the signals. This is how image reconstruction in holography and tomography is usually carried out. The modification of the inverse transformation to allow for random interferences and distortions is implemented as pre-processing of holograms or, correspondingly, image projections before applying the inverse transformation and post-processing after the reconstruction transformation. This technology is illustrated in Sect. 8.7.

As a rule, in image processing applications, end user needs, for solving his particular tasks, apply to images some additional processing to ease visual image analysis. Such a processing is usually called image enhancement. Image enhancement methods are reviewed and illustrated in Sect. 8.8.

In digital holography and digital image processing real images and holograms are represented in computers as arrays of their samples obtained with one or another sampling device in its certain setting. Meanwhile it is very frequently necessary to resample images to obtain, from available array of samples, samples located in position other then those of available samples. The resampling assumes interpolation between available data. Many interpolation methods for sampled data are known. Among them, discrete sinc-interpolation that satisfies discrete sampling theorem described in Ch. 4, has certain advantages. Ch. 9 focuses on properties and efficient computational algorithms for discrete sinc-interpolation and on using it for image zooming, rotation, differentiating, integrating, polar-to-Cartesian coordinate conversion and tomographic reconstruction.

A very important applied problem of image processing is image parameter estimation, and, in particular, localization of objects in images. Methods for solving this problem are discussed in Chs. 10 and 11.

In Ch. 10, a basic mathematical model of observation with additive sensor noise is formulated and applied to the design of optimal device for localization of a target object with the highest possible accuracy and reliability. Analytical formulas are also obtained that characterize potential accuracy and reliability of target localization for single and multi-component images and non-correlated and correlated sensor noise.

Ch. 11 focuses on the problem of reliable localization of a target object in clutter images when the localization is hindered by the presence in images non-target background objects. The main issue in solving this problem is how to reliably discriminate target and non-target objects. In the chapter, a localization device is considered that consists of a linear filter and a unit that finds coordinates of the signal highest maximum at the filter output. The linear filter is optimized so as to secure the highest ratio of the filter response to the target object in its location to standard deviation of the filter response to background non-target image component. This problem is solved for exactly known target objects and for objects that are known to the accuracy of their certain parameters such as, for instance, scale and rotation angle, for spatially homogeneous and inhomogeneous images. The solution called optimal adaptive correlator is extended to the case of localization in color and multi component images. Some practical implication of this solution and its implementation in nonlinear optical and opto-electronic correlators are also discussed and illustrated.

An important place in the arsenal of methods for image denoising, enhancement and segmentation belong to nonlinear filters. Quite a number of nonlinear image processing filters have been reported in the literature. In the Ch. 12, these filters are classified and represented in a unified way in terms of fundamental notions of pixel neighborhood and of a standardized set of estimation operations. The chapter also provides and illustrates practical examples of several unconventional filters for image denoising, enhancement, edge detection and segmentation. In conclusion, possible implementation of the filters in neuro-morphic parallel networks are briefly discussed.

The last, 13-th chapter is devoted to the problem specific for digital holography proper, the digital-to-analog conversion problem of encoding computer generated holograms for recording on physical optical media and to analysis of distortions in reconstructed images associated with encoding methods. As computer generated holograms and optical elements are still recorded using means that are not directly intended for this purpose, there is no unique encoding method. In the chapter, the most simple and known methods are described and analyzed.

In all chapters, exposition is extensively supported by diagrams, graphical and picture illustrations as the author shares the well known Chinese saying that a picture costs more than a thousand words.

REFERENCES

1. D. Gabor, Microscopy by reconstructed waveforms, **1**, *Proc. Royal Society*, A197, 454-487, (1949)
2. http://www.nobel.se/physics/laureates/1971/gabor-lecture.html
3. E. N. Leith, J. Upatnieks, New techniques in wavefront reconstruction, *JOSA*, 51, 1469-1473, (1961)
4. Yu. N. Denisyuk, Photographic reconstruction of the optical properties of an object in its own scattered radiation field, *Dokl. Akad. Nauk SSSR*, **144**,1275-1279 (1962)
5. T. Huang, "Digital Holography", *Proc. of IEEE*, 59, 1335-1346 (1971)
6. B. R. Brown, A. Lohmann, Complex spatial filtering woth binary masks, *Appl. Optics*, **5**, No. 6, 967-969 (1966)
7. T. S. Huang, B. Prasada, Considerations on the generation and processing of holograms by digital computers, *MIT/RLE Quar. Prog. Rep.* 81, Apr. 15, pp. 199-205 (1966)
8. A. W. Lohmann, D. Paris, Binary Fraunhofer holograms generated by computer, *Appl. Optics*, 6, No. 10, 1739-1748 (1967)th. Comp., v. 19, 1965, pp. 297-301
9. J. W. Goodman, R. W. Lawrence, Digital image formation from electronically detected holograms, *Appl. Phys. Lett.*, **11**, pp. 77-79, Aug. 1 (1967)
10. L. B. Lesem, P. M. Hirsch, J. A. Jordan, Kinoform, IBM Journ. Res. Dev., 13, 150, 1969
11. M. .A. Kronrod, N. S. Merzlyakov, L. P. Yaroslavsky, Computer Synthesis of Transparency Holograms, *Soviet Physics-Technical Physics*, v. 13, 1972, p. 414 - 418.
12. M.A. Kronrod, N.S. Merzlyakov, L.P. Yaroslavsky, Reconstruction of a Hologram with a Computer, *Soviet Physics-Technical Physics*, v. 17, no. 2, 1972, p. 419 - 420.
13. L. P. Yaroslavskii, N.S. Merzlyakov, *Methods of Digital Holography*, Consultans Bureau, New York, London (1980) (Russian Edition: L. Yaroslavsky, N. Merzlyakov, *Metody tsifrovoy golographii*, M., Nauka, Moscow, 1977)
14. Wai-Hon Lee, Computer Generated Holograms: Techniques and Applications, In: E. Wolf, Ed., *Progress in Optics*, v. XVI, North Holland, pp. 121-230, (1978)
15. Dallas W. J., Digital Holography, in: Frieden, Ed., *Computers in Optical Research, Topics in Applied Physics*, v. 41, Springer, Berlin, pp.291-366 (1980).
16. O. Bryngdahl, F. Wyrowski, Digital Holography - Computer -generated Holograms, in: E. Wolf, *Progress in Optics*, XXVIII, Elsevier Science Publishers B. V., pp. 3-86 (1990)
17. D. Gabor, Theory of communication, *Journ. Inst. Electr. Eng.*, vol. 93, 429, (1946)
18. D. Gabor, Communication Theory and Physics, *Phil. Mag.*, vol. 41, No. 7, p.1161 (1950)
19. D. Gabor, Light and information, in: *Progress in Optics*, **1**, ed. By E. Wolf, Amsterdam, pp. 109-153, (1961)
20. F. T. S. Yu, *Optics and Information Theory*, Robert E. Krieger Publ. Co, Malabar, Flodira, (1989).
21. J. W. Cooley, J. M. Tukey, An Algorithm for Machine Calculation of Complex Fourier Series, Math.Comp.,v. 19,1965, pp. 297-301

Chapter 2

OPTICAL SIGNALS AND TRANSFORMS

2.1 MATHEMATICAL MODELS OF OPTICAL SIGNALS

2.1.1 Primary definitions and classification

Signals and signal processing methods are treated mathematically through mathematical models. The very basic model is that of a mathematical function. For instance, optical signals are regarded as functions that specify relationship between physical parameters of wave fields such as intensity, and phase and parameters of the physical space such as spatial coordinates and/or of time. Fig. 2.1.1 provides classification of signals as mathematical functions and associated terminology.

Examples of scalar optical signals are monochrome gray scale images. Examples of vectorial (multi-component) signals are two component signals that represent, for instance, orthogonal components of polarized electromagnetic wave fields or left/right stereoscopic images, three component color images represented by their red, green and blue components and multi spectral images that may contain more than three components. Examples of 1-D signals are audio signals or physical measurements as functions of time. Images are modeled as 2-D functions of co-ordinates in image plane while video can be regarded as 3-D functions of image plane co-ordinates and time. Volumetric data are functions of three

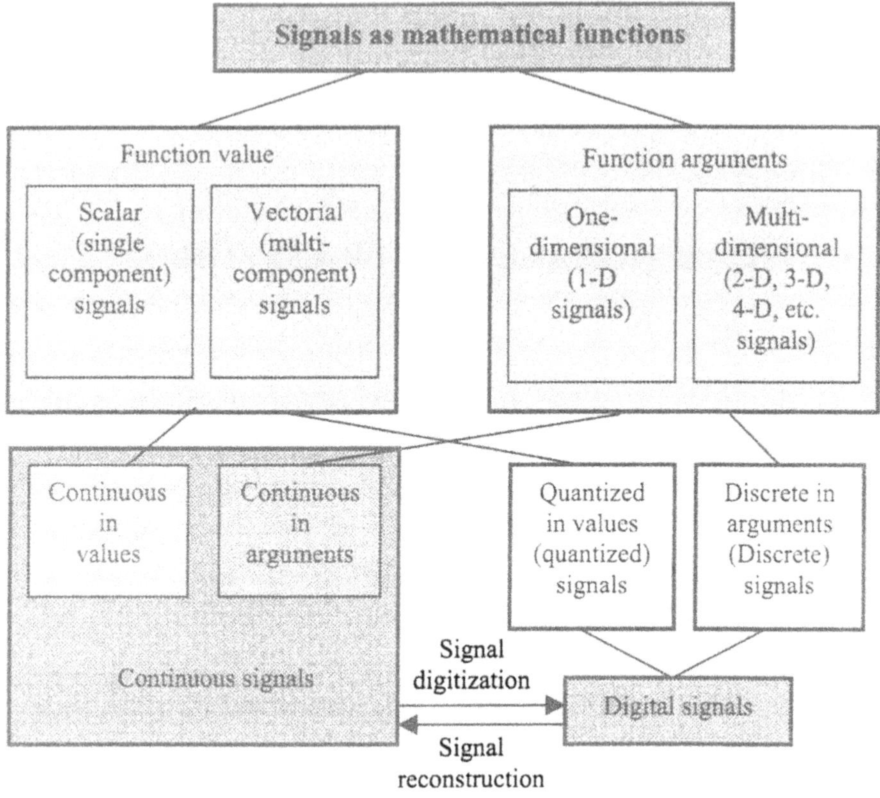

Figure 2-1. Classification of signals as mathematical functions

space co-ordinates. Continuous optical spectrum color images can also be treated as 3-D functions of wavelength and co-ordinates in the image plane.

Discrete signals are signals specified as a list (sequence) of values. Individual entries of this list are called ***signal elements*** or ***signal samples***. ***Quantized signals*** are those that take values from a finite set of possible values (***quantization levels***). Optical signals that carry information on optical properties of real physical objects are ***continuous signals***. They are modeled as continuous functions both in terms of their values and their arguments. Continuous signals are frequently called ***analog signals*** for they are regarded as analogs of the corresponding physical objects.

At the other end of the classification we find ***digital signals***, those that one deals with in computers. In general, each digital signal can be regarded as a single number. In practice, digital signals are represented as a sequence of quantized numbers that may be considered as digits of this single number and are called digital ***signal samples***. In this sense digital signals can be regarded as being both discrete and quantized. For digital signal processing one should convert analog signals into the corresponding digital ones. We will refer to this conversion as ***signal digitization***. Signal digitization always, explicitly or implicitly, assumes that inverse conversion of the digital signal into the corresponding analog signal is possible. We will refer to this digital-to-analog conversion as ***signal reconstruction***.

Throughout the book, we designate continuous signals by lower case roman letters $\boldsymbol{a},\boldsymbol{b}$, etc., and their arguments by $\boldsymbol{x}$. Multi-component signals will be designated by vectors of their components $\boldsymbol{a} = \{\boldsymbol{a}_c\}, \boldsymbol{c} = \mathbf{1,2,...,}\boldsymbol{C}$. For multidimensional continuous signals, argument $\boldsymbol{x}$ is a vector variable with components $\boldsymbol{x} = \{\boldsymbol{x}_d\}, \boldsymbol{d} = \mathbf{1,2,...,}\boldsymbol{D}$, where $\boldsymbol{D}$ is the signal dimensionality.

2.1.2 Signal space

In the signal theory it is very convenient to give signals a geometric interpretation by regarding them as points in some functional space - the ***signal space***. In such a treatment, each signal is represented by a point in the signal space and each point of the signal space corresponds to a certain signal (Fig. 2.1.2). Signal space of continuous signals is a continuum.

Using the notion of the signal space, one can easily portray a general the relationship between continuous and digital signals. Let us split signal space into a set of sub-spaces, or cells that cover the entire space and assign, to each cell, its index number. As a result, each index number will correspond to all signals within the corresponding cell and, for each signal, one can find the index number of the cell to which the signal belongs.

Suppose now that all cells are formed in such a way that all signals within one cell can be regarded identical or indistinguishable for a given application. We will call such cells ***equivalence cells***. In each equivalence cell, one can select a signal that will serve as a ***representative signal*** of all signals within this cell. In such a treatment, conversion of a continuous signal into a corresponding digital signal (analog-to-digital conversion) is assigning to the continuous signal the index number of the equivalence cell to which it belongs. We will call this process ***general digitization***. Inverse process of converting digital signal into analog one (signal reconstruction) is then finding, for each index number (digital signal), a representative signal of the corresponding equivalence cell. In practice, partition of the signal space into a set of equivalence cells is "hard-wired" in analog-to-digital converters and assigning representative signal to a number is "hard-wired" in digital-to-analog converters.

The design of the partition of signal space into a set of equivalence cells is only one of many signal processing tasks that require describing mathematically distinctions between signals. To this goal, the concept of space metrics is introduced which treats distinctions between signals as a distance between the corresponding points in the space.

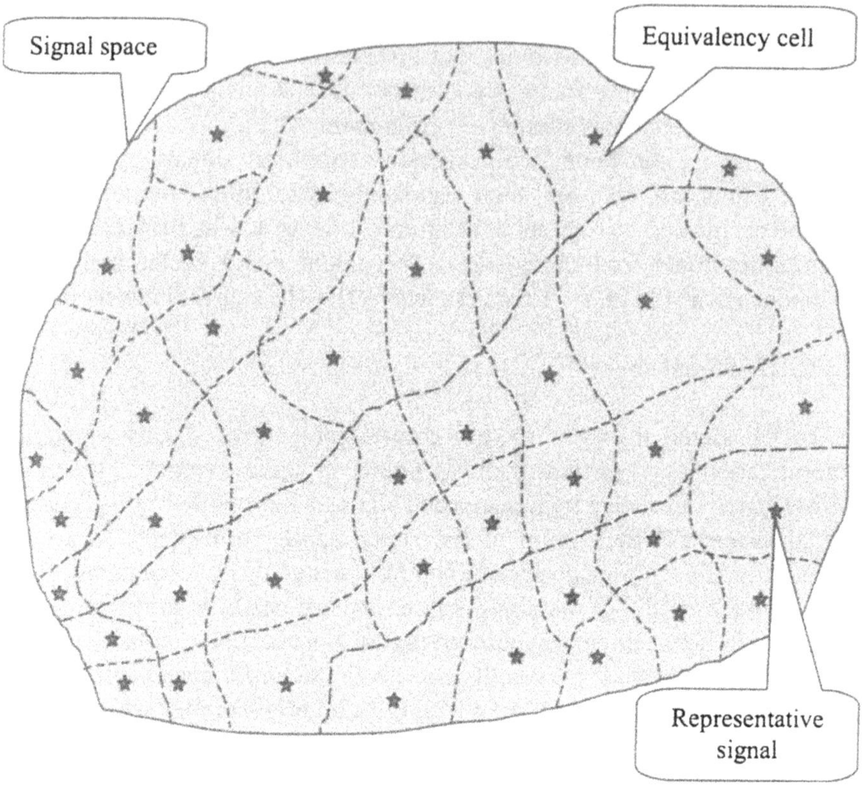

Figure 2-2. Signal space, equivalency cells and representative signals

The ***signal space metrics*** associates with each pair of signals in the space, say, $a^{(1)}$and $a^{(2)}$, a nonnegative real number $d\left(a^{(1)},a^{(2)}\right)$. The method for obtaining this number normally has the following properties:

$$d(a,a)=0\,;\; d\left(a^{(1)},a^{(2)}\right)\geq 0\,; \tag{2.1.1}$$

$$d\left(a^{(1)},a^{(2)}\right)=d\left(a^{(2)},a^{(1)}\right); \tag{2.1.2}$$

$$d\left(a^{(1)},a^{(3)}\right)\leq d\left(a^{(1)},a^{(2)}\right)+d\left(a^{(1)},a^{(3)}\right) \tag{2.1.3}$$

The meaning of the first condition is obvious. The second condition is equally natural, although there are metrics for which this condition does not hold. The idea behind the third condition, referred to as the "triangle rule", formally expresses the following natural requirement of a metrics: if two points are close, in the metrics, to a third point, they must be close to each other. Examples of the metric most commonly used in the theory of signals are presented in the Table 2.1 (second column)

Table 2-1. Typical signal space metrics

Signal type	Designation	Deterministic definition	Statistical definition AV is a statistical averaging operator
Discrete signals	L_N	$\sum_{k=0}^{N-1}\left\|a_k^{(1)}-a_k^{(2)}\right\|$	$AV\left\{\left\|a_k^{(1)}-a_k^{(2)}\right\|\right\}$
	L_N^2	$\sum_{k=0}^{N-1}\left\|a_k^{(1)}-a_k^{(2)}\right\|^2$	$AV\left(\left\|a_k^{(1)}-a_k^{(2)}\right\|^2\right)$
	L_N^P	$\sum_{k=0}^{N-1}\left\|a_k^{(1)}-a_k^{(2)}\right\|^P$	$AV\left(\left\|a_k^{(1)}-a_k^{(2)}\right\|^P\right)$
	LMh_N	$\sum_{k=0}^{N-1}W_k^2\left\|a_k^{(1)}-a_k^{(2)}\right\|^2$ ($\{W_k^2\}$ are scalar weight coefficients)	-
	M_N	$\max_k\left\|a_k^{(1)}-a_k^{(2)}\right\|$	-
Continuous signals	L_x	$\int_X\left\|a_1(x)-a_2(x)\right\|dx$	$AV\left\{\left\|a_1(x)-a_2(x)\right\|\right\}$
	L_X^2	$\int_X\left\|a_1(x)-a_2(x)\right\|^2dx$	$AV\left(\left\|a_1(x)-a_2(x)\right\|^2\right)$
	L_X^P	$\int_X\left\|a_1(x)-a_2(x)\right\|^Pdx$	$AV\left(\left\|a_1(x)-a_2(x)\right\|^P\right)$
	M_X	$\sup_X\left(\left\|a_1(x)-a_2(x)\right\|\right)$, where $\sup(a(x))$ - minimal value that is not exceeded by $a(x)$	-

In many signal and image processing applications, distinctions between signals have to be evaluated statistically, or on average over a set, or ensemble, of signals. This is the case, for instance, when one needs to evaluate performance of a signal processing device or of an algorithm to be used for a variety of signals from a certain class or from a certain data base. The relevant performance measure should characterize the performance for the whole class of signals. Therefore, one or another method of averaging distance measure evaluated for every particular signal of the class is required as it is shown in Table 2.1 (right column).

In order to specify the way of statistical averaging one should introduce characteristics of signal ensemble that are conventionally referred to as ***signal statistical characteristics***. The primary statistical characteristic is probability of signals in the signal space. It defines a relative contribution of signals into the averaging. For continuous signals, probability measure is defined as ***probability density***. For determination of signal space statistical characteristics, mathematical models of signals as ***random processes*** are used. For the theory of random signals readers are referred to dedicated textbooks and monographs such as Refs. [1-4].

From the table one can see that any distance measure involves two items:

- Loss function (we will denote it as ***LOSS***(.,.)), that evaluates numerically the distinctions between the individual signal samples (for discrete signals) or their values for particular values of its arguments (for continuous signals). For instance, for $\boldsymbol{L_N}$ metric, loss functions is defined as an absolute value of the difference between signal samples.
- Method for the averaging loss function values (for instance, by summation over the signal extent or by statistical averaging over a signal ensemble as it is shown in the table).

In image processing it is frequently more adequate to average the loss function over both area of images from an image set and over all images in the set. It is also useful sometimes to weight loss function values with weights that depend on the signal arguments (an example of such weighting one can see in Table 2.1 in the definition of metric $\boldsymbol{LMh_N}$) or other signal parameters. This leads to a family of "local" metrics (***local criteria***) which, for discrete signals, are defined as

$$d\left(a^{(1)},a^{(2)},k\right)=\mathbf{AV}\left\{\sum_n LOC\left(k,n,a_n^{(1)}\right)LOSS\left(a_n^{(1)},a_n^{(2)}\right)\right\}, \tag{2.1.4}$$

where $\mathbf{AV}\{\cdot\}$ designates the averaging over image data set. Local criteria assume evaluation of signal distinctions "locally" around any particular point $\boldsymbol{k}$ of argument of signal's function. Applications of local criteria to the design and substantiating adaptive linear and nonlinear filters will be given in Chapts. 8 and 12 .

2.1.3 Linear signal space, basis functions and signal representation as expansion over a set of basis functions

An important feature of physical signals such as optical ones is that they can be added, subtracted and amplified or attenuated. Mathematically, this is modeled by the property of linearity of the signal space. The signal space is called linear if:

1. An operation of signal summation $(+)$ is defined such that, for any two signals $\boldsymbol{a}^{(1)}$ and $\boldsymbol{a}^{(2)}$, a third signal $\boldsymbol{a}^{(3)} = \boldsymbol{a}^{(1)} + \boldsymbol{a}^{(2)}$ can be uniquely found.
2. The summation operation obeys the commutative and associative laws:
$$\boldsymbol{a}^{(1)} + \boldsymbol{a}^{(2)} = \boldsymbol{a}^{(2)} + \boldsymbol{a}^{(1)}$$
$$\boldsymbol{a}^{(1)} + \boldsymbol{a}^{(2)} + \boldsymbol{a}^{(3)} = \left(\boldsymbol{a}^{(1)} + \boldsymbol{a}^{(2)}\right) + \boldsymbol{a}^{(3)} = \boldsymbol{a}^{(1)} + \left(\boldsymbol{a}^{(2)} + \boldsymbol{a}^{(3)}\right)$$
3. There exists a "zero-signal" $\boldsymbol{\emptyset}$ such that $\boldsymbol{a} + \boldsymbol{\emptyset} = \mathbf{a}$ for all signals of the space.
4. Each element $\boldsymbol{a}$ of the space may be put into the correspondence with a unique opposite signal $(-\boldsymbol{a})$ such that
$$\boldsymbol{a} + (-\boldsymbol{a}) = \boldsymbol{a} - \boldsymbol{a} = \boldsymbol{\emptyset}.$$
5. An operation $(\cdot)$ of multiplying signal by a scalar number is defined such that, for any number α and any signal $\boldsymbol{a}$, a signal $\alpha \cdot \boldsymbol{a}$ can be uniquely found.
6. The following properties hold for the multiplication operation:
$\alpha_1\alpha_2\boldsymbol{a} = \alpha_1(\alpha_2\boldsymbol{a}) = \alpha_2(\alpha_1\boldsymbol{a})$;
$(\alpha_1 + \alpha_2)\boldsymbol{a} = \alpha_1\boldsymbol{a} + \alpha_2\boldsymbol{a}$; $\alpha(\boldsymbol{a}_1 + \boldsymbol{a}_2) = \alpha\,\boldsymbol{a}_1 + \alpha\,\boldsymbol{a}_2$;
$\mathbf{1} \cdot \boldsymbol{a} = \boldsymbol{a}$; $\mathbf{0} \cdot \boldsymbol{a} = \boldsymbol{\emptyset}$;

Usually operations of signal addition and signal multiplication by scalars are treated as the customary operations of point-wise (element-wise) arithmetic summation and multiplication of signal values and the "zero-signal" is considered to be a signal whose values are equal to zero for all its argument values.

In linear signal space, one can specify a set of N signals and use them to generate an arbitrary large set of other signals as a linear combination (weighted sum)

$$\boldsymbol{a} = \sum_{k=0}^{N-1} \alpha_k \varphi_k \tag{2.1.5}$$

of the selected signals that are called basis signals (***basis functions***). Obviously, the set of basis functions should be selected from functions that can't be represented as a linear combination of the rest of the functions from the set, otherwise the set of basis functions will be redundant. The signal space formed of signals generated as a linear combination (2.1.5) of N linearly-independent basis functions is referred to as an ***N-dimensional linear space***.

Each signal $\boldsymbol{a}$ in N-dimensional signal space corresponds to a unique linear combination of basis functions $\{\varphi_k\}$, i.e. is defined by a unique set of scalar coefficients $\{\alpha_k\}$. An (ordered) set of scalar coefficients of decomposition of a given signal over the given basis is the ***signal's representation*** with respect to this basis.

The "zero-signal" may be treated as a reference signal of the signal space. The distinction of signal $\boldsymbol{a}$ from the "zero-signal" displays the individual characteristic of this signal. Mathematically, this characteristic is described through the concept of the signal's norm $\|\boldsymbol{a}\|$:

$$\|\boldsymbol{a}\| = d(\boldsymbol{a}, \boldsymbol{\varnothing}) \tag{2.1.6}$$

The geometrical interpretation of the vector norm is its length.

Signal representation as an expansion over a set of basis functions (2.1.5) is meaningful only when for each signal $\boldsymbol{a}$, its representation $\{\alpha_k\}$ for the given basis φ_k can be found. To this goal, the concept of the ***scalar product*** (***inner product***, ***dot product***) $(\boldsymbol{a}_1, \boldsymbol{a}_2)$ of two signals is introduced as a method for computing a (generally, complex) number that satisfies the following conditions:

$$(\boldsymbol{a}_1, \boldsymbol{a}_2) = (\boldsymbol{a}_1, \boldsymbol{a}_2)^*;$$

$$(\boldsymbol{a}, \boldsymbol{a}) \geq 0; \quad (\boldsymbol{a}, \boldsymbol{a}) = 0, \qquad \textit{iff } \boldsymbol{a} = \boldsymbol{\varnothing};$$

$$(\alpha_1 \boldsymbol{a}_1 + \alpha_2 \boldsymbol{a}_2, \boldsymbol{a}_3) = \alpha_1 (\boldsymbol{a}_1, \boldsymbol{a}_3) + \alpha_2 (\boldsymbol{a}_2, \boldsymbol{a}_3). \tag{2.1.7}$$

where asterisk (*) represents a complex conjugate. Usually the scalar product of two signals is calculated as an integral over their product:

$$(\boldsymbol{a}_1, \boldsymbol{a}_2) = \int_X a_1(x) a_2^*(x) dx \tag{2.1.8}$$

This computation method satisfies the above conditions and, as we will see, is a mathematical model of natural signal transformations such as those in optical systems and signal sensors.

The concepts of signal scalar product and norm may be related by defining the norm through the scalar product :

$$\|a\| = (a,a). \tag{2.1.9}$$

Those functions whose scalar product is zero are called mutually ***orthogonal functions***. If functions are mutually orthogonal, then they are linearly-independent. Therefore mutually orthogonal functions may be used as bases of linear signal spaces.

Having defined a scalar product, one can use it to establish a simple correspondence between signals and their representation for the given basis. Let $\{\varphi_k\}$ be the basis functions and $\{\phi_l\}$ be the reciprocal set of functions orthogonal to $\{\varphi_k\}$ and normalized such that

$$(\varphi_k, \phi_l) = \delta_{k,l} = \begin{cases} \mathbf{0}, & k \neq l \\ \mathbf{1}, & k = l \end{cases} \tag{2.1.10}$$

(symbol $\delta_{k,l}$ is called the ***Kronecker delta***). Then signal representation coefficients $\{\alpha_k\}$ over basis $\{\varphi_k\}$ can be found as

$$\alpha_k = (a, \phi_k) = \sum_{l=0}^{N-1} \alpha_l (\varphi_l, \phi_k) = \sum_{l=0}^{N-1} \alpha_l \delta_{k,l} . \tag{2.1.11}$$

If basis $\{\varphi_k\}$ is composed of mutually orthogonal functions of unit norm:

$$(\varphi_k, \varphi_l) = \delta_{k,l} , \tag{2.1.12}$$

it is reciprocal to itself and is called ***orthonormal basis***. In such a basis signals are representable as

$$a = \sum_{k=0}^{N-1} (a, \varphi_k) \varphi_k , \tag{2.1.13}$$

where

$$(a, \varphi_k) = \alpha_k = \int_X a(x) \varphi_k^*(x) dx . \tag{2.1.14}$$

Knowing the signal representation over the orthonormal basis, one can easily calculate signal scalar product

$$(\boldsymbol{a},\boldsymbol{b})=\left(\sum_{k=0}^{N-1}\alpha_k\varphi_k,\sum_{k=0}^{N-1}\beta_k\varphi_k\right)=\sum_{k=0}^{N-1}\sum_{l=0}^{N-1}(\alpha_k\varphi_k,\beta_l\varphi_l)=\sum_{k=0}^{N-1}\alpha_k\beta_k^* \qquad (2.1.15)$$

and norm

$$\|\boldsymbol{a}\|=\sum_{k=0}^{N-1}|\alpha_k|^2 . \qquad (2.1.16)$$

Representations $\{\alpha_k^{(1)}\}$, $\{\beta_k^{(1)}\}$ and , $\{\alpha_k^{(2)}\}$, $\{\beta_k^{(2)}\}$ of signals $\boldsymbol{a}$ and $\boldsymbol{b}$ over two orthonormal bases $\{\varphi_k^{(1)}\}$ and $\{\varphi_k^{(2)}\}$ are related by means of the ***Parseval's relationship***

$$\sum_{k=0}^{N-1}\alpha_k^{(1)}\left(\beta_k^{(1)}\right)^*=\sum_{k=0}^{N-1}\alpha_k^{(2)}\left(\beta_k^{(2)}\right)^* . \qquad (2.1.17)$$

Up to now, it was assumed that signals are continuous. For discrete signals defined by their elements (samples), a convenient mathematical model is that of vectors.

Let $\{a_k\}$ is a set of samples of a discrete signal, $k=0,1,...,N-1$. They define a vector-column $\mathbf{a}=\{a_k\}$. One can select a set of linearly independent vectors $\{\vec{\varphi}_r\}=\{\{\varphi_{r,k}\}\}$:

$$\sum_{r=0}^{N-1}\lambda_r\vec{\varphi}_r\neq\mathbf{0} \qquad (2.1.18)$$

for arbitrary coefficients $\{\lambda_r\neq\mathbf{0}\}$ as a basis for representing any signal vector:

$$\mathbf{a}_N=\sum_{r=0}^{N-1}\alpha_r\vec{\varphi}_r . \qquad (2.1.19)$$

An ordered set of basis vectors can be regarded as a matrix

$$\mathbf{T}_N=\{\varphi_{k,r}\} \qquad (2.1.20)$$

and signal expansion of Eq. 2.1.18 as a matrix product of the vector-column of coefficients $\boldsymbol{\alpha} = \{\alpha_k\}$ by the basis matrix:

$$\mathbf{a}_N = \mathbf{T}_N \boldsymbol{\alpha}_N . \tag{2.1.21}$$

In signal processing jargon, matrix $\mathbf{T}_N$ is called ***transform matrix*** and signal representation defined by Eq. 2.1.21 is called ***signal transform***. If the transform matrix $\mathbf{T}_N$ has its inverse matrix $\mathbf{T}_N^{-1}$ such that

$$\mathbf{T}_N \cdot \mathbf{T}_N^{-1} = \mathbf{I}_N , \tag{2.1.22}$$

where $\mathbf{I}$ is the identity matrix:

$$\mathbf{I}_N = \{\delta_{k-r}\}, \tag{2.1.23}$$

vector of signal representation coefficients can be found as

$$\boldsymbol{\alpha}_N = \mathbf{T}_N^{-1} \mathbf{a}_N . \tag{2.1.24}$$

Such matrix treatment of discrete signals assumes, that signal scalar product is defined as a matrix product

$$(\mathbf{a},\mathbf{b}) = \mathbf{a}^{tr} \cdot (\mathbf{b})^* = \sum_{k=0}^{N-1} a_k (b_k)^* . \tag{2.1.25}$$

Transforms $\mathbf{T}_N$ and $\mathbf{T}_N^{-1}$ are called mutually orthogonal and signal vector transformations described by Eqs. (2.1.20) and (2.1.23) are called direct and inverse transforms, respectively. Transform matrix that is orthogonal to its transpose

$$\mathbf{T}_N \cdot \mathbf{T}_N^{tr} = \mathbf{I}_N \tag{2.1.26}$$

is called ***orthogonal matrix***. Transform matrix $\mathbf{T}_N$ that is orthogonal to its complex conjugate and transposed copy $(\mathbf{T}_N^*)^{tr}$:

$$\mathbf{T}_N \cdot (\mathbf{T}_N^*)^{tr} = \mathbf{I}_N \tag{2.1.27}$$

is called ***self conjugate matrix*** and the transform is called the ***unitary transform***. For unitary transforms, the Parceval's relationship holds:

$$(\mathbf{a},\mathbf{b}) = (\boldsymbol{\alpha},\boldsymbol{\beta}) = \sum_{k=0}^{N-1} a_k b_k^* = \sum_{r=0}^{N-1} \alpha_r \beta_r^* . \tag{2.1.28}$$

2.1.4 Integral representation of signals

The discrete representation of signals

$$a(x)=\sum_k \alpha_k \varphi_k(x), \tag{2.1.29}$$

described in Sect. 2.1.3 can be extended to a continuous representation obtained were we to substitute in Eqs. 2.1.5 the basis function index k, by a continuous variable, say f, and replace sum by an integral:

$$a(x)=\int_F \alpha(f)\varphi(x,f)df \tag{2.1.30}$$

It is quite natural to extend this approach to the definition of $\alpha(f)$ by introducing reciprocal functions $\phi(f,x)$, to obtain an analog of Eq. 2.1.14:

$$\alpha(f)=\int a(x)\phi(f,x)dx \tag{2.1.31}$$

The function $\alpha(f)$ is called the ***integral transform*** of signal $a(x)$, or its ***spectrum***, over continuous basis $\{\phi(f,x)\}$ called ***transform kernel***. The reciprocity condition for functions $\varphi(x,f)$ and $\phi(f,x)$ or the condition of transform invertibility may be obtained by substituting (2.1.31) into (2.1.30):

$$a(x)=\int_F\left[\int_X a(\xi)\phi(f,\xi)d\xi\right]\varphi(x,f)df=\int_X a(\xi)\delta(x,\xi)d\xi\,, \tag{2.1.32}$$

where

$$\delta(x,\xi)=\int_F \varphi(x,f)\phi(f,\xi)df\,. \tag{2.1.33}$$

The function $\delta(x,\xi)$, defined by Eq.(2.1.33), is called the ***delta-function*** (Dirac delta). It is a continuous analog of Kronecker delta (Eq. 2.1.10) and is a symbol of orthogonality of kernels $\varphi(x,f)$ and $\phi(f,x)$, or, in other words, of invertibility of the integral transform of Eq. (2.1.30).

Formula (2.1.32) may be regarded as a special case of the integral representation with the delta-function as its kernel. One can also treat discrete signal representation (Eq. 2.1.13) as a special case of its integral representation:

$$\alpha(f)=\int_{X} a(x)\phi(f,x)dx=\int_{X}\left[\sum_{k=0}^{N-1}\alpha_k\varphi_k(x)\right]\phi(f,x)dx=\sum_{k=0}^{N-1}\alpha_k\delta(f,f_k)$$

(2.1.34)

where

$$\delta(f,f_k)=\int_{X}\varphi_k(x)\phi(f,x)dx\,. \qquad (2.1.35)$$

Using Eq.(2.1.33) as an analog to Eq.(2.1.12), one can generalize the concept of the orthogonal basis as well.

$$\int_{F}\varphi(x,f)\varphi^*(x,f)df=\delta(x,\xi) \qquad (2.1.36)$$

A basis satisfying this relationship is called ***self-conjugate***. For self-conjugate bases, signal scalar product is invariant to the bases

$$\int_{X} a_1(x)a_2^*(x)dx=\int_{F}\alpha_1(x)\alpha_2^*(f)df\,. \qquad (2.1.37)$$

Eq. (2.1.37) is a continuous analog to the Parseval's relation (2.1.17). In particular,

$$\int_{X}|a_1(x)|^2dx=\int_{F}|\alpha_1(x)|^2df\,. \qquad (2.1.38)$$

2.2 SIGNAL TRANSFORMATIONS

In optical systems, optical signals undergo various transformations. In general, one can mathematically treat signal transformations as a mapping in the signal space. For digital signals, this mapping, in principle, can be specified in a form of a look-up table. However even for digital signals this form of the specification is not practically feasible owing to the enormous volume of the digital signal space (a rough evaluation of it one can see in Sect.3.1). For discrete and continuous signals such a specification is all the more impossible.

Therefore in practice, signal transformations are mathematically described as a combination of certain "elementary" transformations, each of which is specified with the help of a relatively small subset of all feasible input-output pairs. The most important of these are ***point-wise transformations*** and ***linear transformations***.

Signal transformation $\mathbf{PWT}(\cdot)$ is called point-wise if it results in a signal whose value in each point of its argument (coordinate) depends only on the value of the transformed signal in the same point:

$$b(x=\xi)=\mathbf{PWT}[a(x=\xi)]. \tag{2.2.1}$$

For digital signals, ***point-wise transformations*** are applied to individual samples of the signal and are specified by a look-up table with number of entries equal to the number of signal ***quantization levels***. For discrete and continuous signals, it is assumed that point-wise transformations are specified in a form of mathematical functions of a single argument. This function is referred to as system ***transfer function***. If the transfer function does not change with signal coordinates, the transformation is called ***homogeneous transformation***. Otherwise the transformation is ***inhomogeneous transformation***.

A typical example of the element-wise optical transformation is the relation governing the conversion of light radiation into the degree of darkening of a photographic plate. Units responsible for element-wise signal conversion are widely used in models of optical and photographic systems.

Transformation $\mathbf{L}(\cdot)$ is linear if it satisfies the superposition principle:

$$\mathbf{L}\left(\sum_{k=0}^{N-1}\alpha_k a_k(x)\right)=\sum_{k=0}^{N-1}\alpha_k\,\mathbf{L}(a_k(x)); \tag{2.2.2}$$

Obviously, linear signal transformations are defined only for linear signal space and linearly transformed signals also form a linear signal space. Therefore for linear transforms, one may introduce the notion of the sum of transforms, i.e. $\sum_n \mathbf{L}_n$, as well as the product of transforms, $\prod_n \mathbf{L}_n$. The physical equivalent of the product is a series (cascade) connection of units realizing the transform-factors. The physical equivalent of the sum of linear transforms is a parallel connection of units realizing the transform-items. Thanks to the linearity of the transforms, their multiplication is distributive with respect to addition:

$$\begin{aligned} &\mathbf{L}_1(\mathbf{L}_2+\mathbf{L}_3)=\mathbf{L}_1\mathbf{L}_2+\mathbf{L}_1\mathbf{L}_3; \\ &(\mathbf{L}_1+\mathbf{L}_2)\mathbf{L}_3=\mathbf{L}_1\mathbf{L}_3+\mathbf{L}_2\mathbf{L}_3 \end{aligned} \qquad (2.2.3)$$

If a linear transformation $\mathbf{L}$ performs a one-to-one mapping of a signal, then there exists an inverse transformation $\mathbf{L}^{-1}$ such that

$$\mathbf{L}\mathbf{L}^{-1}\mathbf{a}=\mathbf{L}^{-1}\mathbf{L}\mathbf{a}=\mathbf{a} \qquad (2.2.4)$$

A linear transformation of discrete signals is fully characterized by a matrix $\Lambda=\{\lambda_{k,n}\}$ linking the input $\mathbf{a}=\{\alpha_k\}$ and output $\mathbf{b}=\{\beta_k\}$ pairs of the linear transformation $\mathbf{b}=\mathbf{La}$, respectively:

$$\mathbf{b}=\left\{\beta_n=\sum_k \lambda_{k,n}\alpha_k\right\}. \qquad (2.2.5)$$

Thus, in order to describe a linear transformation of a sequence of N numbers, it suffices to specify a matrix of N^2 numbers. Equation (2.2.5) evidences the fact that in the general case, linear transforms are not point-wise, and that they become so only if transformation matrix is a diagonal one ($\{\lambda_{k,n}=\lambda_k\delta_{k-n}\}$).

For continuous signals, integral transforms are the instruments by which one produces a meaningful description of linear transformations. Let a continuous signal $a(x)$ is specified by means of its spectrum $\alpha(f)$ on bases $\{\varphi(x,f);\phi(f,x)\}$ (Eq. 2.1.30, 31 and, in a special case of delta-function base, Eq. 2.1.32). Then the result $b(\xi)$ of applying to this signal a linear transformation $\mathbf{L}$

$$b(\xi)=\mathbf{L}a(x)=\mathbf{L}\int_F \alpha(f)\varphi(x,f)df=\int_F \alpha(f)[\mathbf{L}\varphi(x,f)]df \qquad (2.2.6)$$

is fully specified by the response $h(\xi,f)=\mathbf{L}(\varphi(x,f))$ of the linear transformation to the transform kernel

$$b(\xi) = \int_{F} \alpha(f) h(\xi, f) df . \tag{2.2.7}$$

For the representation of linear transformation output signal $b(\xi)$ also via its spectrum $\beta(f)$ over basis $\phi(f, \xi)$

$$\beta(f) = \int_{\Xi} b(\xi) \phi(f, \xi) d\xi , \tag{2.2.8}$$

spectra $\beta(f)$ and $\alpha(f)$ of a signal and its linear transformation are also related with an integral transform

$$\beta(f) = \int_{F} \alpha(p) H(f, p) dp \tag{2.2.9}$$

whose kernel $H(f, p)$ in its turn is an integral transform, with kernel $\phi(f, \xi)$, of the linear transformation response $h(\xi, f)$ to basis $\varphi(x, f)$ in which spectrum of the signal is specified:

$$H(f, p) = \int h(\xi, p) \phi(f, \xi) d\xi . \tag{2.2.10}$$

Signal integral representation (2.1.32) via delta-function is of especial importance for the characterization of linear transformations. For such a representation, the relationship between output signal $b(x)$ and input signal $a(x)$ of a linear transformation (Eq. 2.2.7) is reduced to

$$b(x) = \int a(\xi) h(x, \xi) d\xi , \tag{2.2.11}$$

where $h(x, \xi) = \mathbf{L}(\delta(x, \xi))$ is linear transformation response to a delta-function called ***impulse response*** or ***point spread function*** of a linear system that implements this transformation.

Cascade combination of units performing linear and point-wise signal transformations (Fig. 2.3) is the basic model used to describe mathematically imaging and holographic systems.

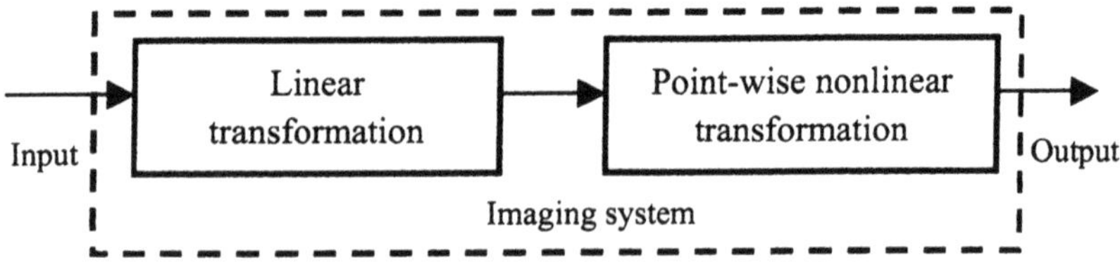

Figure 2-3. Basic model of imaging and holographic systems

2.3 IMAGING SYSTEMS AND INTEGRAL TRANSFORMS

2.3.1 Direct imaging: Convolution integral

Imaging systems are systems that create images of real or phantom objects. In describing imaging systems it is always assumed that there exist a physical or imaginary object plane and an image plane and that the primary goal of imaging systems is reproducing, in image plane, distribution of light reflectance or whatever other physical property of an object in the object plane.

Apparently the very first imaging devices were painters (Fig. 2-4). The first artificial imaging device, was, seemingly, ***camera-obscura*** (pinhole camera, Latin for 'dark room')) that dates back Aristotle (384-322 BCE) and Euclid (365-265 BCE) (Fig. 2-5). Camera obscura was the ancestor of the modern photographic and TV cameras (Fig. 2-6).

Figure 2-4. A woodcut by Albrecht Dürer showing the relationship between the light distribution on an object and image plane (adopted form R. Bracewell, Two-dimensional imaging, Prentice Hall Int. 1995)

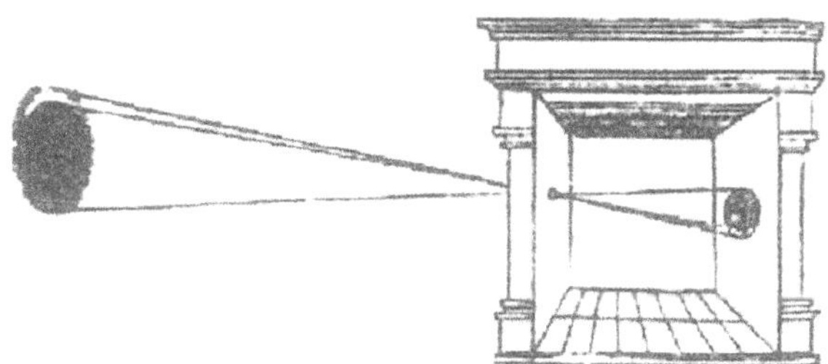

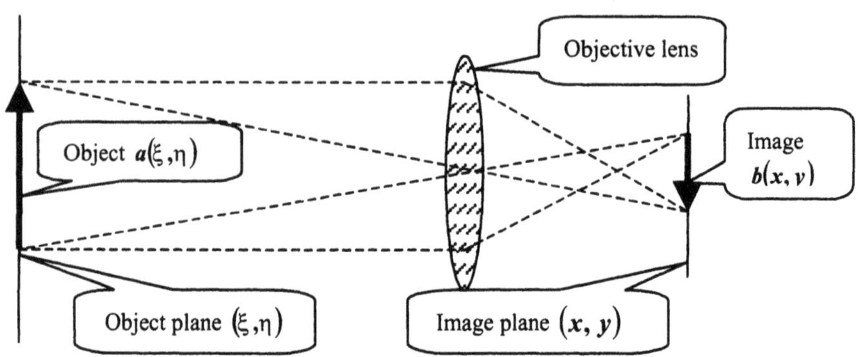

Figure 2-6. Schematic diagram of optics of photographic and TV cameras

Many other modern imaging devices such as X-ray, gamma cameras, scanning electron, acoustic, scanned proximity probe (atomic force, tunnel) microscopes work on the same principle of imaging one point in object plane into a corresponding point in image plane. Being, in their optical part, devices that satisfy the superposition principle, they are most naturally characterized by their point spread function $h(x,y;\xi,\eta)$:

$$b(x,y) = \int_X \int_Y a(\xi,\eta) h(x,y;\xi,\eta) d\xi d\eta \,, \tag{2.3.1}$$

where integration is assumed to be carried out over the object plane. Integral of Eq. (2.3.1) implicitly shows image of a point source located in coordinates (ξ,η).

In general, the point spread function is space variant: image of a point source changes depending on its location in object plane. An important special case is that of space invariant imaging systems in which image of a point source is invariant to the point source location. For space invariant systems, point spread function is a function of difference of coordinates in object and image planes:

$$h(x,y;\xi,\eta) = h(x-\xi, y-\eta). \tag{2.3.2}$$

Imaging equation (2.3.1) for space-invariant systems takes form:

$$b(x,y) = \int_{-\infty}^{\infty} \int_{-\infty}^{\infty} a(\xi,\eta) h(x-\xi, y-\eta) d\xi d\eta \,. \tag{2.3.3, a}$$

called 2-D ***convolution integral***. Correspondingly, 1-D convolution integral is defined as

$$b(x) = \int_{-\infty}^{\infty} a(\xi) h(x-\xi) d\xi \tag{2.3.3, b}$$

Note that integration limits in Eqs. (2.3.3) are infinite. This is a direct consequence of the assumption of space invariance.

Point spread function is the basic specification characteristic of imaging systems. It is used for certification of imaging systems and characterizes the system capability to resolve objects called imaging system ***resolving power***. Given noise level in the signal sensor, it determines variance of error in measuring coordinates of a point source ([5], see also Sect. 10.3). In particular, one can show that, for separable PSF $h(x,y) = h_x(x) h_y(y)$, variances σ_x^2 and σ_y^2 of the error in measuring (x,y) coordinates of a point source are inversely proportional to the energy of the corresponding derivatives of PSF:

$$\sigma_x^2 \propto \left(\int_{-\infty}^{\infty} \left| \frac{\partial}{\partial x} h_x(x) \right|^2 \right)^{-1}; \; \sigma_y^2 \propto \left(\int \left| \frac{\partial}{\partial y} h_y(y) \right|^2 \right)^{-1}. \tag{2.3.4}$$

On a qualitative level it is commonly accepted that resolving power of imaging systems is defined by the ***Rayleigh's criterion***: two point sources are considered resolved if minimum between two corresponding PSF peaks does not exceed 80% or so of the peak maxima (Fig. 2-7).

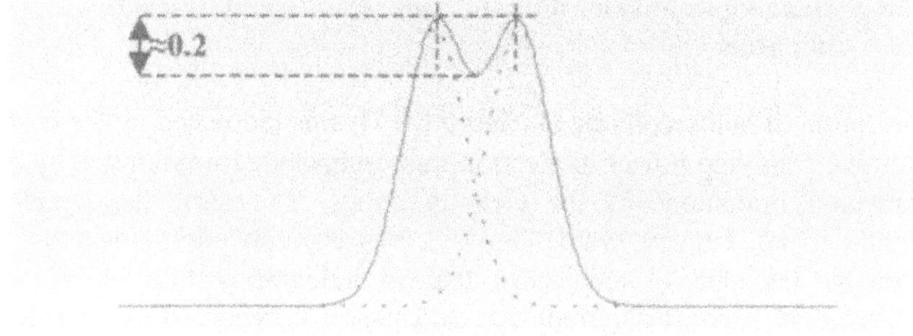

Figure 2-7. Rayleigh's criterion of the resolving power

Imaging systems described by the convolution integral (Eq. 2.3.3) are, in general, not invertible: one can not perfectly restore system input image $a(x,y)$ from the output image $a(x,y)$. The problem of the restoration belongs to so called ***inverse problems***. Methods for image restoration will be discussed in Ch. 8.

In conclusion note that the assumption of imaging system space invariance is most frequently one used in the description and analysis of imaging systems and image restoration. It leads to many simplifications both in system analysis and in image processing. The advantages of space invariance of system PSF can sometimes also be utilized even if the assumption does not satisfactorily hold. This can be achieved if one converts signals from original coordinate system into a in a transformed one in which system PSF becomes space-invariant. After space invariant processing in the transformed coordinate system signal should then be converted back to the original coordinates. This principle is illustrated in Fig.2-8.

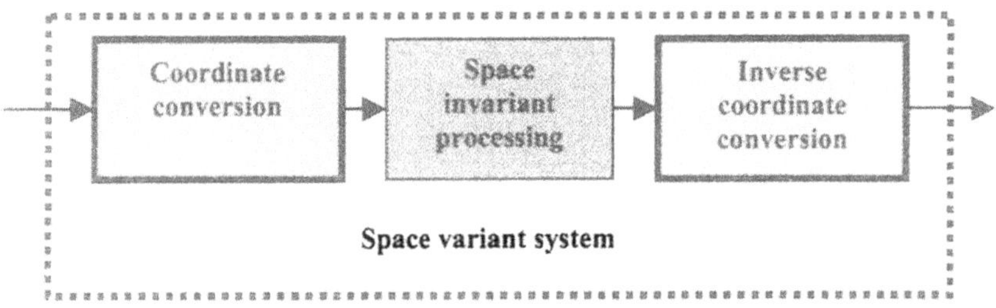

Figure 2-8. Principle of space invariant representation of space variant systems

In digital processing this methods requires signal resampling. Methods for discrete signal resampling will be discussed in Ch. 9.

2.3.2 Imaging in Fourier domain: Holography and diffraction integrals

Invention of holography by D. Gabor ([6,7]) was motivated by the desire to improve resolving power of electron microscopes that was limited by the fundamental limitations of the electron optics. The term "holography" originates from Greece word "holos" (ηωλωσ). By this, inventor of holography intended to emphasize that in holography full information regarding light wave, both amplitude and phase, is recorded by means of interference of two beams, object and reference one. Holography remained an "optical paradox" until the invention of lasers. The first implementations of holography were demonstrated in 1961 by E. Leith and J. Upatnieks ([8])

and by Yu. Denisyuk ([9]). The principle of hologram recording is illustrated in Figs. 2-9 and of hologram reconstruction in Fig. 2-10.

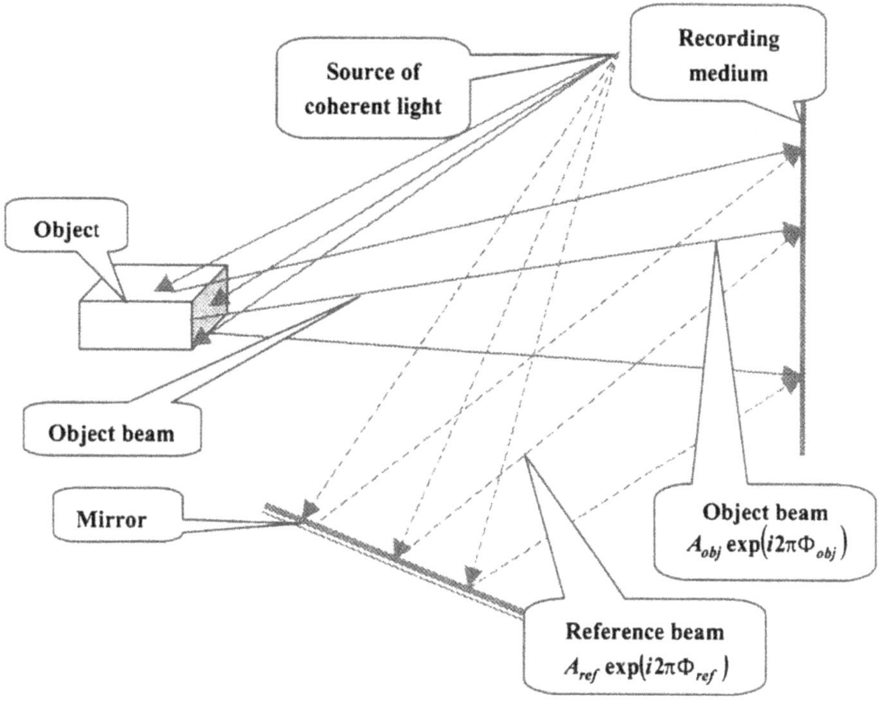

Figure 2-9. Schematic diagram of recording holograms

Radiation scattered by an object (object beam) interferes at a recording medium with a coherent reference beam. The medium records the interference pattern. Hologram is the result of the recording. Physical recording media are sensitive to only energy of radiation. Therefore recorded signal $\boldsymbol{H}(\boldsymbol{x},\boldsymbol{y})$ is proportional to the squared magnitude of the sum of interfering beams:

$$\boldsymbol{H}(\boldsymbol{x},\boldsymbol{y})=\left|A_{obj}\exp\left(i\Phi_{obj}\right)+A_{ref}\exp\left(i\Phi_{ref}\right)\right|^2=A_{obj}^2+A_{ref}^2+$$

$$A_{obj}A_{ref}\exp\left[i\left(\Phi_{obj}-\Phi_{ref}\right)\right]+A_{obj}A_{ref}\exp\left[-i\left(\Phi_{obj}-\Phi_{ref}\right)\right], \qquad (2.3.5)$$

where $A_{obj}\exp\left(i\Phi_{obj}\right)$ and $A_{ref}\exp\left(i\Phi_{ref}\right)$ are complex amplitudes of object and reference beams, respectively.

For reconstruction of the hologram, one should illuminate the hologram with a light beam of the same phase distribution as that of the reference beam $A_0 \exp(i\Phi_{ref})$. The hologram (Fig. 2-10) modulates the reconstruction

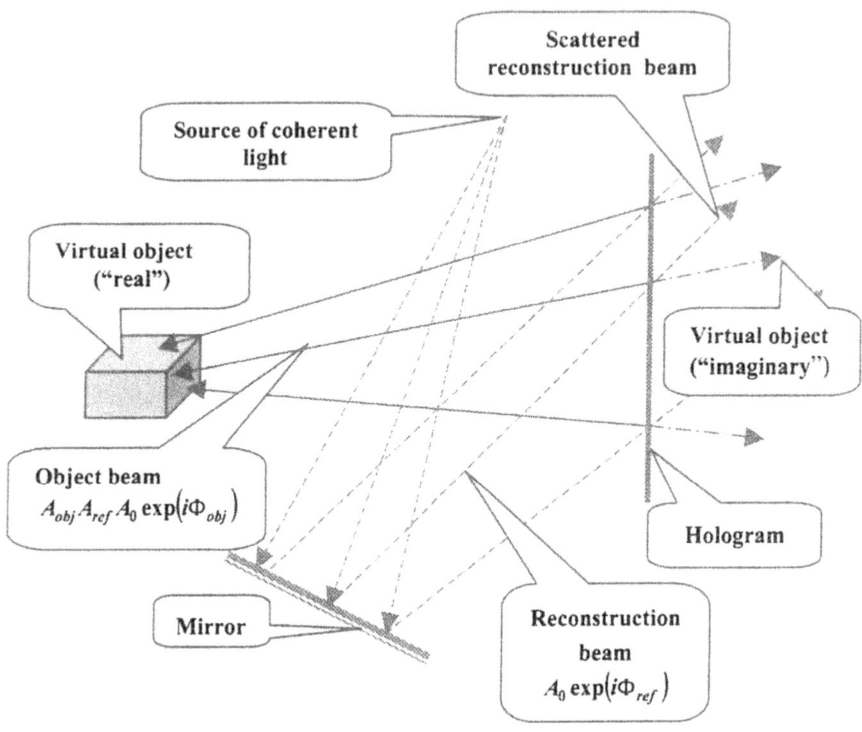

Figure 2-10. Schematic diagram of hologram playback

beam by amplitude and phase. As a result, three beams are generated: a beam $A_{obj}A_{ref}A_0 \exp(i\Phi_{obj})$ that has the same phase distribution as that of the object wave and whose magnitude is proportional to that of the object wave, a scattered reconstruction beam $(A_{obj}^2 + A_{ref}^2)A_0 \exp(i\Phi_{ref})$, and a beam $A_{obj}A_{ref}A_0 \exp(i(2\Phi_{ref} - \Phi_{obj}))$. The former is the beam that reconstructs the object wave front (provided that A_{ref} and A_0 are constant). The latter reconstruct a "conjugate" imaginary object wave front .

The process of recording holograms is a good example of a two-unit imaging system of Fig. 2.3. Point-wise nonlinear transformation unit in this process describes recording interference pattern while linear transformation unit refers to propagation of light from the object to the recording medium.

Consider 1-D model of free space wave propagation (Fig. 2-11). Let complex amplitude of wave front of a monochromatic radiation in coordinate x at the object ('input") plane is $a(x)$. Let also hologram ("output") plane is situated at distance D from the object plane. At the hologram ("output") plane at a point with coordinate f, this wave being traveled a distance $\sqrt{D^2+(x-f)^2}$ will be accordingly attenuated and obtain a phase delay $2\pi\sqrt{D^2+(x-f)^2}/\lambda$, where λ is the coherent radiation wavelength. For the sake of simplicity, we will disregard the effect of wave polarization on thėir superposition that is relevant for electromagnetic radiation and is not for acoustic waves. Then one can conclude that the point spread function of free space wave propagation is

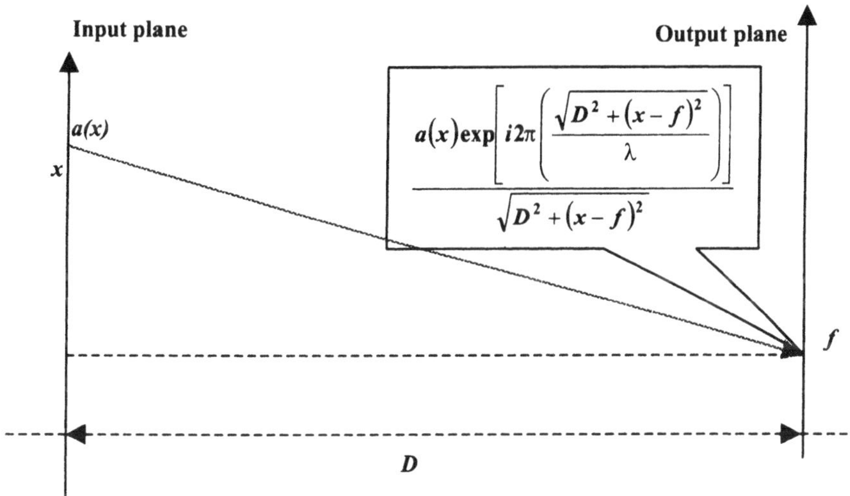

Figure 2-11. Free space wave propagation point spread function

$$h(x,f)=\frac{\exp\left(i2\pi\sqrt{D^2+(x-f)^2}/\lambda\right)}{\sqrt{D^2+(x-f)^2}}. \tag{2.3.6}$$

Obviously, 2-D point spread function of free space wave propagation from input plane (x,y) to output plane (f_x,f_y) is

$$h(x,y;f_x,f_y)=\frac{\exp\left(i2\pi\sqrt{D^2+(x-f_x)^2+(y-f_y)^2}/\lambda\right)}{\sqrt{D^2+(x-f_x)^2+(y-f_y)^2}}, \qquad (2.3.7)$$

The relationship between complex amplitudes $a(x,y)$ and $\alpha(f_x,f_y)$, respectively of the wave front in the input and output planes is then:

$$\alpha(f_x,f_y)=\int_{-\infty}^{\infty}\int_{-\infty}^{\infty}a(x,y)\frac{\exp\left(i2\pi\sqrt{D^2+(x-f_x)^2+(y-f_y)^2}/\lambda\right)}{\sqrt{D^2+(x-f_x)^2+(y-f_y)^2}}dxdy. \qquad (2.3.8)$$

Integral of Eq. 2.3.8 is called ***Kirchhoff's integral*** ([10]).

Assume now that distance D between the input object plane (x,y) and the observation output plane (f_x,f_y) is substantially larger than maximal distances $\left((x-f_x)^2+(y-f_y)^2\right)^{1/2}$ involved in Kirchhoff's integral (2.3.8) so as one can approximate

$$\sqrt{D^2+(x-f_x)^2+(y-f_y)^2}\cong D;$$

$$\exp\left(i2\pi\sqrt{D^2+(x-f_x)^2+(y-f_y)^2}/\lambda\right)\cong \exp\left[i\pi\frac{(x-f_x)^2+(y-f_y)^2}{\lambda D}\right] \qquad (2.3.9)$$

In this approximation, Kirchhoff's integral is reduced, with omitting a scalar multiplier, to its Fresnel approximation:

$$\alpha(f_x,f_y)=\int_{-\infty}^{\infty}\int_{-\infty}^{\infty}a(x,y)\exp\left[i\pi\frac{(x-f_x)^2+(y-f_y)^2}{\lambda D}\right]dxdy. \qquad (2.3.10)$$

that describes the so called ***near zone diffraction***. Eq. 2.3.10 can alternatively be written as

$$\alpha(f_x,f_y)\exp\left(-i\pi\frac{f_x^2+f_y^2}{\lambda D}\right)=$$

$$\int_{-\infty}^{\infty}\int_{-\infty}^{\infty} a(x,y)\exp\left(i\pi\frac{x^2+y^2}{\lambda D}\right)\exp\left[-i2\pi\frac{xf_x+yf_y}{\lambda D}\right]dxdy. \qquad (2.3.11)$$

If, further, distance D is so large that one can neglect phase variations $\left(\pi\frac{f_x^2+f_y^2}{\lambda D}\right)$ and $\left(\pi\frac{x^2+y^2}{\lambda D}\right)$, we arrive at ***far zone diffraction***, or ***Fraunhofer approximation*** of Kirchhoff's integral:

$$\alpha(f_x,f_y)=\int_{-\infty}^{\infty}\int_{-\infty}^{\infty} a(x,y)\exp\left[-i2\pi\frac{xf_x+yf_y}{\lambda D}\right]dxdy. \qquad (2.3.12)$$

Eq. 2.3.12 describes also relationship between wave front complex amplitudes in front and rear focal planes of lenses. Consider 1-D model of a lens (Fig. 2-12). Lens is fully defined by its property to convert light from a point source in the lens focal plane into a parallel beam tilted according to the position of the source with respect to lens optical axis. One can use this property to determine point spread function of a lens.

One can see from the diagram in Fig. 2-12 that, at point f in the lens rear focal plane, the plane wave front originated from a point source of monochromatic light in coordinate $(-x)$ for $x<<F$ (paraxial approximation), is delayed with respect to point at the lens optical axis by distance $\Delta = f\left(-\frac{x}{F}\right)$, where F is lens ***focal length***. From this one can conclude that point spread function of optical systems whose input and output planes are front and rear focal planes of a lens is

$$h(x,f)=\exp(i2\pi\Delta/\lambda)=\exp\left(-i2\pi\frac{fx}{\lambda F}\right) \qquad (2.3.13)$$

and complex amplitudes $a(x)$ and $\alpha(f)$ of monochromatic light in front and rear focal planes of a lens are linked by a relationship:

$$a(f)=\int_X a(x)exp\left(-i2\pi\frac{fx}{\lambda F}\right)dx, \qquad (2.3.14)$$

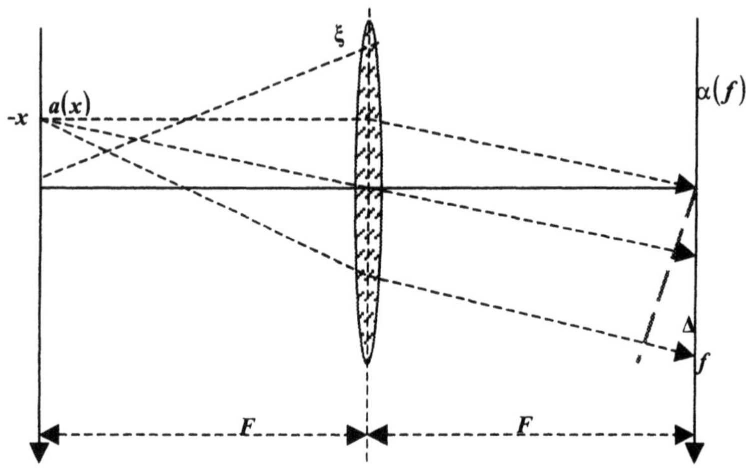

Figure 2-12. Schematic diagram of wave propagation between focal planes of a lens

or, in 2-D denotation,

$$a(f_x, f_y) = \int_X \int_Y a(x, y) \exp\left(-i2\pi \frac{f_x x + f_y y}{\lambda F}\right) dxdy . \quad (2.3.15)$$

It is more convenient to rewrite Eqs. 2.3.8, 2.3.10, 2.3.12, 2.3.14 and 2.3.15 in dimensionless coordinates normalized, correspondingly, by $\sqrt{\lambda D}$ or by $\sqrt{\lambda F}$. In such normalized coordinates, obtain, neglecting irrelevant constant factors:

$$\alpha_K(f_x, f_y) = \int_{-\infty}^{\infty} \int_{-\infty}^{\infty} a(x, y) \frac{\exp\left\{-i\pi \sqrt{d_\lambda^2 + d_\lambda \left[(x - f_x)^2 + (y - f_y)\right]^2}\right\}}{\sqrt{d_\lambda^2 + d_\lambda \left[(x - f_x)^2 + (y - f_y)\right]^2}} dxdy \quad (2.3.16)$$

where $d_\lambda = D/\lambda$ is a dimensionless distance parameter,

$$\alpha_{Fr}(f_x, f_y) = \int_{-\infty}^{\infty} \int_{-\infty}^{\infty} a(x, y) \exp\left\{-i\pi \left[(x - f_x)^2 + (y - f_y)^2\right]\right\} dxdy ; \quad (2.3.17)$$

and

$$\alpha_F(f) = \int_{-\infty}^{\infty} a(x)\exp(i2\pi fx)dx\,; \tag{2.3.18}$$

$$a_F(f_x, f_y) = \int_{-\infty}^{\infty}\int_{-\infty}^{\infty} a(x,y)\exp[i2\pi(f_x x + f_y y)]dxdy\,. \tag{2.3.19}$$

We will refer to Eq. 2.3.16 to as ***integral Kirchhof transform***. Integral transform defined by Eq. 2.3.17 is called ***integral Fresnel transform***. Integral transforms defined by Eq. 2.3.18 and 19 are 1-D and 2-D ***integral Fourier transforms***.

It follows from the above that optical system that consists of a lens and input and output planes in the lens focal planes and that uses monochromatic radiation acts as a Fourier transformer, 1-D with a cylindrical lens, and 2-D with spherical one, of the wave front at the lens front focal plane. One can also use two such Fourier transform systems in cascade and place into their common focal plane a ***spatial light modulator*** that is capable of modification of the incident wave front magnitude or/and phase (Fig. 2-13). An optical system built in this way is a simplest coherent optical image processing system. Image processing and target location methods that can be implemented in such systems are discussed in Chs. 8, 10 and 11.

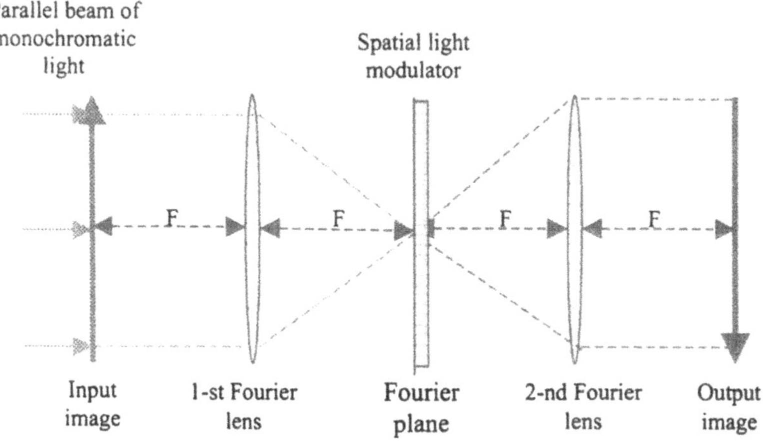

Figure 2-13. Schematic diagram of an optical image processing system

From schematic diagram in Fig. 2-11 it follows also that lens itself can be considered as a spatial light modulator that adds to the phase of the incident light a component that is a quadratic function of a distance from the lens optical axis. Indeed, a spherical wave at point ξ in lens plain from a point source at the optical axis in the input plane has complex amplitude

$$A \propto \exp\left(i2\pi \frac{\sqrt{F^2+\xi^2}}{\lambda} \right). \tag{2.3.20}$$

In paraxial approximation, when size of the lens is much less than its focal distance, wave amplitude is approximately

$$A \propto A_0 \exp\left(i\pi \frac{\xi^2}{\lambda F} \right). \tag{2.3.21}$$

Because this spherical wave is converted by the lens into a plain wave, complex transparency of the lens is then

$$T = \exp\left(-i\pi \frac{\xi^2}{\lambda F} \right). \tag{2.3.22}$$

This property of lens allows to regard an optical system of three lenses shown in Fig.2-14 as an optical Fresnel transformer, 1-D with cylindrical and 2-D with spherical lens.

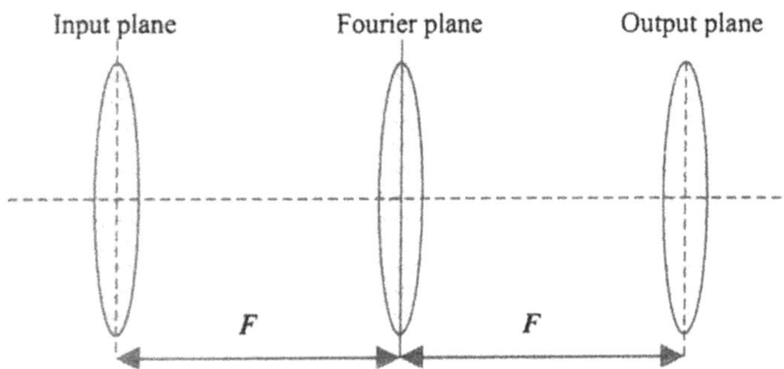

Figure 2-14. Schematic diagram of an optical Fresnel transformer

2.4 FOURIER TRANSFORM AND ITS DERIVATIVES

Fourier transform plays an exceptional role in digital holography and image processing. Apart of its fundamental role in mathematical representation of signal transformations in wave propagation and image formation described in the previous section, it has, in its discrete representation as Discrete Fourier Transform (DFT, see Chapt. 4), a remarkable property of efficient computation via Fast Fourier Transform (see Chapt. 6). As we will see it later in this and next sections, integral Fourier transform represents also a very convenient tool for linking all linear transforms. In this section, we discuss main properties of integral Fourier transform and such its derivatives as Cosine Fourier Transform, Hankel transform, Hartley transform and Mellin transform. We will describe here also Fresnel, Fractional Fourier and Hilbert transforms that are closely associated with Fourier transform.

In order to make derivations more compact, we will, when it will be convenient, treat two-dimensional coordinates (x, y) and (f_x, f_y) as two-component vectors $\mathbf{x} = \begin{bmatrix} x \\ y \end{bmatrix}$ and $\mathbf{f} = \begin{bmatrix} f_x \\ f_y \end{bmatrix}$. In such denotation, 2-D integral Fourier transform is written as

$$\alpha(\mathbf{f}) = \int_{-\infty}^{\infty} a(\mathbf{x}) \exp(i2\pi \mathbf{x}^{tr} \mathbf{f}) d\mathbf{x} . \tag{2.4.1}$$

where $\mathbf{x}^{tr}$ is transposed vector $\mathbf{x}$. Eq. 2.4.1 can also be generally regarded as a multi-dimensional integral transform assuming that $\mathbf{x}$ and $\mathbf{f}$ can be, in general, multi component vectors.

2.4.1 Properties of integral Fourier Transform

The most important properties of integral Fourier transform are the following.

Invertibility

Integral Fourier transform is invertible. According to the integral Fourier theorem ([11]) signal $a(\mathbf{x})$ can be found from its ***Fourier spectrum*** $\alpha(\mathbf{f})$ as

$$a(\mathbf{x}) = \int_{-\infty}^{\infty} \alpha(\mathbf{f}) \exp(-i2\pi \mathbf{f}^{tr} \cdot \mathbf{x}) d\mathbf{f} = \lim_{\mathbf{F} \to \infty} \int_{\mathbf{F}} \alpha(\mathbf{f}) \exp(-i2\pi \mathbf{f}^{tr} \cdot \mathbf{x}) d\mathbf{f} , \tag{2.4.2}$$

where $\mathbf{F}$ is the area of integration and

$$a(\mathbf{x}) = \int_{-\infty}^{\infty} \alpha(\mathbf{f}) \exp\left[-i2\pi\mathbf{f}^{tr} \cdot \mathbf{x}\right] d\mathbf{f} . \qquad (2.4.3)$$

is called ***inverse Fourier transform***. Therefore, the kernel of the integral Fourier transform is self-conjugate with delta function that symbolizes the self- conjugacy defined as

$$\delta(\mathbf{x},\boldsymbol{\xi}) = \lim_{\mathbf{F}\to\infty} \int_{\mathbf{F}} \exp\left[i2\pi\mathbf{f}^{tr} \cdot (\mathbf{x}-\boldsymbol{\xi})\right] d\mathbf{f} . \qquad (2.4.4)$$

In the special case of 1-D integral Fourier transform:

$$\delta(x,\xi) = \lim_{F\to\infty} \int_{-F}^{F} \exp\left[i2\pi f(x-\xi)\right] df = \lim_{F\to\infty} \left\{2F \,\mathrm{sinc}\left[2\pi F(x-\xi)\right]\right\}, \qquad (2.4.5)$$

where

$$\mathrm{sinc}(x) = \frac{\sin x}{x} \qquad (2.4.6)$$

is so called ***sinc-function***. Sinc-function can be regarded as Fourier transform

$$\mathrm{sinc}(2\pi Fx) = \int_{-\infty}^{\infty} \mathrm{rect}\left(\frac{f+F}{2F}\right) \exp(i2\pi fx) dx \qquad (2.4.7)$$

of a ***rect-function***:

$$\mathrm{rect}(x) = \begin{cases} 1, \ 0 \le x < 1 \\ 0, \ \text{otherwise} \end{cases} . \qquad (2.4.8)$$

It is important to note that the invertibility of integral Fourier Transform is a mathematical idealization. In practice, integration in Fourier transform is always implemented in finite limits defined by aperture of optical systems, dimensions of lenses and similar factors. These limitations are referred to as ***bandwidth limitations***. Owing to the bandwidth limitations, artifacts known as ***Gibb's effects*** may appear in form of oscillations at signal sharp edges whenever Fourier transform is carried out. More on Gibb's effects see in Sect. 3.2.2.1.

Separability. It follows from the definition of multi-dimensional integral Fourier transform (Eq. 2.4.1) that it is separable to separate one-dimensional transforms.

$$\alpha(f_1,\dots,f_M)=\int_{-\infty}^{\infty}\dots\int_{-\infty}^{\infty}a(x_1,\dots,x_M)\exp\left(i2\pi\sum_{m=1}^{M}x_m f_m\right)dx_1,\dots,dx_M =$$

$$\int_{-\infty}^{\infty}\exp(i2\pi x_1 f_1)dx_1\int_{-\infty}^{\infty}\exp(i2\pi x_2 f_2)dx_2\dots\int_{-\infty}^{\infty}a(x_1,\dots,x_M)\exp(i2\pi x_M f_M)dx_M \quad (2.4.9)$$

Shift theorem Absolute value of signal Fourier spectrum is invariant to signal shift. Signal shift causes only linear modulation of the phase of the spectrum.

Proof. From Eq. 2.4.3 it follows that

$$a(\mathbf{x}-\mathbf{x}_0)=\int_{-\infty}^{\infty}\alpha(\mathbf{f})\exp\left[-i2\pi\mathbf{f}^{tr}\cdot(\mathbf{x}-\mathbf{x}_0)\right]d\mathbf{f}= \\ \int_{-\infty}^{\infty}\left[\alpha(\mathbf{f})\exp(i2\pi\mathbf{f}^{tr}\cdot\mathbf{x}_0)\right]\exp\left[-i2\pi\mathbf{f}^{tr}\cdot\mathbf{x}\right]d\mathbf{f}, \quad (2.4.10)$$

which implies that if Fourier spectrum of $a(\mathbf{x})$ is $\alpha(\mathbf{f})$, spectrum of $a(\mathbf{x}-\mathbf{x}_0)$ is $\alpha(\mathbf{f})\exp(i2\pi\mathbf{f}^{tr}\cdot\mathbf{x}_0)$ and that absolute value of signal Fourier spectrum is invariant to the signal shift.

Rotation theorem: Signal rotation results in the same rotation of its spectrum.

Proof. In vector-matrix denotation of signal and spectrum coordinates, signal rotation by an angle ϑ can be represented as multiplication of coordinate vector $\mathbf{x}$ by a rotation matrix $\mathbf{Rot}_\vartheta$:

$$\mathbf{Rot}_\vartheta=\begin{bmatrix}\cos\vartheta & -\sin\vartheta\\ \sin\vartheta & \cos\vartheta\end{bmatrix}. \quad (2.4.11)$$

From Eq. 2.4.3 obtain:

$$a(\mathbf{Rot}_\vartheta \cdot \mathbf{x}) = \int_{-\infty}^{\infty} \alpha(\mathbf{f}) \exp(-i2\pi \mathbf{f}^{tr} \cdot (\mathbf{Rot}_\vartheta \cdot \mathbf{x})) d\mathbf{f} =$$

$$\int_{-\infty}^{\infty} \alpha(\mathbf{f}) \exp(-i2\pi (\mathbf{f}^{tr} \cdot \mathbf{Rot}_\vartheta) \cdot \mathbf{x}) d\mathbf{f} = \int_{-\infty}^{\infty} \alpha(\mathbf{f}) \exp\left(-i2\pi \left((\mathbf{Rot}_\vartheta)^{tr} \cdot \mathbf{f}\right)^{tr} \cdot \mathbf{x}\right) d\mathbf{f} =$$

$$\int_{-\infty}^{\infty} \alpha(\mathbf{f}) \exp(-i2\pi (\mathbf{Rot}_{-\vartheta} \cdot \mathbf{f})^{tr} \cdot \mathbf{x}) d\mathbf{f}.$$

(2.4.12)

Changing variables $\mathbf{Rot}_{-\vartheta} \cdot \mathbf{f} \to \tilde{\mathbf{f}}$ and by virtue of the fact that the Jacobian of this coordinate conversion is

$$J(\mathbf{Rot}_{-\vartheta}) = \cos^2 \vartheta + \sin^2 \vartheta = 1, \tag{2.4.13}$$

obtain finally:

$$a(\mathbf{Rot}_\vartheta \cdot \mathbf{x}) = \int_{-\infty}^{\infty} \alpha\left(\mathbf{Rot}_\vartheta \cdot \tilde{\mathbf{f}}\right) \exp(-i2\pi \tilde{\mathbf{f}} \cdot \mathbf{x}) d\tilde{\mathbf{f}}. \tag{2.4.14}$$

Therefore if Fourier spectrum of $a(\mathbf{x})$ is $\alpha(\mathbf{f})$, spectrum of $a(\mathbf{Rot}_\vartheta \cdot \mathbf{x})$ is $\alpha(\mathbf{Rot}_\vartheta \cdot \mathbf{f})$.

Scaling theorem: Scaling signal coordinates S times causes $1/S$ scaling of its spectrum magnitude and spectrum coordinates.

Proof.

$$a(S\mathbf{x}) = \int_{-\infty}^{\infty} \alpha(\mathbf{f}) \exp[-i2\pi \mathbf{f} \cdot (S\mathbf{x})] d\mathbf{f} = \int_{-\infty}^{\infty} \alpha(\mathbf{f}) \exp[-i2\pi (S\mathbf{f}) \cdot \mathbf{x}] d\mathbf{f} =$$

$$\frac{1}{S} \int_{-\infty}^{\infty} \alpha\left(\frac{1}{S}\mathbf{f}\right) \exp[-i2\pi \tilde{\mathbf{f}} \cdot \mathbf{x}] d\tilde{\mathbf{f}}.$$

(2.4.15)

Therefore, if Fourier spectrum of $a(\mathbf{x})$ is $\alpha(\mathbf{f})$, Fourier spectrum of $a(S\mathbf{x})$ is $\frac{1}{S}\alpha\left(\frac{1}{S}\mathbf{f}\right)$.

Convolution theorem: Inverse Fourier transform of a product $\alpha(\mathbf{f})\beta(\mathbf{f})$ of spectra of two signals $a(\mathbf{x})$ and $b(\mathbf{x})$ is convolution $\int_{-\infty}^{\infty} a(\boldsymbol{\xi})b(\mathbf{x}-\boldsymbol{\xi})d\boldsymbol{\xi}$ of these signals.

Proof.

$$\int_{-\infty}^{\infty}\alpha(\mathbf{f})\beta(\mathbf{f})\exp(-i2\pi\mathbf{f}''\mathbf{x})d\mathbf{f} =$$
$$\int_{-\infty}^{\infty}\left[\int_{-\infty}^{\infty}a(\boldsymbol{\xi})\exp(i2\pi\mathbf{f}''\boldsymbol{\xi})d\boldsymbol{\xi}\right]\beta(\mathbf{f})\exp(-i2\pi\mathbf{f}''\mathbf{x})d\mathbf{f} = \quad . \tag{2.4.16}$$
$$\int_{-\infty}^{\infty}a(\boldsymbol{\xi})\left[\int_{-\infty}^{\infty}\beta(\mathbf{f})\exp(-i2\pi\mathbf{f}''(\mathbf{x}-\boldsymbol{\xi}))d\mathbf{f}\right]d\boldsymbol{\xi} = \int_{-\infty}^{\infty}a(\boldsymbol{\xi})b(\mathbf{x}-\boldsymbol{\xi})d\boldsymbol{\xi}$$

Convolution theorem links Fourier transform and convolution integral. As it was shown in Sect. 2.3.1, convolution integral describes image formation in space invariant imaging systems. Convolution theorem shows that image formation processes can also be treated in terms of modulation of image Fourier spectra by Fourier transform $H(\mathbf{f})$ of convolution integral's kernel, impulse response $h(\mathbf{x})$ of the space invariant system:

$$H(\mathbf{f}) = \int_{-\infty}^{\infty}h(\mathbf{x})\exp(i2\pi\mathbf{f}\mathbf{x})d\mathbf{x}. \tag{2.4.17}$$

Function $H(\mathbf{f})$ is called system ***frequency response*** (the term customary in signal processing and communication) or ***modulation transfer function*** (MTF, the term customary within optical community).

Symmetry properties

Signal and their spectra symmetry properties are summarized in the following equations in which symbol $*$ means complex conjugation.

$$a(\mathbf{x}) = \pm a(-\mathbf{x}) \xleftarrow{\text{FourierTransform}}\!\!\rightarrow \alpha(\mathbf{f}) = \pm\alpha(-\mathbf{f}); \tag{2.4.18}$$

$$a(\mathbf{x}) = \pm a^*(\mathbf{x}) \xleftarrow{\text{FourierTransform}}\!\!\rightarrow \alpha(\mathbf{f}) = \pm\alpha^*(-\mathbf{f}); \tag{2.4.19}$$

$$a(\mathbf{x}) = a(-\mathbf{x}) = \pm a^*(\mathbf{x}) \xleftarrow{\text{FourierTransform}}\!\!\rightarrow \alpha(\mathbf{f}) = \alpha(-\mathbf{f}) = \pm\alpha^*(-\mathbf{f}); \tag{2.4.20}$$

$$a(\mathbf{x}) = -a(-\mathbf{x}) = \pm a^*(\mathbf{x}) \xleftarrow{\text{FourierTransform}}\!\!\rightarrow \alpha(\mathbf{f}) = -\alpha(-\mathbf{f}) = \pm\alpha^*(-\mathbf{f}); \tag{2.4.21}$$

2.4.2 Special cases of the integral Fourier transform: integral Cosine, Hartley, Hankel and Mellin transforms

2.4.2.1 Integral Cosine transform

When $a(\mathbf{x})$ is an even function ($a(\mathbf{x}) = a(-\mathbf{x})$), integral Fourier Transform in reduced to ***integral Cosine transform***:

$$\alpha(\mathbf{f}) = \int_{-\infty}^{\infty} a(\mathbf{x}) \cos(2\pi \mathbf{x}'' \mathbf{f}) d\mathbf{x}. \tag{2.4.22}$$

Proof:

$$\alpha(\mathbf{f}) = \int_{-\infty}^{\infty} a(\mathbf{x}) \exp(i2\pi \mathbf{x}'' \mathbf{f}) d\mathbf{x} = \int_{-\infty}^{\infty} a(-\mathbf{x}) \exp(i2\pi \mathbf{x}'' \mathbf{f}) d\mathbf{x} =$$
$$\int_{-\infty}^{\infty} a(\mathbf{x}) \exp(-i2\pi \mathbf{x}'' \mathbf{f}) d\mathbf{x} = \tilde{\alpha}(\mathbf{f}) = \frac{\alpha(\mathbf{f}) + \tilde{\alpha}(\mathbf{f})}{2} =$$
$$\int_{-\infty}^{\infty} a(\mathbf{x}) \cos(2\pi \mathbf{x}'' \mathbf{f}) d\mathbf{x} \tag{2.4.23}$$

Such a transformation is carried out in incoherent optical information processing and holographic systems ([12-15]).

By virtue of symmetry property of integral Fourier transform (Eq.2.4.15), inverse cosine transform is identical to the direct one:

$$a(\mathbf{x}) = \int_{-\infty}^{\infty} \alpha(\mathbf{f}) \cos(2\pi \mathbf{x}'' \mathbf{f}) d\mathbf{x}. \tag{2.4.24}$$

Multi-dimensional cosine transform is not separable to 1-D transforms. It is separable if $a(\mathbf{x})$ is even in all dimensions separately. In particular, 2-D separable cosine transform

$$\alpha(f_x, f_y) = \int_0^{\infty}\int_0^{\infty} a(x, y) \cos(2\pi f_x x) \cos(2\pi f_y y) dx fy. \tag{2.4.25}$$

Fig. 2-15 illustrates 2 types of symmetry that correspond to inseparable (left image) and separable (right image) integral Cosine transforms.

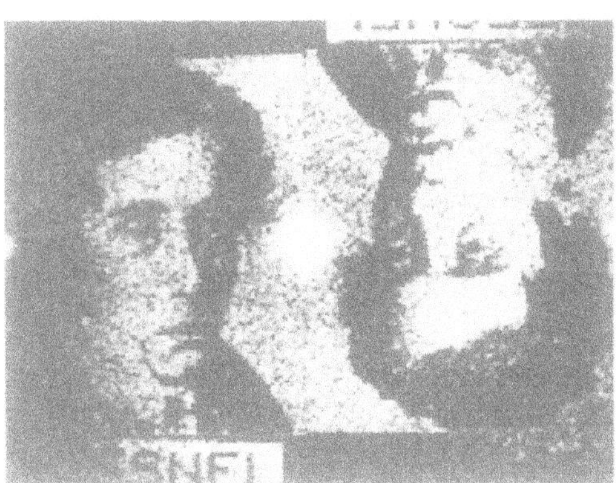
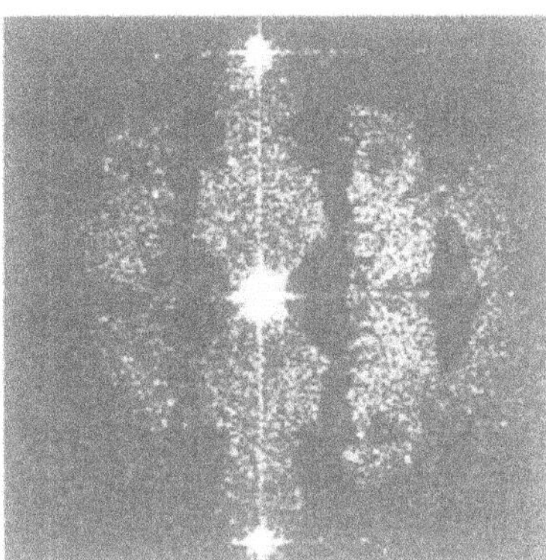

Figure 2-15. Two types of image even symmetry observed in images reconstructed from computer generated Fourier holograms.

An important characteristic properties of integral Cosine transform is that it has a real valued kernel.

2.4.2.2 Hartley transform

Yet another derivative of integral Fourier transform with real valued kernel is ***Hartley transform*** ([16]). 1-D Hartley transform is defined as

$$\alpha(f) = \int_{-\infty}^{\infty} a(x)[\cos(2\pi fx) + \sin(2\pi fx)]dx . \tag{2.4.26}$$

With a multiplier $\sqrt{2}$, it is equivalent to

$$\alpha(f) = \int_{-\infty}^{\infty} a(x)\cos(2\pi fx - \pi / 4)dx . \tag{2.4.27}$$

Similarly to the Cosine transform, Hartley transform and its inverse are identical. For real valued signals, the following relationships linking their Fourier $\alpha_F(f)$ and Hartley $\alpha_H(f)$ spectra may be readily derived as

$$\alpha_H(f) = \mathbf{Re}(\alpha_F(f)) + \mathbf{Im}(\alpha_F(f)) = \frac{(i+1)\alpha(f) + (i-1)\alpha(-f)}{2i};$$
$$\alpha_F(f) = \frac{\alpha_H(f) + \alpha_H(-f)}{2} + i\frac{\alpha_H(f) - \alpha_H(-f)}{2}. \quad (2.4.28)$$

where $\mathbf{Re}(\alpha)$ and $\mathbf{Im}(\alpha)$ are real and imaginary parts of α.
As in the case of the Cosine transform, separable

$$\alpha(f_x, f_y) = \int_{-\infty}^{\infty}\int_{-\infty}^{\infty} a(x,y)\cos(2\pi f_x x - \pi/4)\cos(2\pi f_y y - \pi/4)dxdy \quad (2.4.29)$$

and inseparable

$$\alpha(f_x, f_y) = \int_{-\infty}^{\infty}\int_{-\infty}^{\infty} a(x,y)\cos[2\pi(f_x x + f_y y) - \pi/4]dxdy \quad (2.4.30)$$

versions of 2-D Hartley transforms are possible.

2.4.2.3 Hankel transform

If a 2-D signal $a(x,y)$ is circularly symmetric that is, in polar coordinate system (r,ϑ), ($x = r\cos\vartheta;\quad y = r\sin\vartheta$), it is a function of single variable r:

$$a(x,y) = a(r), \quad (2.4.31)$$

2-D integral Fourier transform is reduced the so called Hankel transform:

$$\alpha(f) = 2\pi\int_0^{\infty} r a(r) J_0(2\pi f r)dr, \quad (2.4.32)$$

where $J_0(\cdot)$ is zero-order Bessel function of the first kind:

$$J_0(f) = \frac{1}{2\pi}\int_{-\pi}^{\pi} \exp(if\cos\vartheta)d\vartheta. \quad (2.4.33)$$

One can show this by replacing Cartesian coordinates (x,y) and (f_x,f_y) in 2-D Fourier transform by polar coordinates (r,ϑ) and (f,θ), respectively:

$$\alpha(f_x,f_y)=\int_{-\infty}^{\infty}\int_{-\infty}^{\infty}a(x,y)\exp[i2\pi(f_x x+f_y y)]dxdy=$$
$$\int_{0}^{\infty}\int_{-\pi}^{\pi}ra(r)\exp[i2\pi(f_x r\cos\vartheta+f_y r\sin\vartheta)]dxdy= \qquad (2.4.34)$$
$$\alpha(f,\theta)=\int_{0}^{\infty}\int_{-\pi}^{\pi}a(r)\exp[i2\pi(\cos\theta\cos\vartheta+\sin\theta\sin\vartheta)rf]rdrd\vartheta=$$
$$\int_0^{\infty}a(r)rdr\int_{-\pi}^{\pi}\exp(i2\pi\cos(\vartheta-\theta))d\theta=2\pi\int_0^{\infty}a(r)J_0(2\pi rf)rdr$$

Integral Fourier transform $\alpha_F(f_x,f_y)$ and Hankel transform $\alpha_{Hn}(f)$ of a radial symmetric signal $a(r)$ (Eq. 2.4.29) are related as follows:

$$\alpha_F(f_x,f_y)=\alpha_{H_n}\left(\sqrt{f_x^2+f_y^2}\right). \qquad (2.4.35)$$

From inverse 2-D Fourier transform one can derive that inverse Hankel transform coincides with the direct one:

$$a(r)=2\pi\int_0^{\infty}\alpha(f)J_0(2\pi fr)fdf. \qquad (2.4.36)$$

As one can see, Bessel function plays in Hankel transform role which complex exponential function plays in Fourier transform. The role played by sinc-function in Fourier transform, in Hankel transform is played by a ***jinc-function*** ([17]) that, similarly to sinc-function as a Fourier transform of a rect-function (Eq. 2.4.8), is Hankel transform of the rect-function:

$$\mathrm{jinc}(r)=2\pi\int_0^{\infty}\mathrm{rect}(f)J_0(2\pi fr)df. \qquad (2.4.38)$$

Sinc- and ***jinc*** -functions are compared in Fig. 2-16. As one can see from the figure, jinc function oscillates much less and converges to zero much faster then sinc function.

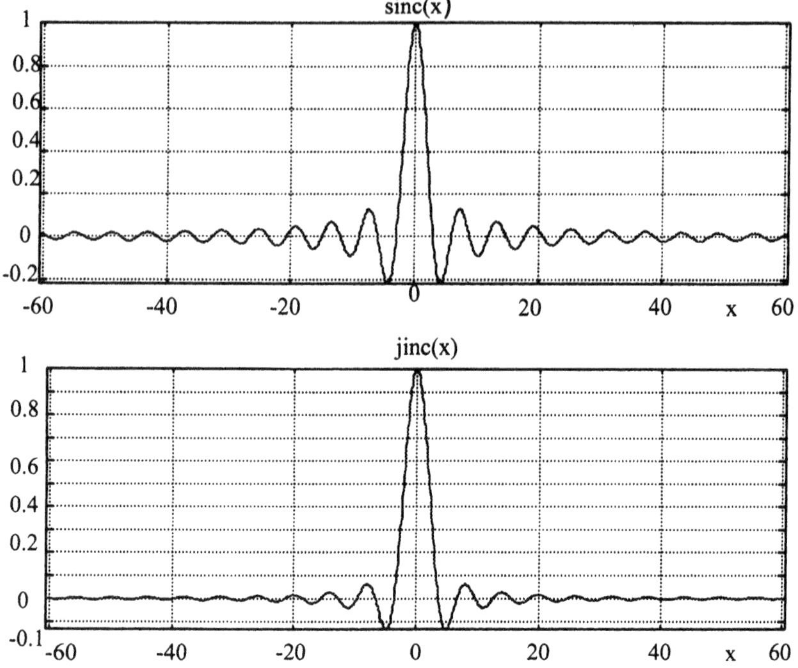

Figure 2-16. **Sinc-** and **jinc-** functions

2.4.2.4 Mellin transform

The shift invariance of the Fourier transform may be turned into a scale invariance property if one introduces a logarithm scaling of the signal coordinate:

$$\boldsymbol{x} = \ln z \tag{2.4.39}$$

Substituting Eq 2.4.39 in Eq. 2.3.17a yields:

$$\alpha(f) = \int_0^\infty a(\ln z)\exp(i2\pi f \ln z)\frac{dz}{z} = \int_0^\infty a(\ln z) z^{i2\pi f - 1} dz\,. \tag{2.4.40}$$

This brings us to a transform:

$$\mu(f) = \int_0^\infty a(z) z^{i2\pi f - 1} dz\,. \tag{2.4.41}$$

referred to as the ***Mellin transform***. For Mellin transform, scaling signal coordinate $z = k\tilde{z}$ modifies its Mellin spectrum in the same way as signal shift modifies Fourier spectrum (see Eq.2.4.10):

$$\tilde{\mu}(f) = k\int_0^{\infty} a(k\tilde{z})(k\tilde{z})^{i2\pi f-1}d\tilde{z} = k^{i2\pi f}\int_0^{\infty} a(k\tilde{z})(\tilde{z})^{i2\pi f-1}d\tilde{z} = k^{i2\pi f}\mu(f). \tag{2.4.42}$$

Therefore, Mellin spectrum of a signal, by its absolute value, is invariant to scaling signal coordinate.

Two dimensional Mellin Transform is generated by the 2-D Fourier transform where coordinate conversion (2.4.40) is applied to both coordinates and, by this definition, is separable over two variables:

$$\mu(f_1, f_2) = \int_0^{\infty}\int_0^{\infty} a(z_1, z_2) z_1^{i2\pi f_1-1} z_2^{i2\pi f_2-1} dz_1 dz_2 . \tag{2.4.43}$$

More details regarding Mellin transform one can find in [18].

2.4.3 Integral Fresnel and related transform

2-D integral Fresnel transform was introduced in Sect. 2.3.2 (Eq. 2.3.17) as a mathematical model that approximates Kirchhoff integral in near zone diffraction. 1-D Fresnel transform is defined respectively as

$$\alpha_{Fr}(f) = \int_{-\infty}^{\infty} a(x)\exp\left[-i\pi(x-f)^2\right]dx . \tag{2.4.44}$$

Fresnel transform and Fourier transform are very closely connected and may be reduced one to another by mean of phase modulation of signal and its spectrum with ***chirp-functions*** functions $\mathbf{chirp}(x) = \exp(-i\pi x^2)$ and $\mathbf{chirp}(f) = \exp(-i\pi f^2)$:

$$\alpha_{Fr}(f) = \exp(-i\pi f^2)\int_{-\infty}^{\infty} a(x)\exp(-i\pi x^2)\exp(i2\pi f x)dx \tag{2.4.45}$$

From the invertibility of the Fourier transform it follows that inverse Fresnel transform can be obtained as:

$$a(x)=\exp(i\pi x^2)\int_{-\infty}^{\infty}\alpha_{Fr}(f)\exp(i\pi f^2)\exp(-i2fx)df =$$

$$\lim_{F\to\infty}\int_{-F}^{F}\alpha_{Fr}(f)\exp\left[i\pi(x-f)^2\right]df = \int_{-\infty}^{\infty}\alpha_{Fr}(f)\exp\left[i\pi(x-f)^2\right]df\,. \tag{2.4.46}$$

Therefore, the kernel of the Fresnel transform is self conjugate and his corresponding delta function is:

$$\begin{aligned}\delta(x,\xi) &= \lim_{F\to\infty}\exp\left[i\pi\left(x^2-\xi^2\right)\right]\int_{-F}^{F}\exp\left[-i2\pi f(x-\xi)\right]df = \\ &\exp\left[i\pi\left(x^2-\xi^2\right)\right]\cdot\lim_{F\to\infty}\left\{2F\,\mathrm{sinc}\left[2\pi F(x-\xi)\right]\right\}.\end{aligned} \tag{2.4.47}$$

The result of Fresnel transform of a rect-function

$$\begin{aligned}\mathrm{frinc}(x;F) &= \int_{-\infty}^{\infty} rect\left(\frac{f+F}{2F}\right)\exp\left[i\pi(x-f)^2\right]df = \\ &\exp(i\pi x^2)\int_{-F}^{F}\exp(i\pi f^2)\exp(-i2\pi fx)df\end{aligned} \tag{2.4.48}$$

is an analog of sinc-function of Fourier transform (Eq. 2.4.7). We will refer to it as to ***frinc-function***. Frinc-function is a version of a function

$$\mathrm{Fr}^{\pm}(z)=\mathrm{FrC}(z)\pm\mathrm{FrS}(z)=\frac{1}{\sqrt{2\pi}}\int_{0}^{\sqrt{z}}\exp(\pm ix^2)dx \tag{2.4.49}$$

known in mathematical literature as ***Fresnel integral*** ([19]). The relationship between frinc-function and Fresnel integral is described by the formula:

$$\begin{aligned}\mathrm{frinc}(x;F) &= \int_{-F}^{F}\exp\left[i\pi(x-f)^2\right]df = \frac{\sqrt{2}}{\sqrt{2\pi}}\int_{\sqrt{\pi}(x-F)}^{\sqrt{\pi}(x-F)}\exp(ix^2)dx = \\ &= \sqrt{2}\,\mathrm{Fr}^{+}\left(\pi(x+F)^2\right)-\sqrt{2}\,\mathrm{Fr}^{+}\left(\pi(x-F)^2\right)\end{aligned} \tag{2.4.50}$$

Fresnel transform can also be regarded as a convolution transform and it can also be linked with Fourier transform through the convolution theorem. Fresnel transform $\alpha_{Fr}(f)$ of a signal $a(x)$ can be found as inverse Fourier transform

$$\alpha_{Fr}(f)=\int_{-\infty}^{\infty}\alpha_F(p)\mathbf{CHIRP}(p)\exp(-i2\pi fp)dp \tag{2.4.51}$$

of a product of signal Fourier spectrum $\alpha_F(p)$ and Fourier transform $\mathbf{CHIRP}(p)$ of a chirp-function which is, to the accuracy of irrelevant constant, also a chirp-function ([19]):

$$\mathbf{CHIRP}(p)=\int_{-\infty}^{\infty}\exp(-i\pi x^2)\exp(i2\pi px)dx=\frac{1+i}{\sqrt{2}}\exp(i\pi p^2)=\frac{1+i}{\sqrt{2}}\mathbf{chirp}(ip). \tag{2.4.52}$$

Two dimensional analogs of Eqs. 2.4.45 and 46 follow straightforwardly. Integral Kirchhoff transform (Eq. 2.3.16) whose approximation is Fresnel transform is also a convolution and can also be represented via Fourier transform:

$$\alpha_K(f_x,f_y)=\int_{-\infty}^{\infty}\int_{-\infty}^{\infty}\alpha_F(p_x,p_y)\mathbf{K}(p_x,p_y)\exp[-i2\pi(f_xp_x+f_yp_y)]dp_xdp_y \text{ ,} \tag{2.4.53}$$

where

$$\mathbf{K}(p_x,p_y)=\int_{-\infty}^{\infty}\int_{-\infty}^{\infty}\frac{\exp\left\{-i\pi\sqrt{d_\lambda^2+d_\lambda\left[(x-f_x)^2+(y-f_y)\right]^2}\right\}}{\sqrt{d_\lambda^2+d_\lambda\left[(x-f_x)^2+(y-f_y)\right]^2}}\times \exp[-i2\pi(f_xp_x+f_yp_y)]dxdy \tag{2.4.54}$$

The result of Fresnel transform of an arbitrary function $a(x)$ is, generally, a complex valued function. Its absolute value can be interpreted as the magnitude of the wave front and its phase as the phase of the wave front in near zone of diffraction of wave front $a(x)$. Physical sensors are insensitive to the phase of wave fronts and measure only its magnitude. Therefore one may want to consider a reduced, or ***partial Fresnel transform***

$$\tilde{\alpha}_{Fr}(f)=\int_{-\infty}^{\infty}a(x)\exp(-i\pi x^2)\exp(i2\pi fx)dx \tag{2.4.55}$$

Partial Fresnel transform is Fourier transform of a signal modulated by a chirp-function. Its inverse transform is simply the inverse Fourier transform phase modulated with a ***chirp-function***:

$$a(x) = \exp(i\pi x^2) \int_{-\infty}^{\infty} \tilde{\alpha}_{Fr}(f) \exp(-i2\pi f x) df . \quad (2.4.56)$$

Fresnel transform as a mathematical model of transformation in optical systems can be regarded as a special case of the so called ***ABCD-transform*** ([20-22]):

$$\alpha_{ABCD}(f) = C \int_{-\infty}^{\infty} a(x) \exp\left[\frac{i\pi}{B}\left(Ax^2 - 2fx + Df^2\right)\right] dx . \quad (2.4.57)$$

When $A = D = 0$ and $B = 1$, ABCD-transform is reduced to Fourier transform, when $A = B = D = 1$, ABCD-transform is reduced to Fresnel transform. The special case of $A = D = \cos\varphi$ and $B = \sin\varphi$ is called ***Fractional Fourier transform***:

$$\alpha_{FrF}(f) = C \exp\left(i\pi \frac{f^2}{\tan\varphi}\right) \int_{-\infty}^{\infty} a(x) \exp\left(i\pi \frac{x^2}{\tan\varphi}\right) \exp\left[\frac{i2\pi f x}{\sin\varphi}\right] dx . \quad (2.4.58)$$

Originally introduced in quantum physics ([23]), it was later found to be convenient for describing optical phenomena such as wave propagation in graded index optical fibers ([24-27]).

From the fact that ABCD-transform is reducible to Fourier transform:

$$\alpha_{ABCD}(f) = C \exp\left(i\pi \frac{Df^2}{B}\right) \int_{-\infty}^{\infty} a(x) \exp\left(i\pi \frac{Ax^2}{B}\right) \exp\left[-\frac{i2\pi f x}{B}\right] dx =$$

$$BC \exp\left(i\pi \frac{Df^2}{B}\right) \int_{-\infty}^{\infty} a(B\tilde{x}) \exp(i\pi AB\tilde{x}^2) \exp(-i2\pi f\tilde{x}) d\tilde{x} \quad (2.4.59)$$

it follows that its is invertible and his kernel is self conjugate:

$$a(x) = \frac{1}{BC} \int_{-\infty}^{\infty} \alpha_{ABCD}(f) \exp\left[-i\pi \frac{Ax^2 - 2fx + Df^2}{B}\right] df . \quad (2.4.60)$$

2.4.4 Hilbert transform

Hilbert transform represents yet another signal transform of the convolution type closely linked with Fourier transform. Hilbert transform deals with the redundancy associated with the symmetry of Fourier spectra of real valued signals (Eq. 2.4.18).

Let $\alpha(f)=\alpha^*(-f)$ is Fourier spectrum of a signal $a(x)$. Form an auxiliary signal $\tilde{a}(x)$ by cutting out frequency components of signal $a(x)$ for negative frequencies and, in order to preserve signal energy, multiplying its spectrum on positive frequencies by a factor of 2:

$$\tilde{a}(x)=2\int_0^\infty \alpha(f)\exp(-i2\pi fx)df=\int_{-\infty}^{\infty}\alpha(f)[\mathrm{sign}(f)+1]\exp(-i2\pi fx)df, \tag{2.4.61}$$

where $\mathrm{sign}(f)=\frac{|f|}{f}=\begin{cases}+1, & f\geq 0\\ -1, & f<0\end{cases}$.

Signal $\tilde{a}(x)$ can be represented in a form of two components: $\tilde{a}(x)=a(x)+\tilde{\tilde{a}}(x)$, where $\tilde{\tilde{a}}(x)$ is

$$\tilde{\tilde{a}}(x)=\int_{-\infty}^{\infty}\alpha(f)\mathrm{sign}(f)\exp(-i2\pi fx)df. \tag{2.4.62}$$

Because spectrum $\alpha(f)sign(f)$ is anti-symmetrical: $\alpha(f)sign(f)=\alpha^*(-f)sign(-f)$, for real valued signal $a(x)$, real part of $\tilde{a}(x)$ is equal to $a(x)$ and $\tilde{\tilde{a}}(x)$ is pure imaginary. Real valued signal $\hat{a}(x)$

$$\hat{a}(x)=\frac{1}{i}\tilde{\tilde{a}}(x)=\frac{1}{i}\int_{-\infty}^{\infty}\alpha(f)\mathrm{sign}(f)\exp(-i2\pi fx)df. \tag{2.4.63}$$

is called Hilbert transform of $a(x)$ ([3]).

Signal

$$\tilde{a}(x)=a(x)+i\hat{a}(x) \tag{2.4.64}$$

is called ***analytical signal*** for real valued signal $a(x)$. Fourier spectrum of analytical signals is nonzero only for positive frequencies.

From Eq. 2.4.64 and by virtue of the convolution theorem it follows that Hilbert transform is a signal convolution with PSF defined as

$$h(x)=\frac{1}{i}\int_{-\infty}^{\infty}\mathrm{sign}(f)\exp(-i2\pi fx)df\,. \qquad (2.4.65)$$

To find this integral, one can use identity ([28]):

$$\int_{0}^{\infty}\frac{\sin mx}{x}dx=\frac{\pi}{2}\mathrm{sign}(m) \qquad (2.4.66)$$

and modify it in the following way:

$$\int_{0}^{\infty}\frac{\sin 2\pi fx}{x}dx=\int_{-\infty}^{0}\frac{\sin 2\pi fx}{x}dx=\frac{1}{2}\int_{-\infty}^{\infty}\frac{\sin 2\pi fx}{x}dx+\frac{1}{2}\int_{-\infty}^{\infty}\frac{\cos 2\pi fx}{x}dx=$$
$$\frac{1}{2i}\int_{-\infty}^{\infty}\frac{\exp(i2\pi fx)}{x}dx=\frac{\pi}{2}\mathrm{sign}(f)\,. \qquad (2.4.67)$$

Then taking inverse Fourier transform, obtain:

$$\frac{1}{i}\int_{-\infty}^{\infty}\mathrm{sign}(f)\exp(-i2\pi fx)df=-\frac{1}{\pi x}\,. \qquad (2.4.68)$$

Therefore, Hilbert transform, as a convolution integral, is defined as

$$\hat{a}(x)=\frac{1}{\pi}\int_{-\infty}^{\infty}\frac{a(\xi)}{\xi-x}d\xi\,. \qquad (2.4.69)$$

Because $(i^{-1}\,\mathrm{sign}\,f)(i^{-1}\,\mathrm{sign}\,f)=-1$, inverse Hilbert transform is identical to the direct one except for the sign:

$$a(x)=\frac{1}{\pi}\int_{-\infty}^{\infty}\frac{\hat{a}(\xi)}{x-\xi}d\xi\,. \qquad (2.4.70)$$

Note that these definitions are directly linked with the definitions of the direct and inverse Fourier transform (Eqs. 2.4.1 and 2.4.3). If in these definitions sign of exponents are opposite (see, for instance, [3]) , direct and inverse Hilbert transforms also will also change their signs.

One of the most immediate applications of Hilbert transform is determination of ***signal's complex amplitude, signal's envelope*** and phase. Signal's envelope is absolute value of the corresponding analytical signal:

$$\tilde{A}(x) = \sqrt{a^2(x) + \hat{a}^2(x)}. \qquad (2.4.71)$$

Such name is justified by the fact that $\tilde{A}(x) \geq |a(x)|$ and that functions $\tilde{A}(x)$ and $a(x)$ never intersect and have common tangents wherever they touch each other:

$$\left.\frac{d}{dx}\tilde{A}(x)\right|_{A(x)=a(x)} = \frac{d}{dx}a(x) \qquad (2.4.72)$$

The phase of the analytical signal is defined as

$$\theta(x) = \mathbf{Im}(\ln \tilde{a}(x)) = \arctan\left(\frac{a(x)}{\hat{a}(x)}\right) \qquad (2.4.73)$$

and complex amplitude is defined as $A(x) = \tilde{A}(x)\exp(i\theta(x))$.

The physical meaning of the envelope and phase of analytical signal can be illustrated on an example of sinusoidal functions

$$a(x) = a_0(x)\cos[2\pi f_0 x + \theta(x)] \qquad (2.4.74)$$

with a carrier frequency f_0 variable amplitude $a_0(x)$ and phase θ. Such functions are, for instance, mathematical models of signals in holography and interferometry. Rewrite Eq. 2.4.74 in the form:

$$a(x) = a_0(x)\cos\theta(x)\cos 2\pi f_0 x - a_0(x)\sin\theta(x)\sin 2\pi f_0 x. \qquad (2.4.75)$$

Because Fourier spectra of functions $\cos 2\pi f_0 x$ and $\sin 2\pi f_0 x$ are, respectively, $[\delta(f - f_0) + \delta(f + f_0)]/2$ and $[\delta(f + f_0) - \delta(f - f_0)]/2i$, their Hilbert transforms are, respectively, $-\sin 2\pi f_0 x$ and $\cos 2\pi f_0 x$. Then, in the assumption that Fourier spectra of functions $a_0(x)\cos\theta(x)$ and $a_0(x)\sin\theta(x)$ do not extend outside interval $[-f_0, f_0]$, Hilbert transform of $a(x)$ is:

$$\hat{a}(x) = -a_0(x)\cos\theta(x)\sin 2\pi f_0 x - a_0(x)\sin\theta(x)\cos 2\pi f_0 x = -a_0(x)\sin[2\pi f_0 x + \theta(x)] \quad (2.4.76)$$

and its analytical signal is

$$\begin{aligned} a(x) &= a_0(x)\cos[2\pi f_0 x - \theta(x)] - ia_0(x)\sin[2\pi f_0 x - \theta(x)] = \\ & a_0(x)\exp[-(i2\pi f_0 x - \theta(x))] = \\ & a_0(x)\exp(i\theta(x))\exp[-i2\pi f_0 x] = A(x)\exp[-i2\pi f_0 x] \end{aligned} \quad (2.4.77)$$

where $A(x) = a(x)\exp(i\theta(x))$, the form adopted for representation of sinusoidal signals on a complex plain.

2.5 IMAGING FROM PROJECTIONS: RADON AND ABEL TRANSFORMS

One of the most important recent findings in image science resulted in various methods of tomography is the discovery that image can be reconstructed from its projections. Signal formation by the procedure of image projecting with parallel beams is described by ***Radon transform***.

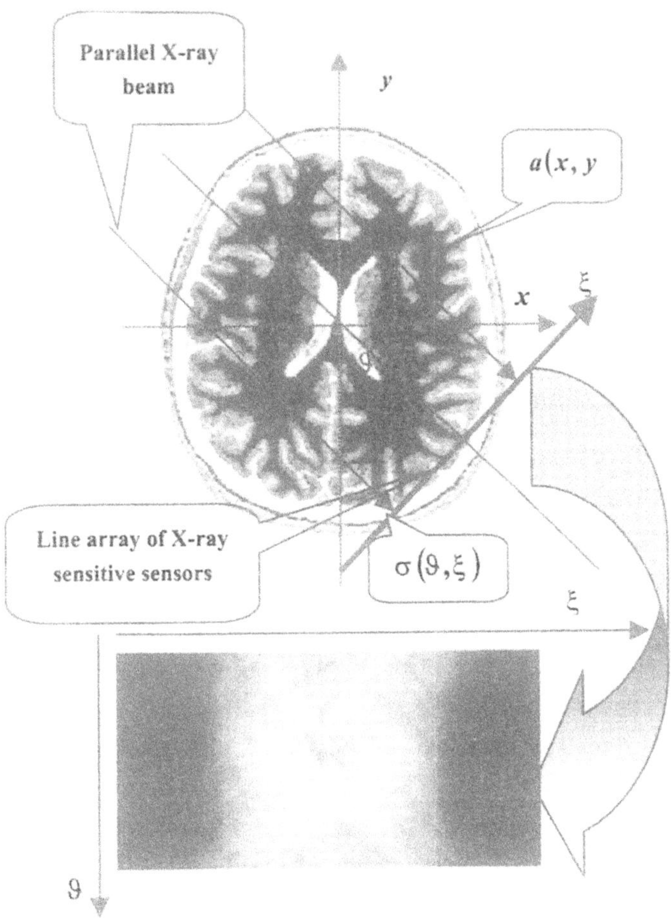

Figure 2-17. Parallel beam projection tomography

Radon transform defines projection of 2-D signal $a(x,y)$ given in a rectangle coordinate system (x,y) onto an axis ξ rotated with respect to axis x by the angle ϑ (Fig. 2-17) as:

$$\sigma(\vartheta,\xi) = \int_X \int_Y a(x,y)\delta(\xi - x\cos\vartheta - y\sin\vartheta)dxdy \tag{2.5.1}$$

Projection signal $\sigma(\vartheta,\xi)$ is called ***sinogram***.

Radon transform and Fourier transforms of projections are linked with ***projection theorem***:

1-D Fourier transforms $\alpha_\vartheta(f)$ of object projections $\sigma(\vartheta,\xi)$ are cross-sections of 2-D object spectrum taken at the corresponding angles.

Proof:

$$\alpha_\vartheta(f) = \int_{-\infty}^{\infty} \sigma(\vartheta,\xi)\exp(i2\pi f\xi)d\xi =$$

$$\int_{-\infty}^{\infty}\left\{\int_X\int_Y a(x,y)\delta(\xi - x\cos\vartheta - y\sin\vartheta)dxdy\right\}\exp(i2\pi f\xi)d\xi = \tag{2.5.2}$$

$$\int_X\int_Y a(x,y)\exp[i2\pi(fx\cos\vartheta + fy\sin\vartheta)]dxdy =$$

$$\alpha(f\cos\vartheta, f\sin\vartheta)$$

where $\alpha(f_x, f_y)$ is 2-D Fourier spectrum of $a(x,y)$.

From the projection theorem it follows that inverse Radon transform can be computed as 2-D inverse Fourier transform of projections' spectra $\alpha_\vartheta(f)$ in polar coordinate system (f,ϑ):

$$a(x,y) = \int_{-\pi}^{\pi}\int_{-\infty}^{\infty}\alpha_\vartheta(f)\exp[-i2\pi(f\cos\vartheta + f\sin\vartheta)]d\vartheta df =$$

$$\int_{-\pi}^{\pi}\int_{-\infty}^{\infty}\alpha(f\cos\vartheta, f\sin\vartheta)\exp[-i2\pi(f\cos\vartheta + f\sin\vartheta)]d\vartheta df \tag{2.5.3}$$

This method of computing inverse Radon transform is called ***Fourier reconstruction method***. Yet another method of inverting Radon transform, ***filtered back projection method,*** is also derived from the interrelation between Fourier and Radon transforms.

2-D function $a(x,y)$ can be found from its 2-D Fourier spectrum $\alpha(f_x, f_y)$ as

$$a(x,y) = \int_{-\infty}^{\infty}\int_{-\infty}^{\infty}\alpha(f_x, f_y)\exp[-i2\pi(f_x x + f_y x)]df_x df_y \tag{2.5.4}$$

In polar co-ordinate system (w,θ) in frequency domain (Fig. 2-18),

$$f_x = w\cos\theta;\ f_y = w\sin\theta;\ df_x df_y = |w| dw d\theta\ ;\ \theta \in [0,\pi], \qquad (2.5.5)$$

obtain:

$$\int_0^{\pi} d\theta \int_{-\infty}^{\infty} |w| \alpha(w,\theta) \exp[-i2\pi w(x\cos\theta + y\sin\theta)] dw = \int_0^{\pi} FBproj(x\cos\theta + y\sin\theta) d\theta, \qquad (2.5.6)$$

where

$$FBproj(x\cos\theta + y\sin\theta) = \int_{-\infty}^{\infty} |w| \alpha(w,\theta) \exp[-i2\pi w(x\cos\theta + y\sin\theta)] dw. \qquad (2.5.7)$$

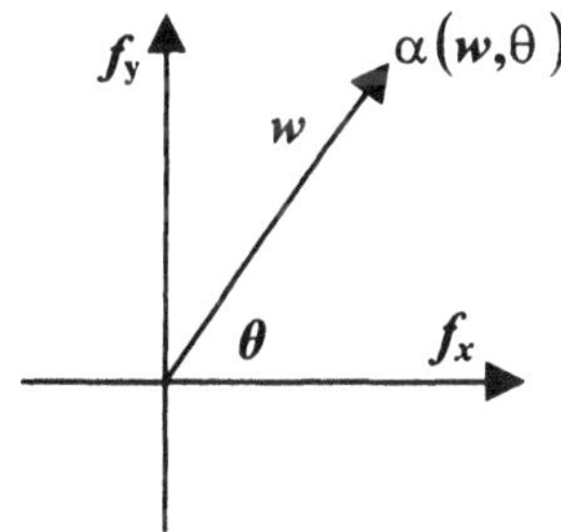

Figure 2-18. Cartesian and polar coordinate systems in Fourier transform domain

Eq. (2.5.6) describes reconstruction of object signal $a(x,y)$ as a "back projection" of signal $FBproj(x\cos\theta + y\sin\theta)$ of Eq. (2.5.7) that is obtained as inverse Fourier Transform of 1-D spectrum $\alpha(w,\theta)$ modified by a filter with frequency response $|w|$ (***"ramp filter"***). Spectrum $\alpha(w,\theta)$ is a cros-section of the object's 2-D spectrum along straight line that goes under angle θ . According to the projection theorem, it is 1-D Fourier Transform the corresponding projection of the object. Therefore, Eq. (2.5.6) represents reconstruction of object signal $a(x,y)$ as a result of accumulating "back projections" of its "filtered projections" $FBproj(x\cos\theta + y\sin\theta)$.

Obviously, exact inversion of Radon transform assumes infinite number of projections. In practice, only finite number of projections is always obtained. Although reconstruction methods can, in principle, be implemented in optical setups, the reconstruction in most applications is carried out in a digital form in computers. Some computer algorithms of tomographic reconstruction are described in Chs. 8 and 9.

For circularly symmetric signals $a(x,y) = a\left(\sqrt{x^2 + y^2}\right)$, Radon transform is reduced to ***Abel transform***:

$$\bar{a}(x) = \int_{-\infty}^{\infty} a\left(\sqrt{x^2 + y^2}\right) dy, \tag{2.5.8}$$

or, in polar coordinates (r, ϑ)

$$\bar{a}(x) = 2\int_{x}^{\infty} \frac{a(r)}{\sqrt{r^2 - x^2}} r dr. \tag{2.5.9}$$

Abel transform is naturally linked with Hankel and Fourier transforms. In particular, Abel transform of the jinc-function is the sinc-function

$$2\int_{x}^{\infty} \frac{\text{jinc}(r)}{\sqrt{r^2 - x^2}} r dr = \text{sinc}(x). \tag{2.5.10}$$

More details regarding properties of Radon and Abel transforms can be found in Refs. [17, 29,30].

2.6 MULTI RESOLUTION IMAGING: WAVELET TRANSFORMS

Above described convolution and Fourier and related orthogonal transforms describe wave propagation and image formation in a large variety of imaging devices. It is frequently convenient to regard applying these signal transforms to signals $a(x)$ taken at an arbitrary scale s in coordinate sx. Because kernel of Fourier and related transforms is a function of a product of coordinates in signal and transform domain, these transforms can readily be treated as transforms of signals in scaled coordinates with the scale factor being a coordinate in transform domain. This is not the case for convolution transforms. Applying the idea of signal treatment in scaled coordinates to convolution leads to a new type of integral transforms that was given name of ***wavelet transforms*** ([31-34]).

Consider convolution integral in a form

$$\bar{a}(x) = \int_{-\infty}^{\infty} a(\xi)h(x-\xi)d\xi \,. \tag{2.6.1}$$

Let now signal $a(x)$ is considered in a scaled coordinate system sx. Then its convolution with PSF $h(\cdot)$ depends on both s and x:

$$\bar{a}(s;x) = \int_{-\infty}^{\infty} a(s\xi)h(x-\xi)d\xi \,. \tag{2.6.2}$$

In Eq. 2.6.2 signal is taken at different scales and convolution kernel is kept the same. One can now translate signal scaling into convolution kernel scaling by changing variables $s\xi \rightarrow \xi$ and obtain:

$$\bar{a}(s;x) = \frac{1}{s}\int_{-\infty}^{\infty} a(\xi)h\left(\frac{sx-\xi}{s}\right)d\xi \tag{2.6.3}$$

or, in scaled coordinates $sx \rightarrow x$:

$$\bar{a}(s;x) = \frac{1}{s}\int_{-\infty}^{\infty} a(\xi)h\left(\frac{x-\xi}{s}\right)d\xi \,. \tag{2.6.4}$$

Eq. 2.6.4 is commonly accepted as a definition of wavelet transform. In such a representation it can be regarded as signal analysis with PSF taken at different scales, or signal ***mutiresolution*** analysis.

From the point of view of classification of integral transforms, wavelet transform can be given yet another interpretation. Different transforms can be classified in terms of transform kernel. All transform kernels are built from a single mother function by its modification parameterized in terms of transform domain coordinates. Convolution integrals represent transforms with kernels obtained from a mother function by its translation or shift. In Fourier and related transforms kernels are built by means of scaling coordinates of the mother function. Wavelet transform are built on the principle of combination of shift and scaling. Such classification helps to understand approaches to the design of transforms for signal discretization and digital processing (see Sect. 3.2).

In 1-D wavelet transforms, transform domain argument of the kernel function has two components: shift ξ and scale s. There might be many wavelet transforms that differ one from another in selection of the mother wavelet $h(\cdot)$. This selection is the main issue in the design of wavelet transforms and is governed by the requirement of the invertibility of the transform and by implementation issues.

As one of the immediate examples of the wavelet transforms one can consider Fresnel Transform in non-reduced coordinates

$$\alpha\left(\sqrt{\lambda D}; f\right) = \int_{-\infty}^{\infty} a(x) \exp\left[i\pi \frac{(x-f)^2}{\lambda D} \right] dx \qquad (2.6.5)$$

with scale parameter $\sqrt{\lambda D}$ (see also Eq. 2.3.10). Haar Transform suggested in 1910 ([35]) is yet another example of a wavelet transform. Kernel of Haar transform is a bipolar step function

$$h(s; x) = rect\left(\frac{2x}{s}\right) - rect\left(\frac{2x-1}{s}\right). \qquad (2.6.6)$$

By the present time, wavelet transforms have become a very popular tool for signal processing and quite a few of different wavelet transforms have been suggested ([33]). Two dimensional and multi-dimensional wavelets are usually defined as separable functions.

2.7 SLIDING WINDOW TRANSFORMS AND "TIME-FREQUENCY" (SPACE-TRANSFORM) SIGNAL REPRESENTATION

2.7.1 Transforms in sliding window (windowed transforms)

Fourier integral transform and its derivatives are all defined with respect to signals defined on infinite intervals. In reality, however, only a finite fraction of the signal selected by the signal sensor aperture is or should be analyzed. This can mathematically be modeled by signal windowing. In this way we arrive at ***windowed transforms***:

$$\alpha(x,f)=\int_{-\infty}^{\infty} w(\xi)a(x-\xi)\varphi(f,\xi)d\xi \qquad (2.7.1)$$

where $\varphi(f,\xi)$ is a transform kernel and $w(\xi)$ is a window function such that it is close to 1 within a certain vicinity to $x=0$ and then more or less rapidly decays to zero:

$$w(0)=1; \quad \lim_{x\to\infty} w(x)=0 \qquad (2.7.2)$$

Windowed transforms of signal $a(x)$ are defined for every point x of signal coordinates and usually are computed in a process of regular signal scanning along its coordinate. In this sense they will be referred to as ***sliding window transforms***.

2.7.2 Sliding window Fourier Transform

Sliding window Fourier transform (SWFT) is one of the most important sliding window transforms. It is defined as

$$\alpha(x,f)=\int_{-\infty}^{\infty} w(\xi)a(x-\xi)\exp(i2\pi f\xi)d\xi\,. \qquad (2.7.3)$$

SWFT of 1-D signals is a function of two variables: coordinate in signal domain and frequency in Fourier domain. It represents what one can call signal ***local spectrum***; that is, spectrum of a (windowed) signal segment centered at the center of the window. This spectrum regarded as a function of coordinate x that defines running position of the window and frequency

f is called signal ***space-frequency representation*** (***time-frequency representation***, in communication theory jargon).

From inverse Fourier transform of the windowed Fourier transform spectrum $\alpha(x, f)$ obtain:

$$w(\xi)a(x-\xi) = \int_{-\infty}^{\infty} \alpha(x, f)\exp(-i2\pi f\xi)df\,. \tag{2.7.4}$$

By virtue of Eq. 2.7.3 it follows from Eq. 2.7.4 that

$$a(x) = \int_{-\infty}^{\infty} \alpha(x, f)df\,. \tag{2.7.5}$$

Eq. 2.7.5 can be regarded as inverse SWFT.

Window function of SWFT can be an arbitrary function that satisfies Eq. 2.7.4. Selection of the window function is governed by the requirements to local spectrum analysis. SWFT with window function

$$w(\xi) = \exp\left(-\frac{\xi^2}{2\sigma^2}\right) \tag{2.7.6}$$

with σ as a window width parameter is known as ***Gabor transform*** ([7]).

2.7.3 Sliding window and wavelet transforms as signal sub-band decomposition

One can give to Windowed Fourier Transform one more and very instructive interpretation. One can regard Eq. 2.7.3 as a convolution integral of $a(x)$ with kernel $w(\xi)\exp(i2\pi f\xi)$. Therefore, according to the convolution theorem, Fourier transform of the signal time-frequency representation $\alpha(x, f)$ at frequency f

$$\mathrm{A}(p, f) = \int_{-\infty}^{\infty} \alpha(x, f)\exp(i2\pi px)dx \tag{2.7.7}$$

is a product $\mathrm{A}(p, f) = \alpha(p)W_f(p)$ of signal Fourier spectrum

$$\alpha(p) = \int_{-\infty}^{\infty} a(x)\exp(i2\pi px)dx \tag{2.7.8}$$

and Fourier Transform $W_f(p)$ of the convolution kernel $w(\xi)\exp(i2\pi f\xi)$. The latter is nothing but shifted by f Fourier spectrum of the window function $w(\xi)$:

$$W_f(p) = \int_{-\infty}^{\infty} [w(\xi)\exp(i2\pi f\xi)]\exp(i2\pi p\xi)d\xi = \int_{-\infty}^{\infty} w(\xi)\exp[i2\pi(p+f)]d\xi = W(p+f) \tag{2.7.9}$$

Therefore, signal time-frequency representation $\alpha(x,f)$ at frequency f can be regarded as signal's sub band formed by signal filtering by a filter with frequency response $W(p+f)$.

Similar is true for a general sliding window transformation (Eq. 2.7.1): sliding window signal transformation is a signal's sub-band formed by a filter with frequency response equal to Fourier Transform of the corresponding windowed basis functions:

$$W_f(p) = \int_{-\infty}^{\infty} w(\xi)\varphi(f,\xi)\exp(i2\pi p\xi)d\xi\,. \tag{2.7.10}$$

Because wavelet transforms are, for every scale, signal convolution they can also be treated as signal sub-band decompositions formed by signal filtering by filters with frequency response equal to Fourier Transform of the mother function on the corresponding scale:

$$W_s(p) = \int_{-\infty}^{\infty} \varphi\left(\frac{\xi}{s}\right)\exp(i2\pi p\xi)d\xi\,. \tag{2.7.11}$$

This property establishes a link between wavelet and sliding window transforms. Figs. 2.19 and 20 illustrate arrangements of sub-bands for sliding window Fourier transform and wavelet base known as Binom5, respectively. As one can see from the figures the main distinction between these two sub-ban decompositions is that sliding window Fourier transform decomposes signal into sub-bands of the same width uniformly arranged in the base band while sub-bands for wavelet decomposition are arranged in logarithm scale and have width that is doubled with the increase of its frequency.

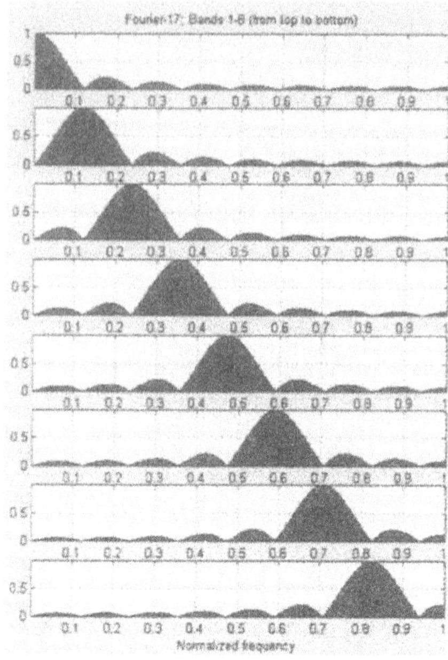

Figure 2-19. Arrangement of signal sub-bands for sliding window Fourier Transform

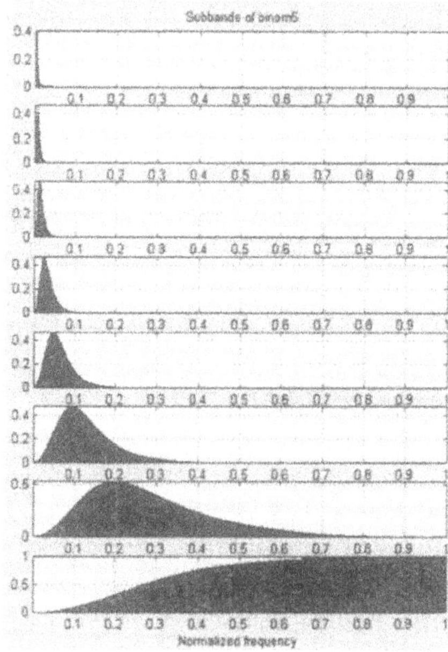

Figure 2-20. Arrangement of signal sub-bands in wavelet Binom5 decomposition

2.8 STOCHASTIC TRANSFORMATIONS AND STATISTICAL MODELS

Description of mathematical models of signal transformations will not be complete without considering statistical approach to signals and transformations. As it was mentioned in Sect. 2.1.2, results of signal processing and performance of imaging and measurement systems is usually evaluated on average. For statistical averaging, signal transformation characteristics such as, for instance, point spread function and frequency response, are, similarly to the signals, also modeled as random values or functions. In addition, one would need to statistically describe interaction between signals and transformation systems. The most useful models to describe the interaction are those of additive, multiplicative, Poisson noise, impulse and speckle noise models.

2.8.1 Additive, multiplicative, Poisson noise and impulse noise models

The model of ***additive signal independent noise model*** (ASIN-model) assumes signal transformation described as

$$b(x) = a(x) + n(x), \tag{2.8.1}$$

where $a(x)$ is input signal, $b(x)$ is output signal and $n(x)$ is a random process statistically independent on signal $a(x)$ and referred to as ***additive noise***. The assumption of statistical independence assumes that statistical averaging applied to signal $b(x)$ and to results of its transformations can be carried out on statistical ensembles of signals $\{a(x)\}$ and noise $\{n(x)\}$ separately:

$$AV_{\{a(x)\},\{n(x)\}}\{b(x)\} = AV_{\{a(x)\}}\{a(x)\} + AV_{\{n(x)\}}\{n(x)\}. \tag{2.8.2}$$

For ASIN-model, mean value $\overline{n} = AV_{\{n(x)\}}\{n(x)\}$ of noise is of no concern and it is natural to assume that $\overline{n} = 0$. With this assumption, mean value of the signal corrupted with additive zero mean noise is equal to the uncorrupted signal and standard deviation of its fluctuations is equal to standard deviation of noise:

$$AV_{\{n(x)\}}\{b(x)\} = a(x)\}. \tag{2.8.3}$$

$$\sqrt{AV_{\{n(x)\}}\{|b(x)-a(x)|^2\}}=\sqrt{AV_{\{n(x)\}}\{|n(x)|^2\}}. \tag{2.8.4}$$

For ASIN-model, ratio of signal mean value to standard that characterizes intensity of signal fluctuations with respect to signal mean value is proportional to signal mean value.

Typical statistical model of additive noise is Gaussian random process ([36]). Gaussian random process is completely statistically determined by its probability density called ***normal distribution***

$$p(n)=\frac{1}{\sqrt{2\pi}\sigma_n}\exp\left[-\frac{(n-\bar{n})^2}{2\sigma_n^2}\right], \tag{2.8.5}$$

where σ_n is its standard deviation and $\bar{n}$ is its mean value, which, for ASIN-model, is usually assumed to be zero, and by its ***correlation function***

$$R_n(x_1,x_2)=AV_{\{n(x)\}}\{n(x_1)\cdot n(x_2)\}. \tag{2.8.6}$$

According to the ***central limit theorem*** of the probability theory, normal distribution is a good model for probability density distribution of random values that are obtained as a sum of very many statistically independent random variables. For instance, it is a good model of the distribution of ***thermal noise*** caused by random fluctuations of velocities of electrons in electron current in signal sensors.

Random process is a ***stationary process***, if its correlation function depends only on distance between points x_1 and x_2: $R_n(x_1,x_2)=R_n(x_1-x_2)$. Fourier transform of correlation function of a stationary random process is called its ***spectral density***, or ***power spectrum***:

$$N(f)=\int_{-\infty}^{\infty}R_n(x)\exp(i2\pi fx)dx. \tag{2.8.7}$$

Alternatively, spectral density of a stationary random process can be defined as average, over all realizations of the process $\{n(x)\}$, of squared module of Fourier spectra of the realizations

$$N(f)=AV_{\{n(x)\}}\left\{\left|\int_{-\infty}^{\infty}n(x)\exp(i2\pi fx)dx\right|^2\right\}. \tag{2.8.8}$$

Random process is called ***uncorrelated*** if its spectral density is uniform function, or respectively, its correlation function is delta-function:

$$N(f) = N_0 = \sigma_n^2;$$
$$R_n(x) = N_0 \delta(x) = \sigma_n^2 \delta(x). \quad (2.8.9)$$

Uncorrelated Gaussian random process is conventionally called ***white noise***.

In a number of applications, additive noise is highly correlated. The correlation is most convenient to describe as concentration of noise power spectrum in certain areas of frequency domain. If noise power spectrum is concentrated only around a few isolated points in frequency domain, this type of additive noise is called ***moiré noise***.

Multiplicative noise (MN-) model assumes signal transformation defined as

$$b(x) = m(x) \cdot a(x), \quad (2.8.10)$$

where $m(x)$ is signal independent random process with mean equal to 1. Similarly to mean value of signal corrupted by additive zero mean noise, mean value of a signal corrupted by multiplicative noise is also equal to the mean value of non-corrupted signal:

$$AV\{b(x)\} = AV_{\{a(x)\}}\{a(x)\} \cdot AV_{\{n(x)\}}\{n(x)\} = AV_{\{a(x)\}}\{a(x)\}. \quad (2.8.11)$$

Variance of the corrupted signal is proportional to the variance of non-corrupted signal and variance of noise:

$$AV\{b^2(x)\} = AV_{\{a(x)\}}\{a^2(x)\} \cdot AV_{\{n(x)\}}\{n^2(x)\}. \quad (2.8.12)$$

For multiplicative noise, ratio of corrupted signal mean value to its standard deviation is equal to standard deviation of the multiplicative noise and does not depend on the signal mean value. This is a characteristic property of multiplicative noise.

Poisson noise model

Poisson noise model describes quantum fluctuations in sensors such as fluctuations of number of electrons in CCD or CMOS photo and X-ray sensitive sensors. For Poisson noise model, signal values are nonnegative integer numbers that have ***Poisson distribution*** probability density:

$$P(q) = \frac{\bar{q}^q}{q!}\exp(-\bar{q}), \tag{2.8.13}$$

where $\bar{q}$ is mean value of q. It is the property of Poisson distribution that standard deviation of q is also equal to $\bar{q}$:

$$\left(AV_q\left\{(q - \bar{q})^2\right\}\right)^{1/2} = \bar{q}. \tag{2.8.14}$$

For $\bar{q} \to \infty$, Poisson distribution, according to Moivre-Laplace theorem, tends to normal distribution with standard deviation $\sqrt{\bar{q}}$

$$P(q) = \frac{1}{\sqrt{2\pi\bar{q}}}\exp\left[-\frac{(q - \bar{q})^2}{2\bar{q}}\right]. \tag{2.8.15}$$

For Poisson distributions, ratio of standard deviation to mean value tends to zero with growth of the mean value:

$$\frac{\left[AV_q\left|q - AV_q(q)\right|^2\right]^{1/2}}{AV_q(q)} = \frac{\sqrt{\bar{q}}}{\bar{q}} \xrightarrow[\bar{q}\to\infty]{} 0, \tag{2.8.16}$$

which means that with growth of $\bar{q}$ fluctuations are becoming less and less noticeable.

Impulse noise model (ImpN-) model

$$b(x) = [1 - e(x)] \cdot a(x) + e(x)n(x) \tag{2.8.17}$$

assumes that noise is caused by two random factors: random binary process $e(x)$ that takes two values, zero and one, and an additive noise $n(x)$. ImpN-model is applicable to discrete signals. It is characteristic for signal transmission systems with random faults. The process $e(x)$ is specified by the probability that $e(x) = 1$ (the probability of error, or the probability of signal loss). Statistical properties of noise $n(x)$ can be arbitrary. If random process $n(x)$ takes only two values equal to minimum and maximum of $a(x)$, it is, in image processing jargon, referred to as ***pepper and salt noise***.

Justification for using ASIN-, MN- and ImpN-models and statistical characteristics of noise ensemble should be based on the design characteristics of the sensors, imaging and image transmission systems.

Examples of images distorted with additive and impulse noise are shown in Fig. 2-21. The noise level in images is characterized by the ratio PSNR of image dynamic range to standard deviation of noise. For impulse noise, standard deviation of noise is computed as standard deviation of difference between initial and noisy images.

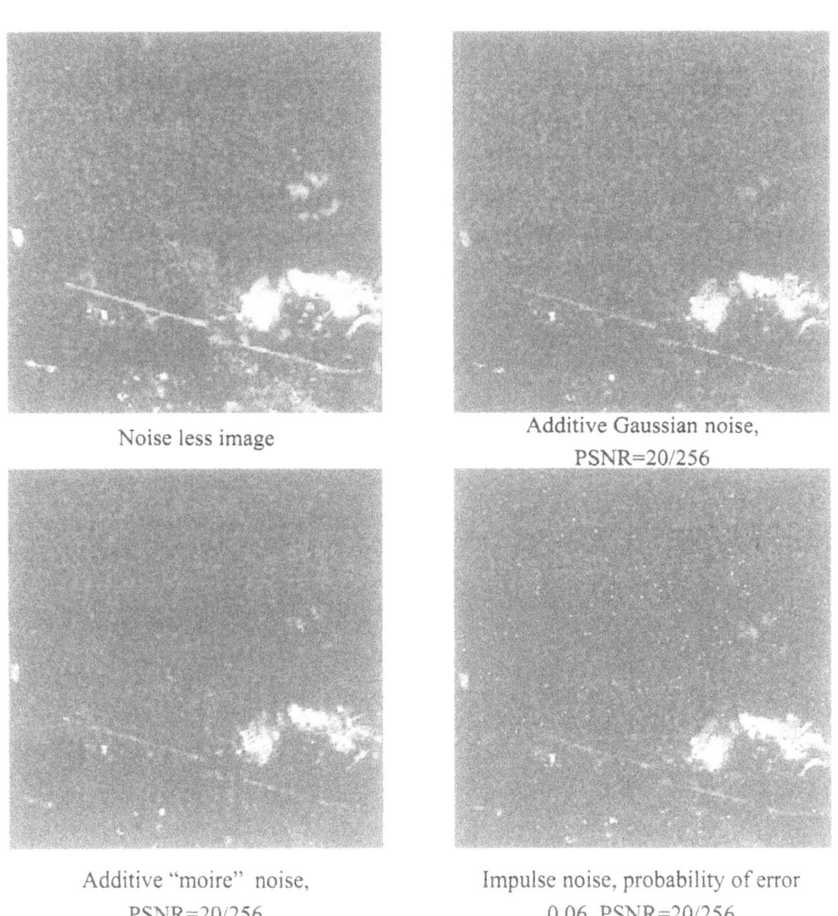

Figure 2-21. Examples of images with additive and impulse noise. Observe different visibility of different types of noise with the same standard deviation

2.8.2 Speckle noise model

Phenomenon of ***speckle noise*** is characteristic for coherent imaging systems such as holographic ones, synthetic aperture radar and ultrasound imaging systems. Speckle noise originates from property of objects to diffusely scatter irradiation and is caused by distortions introduced by wave front sensors. Conventionally it is associated with finite resolving power of sensors, or, in holographic systems, by the limitation of the area in which wave front is measured by sensors or recorded on holograms ([37]).

Let $a(\xi,\eta)\exp[i\theta(\xi,\eta)]$ be complex amplitude of wave front in the object plane that describes object reflectance/transmittance properties. For describing property of objects to diffusely scatter irradiation, $\theta(\xi,\eta)$ can be modeled as a random process. Diffuse uniform in all direction scattering correspond to spatially uncorrelated process $\theta(\xi,\eta)$. Specular scattering takes place if $\theta(\xi,\eta)$ is highly spatially correlated process.

Consider coherent imaging system with point spread function $h(x,y;\xi,\eta)$. Then, for input object $a(\xi,\eta)\exp(i\theta(\xi,\eta))$, output image formed by a sensor that is sensitive to squared magnitude of the output wave front is

$$b^2(x,y)=\left|\int_{-\infty}^{\infty}\int_{-\infty}^{\infty}a(\xi,\eta)\exp[i\theta(\xi,\eta)]h(x,y;\xi,\eta)d\xi\, d\eta\right|^2 . \qquad (2.8.18)$$

Let the object has a uniformly painted surface such that $a(\xi,\eta)=A_0$. Then:

$$b^2(x,y)=\left|A_0\int_{-\infty}^{\infty}\int_{-\infty}^{\infty}\exp[i\theta(\xi,\eta)]h(x,y;\xi,\eta)d\xi\, d\eta\right|^2=\left|b^{re}\right|^2+\left|b^{im}\right|^2 \qquad (2.8.19)$$

where

$$b^{re}=A_0\int_{-\infty}^{\infty}\int_{-\infty}^{\infty}\cos[\theta(\xi,\eta)]h(x,y;\xi,\eta)d\xi\, d\eta , \qquad (2.8.20)$$

$$b^{im}=A_0\int_{-\infty}^{\infty}\int_{-\infty}^{\infty}\sin[\theta(\xi,\eta)]\, h(x,y;\xi,\eta)\, d\xi\, d\eta \qquad (2.8.21)$$

If functions $\cos[\theta(\xi,\eta)]$ and $\sin[\theta(\xi,\eta)]$ are changing much faster then point spread function $h(x,y;\xi,\eta)$, or, in other words, many uncorrelated values of $\exp[i\theta(\xi,\eta)]$ are observed within aperture of the imaging system, the central limit theorem of the probability theory can be used to assert that b^{re} and b^{im} are normally distributed with zero mean:

$$\overline{b^{re}} = AV_\theta\left(b^{re}\right) = A_0 \int_{-\infty}^{\infty}\int_{-\infty}^{\infty} AV_\theta\left\{\cos[\theta(\xi,\eta)]\right\} h(x,y;\xi,\eta)\,d\xi\,d\eta = 0\,;$$

$$\overline{b^{im}} = AV_\theta\left(b^{im}\right) = A_0 \int_{-\infty}^{\infty}\int_{-\infty}^{\infty} AV_\theta\left\{\sin[\theta(\xi,\eta)]\right\} h(x,y;\xi,\eta)\,d\xi\,d\eta = 0\,.$$

(2.8.22)

Find correlation function of orthogonal components b^{re} and b^{im}. For b^{re},

$$R_{b^{re}}(\mathbf{x}_1,\mathbf{x}_2) = AV_\theta\left\{b^{re}(\mathbf{x}_1)\left[b^{re}(\mathbf{x}_2)\right]^*\right\} =$$

$$A_0^2 \int_{-\infty}^{\infty}\int_{-\infty}^{\infty} AV_\theta\left\{\cos[\theta(\boldsymbol{\xi}_1)]\cos[\theta(\boldsymbol{\xi}_2)]\right\} h(\mathbf{x}_1;\boldsymbol{\xi}_1)h^*(\mathbf{x}_2;\boldsymbol{\xi}_2)\,d\boldsymbol{\xi}_1\,d\boldsymbol{\xi}_2 =$$

$$A_0^2 \int_{-\infty}^{\infty}\int_{-\infty}^{\infty} R_\theta(\boldsymbol{\xi}_1,\boldsymbol{\xi}_2) h(\mathbf{x}_1;\boldsymbol{\xi}_1)h^*(\mathbf{x}_2;\boldsymbol{\xi}_2)\,d\boldsymbol{\xi}_1\,d\boldsymbol{\xi}_2\,, \qquad (2.8.23)$$

where $\mathbf{x} = (x,y)$, $\boldsymbol{\xi} = (\xi,\eta)$ and $R_\theta(\boldsymbol{\xi}_1,\boldsymbol{\xi}_2)$ is correlation function of $\cos[\theta(\xi,\eta)]$. In the above accepted assumption of applicability of the central limit theorem, $R_\theta(\boldsymbol{\xi}_1,\boldsymbol{\xi}_2)$ can be regarded, with respect to integration with $h(x,y;\xi,\eta)$, delta function which implies that

$$R_{b^{re}}(\mathbf{x}_1,\mathbf{x}_2) = \frac{1}{2}A_0^2 \int_{-\infty}^{\infty} h(\mathbf{x}_1;\boldsymbol{\xi})h^*(\mathbf{x}_2;\boldsymbol{\xi})\,d\boldsymbol{\xi}\,\cdot \qquad (2.8.24)$$

In the same way one can obtain that:

$$R_{b^{im}}(\mathbf{x}_1,\mathbf{x}_2) = R_{b^{re}}(\mathbf{x}_1,\mathbf{x}_2) = \frac{1}{2}A_0^2 \int_{-\infty}^{\infty} h(\mathbf{x}_1;\boldsymbol{\xi})h^*(\mathbf{x}_2;\boldsymbol{\xi})\,d\boldsymbol{\xi}\,\cdot \qquad (2.8.25)$$

From Eqs. 2.8.25 and 26 it follows that variances of orthogonal components b^{re} and b^{im} are

$$\sigma_b^2 = \overline{\left(b^{re}\right)^2} = \overline{\left(b^{im}\right)^2} = \frac{1}{2} A_0^2 \int_{-\infty}^{\infty}\int_{-\infty}^{\infty} \left|h(x,y;\xi,\eta)\right|^2 d\xi\, d\eta . \qquad (2.8.26)$$

In general, they are functions of coordinate (x,y) in image plane. For space invariant imaging system ($h(x,y;\xi,\eta) = h(x-\xi, y-\eta)$) they are coordinate independent:

$$\sigma_b^2 = \frac{1}{2} A_0^2 \int_{-\infty}^{\infty}\int_{-\infty}^{\infty} \left|h(x,y)\right|^2 dxdy \qquad (2.8.27)$$

Having defined probability density function of orthogonal components of $b^2(x,y)$ one can now find probability density of $b^2(x,y)$ as the probability density of sum of squared independent variables b^{re} and b^{im} with normal distribution. To this end, introduce random variables R and ϑ such that

$$b^{re} = R\cos\vartheta ; \quad b^{im} = R\sin\vartheta ; \quad b(x,y) = R^2 . \qquad (2.8.28)$$

Probability that b^{re} and b^{im} take values within a rectangle $d(b^{re})\mathrm{d}(b^{im})$ is

$$\frac{1}{2\pi\sigma_b^2}\exp\left[-\frac{\left(b^{re}\right)^2 + \left(b^{im}\right)^2}{2\sigma_b^2}\right] db^{re}\, db^{im} =$$

$$\frac{d\theta}{2\pi}\frac{R}{\sigma_b^2}\exp\left(-\frac{R^2}{2\sigma_b^2}\right) dR = \frac{d\theta}{2\pi}\frac{1}{2\sigma_b^2}\exp\left(-\frac{R^2}{2\sigma_b^2}\right) dR^2 . \qquad (2.8.29)$$

This is joint probability of variables $\left\{R^2, \vartheta\right\}$. It follows, therefore, that $\left\{R^2 = b(x,y)\right\}$ and ϑ are statistically independent, ϑ is uniformly distributed in the range $\left\{0, 2\pi\right\}$ and $b(x,y)$ has exponential distribution density:

$$P(b) = \frac{1}{2\sigma_b^2}\exp\left(-\frac{b}{2\sigma_b^2}\right), \qquad (2.8.30)$$

where σ_b^2 is defined by Eq. (2.8.27).

From Eqs.2.8.19 and 27 one can immediately see that mean value of $b^2(x,y)$ is:

$$\overline{b^2(x,y)} = 2\sigma_b^2 = A_0^2 \int_{-\infty}^{\infty}\int_{-\infty}^{\infty} |h(x,y)|^2 \, dxdy \quad (2.8.31)$$

It is the property of the exponential distribution that its standard deviation is equal to its mean value

$$\sigma_{b^2} = \left(\overline{\left[b^2(x,y) \right]^2} \right)^{1/2} = A_0^2 \int_{-\infty}^{\infty}\int_{-\infty}^{\infty} |h(x,y)|^2 \, dxdy \; . \quad (2.8.32)$$

Fluctuation of $b^2(x,y)$ around its mean value are called speckle noise. It is customary to characterize intensity of speckle noise by ration of its standard deviation to mean value called "***speckle contrast***":

$$Spckl\_contrast = \frac{\sigma_b^2}{\overline{b^2(x,y)}} \; . \quad (2.8.33)$$

It follows from Eqs. 2.8.31 and 32 that, for objects that scatter irradiation almost uniformly in the space, speckle contrast of their image obtained in coherent imaging systems is unity:

$$Spckl\_contrast = 1 \, . \quad (2.8.34)$$

Therefore speckle noise can be regarded to be multiplicative with respect to image mean value. Fig. 2-22 illustrates appearance of speckle noise and its spatial correlations that depend on imaging system point spread function (Eqs. 2.8.24 and 25).

In general, speckle noise in coherent imaging systems appears not only owing to insufficient resolving power of the systems. In fact, any distortions of the wave field that may happen in the sensor such as limitation of the signal dynamic range or signal quantization for its conversion into a digital signal for signal processing also cause appearance of speckle noise. These phenomena are discussed in Sect. 7.3.

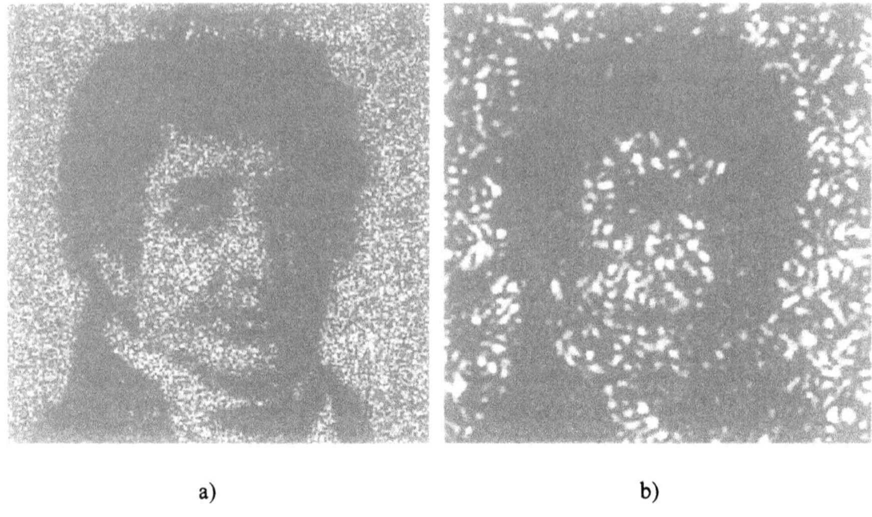

a) b)

Figure 2-22. Images with speckle noise for imaging systems with high (a) and low (b) resolving power.

References

1. W. B. Davenport, W. L. Root, An introduction to the theory of random signals and noise, McGraw-Hill, N.Y., 1958
2. B. R. Frieden, Numerical Methods in Probability Theory and Statistics, in: Computers in Optical Research, Methods in Applications, B. R. Frieden, ed., Springer Verlag, N.Y., 1980
3. A. Papoulis, Signal Analysis, McGraw-Hill, N.Y., 1977
4. J. M. Wozencraft, I. M. Jacobs, Principles of Communication Engineering, J. Wiley & Sons, N. Y., 1965
5. L.P. Yaroslavsky, The Theory of Optimal Methods for Localization of Objects in Pictures, In: Progress in Optics, Ed. E. Wolf, v.XXXII, Elsevier Science Publishers, Amsterdam, 1993
6. D. Gabor, A new microscopic principle, Nature, v. 161, 777-778, 1948
7. D. Gabor, Theory of Communication, Journ. Of IEE, vol. 93, 1946, pp. 429-457
8. E.N. Leith, J. Upatnieks, New techniques in wavefront reconstruction, JOSA, v. 51, 1469-1473, 1961
9. Yu. N. Denisyuk, Photographic reconstruction of the optical properties of an object in its own scattered radiation field, Dokl. Akad. Nauk SSSR, v. 1444, 1275-1279,
10. J.W. Goodman, Introduction to Fourier Optics, Second edition, McGraw-Hill, 1988
11. E.C. Titchmarsh, Introduction to the Theory of Fourier Integrals, Clarendon Press, Oxford, 1948
12. L. Mertz, Transformation in Optics, Wiley, N.Y., 1965
13. G. L. Rogers, Noncoherent Optical Processing, Wiley, N.Y., 1977
14. A. Lohmann, J. Opt. Soc. Am., vol. 55, 1555-1556, 1965
15. A.S. Marathay, J. Opt. Soc. Am., vol. 4, 1861-1868
16. R. V. L. Hartley, "A more symmetrical Fourier analysis applied to transmission problems," *Proc. IRE, v.* 30, 144-150 (1942)
17. R. Bracewell, Two-dimensional Imaging, Prentice Hall, 1995
18. J. Bertrand, P. Bertrand, J.-Ph. Ovarlez, The Mellin Transform, in: The Transforms and Applications Handbook, Ed. A.D. Poularikas, CRC Press, 1966
19. Janke, Emde, Loesh, Tafeln Hoeherer Functionen, 6 Aufgabe, B. G. Teubner Verlaggesellschaft, Stuttgart, 1960 ((see also http://functions.wolframcom)
20. P. Baues, Optoelectronics, v. 1, 37 (1969)
21. S. A. Collins, J. Opt. Soc. Am., vol. 60, p. 1168 (1970)
22. S. Abe, J.T. Sheridan, Opt. Commun., vol. 113, p. 385 (1995)
23. V. Namias, The fractional order Foutier transform and its applications in quantum mechanics, J. Inst. Maths. Applics., v. 25, pp. 241-265, 1980
24. A. Lohmann, Fourier curious, in: International trends in Optics, vol. 2, ICO series, J. C, Dainty, ed., 1994
25. A. Lohmann, Image rotation, Wigner rotation and the fractional Fourier transform, J. Opt. Soc. Am., vol. A10, pp. 2181-2186 (1993)
26. P. Pellat-Finet, G. Bonnet, Fractional order Fourier transform and Fourier optics, Opt. Comm., 111, 141 (1994)
27. H. M. Ozaktas, Z. Zalevsky, M. A. Kutay, The Fractional Fourier Transform, with application in Optics and Signal Processing, Wiley, N. Y., 2000
28. CRC Handbook of Mathematical Sciences, 6th Edition, W. H. Beyer, Editor, CRC Press, Roca Baton, 1978

29. H. Barrett, The Radon Transform and its Applications, In: Progress in Optics, ed. By E. Wolf, Elsevier, Amsterdam, vol. 21, pp. 219-286, 1984,.
30. S. R. Deans, Radon and Abel Transforms, in: The Transforms and Applications Handbook, Ed. A.D. Poularikas, CRC Press, 1966
31. Y. Meyer, Wavelets: Algorithms and Applications, SIAM, 1993
32. S. G. Mallat, A theory for multiresolution signal decomposition: The wavelet representation, IEEE Trans. Pattern Anal. Machine Intell., PAMI-31, pp. 674-693, 1989
33. S. Mallat, A Wavelet Tour of Signal Processing, 2nd edition, Academic Press, 1999
34. M. Vetterli, J. Kovacevic, Wavelets and subband coding, Prentice Hall, Englewood Cliffs, N.J., 1995
35. A. Haar, Zur theorie der orthogonalen funktionensysteme, Math. Annal., Vol. 69,pp. 331-371, 1910
36. A. Papoulis, Probability, Random Variables and Stochastic Processes, McGrow-Hill, N.Y., 1984
37. J.W. Goodman. Statistical Properties of Laser Speckle Patterns, In: Laser Speckle and related Phenomena, J. C. Dainty, Ed., Springer Verlag, Berlin, 1975

Chapter 3

DIGITAL REPRESENTATION OF SIGNALS

3.1 PRINCIPLES OF SIGNAL DIGITIZATION

As it was discussed in Sect. 2.1, signal digitization can be treated in general terms as determination, for each particular signal, of an index of the equivalency cell to which the signal belongs in the signal space and signal reconstruction can be treated as generating a representative signal of the cell from its index. This is, for instance, what we do when we describe everything in our life with words in speaking or writing. In this case this is our brain that makes the job of subdividing "signal space" into the "equivalency cells" of notions and recognizing which word (cell index) corresponds to what we want to describe. The volume of our vocabulary is about $\mathbf{10^5 \div 10^6}$ words. The variety of signals we have to deal with in signal processing and especially in optics and holography is immeasurably larger. One can see this from a simple example of the number of different images of, for instance, $\mathbf{500 \times 500}$ pixels with $\mathbf{256}$ gray levels. This number is $\mathbf{256^{500\times500}}$. No technical device will ever be capable of storing so many images for comparing them with input images to be digitized.

A solution of this problem of the digitization complexity is found in a two step digitization procedure. At the first step of the digitization, called ***signal discretization***, continuous signals are converted into discrete ones defined as a set (sequence) of numbers, that constitute signal discrete representation. At the second stage, called ***scalar*** (***element wise***) ***quantization***, this set of numbers is entry wise converted into a set of quantized numbers, which finally results in a digital signal that corresponds to the initial continuous one. In the same way as written speech is a sequence of letters selected from an alphabet, digital representation of signals is a sequence of samples that can take one of a finite set of quantized values. In terms of signal space, discretization can be regarded as introducing a coordinate system in the signal space and determining signal coordinates. Element wise quantization is then quantization of those coordinates. Equivalence cells of the signal space are therefore approximated be hypercubes with edges parallel to the coordinate axis.

3.2 SIGNAL DISCRETIZATION AS EXPANSION OVER A SET OF BASIS FUNCTIONS. TYPICAL BASIS FUNCTIONS AND CLASSIFICATION

In principle, there might be many different ways of converting continuous signals into discrete ones represented by a set of numbers. However, entire technological tradition and all technical devices that are used at present for such a conversion implement a method that can mathematically be modeled as computing coefficients that represent signals in their expansion over a set of basis functions. Let $\{\varphi_k^{(d)}(x)\}$ be the set of basis functions used for signal discretization. Then coefficients $\{\alpha_k\}$ of signal discrete representations are computed as

$$\alpha_k = \int_X a(x)\varphi_k^{(d)}(x)dx \tag{3.2.1}$$

This equation is a mathematical formulation of operations performed by signal sensors. Functions $\{\varphi_k^{(d)}(x)\}$ describe spatial sensitivity of sensors, that is sensor point spread function, or ***sensor aperture functions***.

Such a discretization with discretization basis $\{\varphi_k^{(d)}(x)\}$ assumes that signal can be reconstructed from its discrete representation $\{\alpha_k\}$ with the use of a reciprocal set $\{\varphi_k^{(r)}(x)\}$ of reconstructing basis functions:

$$a(x) \approx \sum_{k=0}^{N-1} \alpha_k \varphi_k^{(r)}(x) \tag{3.2.2}$$

Functions $\{\varphi_k^{(r)}(x)\}$ characterize point spread functions of signal reconstruction devices (such as, for instance, computer display), or their ***reconstruction aperture functions***. Approximation symbol in Eq. (3.2.2) is used to indicate that discretization $\{\varphi_k^{(d)}(x)\}$ and reconstruction $\{\varphi_k^{(r)}(x)\}$

basis functions are in practice not necessarily mutually orthogonal as they should be for the perfect signal reconstruction. In addition, signal reconstructed from its discrete representation is practically always only a more or less accurate copy of the initial signal. Even if discretization and reconstruction bases are orthogonal perfect reconstruction requires, in general, infinite number of terms N in the reconstruction equation (3.2.2) which is of cause not possible.

For different bases, the signal approximation accuracy for a given N might be different. Obviously, the bases that provide better approximation accuracy for a given N are preferable. Therefore the convergence speed of the expansion (3.2.2) with the number of terms N is the first issue one should check when selecting a basis. Approaches to the selection of bases optimal in this respect are discussed in Sect. 3.2.5.

The second issue is that of basis complexity. In principle, discretization and reconstruction basis functions can be implemented in the signal discretization and restoration devices (or, in digital processing, in computer memory), correspondingly, in form of templates. However, practically this is almost never possible because the number N of the required template functions is very large (for instance for images it is of order of magnitude of $10^5 \div 10^7$). This why these functions are usually generated (in hardware or software) from a "mother" function rather then "prefabricated" and stored.

3.2.1 Shift (convolution) basis functions

The simplest method for generating set of functions from one "mother" function is that of translation (coordinate shift):

$$\varphi_k(x) = \varphi(x - k\Delta x) \tag{3.2.3}$$

Functions generated in this way are called ***shift basis functions***. Elementary shift interval Δx is called ***the discretization interval***. It is usually, though not always, the same for the entire range of signal coordinates x. Positions $\{x = k\Delta x\}$ are called ***sampling points*** (for images, raster points).

Two family of functions exemplify shift basis functions: ***rectangular impulse functions***

$$\varphi_k(x) = rect\left(\frac{x - k\Delta x}{\Delta x}\right), \tag{3.2.4}$$

where

$$rect(\xi)=\begin{cases}1, & 0\le\xi<1\\ 0, & otherwise\end{cases} \tag{3.2.5}$$

and ***sinc-functions***

$$\varphi_k(x)=\mathrm{sinc}\left(\pi\,\frac{x-k\Delta}{\Delta x}\right), \tag{3.2.6}$$

where

$$\mathrm{sinc}(x)=\frac{\sin x}{x} \tag{3.2.7}$$

As it is clear from the definition (Eqs. (3.2.4-5)), rectangle impulse basis functions are orthogonal. For rectangle impulse basis functions

$$\left\{\varphi_k^{(d)}(x)=\frac{1}{\Delta x}rect\left(\frac{x-k\Delta x}{\Delta x}\right)\right\}, \tag{3.2.8}$$

used as the discretization basis, signals are represented their mean values over discretization intervals:

$$\alpha_k=\frac{1}{\Delta x}\int_X a(x)rect\left(\frac{x-k\Delta x}{\Delta x}\right)dx=\frac{1}{\Delta x}\int_{k\Delta x}^{(k+1)\Delta x}a(x)dx=\overline{a(k\Delta x)}. \tag{3.2.9}$$

Reciprocal signal reconstruction basis in this case is composed of functions

$$\left\{\varphi_k^{(r)}(x,k)=rect\left(\frac{x-k\Delta x}{\Delta x}\right)\right\} \tag{3.2.10}$$

and signals are reconstructed with this basis in a form of piece-wise constant functions (Figure 3-1):

$$a(x)\approx\tilde{a}(x)=\sum_{k=0}^{N-1}\overline{a(k\Delta x)}rect\left(\frac{x-k\Delta x}{\Delta x}\right) \tag{3.2.11}$$

Sinc- functions (sampling functions) are in a sense dual to rectangle impulse ones. As it follows from the equation:

$$\int_{-\infty}^{\infty} rect\left[\left(f + 1/2\Delta x\right)\Delta x\right]\exp\left(-i2\pi fx\right)dx =$$

$$\int_{-1/2\Delta x}^{1/2\Delta x} \exp\left(-i2\pi fx\right)df = \frac{1}{\Delta x}\text{sinc}\left(\frac{\pi x}{\Delta x}\right), \tag{3.2.12}$$

their Fourier spectrum is a rectangle impulse function (see Figure 3-2).

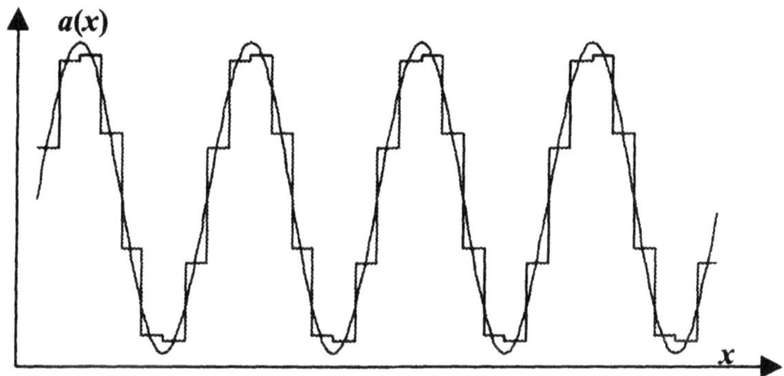

Figure 3-1. Continuous sinusoidal signal and its copy reconstructed after discretization with rectangle impulse discretization and reconstruction basis functions

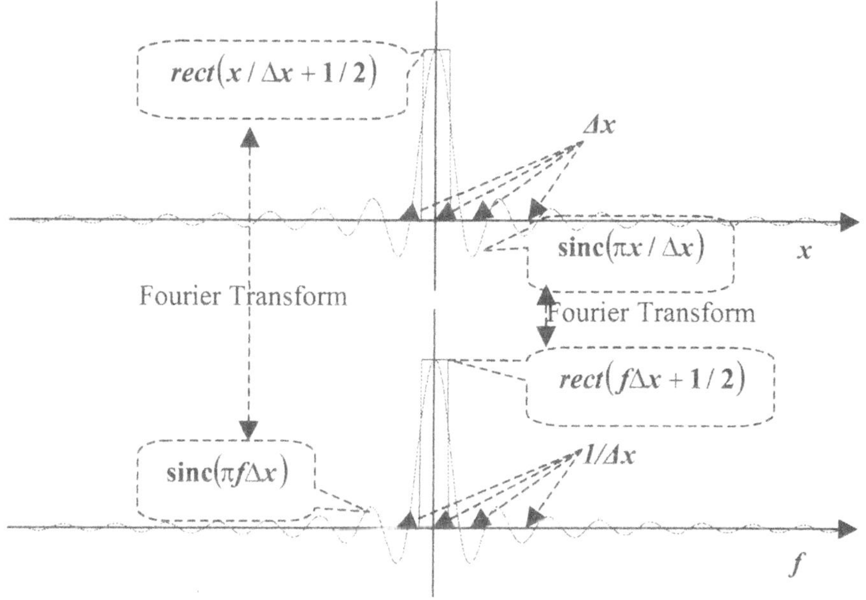

Figure 3-2. Duality of rectangle impulse and sinc basis functions

Functions $\left\{ \frac{1}{\Delta x} \mathrm{sinc}\left(\frac{\pi (x - k\Delta x)}{\Delta x} \right) \right\}$ can be used as the discretization basis:

$$\alpha_k = \frac{1}{\Delta x} \int_{-\infty}^{\infty} a(x) \mathrm{sinc}\left(\frac{\pi (x - k\Delta x)}{\Delta x} \right) dx \; . \tag{3.2.13}$$

Using ***shift theorem*** and ***convolution theorem*** of the theory of Fourier Transform, (Sect. 2.4) and Eq. (3.2.12) one can show that sinc-functions are mutually orthogonal:

$$\frac{1}{\Delta x} \int_{-\infty}^{\infty} \mathrm{sinc}\left(\frac{\pi (x - k\Delta x)}{\Delta x} \right) \mathrm{sinc}\left(\frac{\pi (x - l\Delta x)}{\Delta x} \right) dx =$$
$$\frac{1}{\Delta x} \int_{-\infty}^{\infty} \mathrm{sinc}\left(\frac{\pi x}{\Delta x} \right) \mathrm{sinc}\left(\frac{\pi (x - (k - l)\Delta x)}{\Delta x} \right) dx =$$
$$\int_{-1/2\Delta x}^{1/2\Delta x} \exp(-i2\pi f (k - l)\Delta x) df = \mathrm{sinc}(\pi (k - l)) = \delta(k - l) = \begin{cases} 1, k = l \\ 0, k \neq l \end{cases} \tag{3.2.14}$$

Therefore functions $\left\{ \mathrm{sinc}\left(\frac{\pi (x - k\Delta x)}{\Delta x} \right) \right\}$ are reconstruction functions for this basis:

$$a(x) \approx \tilde{a}(x) = \sum_{k=0}^{N-1} \alpha_k \, \mathrm{sinc}\left(\pi \frac{x - k\Delta x}{\Delta x} \right). \tag{3.2.15}$$

Because Fourier spectrum of functions $\left\{ \mathrm{sinc}\left(\frac{\pi (x - k\Delta x)}{\Delta x} \right) \right\}$ is bounded in the range $[-1/2\Delta x, 1/2\Delta x]$ (see Eq.3.2.12), basis composed of them generates signals that belong to the class of ***band limited functions***. It also follows from Eq. (3.2.15) that coefficients $\{\alpha_k\}$ of discrete representation of band limited functions are equal to functions' samples at equidistant points $\{x = k\Delta x\}$:

$$\alpha_k = \tilde{a}(k\Delta x) \tag{3.2.16}$$

This is where the name "sampling functions" derives from. According with the sampling theorem (see Section 3.3), the sampling functions are a gold standard for the discrete representation of signals with shift basis functions.

Eqs. 3.2.11 and 3.2.15 can also be regarded as convolution $\mathbf{conv}(.,.)$ of signals

$$\tilde{\tilde{a}}(x) = \sum_{k=0}^{N-1} \alpha_k \delta(x - k\Delta x) \tag{3.2.17a}$$

with reconstruction basis functions:

$$\tilde{a}(x) = \mathbf{conv}\left(\tilde{\tilde{a}}(x), \varphi_d(x)\right) = \mathbf{conv}\left(\sum_{k=0}^{N-1} \alpha_k \delta(x - k\Delta x), \varphi_d(x)\right) = \sum_{k=0}^{N-1} \alpha_k \varphi_d(x - k\Delta x). \tag{3.2.17b}$$

This justifies yet another name for shift basis functions, the ***convolution basis functions***, and treatment of Eq. (3.2.17, b) for reconstruction of continuous signals from their discrete representation as ***convolution based interpolation***.

3.2.2 Scale (multiplicative) basis functions

3.2.2.1 Sinusoidal basis functions

Scaling function coordinates is yet another method for generating basis functions from one mother function:

$$\varphi_k(x) = \varphi(kx) \tag{3.2.18}$$

Apparently, there are only a few examples of such ***scaling basis functions*** also called Fourier kernels [1]. All of them are derivatives of sinusoidal functions $\{\cos(2\pi kx / X)\}$ and $\{\sin(2\pi kx / X)\}$. These functions are orthogonal. Functions $\{\cos(2\pi kx / X)\}$ when used as a reconstruction basis, reconstruct signals

$$a(x) \approx \tilde{a}^{(c)} = \sum_{k=0}^{N-1} \alpha_k^{(c)} \cos(2\pi kx / X) \tag{3.2.19}$$

that are periodical, with period X, and even functions ($a(x)=a(-x)$). Functions $\{\sin(2\pi kx/X)\}$ used as a reconstruction basis reconstruct signals

$$a(x)\approx \tilde{a}^{(s)}=\sum_{k=0}^{N-1}\alpha_k^{(s)}\sin(2\pi kx/X) \tag{3.2.20}$$

that are also periodical, with period X, and odd functions $a(x)=-a(-x)$. Signal representation coefficients for these bases are found, respectively, as,

$$\alpha_k^c=\frac{1}{X}\int_{-X/2}^{X/2}a(x)\cos(2\pi kx/X)dx \tag{3.2.21}$$

$$\alpha_k^s=\frac{1}{X}\int_{-X/2}^{X/2}a(x)\sin(2\pi kx/X)dx \tag{3.2.22}$$

In reality, signals are, in general, neither even nor odd functions. There are two ways to avoid this obstacle in using sinusoidal basis functions. One way is form, from signals taken at interval $[0,X]$, auxiliary even or odd signals defined on interval $[0,2X]$ of double length as, respectively:

$$\tilde{\tilde{a}}^{(c)}(x)=\begin{cases} a(x), & 0\le x<X \\ a(2X-x), & X\le x<2X \end{cases} \tag{3.2.23}$$

and

$$\tilde{\tilde{a}}^{(s)}(x)=\begin{cases} a(x), & 0\le x<X \\ -a(2X-x), & X\le x<2X \end{cases} \tag{3.2.24}$$

With this signal "***symmetrization***", bases $\{\cos(\pi kx/X)\}$ and $\{\sin(\pi kx/X)\}$ can be used for discretization and reconstruction of arbitrary signals. Signal representation coefficients should be computed in this case as, respectively,

$$\tilde{\tilde{\alpha}}_k^{(c)}=\frac{1}{2X}\int_0^{2X}\tilde{\tilde{a}}(x)\cos(\pi kx/X)dx=\frac{1}{X}\int_0^{X}a(x)\cos(\pi kx/X)dx \tag{3.2.25}$$

$$\tilde{\tilde{\alpha}}_k^{(s)} = \frac{1}{2X}\int_0^{2X}\tilde{\tilde{a}}(x)\sin(\pi kx/X)dx = \frac{1}{X}\int_0^{X}a(x)\sin(\pi kx/X)dx \qquad (3.2.26)$$

and signals should be reconstructed from their discrete representations $\left\{\tilde{\tilde{\alpha}}_k^{(c)}\right\}$ and $\left\{\tilde{\tilde{\alpha}}_k^{(s)}\right\}$ as, respectively,

$$a(x) \approx \tilde{a} = \left(\sum_{k=0}^{N-1}\alpha_k^{(c)}\cos(\pi kx/X)\right)rect\left(\frac{x}{X}\right) \qquad (3.2.27)$$

or

$$a(x) \approx \tilde{a} = \left(\sum_{k=0}^{N-1}\alpha_k^{(s)}\sin(\pi kx/X)\right)rect\left(\frac{x}{X}\right) \qquad (3.2.28)$$

The second way to avoid the above mentioned limitations of sinusoidal basis functions $\{\cos(2\pi kx/X)\}$ and $\{\sin(2\pi kx/X)\}$ is to use them in combination:

$$a(x) = \sum_{k=0}^{N-1}\left(\alpha_k^{(c)}\cos(2\pi kx/X) + \alpha_k^{(s)}\sin(2\pi kx/X)\right), \qquad (3.2.29)$$

where the representation coefficients $\{\alpha_k^{(c)}\}$ and $\{\alpha_k^{(s)}\}$ are defined by Eqs. (3.2.21) and (3.2.22).

All these methods of signal expansion over set of sinusoidal basis functions are different versions of well known Fourier series expansion of a signal in the interval of the length X. Mathematical treatment of Fourier series expansion is much simplified if pairs of functions $\{\cos(2\pi kx/X)\}$ and $\{\sin(2\pi kx/X)\}$ are replaced by a complex exponential basis functions $\{\exp(i2\pi kx/X)\}$:

$$a(x) = \sum_k \alpha_k \exp(i2\pi kx/X) \qquad (3.2.30)$$

where coefficients $\{\alpha_k\}$ are complex numbers:

$$\alpha_k = \frac{1}{X}\int_{-X/2}^{X/2}a(x)\exp(-i2\pi kx/X)dx \qquad (3.2.31)$$

that can be regarded as $(1/X)$-normalized samples of signal Fourier transform spectrum taken at frequencies $\{k/X\}$.

While in signal discrete representation the number of terms in the expansion (3.2.31) is always finite, in classical Fourier series expansion assumes it is infinite. Limitation of the number of terms in Fourier series (3.2.20) is equivalent to dropping all signal components with frequencies higher than $(N-1)/X$, where N is the number of terms. This frequency band limitation can mathematically be explained as multiplying signal Fourier spectrum by a rectangle function $rect\left(\frac{f+(N-1)/2X}{(N-1)/X}\right)$ which is equivalent to the signal convolution with a corresponding sinc-function, Fourier transform of the rectangle function. As a result of the convolution, sinc-function originated oscillations may appear at signal edges (Figure 3-3). As it was already mentioned in Sect. 2.4.1, these specific distortions in reconstructed signals are called ***Gibb's effects***.

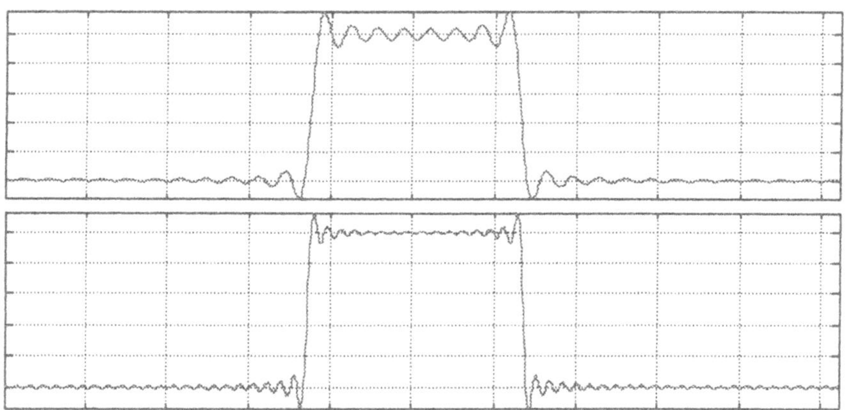

Figure 3-3. Gibb's effect oscillations at edges of a rectangular impulse for two different, lower (upper graph) and higher (lower graph) band limitation

It is remarkable that only frequency of Gibb's oscillations depend on band limitation, not their magnitude. With the increase of signal's bandwidth oscillation frequency grows while the level of the highest oscillation overshoot remains on the level of approximately **0.21** of the impulse height (for oscillations at edges of a rectangle impulse).

The fundamental property of sinusoidal basis functions is that, owing to their periodicity, they generate signals that are periodical functions. Periodical signals are dual to signals composed from shifted delta-functions introduced in Sect.3.2.1 (Eqs. 3.2.17, a,b). This duality will be discussed in some more details in Sect. 3.3.

Basis of sinusoidal functions is of especial importance for digital holography because of its direct connections with fundamental optical transforms such as Fourier and Fresnel transforms.

3.2.2.2 Walsh functions

For basis of exponential functions $\{\exp(i2\pi kx/X)\}$, generating basis functions by scaling their argument can be treated also as generating by means of multiplying mother function:

$$\exp(i2\pi kx/X) = \prod_{l=1}^{k} \exp(i2\pi x/X) \qquad (3.2.32)$$

There exists yet another family of orthogonal functions built with the same principle, ***Walsh functions***. Walsh functions are distinguished by the fact that they can assume only two values; either 1, or -1. They are generated by multiplication of clipped sinusoidal functions called the ***Rademacher functions***

$$rad_k(x) = sign\left(\sin\left(2^k \pi x/X\right)\right) \qquad (3.2.33)$$

Any two Rademacher functions are mutually orthogonal. However, the system of functions $\{rad_k(x)\}$ is incomplete in the same sense as families of functions $\{\cos(2\pi kx/X)\}$ and $\{\sin(2\pi kx/X)\}$ are incomplete. The Walsh functions are actually an extension of Rademacher functions to a complete system. These are defined as

$$wal_k(x) = \prod_{m=0}^{\infty} \left(rad_{m+1}(x)\right)^{k_m^{GC}}, \qquad (3.2.34)$$

where k_m^{GC} is the m-th digit of the so-called Gray code of number k. Gray code digits are generated from digits $\{k_m\}$ of binary representation of the number

$$k = \sum_{m=0}^{\infty} k_m 2^m; \quad k_m = 0,1 \qquad (3.2.35)$$

according to the following rule:

$$k_m^{GC} = k_m \oplus k_{m+1}, \qquad (3.2.36)$$

where $\oplus$ stands for modulo 2 addition.

Formula (3.2.34) reveals the origin of the Walsh functions. However, for the purpose of calculating their values another representation of the Walsh functions::

$$wal_k(\xi) = \prod_{m=0}^{\infty}\left[(-1)^{\xi_{m+1}}\right]^{k_m^{GC}} = (-1)^{\sum_{m=0}^{\infty} k_m^{GC}\xi_{m+1}} \tag{3.2.37}$$

is useful, where

$$\xi = x/X = \sum_{m=0}^{\infty}\xi_m 2^{-m};\ \xi_0 = 0; \qquad \xi_{m>0} = 0,1 \tag{3.2.38}$$

The Walsh functions are orthogonal on the interval $[0, X]$. Because of their origin as a product of Rademacher functions, multiplication of two Walsh functions results in another Walsh function with shifted an index or argument. The shift, called ***dyadic shift***, is determined through the bit-by-bit modulo 2 addition of functions' indices or, respectively, arguments:

$$wal_k(\xi)wal_l(\xi) = wal_{k\oplus l}(\xi); \tag{3.2.39}$$

$$wal_k(\xi)wal_k(\zeta) = wal_k(\xi \oplus \zeta) \tag{3.2.40}$$

One can regard $\sum_{m=0}^{\infty} k_m^{GC}\xi_{m+1}$ in the definition of Walsh as a scalar product of vectors $\{k_m^{GC}\}$ and $\{\xi_{m+1}\}$. In this sense one can treat basis of Walsh functions as a scale basis. Multiplicative nature of Walsh functions and exponential ones justifies yet another name for these families of bases, multiplicative bases.

Walsh function are akin to complex exponential functions $\{\exp(i2\pi kx/X)\}$ in one more respect. This can be seen if (-1) in Eq. (3.2.27) is replaced by $\exp(i\pi)$:

$$wal_k(\xi) == \exp\left(i\pi \sum_{m=0}^{\infty} k_m^{GC}\xi_{m+1}\right) \tag{3.2.41}$$

and from comparison of graphs of sinusoidal and Walsh functions shown in figure 3-4

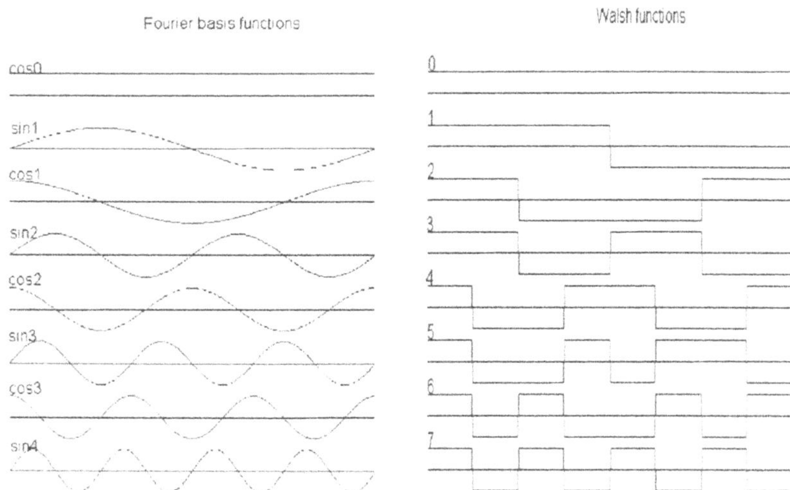

Figure 3-3. Sinusoidal and Walsh functions

As one can see from the figure, an important parameter that unites both families of functions is the number of zero crossing which coincides, for these functions, with their index k. As it was mentioned above, for sinusoidal functions, this parameter has an association with the notion of frequency. For Walsh functions, this parameter obtained name of "***sequency***"([2]).

Sequency wise ordering of Walsh functions is not accidental. In principle, ordering basis functions can be arbitrary. The only condition one should obey is matching between signal representation coefficients $\{\alpha_k\}$ and discretization and reconstruction basis functions $\{\varphi_k^{(d)}(x)\}$ and $\{\varphi_k^{(r)}(x)\}$. However almost always natural ways of natural ordering exist. For instance, for shift (convolution) bases the natural ordering is according to successive coordinate shifts. Sinusoidal basis functions are naturally ordered in frequency of signal Fourier transform. This why sequence (number of zero crossing) wise ordering of Walsh functions can also be regarded natural.

However, the advantage of such an ordering is more fundamental. The most natural requirement to ordering basis functions for signal representation is the ordering for which signal approximation error for finite number N of terms in signal expansion over the basis monotonically decreases with the growth of N. For Walsh functions, it happens that, for conventional analog signals, sequency wise ordering satisfies this requirement much better than other methods of ordering. Figure 3-5 illustrates this feature of Walsh function signal spectra. Graphs on the figure show averaged squared Walsh spectral coefficients of image rows as functions of coefficients' indices for sequence wise (Walsh) ordering (left) and for Hadamard ordering (right). In

Hadamard ordering, digits $\{k_m\}$ of binary representation of index k (3.2.35) rather than digits $\{k_m^{GC}\}$ are used for generating Walsh functions:

$$walhad_k(x) = (-1)^{\sum_{m=0}^{\infty} k_m \xi_{m+1}} \tag{3.2.42}$$

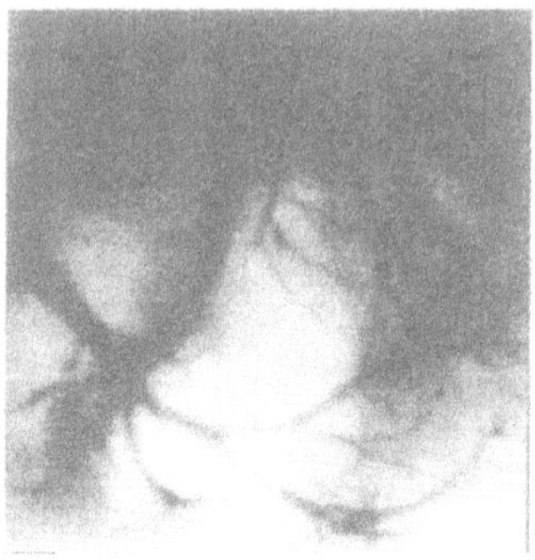

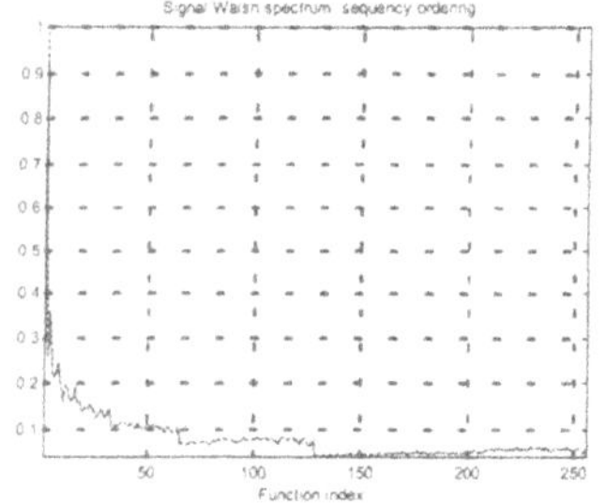

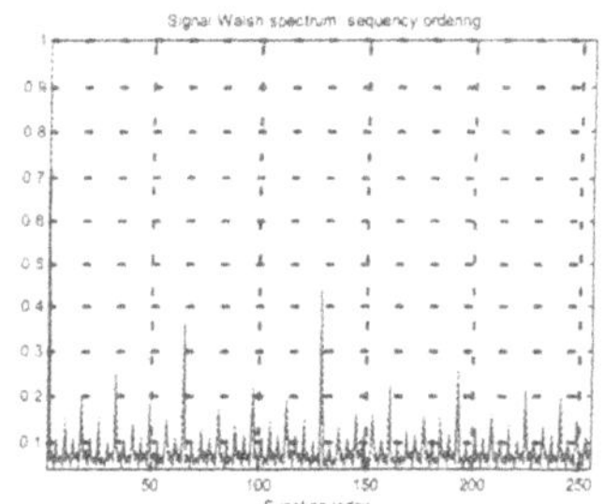

Figure 3-5. Comparison of image Walsh function spectra in Walsh (left) and Hadamard (right) ordering

In conclusion note that signal expansion over Walsh function basis approximates signals with piecewise functions just as it is the case for rectangle impulse basis functions.

3.2.3 Wavelets

A distinctive feature of shift (convolution) bases is that, for them, signal representation coefficients depend on signal values in the vicinity of the corresponding sampling point, and therefore they carry local information about signals. In contrast to them, signal discrete representation for scale (multiplicative) bases is "global": signal representation coefficients depend on the entire signal. Sometimes it is useful to have a combination of these two features in the signal discrete representation. This is achieved with wavelet basis functions built using a combination of shift and scaling of a mother function. A most immediate example of such a combination are ***Haar functions***. Haar functions are generated from from shift basis of rectangle impulse functions $\{rect(x - k\Delta x / \Delta x)\}$ and Rademacher functions $\{sign(\sin(2^k \pi x / X))\}$ in the following way:

$$har_k(x) = 2^{\hat{m}} rad_{\hat{m}+1}(x) rect\left(2^{\hat{m}} \frac{x}{X} - [k] \bmod 2^{\hat{m}}\right), \qquad (3.2.43)$$

where $\hat{m}$ is index of the most significant non-zero digit (bit) in binary representation of k (Eq. (3.2.35)) and $[k] \bmod 2^{\hat{m}}$ is residual from division of k by $2^{\hat{m}}$.

The Haar functions are orthonormal on the interval $[0, X]$. Graphs of first eight Haar functions are shown in Fig. 3-6.

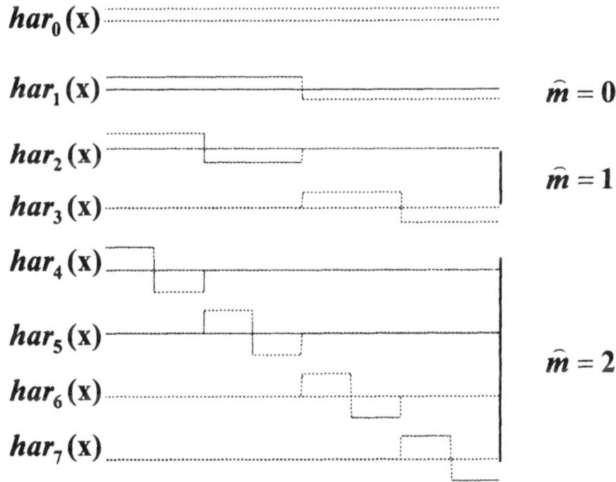

Figure 3-6. First eight Haar functions

The figure clearly reveals the way in which Haar functions are built. One can see from the figure, that Haar functions form groups in scale and that functions within each group are generated by shift in coordinates with shift interval coordinated with the scale. One can also see that signals reconstructed from their discrete representation over Haar basis functions belong to the same family of piece-wise constant functions as signals reconstructed in the base of rectangle impulse basis functions.

Numerical values of Haar functions may be found at each point by expressing them, as in the case of Walsh functions, via a binary code argument representations (3.2.35 and 38):

$$\boldsymbol{har}_k(\boldsymbol{x}) = \mathbf{2}^{\hat{m}}(-\mathbf{1})^{\xi_{\hat{m}+1}}\,\delta\left([\xi]_{\hat{m}} - (\boldsymbol{k})\,\mathbf{mod}\,\mathbf{2}^{\hat{m}}\right). \tag{3.2.44}$$

At present time, numerous other wavelet bases are known ([3]).

3.2.4 Discrete bases

In digital processing signals are represented in a discrete form by their samples obtained as a result of the discretization. For discrete signals, signal discretization and signal reconstruction procedures that take place in signal data compression carried out in digital computers (see Sect. 3.6) are implemented as decimation or pruning coefficients of signal transforms, and as, respectively, inverse transform of decimated or pruned vectors of the coefficients. Basis vectors that form matrices of orthogonal transforms are most usually derived from continuous basis functions described in Sect. 3.2. In Chapt. 4 we will show the derivation of discrete orthogonal transforms that are discrete analogs of such integral transforms as convolution, Fourier and Fresnel Transforms.

Main demands to discrete orthogonal transforms are basically similar to those to continuous bases. These are good ***energy compaction property*** (low MSE of signal reconstruction from pruned set of transform coefficients, see Sect. 3.2.5) and low computational complexity. They were the driving motives in the developments of fast orthogonal transforms in 1970-1980-th ([4-6]) when most of the transforms that are now in use were invented. In Table 3-1 most frequently used orthogonal transforms are listed and their definitions is provided. Computational complexity of transforms listed in the table, except that of Haar Transform, is of the order of magnitude $\boldsymbol{O}(\boldsymbol{N}\log\boldsymbol{N})$, where $\boldsymbol{N}$ is the transform dimensionality. Computational complexity of Haar Transform is even lower: $\boldsymbol{O}(\boldsymbol{N})$. Illustrative comparison of energy compaction properties of most frequently used orthogonal transforms will be given in Sect. 3.2.6.

More recently, discrete wavelet transforms gained considerable popularity because of their relatively good energy compaction properties, low computational complexity (similar to that of Haar Transform that is the simplest binary example of wavelets) and their multi resolution property (as it was mentioned, wavelet signal decomposition is hybrid, local and global).

Table 3-1. Discrete orthogonal transforms

Transform	Transform matrix $\boldsymbol{N} \times \boldsymbol{N}$

Transform	Transform matrix $N \times N$
Discrete Fourier Transforms (DFT)	Direct: $\mathbf{DFT} = \left\{ \frac{1}{\sqrt{N}} \sum_{k=0}^{N-1} a_k \exp\left(i2\pi \frac{kr}{N} \right) \right\}$
	Inverse: $\mathbf{IDFT} = \left\{ \frac{1}{\sqrt{N}} \sum_{k=0}^{N-1} a_k \exp\left(-i2\pi \frac{kr}{N} \right) \right\}$
Discrete cosine Transforms (DCT)	Direct: $\mathbf{DCT} = \left\{ \frac{2}{\sqrt{2N}} \cos\left(\pi \frac{(k+1/2)r}{N} \right) \right\}$
	Inverse: $\mathbf{IDCT} = \left\{ \frac{2}{(1+\delta(r))\sqrt{2N}} \cos\left(\pi \frac{(k+1/2)r}{N} \right) \right\}$
DCT-IV	$\mathbf{DCTIV} = \left\{ \frac{\sqrt{2}}{\sqrt{N}} \cos\left(\pi \frac{(k+1/2)(r+1/2)}{N} \right) \right\}$
	$\mathbf{IDCTIV} = \mathbf{DCTIV}$
Discrete Sine Transform (DST)	$\mathbf{DST} = \left\{ \frac{1}{\sqrt{2N}} \sin\left(2\pi \frac{(k+1)(r+1)}{2N+1} \right) \right\}$
	$\mathbf{IDST} = \mathbf{DST}$
DST-IV	$\mathbf{DSTIV} = \left\{ \frac{1}{\sqrt{2N}} \sin\left(2\pi \frac{(k+1/2)(r+1/2)}{2N} \right) \right\}$
	$\mathbf{IDSTIV} = \mathbf{DSTIV}$
Walsh Transform	$\mathbf{WAL} = \left\{ \frac{1}{\sqrt{N}} (-1)^{\sum_{m=0}^{n-1} k_{n-m-1}^{GC} r_m} \right\}$; $n = \log_2 N$; $\{r_m\}$ are digits of binary representation of row index r; $\{k_{n-m-1}^{GC}\}$ are digits of Gray code of bit reversed digits $\{k_{n-m-1}\}$ of binary representation $\{k_m\}$ of column index; $m = 0,1,\ldots,n-1$; $\{k_s^{GC} = k_s \oplus k_{s+1}\}$; $\oplus$ - modulo 2 binary addition.
Haar Transform	$\mathbf{HAAR} = \left\{ \frac{2^{\hat{m}/2}}{\sqrt{N}} (-1)^{k_{n-\hat{m}-1}} \prod_{m=0}^{\hat{m}-1} \delta(k_{n-m} \oplus r_m) \right\}$; $n = \log_2 N$; $\hat{m}$ - index of the most significant nonzero digit in binary representation of row index r;

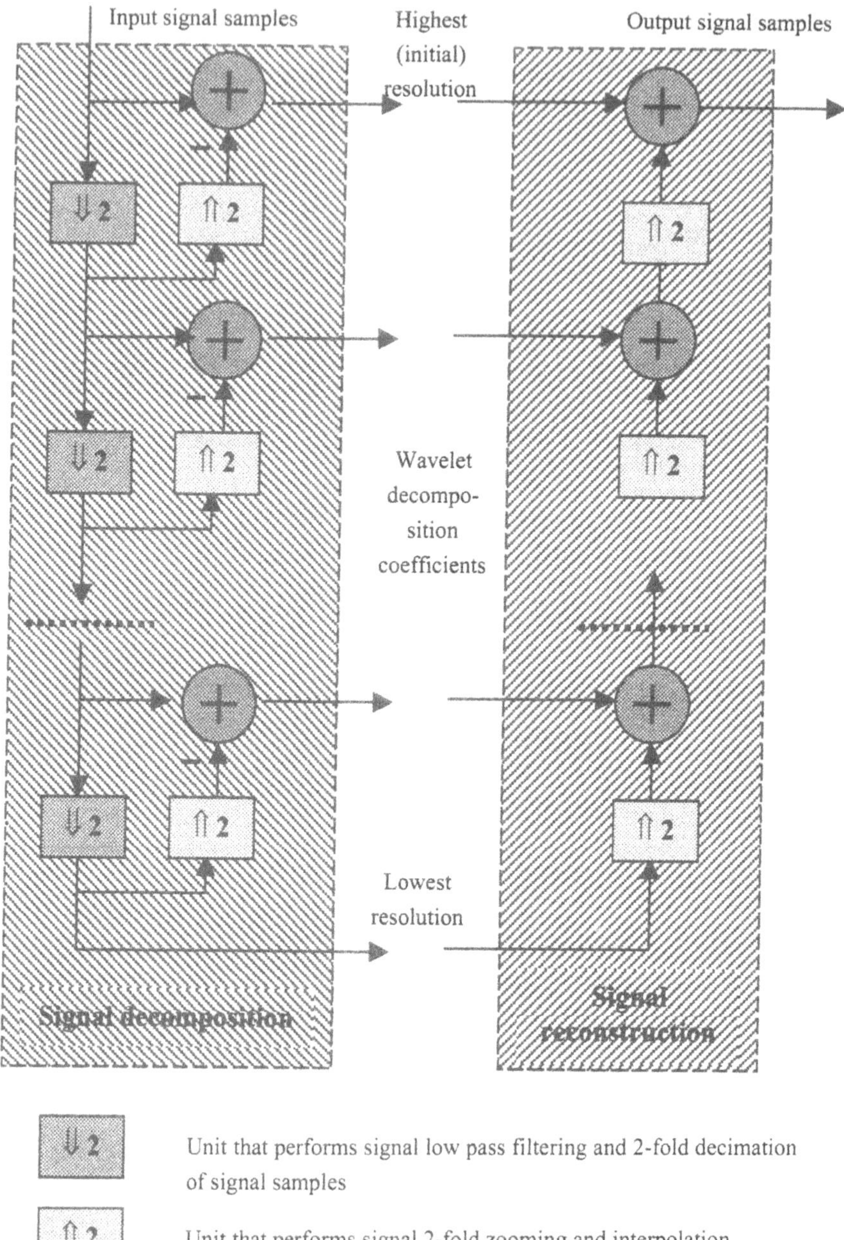

Figure 3-6. Schematic diagram of discrete wavelet (multi resolution) signal decomposition and reconstruction

There are many ways to introduce and define wavelet transform as well as there is unlimited number of modifications of wavelet bases. Fig. 3-6 represents one of the most simple and straightforward method, the algorithmic one. As it is shown in the figure, signal wavelet decomposition is made in several steps using two basic signal processing procedures: decimation and interpolation. As a result, signal of N samples is decomposed into $\log_2 N$ sub-signals with descending resolution (number of samples) each of which represents different sub-bands of the signal, from high to low frequencies with the lowest level representing signal dc component The way how low pass filtering for decimation and how signal interpolation from its decimated copy are implemented defines type of the wavelets. This method of signal representation is known also by names multiresolution, sub-band representation and pyramidal coding (see, for instance, [3,7-9]).

3.2.5 Optimality of bases. Karhunen-Loeve, Hotelling and Singular Values Decomposition Transforms

In this section, we address the problem of optimality of bases in terms of the accuracy of signal approximation with a limited number of terms in their expansion over the bases. For generality, we will assume that signals belong to a certain family (ensemble) of signals and that the accuracy of signal representation is evaluated on average over this family.

Let $\{a(x,\omega)\}$ be a set (in terms of a parameter $\omega \in \Omega$) of signals to be represented by their approximations in a form of finite expansions:

$$a(x,\omega) \cong \tilde{a}(x,\omega) = \sum_{k=0}^{N-1} \alpha_k(\omega)\varphi_k(x) \tag{3.2.45}$$

over orthonormal basis functions $\{\varphi_k(x)\}$ such that

$$\int_X \varphi_k(x)\varphi_l^*(x)dx = \delta(k,l) = 0^{|k-l|}. \tag{3.2.46}$$

Let also evaluate the approximation accuracy in terms of mean squared approximation error (MSE):

$$|\varepsilon(\omega)|^2 = \int_X |a(x,\omega) - \tilde{a}(x,\omega)|^2 dx = \int_X \left| a(x,\omega) - \sum_{k=0}^{N-1} \alpha_k(\omega)\varphi_k(x) \right|^2 dx \tag{3.2.47}$$

Optimal values of $\{\alpha_k(\omega)\}$ that minimize MSE can be found by equaling to zero derivatives of Eq. (3.2.47) over $\{\alpha_k(\omega)\}$:

$$\frac{\partial}{\partial \alpha_k(\omega)^*} \int_X \left| a(x,\omega) - \sum_{k=0}^{N-1} \alpha_k(\omega)\varphi_k(x) \right|^2 dx = 0, \tag{3.2.48}$$

from which one can obtain:

$$\alpha_k(\omega) = \int_X a(x,\omega)\varphi_k^*(x)dx, \tag{3.2.49}$$

where $*$ stands for a complex conjugate. This means that, for basis $\{\varphi_k(x)\}$ used as a reconstruction basis, the reciprocal discretization basis is composed of functions $\{\varphi_k^*(x)\}$.

From Eq. (3.2.48) it follows that minimal MSE is equal to

$$|\varepsilon(\omega)|^2_{\min} = \int_X |a(x,\omega)|^2 dx - \sum_{k=0}^{N-1} |\alpha_k(\omega)|^2 \tag{3.2.50}$$

By an appropriate selection of bases functions $\{\varphi_k(x)\}$ one can further minimize average MSE over the given set Ω of signals as defined by parameter ω :

$$\begin{aligned}
\{\varphi_k(x)\}_{opt} &= \arg\min_{\{\varphi_k(x)\}} \left\{ AV_\Omega \left(\int_X |a(x,\omega)|^2 dx - \sum_{k=0}^{N-1} |\alpha_k(\omega)|^2 \right) \right\} = \\
&\arg\min_{\{\varphi_k(x)\}} \left\{ AV_\Omega \left(\int_X |a(x,\omega)|^2 dx \right) - AV_\Omega \left(\sum_{k=0}^{N-1} |\alpha_k(\omega)|^2 \right) \right\} = \\
&\arg\max_{\{\varphi_k(x)\}} \left\{ AV_\Omega \left(\sum_{k=0}^{N-1} |\alpha_k(\omega)|^2 \right) \right\} = \\
&\arg\max_{\{\varphi_k(x)\}} \left\{ AV_\Omega \left(\sum_{k=0}^{N-1} \iint_X a(x,\omega)a^*(y,\omega)\varphi_k^*(x)\varphi_k(y)dxdy \right) \right\} = \\
&\arg\max_{\{\varphi_k(x)\}} \left\{ \sum_{k=0}^{N-1} \iint_X AV_\Omega \left(a(x,\omega)a^*(y,\omega) \right) \varphi_k^*(x)\varphi_k(y)dxdy \right\} = \\
&\arg\max_{\{\varphi_k(x)\}} \left\{ \sum_{k=0}^{N-1} \iint_X R_a(x,y)\varphi_k^*(x)\varphi_k(y)dxdy \right\}
\end{aligned} \tag{3.2.51}$$

where AV_Ω means averaging over the signal set Ω. Function $R_a(x,y)$

$$R_a(x,y) = AV_\Omega\left(a(x,\omega)a^*(y,\omega)\right) \tag{3.2.52}$$

is called "correlation function" of the set of signals $\{a(x,\omega)\}$. Denote

$$R\Phi(x) = \int_X R_a(x,y)\varphi_k(y)dy \tag{3.2.53}$$

Then optimal bases functions $\{\varphi_k(x)\}$ are found from

$$\{\varphi_k(x)\} = \underset{\{\varphi_k(x)\}}{\arg\max}\left\{\sum_{k=0}^{N-1}\int_X R\Phi(x)\varphi_k^*(x)dx\right\} \tag{3.2.54}$$

From Cauchy-Schwarz-Buniakowsky inequality

$$\int_X f_1(x)f_2^*(x)dx \le \left(\int_X |f_1(x)|^2 dx\right)^{1/2}\left(\int_X |f_2(x)|^2 dx\right)^{1/2} \tag{3.2.55}$$

it follows that solution of Eq. (3.2.54) is

$$R\Phi(x) = \int_X R_a(x,y)\varphi_k(y)dy = \lambda_k\varphi_k(x) \tag{3.2.56}$$

This means that optimal bases functions are eigen functions of integral equation Eq.(3.2.56) with kernel defined by the correlation function (3.2.52) of the set of signals ([10,11]). Such bases are called ***Karhunen-Loeve bases***.

Similar reasoning can be applied to integral representation of signals. Let signals $a(x,\omega)$ are approximated by their integral representation over self-conjugate basis $\varphi(x,f)$:

$$a(x,\omega) \cong \tilde{a}(x,\omega) = \int_F \alpha(f,\omega)\varphi(x,f)df. \tag{3.2.57}$$

Then mean squared approximation error

$$AV_{\Omega}|\varepsilon(\omega)|^2 = AV_{\Omega}\left\{\left|a(x,\omega) - \int_F \alpha(f,\omega)\varphi(x,f)df\right|^2\right\} \tag{3.2.58}$$

will be minimal if signal integral representation $\alpha(f,\omega)$ is obtained as:

$$\alpha(f,\omega) = \int_X a(x,\omega)\varphi^*(x,f)dx \tag{3.2.59}$$

and basis $\alpha(x,f)$ satisfies integral equation

$$\int_X R_a(x,y)\varphi(y,f)dy = \lambda(f)\varphi^*(x,f), \tag{3.2.60}$$

where $R_a(x,y)$, as before, is correlation function (Eq. 3.2.52) of the set of signals $a(x,\omega)$.

An important special case is that of signals with correlation function that depends only on the difference of its arguments:

$$R_a(x,y) = R_a(x-y) \tag{3.2.61}$$

In this case the basis optimality condition of Eq. 3.2.60 takes the form:

$$\int_{-\infty}^{\infty} R_a(x-y)\varphi(y,f)dy = \lambda(f)\varphi^*(x,f). \tag{3.2.62}$$

As one can see this condition is satisfied with exponential basis functions

$$\varphi(x,f) = \exp(i2\pi fx). \tag{3.2.63}$$

Two important conclusions follow from this result:

- Integral Fourier transform is an MSE optimal Karhunen-Loeve transform for signals with correlation function that depends only on difference of its arguments;
- The best, in MSE sense, approximation of such signals is their approximation with functions whose Fourier spectrum is equal to zero outside a certain bandwidth [-*F*,*F*] (***band limited functions***):

$$\tilde{a}(x,\omega)=\int_{-F}^{F}\alpha(f,\omega)\varphi(x,f)df\ . \qquad (3.2.64)$$

that defines minimal approximation MSE:

$$\left[AV_{\Omega}|\varepsilon(\omega)|^{2}\right]_{min}=\int_{-\infty}^{-F}AV_{\Omega}\left\{|\alpha(f,\omega)|^{2}\right\}df+\int_{F}^{\infty}AV_{\Omega}\left\{|\alpha(f,\omega)|^{2}\right\}df \qquad (3.2.65)$$

The discrete equivalent of the Karhunen-Loeve expansion is ***Hotelling expansion*** or method of ***principal components*** ([12]).

Let $\mathbf{a}_{\omega}=\{a_k\},\ k=0,1,...,N-1$ is a discrete signal from an ensemble of signals Ω defined by parameter ω. Find an orthogonal transform $\mathbf{T}$

$$\mathbf{T}_N=\{\varphi_{k,r}\};\qquad \mathbf{T}_N\left(\mathbf{T}_N^{*}\right)^{tr}=\mathbf{I}_N; \qquad (3.2.66)$$

$$\boldsymbol{\alpha}=\mathbf{T}_N\mathbf{a}=\left\{\alpha_r^{(\omega)}=\sum_{k=0}^{N-1}a_k\varphi_{k,r}\right\} \qquad (3.2.67)$$

that minimizes, on average over the signal ensemble Ω, Euclidean distance

$$AV_{\Omega}\left[d\left(\mathbf{a}_{\omega},\tilde{\mathbf{a}}_{\omega}\right)\right]=AV_{\Omega}\left[\sum_{k=0}^{N-1}\left(a_k^{\omega}-\tilde{a}_k^{\omega}\right)^2\right] \qquad (3.2.68)$$

between signals and their approximation

$$\tilde{\mathbf{a}}_{\omega}=\left\{a_k=\sum_{r=0}^{\tilde{N}-1}\alpha_r^{(\omega)}\varphi_{k,r}^{*}\right\}. \qquad (3.2.69)$$

By virtue of Parseval's relation, the transform is a solution of the equation

$$\mathbf{T}=\underset{\{\varphi_{k,r}\}}{\arg\max}\,AV_{\Omega}\left[\sum_{r=0}^{\tilde{N}-1}\left|\alpha_r^{(\omega)}\right|^2\right]=$$

$$\underset{\{\varphi_{k,r}\}}{\arg\max}\,AV_{\Omega}\left\{\sum_{r=0}^{\tilde{N}-1}\sum_{k=0}^{N-1}\sum_{\substack{n=0\\ \{\varphi_{k,r}\}}}^{N-1}a_k^{\omega}\left(a_n^{\omega}\right)^{*}\varphi_{k,r}\left(\varphi_{n,r}\right)^{*}\right\}. \qquad (3.2.70)$$

Denote

$$R_a(k,n) = AV_{\Omega}\left[a_k^{(\omega)}\left(a_n^{(\omega)}\right)^*\right] \tag{3.2.71}$$

Then transform optimization equation (3.2.54) becomes:

$$\mathbf{T} = \underset{\{\varphi_{k,r}\}}{\arg\max}\left\{\sum_{r=0}^{\tilde{N}-1}\sum_{k=0}^{N-1}\sum_{l=0}^{N-1} R_a(k,n)\varphi_{k,r}\left(\varphi_{n,r}\right)^*\right\} =$$
$$\underset{\{\varphi_{k,r}\}}{\arg\max}\left\{\sum_{r=0}^{\tilde{N}-1}\left(\sum_{n=0}^{N-1}\left(\varphi_{n,r}\right)^*\sum_{k=0}^{N-1} R_a(k,n)\varphi_{k,r}\right)\right\}. \tag{3.2.72}$$

From Cauchy-Schwarz-Buniakowsky inequality

$$\left(\sum_{k=0}^{N-1} a_k b_k\right)^2 \le \sum_{k=0}^{N-1}\left|a_k\right|^2 \cdot \sum_{k=0}^{N-1}\left|b_k\right|^2 \tag{3.2.73}$$

that become equality iff $\{a_k = \lambda b_k\}$, where λ is an arbitrary constant, it follows that the transform optimization equation is a transform is reduced to the equation

$$\sum_{k=0}^{N-1} R_a(k,n)\varphi_{k,r} = \lambda_r \varphi_{n,r}. \tag{3.2.74}$$

If no ensemble averaging is required when evaluating pruned transform coefficients reconstruction, Eq. 3.2.74 becomes

$$\sum_{k=0}^{N-1} a_k\left(a_n\right)^* \varphi_{k,r} = \lambda_r \varphi_{n,r}. \tag{3.2.75}$$

Signal decomposition with a transform defined by basis functions that are solutions of Eq. 3.2.75 is called ***singular value decomposition*** (SVD.
In matrix denotations, Eqs. 3.2.74 and 75 are:

$$\mathbf{R}_a \mathbf{T}_N = \mathbf{T}_N \Lambda_N \tag{3.2.76}$$

and

$$\mathbf{a} \cdot \left(\mathbf{a}^*\right)^{tr} \mathbf{T}_N = \mathbf{T}_N \Lambda_N, \tag{3.2.77}$$

where $\mathbf{R}_a = \{R_a(k,n)\} = AV_\Omega\left[\mathbf{a}\cdot(\mathbf{a}^*)^{tr}\right]$ is signal correlation matrix, $\mathbf{a} = \{a_k\}$ - vector-column of signal samples, $(\mathbf{a}^*)^{tr}$ - complex conjugate transposed vector (vector-row), $\Lambda_N = (\lambda_r\delta(k,r))$ - a diagonal matrix. From Eqs. (3.2.76-77) one can see that one of distinctive features of Hotelling and SVD transforms is that they diagonalize signal correlation matrix:

$$(\mathbf{T}_N^*)^{tr}\mathbf{R}_a\mathbf{T}_N = \Lambda_N \tag{3.2.78}$$

or, respectively, matrix obtained as “outer product” $\left[\mathbf{a}\cdot(\mathbf{a}^*)^{tr}\right]$ of signal vectors:

$$(\mathbf{T}_N^*)^{tr}\left[\mathbf{a}\cdot(\mathbf{a}^*)^{tr}\right]\mathbf{T}_N = \Lambda_N \tag{3.2.79}$$

Eqs. (3.2.55, 73-76) determine MSE optimal bases and transforms via signal correlation function, correlation matrix or outer product. One can also deduce from the above reasoning that each particular orthogonal transform can be regarded to be MSE optimal for a certain class of signals whose correlation function (matrix) satisfies these equations.

The theoretical value of Karhunen-Loeve, Hotelling and SVD transforms in signal processing is that they provide the lower bound for compact (in terms of number of terms in signal decomposition) signal representation for MSE criterion of signal approximation. Their practical value is, however, quite limited because of high complexity of generating transform basis functions and matrices and computing signal representation coefficients.

3.2.6 Two dimensional and multi dimensional bases

Two-dimensional and multi-dimensional basis functions are most frequently (although not always) generated as separable functions of two variables by the product of two one-dimensional ones. The underlying reason is the easiness of hardware and software generating the basis functions and lower computational complexity of computing signal representation coefficients

In the case of separable basis functions, which are the product of two one-dimensional functions, the job of evaluating of a scalar product is reduced to performing two one-dimensional summations rather then one two-dimensional:

$$\alpha_{r,s} = \sum_{k=0}^{N_1-1}\sum_{l=0}^{N_2-1} a_{k,l}\varphi_{r,s}^{(d)}(k,l) = \sum_{k=0}^{N_1-1}\sum_{l=0}^{N_2-1} a_{k,l}\varphi_r^{(d1)}(k)\varphi_s^{(d2)}(l) =$$

$$\sum_{k=0}^{N_1-1}\varphi_s^{(d2)}(l)\sum_{l=0}^{N_2-1} a_{k,l}\varphi_r^{(d1)}(k) \tag{3.2.80}$$

As one can see from Eq. 3.2.80, direct computation of 2-D sum requires $N_1^2 N_2^2$ multiplication/summation operations while two separable 1-D summations require $N_1 N_2 (N_1 + N_2)$ operations. For large 2_D arrays, saving is very substantial. Note that in separable two- and multi-dimensional bases multiplicand 1-D basis functions are not necessarily should be identical, though most frequently they are.

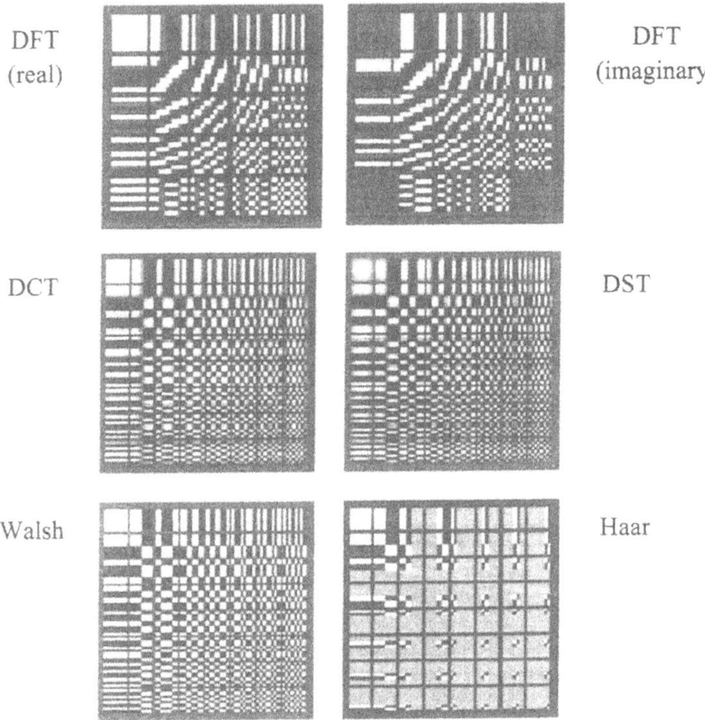

Figure 3-7. Examples of 8x8 sets of 2-D basis functions for DFT, DST, Walsh and Haar transforms displayed as images

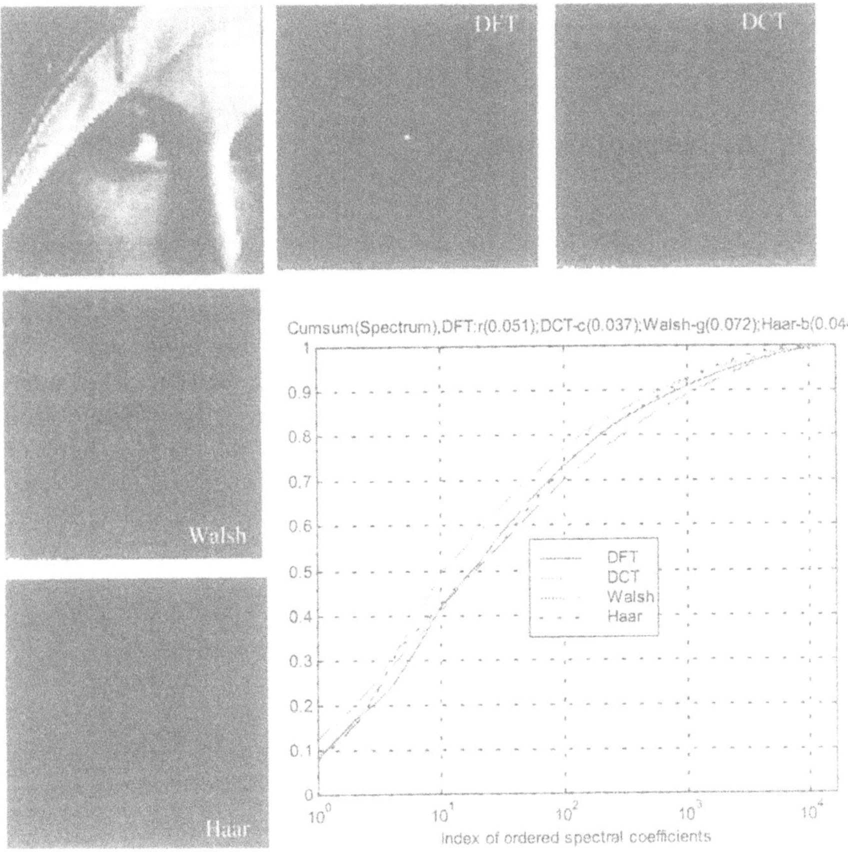

Figure 3-8. Comparison of image DFT,DCT, Walsh and Haar spectra. Graph illustrates the speed with which magnitude wise sorted spectral coefficients converge to the entire image signal energy.

Figures 3-7 demonstrates examples of sets of 2-D discrete bases functions for DFT, DCT, Walsh and Haar transforms and Figure 3-8 shows image spectra in these bases and how rapidly spectrum pruning error converges to zero for different bases. One can see that, for this particular image, fraction of the coefficients that contain 90% of the signal energy is 0,051 for DFT, 0.037 for DCT, 0.072 for Walsh transform and 0.044 for Haar transform which evidences that DCT spectrum truncation error converges to zero faster than that for other transforms. In Sect. 3.6, we will review applications of. signal discrete orthogonal transforms to data compression, and in Chapt. 8 we will show their use for signal denoising and deblurring.

3.3 SHIFT (CONVOLUTION) BASES FUNCTIONS AND SAMPLING THEOREM

3.3.1 1-D sampling theorem

The most widely used method for signal/image discretization is signal/image ***sampling*** by sensing them with a set of sensors that are equidistantly placed in signal/image coordinates. The sampling is carried out either by a mechanical or electronic signal/image scanning with a single sensor or with an array of sensors that work in parallel. The latter is exemplified by modern CCD and CMOS TV cameras. Remarkable enough that image sampling was first invented by the Nature in a form of a compound and retinal eyes (Fig. 3-9)

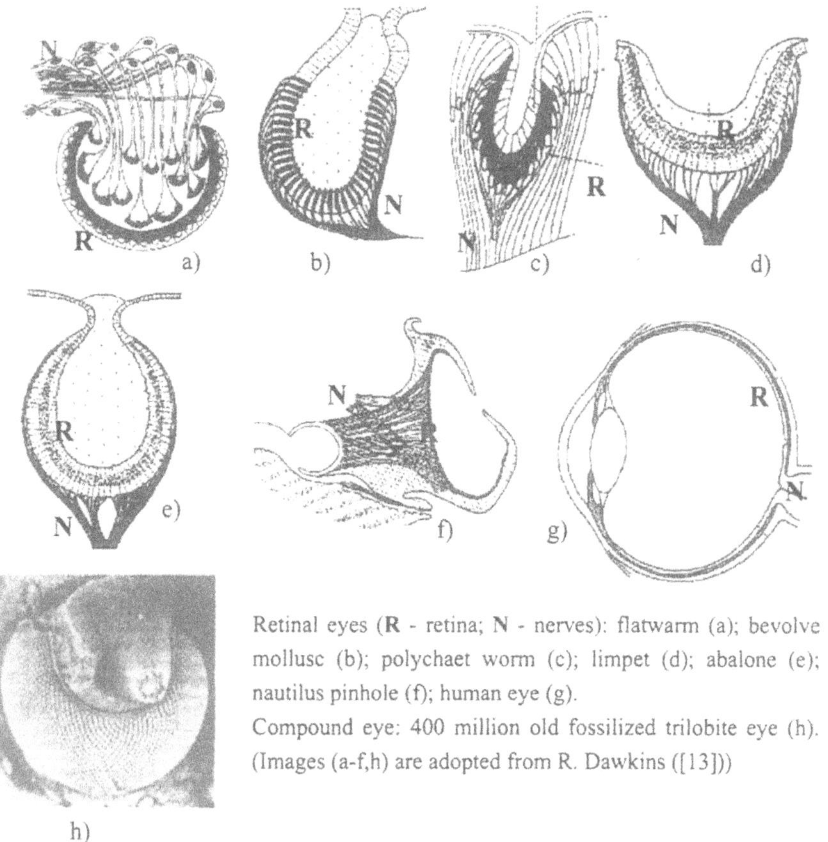

Figure 3-9. Different types of vision organs in the animal kingdom. In all types of eyes one can see devices for image discretization by sampling

Signal sampling can mathematically be treated as signal representation over shifted bases. Let $\varphi_r(x)$ be a point spread function of a signal/image reconstruction/display device and $\varphi_d(x)$ be a point spread function of a signal discretization device given in the same measurement units. Then discrete representation of a signal $a(x)$ over a basis generated by shifts of the discretization device point spread function by multiple of the discretization interval Δx is:

$$\alpha_k = \int_X a(x)\varphi_d(x - k\Delta x)dx\,, \tag{3.3.1a}$$

and its reconstruction over basis formed by the corresponding shifts of point spread function of the reconstruction device is

$$a(x) \cong a_r(x) = \sum_k \alpha_k \varphi_r(x - k\Delta x). \tag{3.3.1b}$$

Integration over x in Eq. 3.3.1a and summation over k in Eq.3.3.1b are performed within signal boundaries.

Although usually signals are finite functions, mathematical analysis of the discretization by sampling is much simplified if one assumes that signals are infinite and integer index k of shift bases functions also runs in infinite limits:

$$\alpha_k = \int_{-\infty}^{\infty} a(x)\varphi_d(x - k\Delta x)dx \tag{3.3.2a}$$

$$a(x) \cong a_r(x) = \sum_{k=-\infty}^{\infty} \alpha_k \varphi_r(x - k\Delta x) \tag{3.3.2b}$$

Eq.(3.3.2a) means that coefficients $\{\alpha_k\}$ of signal discrete representation over shift basis functions $\{\varphi_d(x - k\Delta x); \varphi_r(x - k\Delta x)\}$ are equidistant samples

$$\alpha_k = a_{df}(k\Delta x) = \int_{-\infty}^{\infty} a_{df}(x)\delta(x - k\Delta x)dx \tag{3.3.3}$$

of signal $a_{df}(x)$

$$a_{df}(x) = \int_{-\infty}^{\infty} a(\xi)\varphi_d(x-\xi)d\xi \qquad (3.3.3)$$

obtained from signal $a(x)$ by its convolution with a point spread function of the discretization device. A model of the discretization device that corresponds to such a representation is shown in Fig. 3-10.

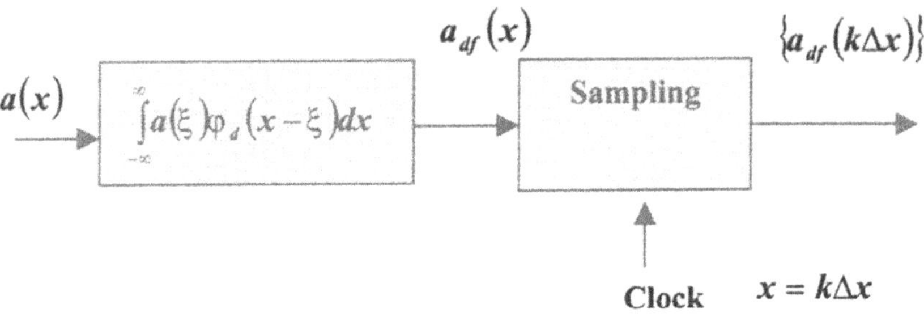

Figure 3-10. Mathematical model of the sampling device

Signal reconstruction from its sampled representation described by Eq. (3.3.1a) can, in its turn, be also regarded as filtering a discrete signal

$$\tilde{a}(x) = \sum_{k=-\infty}^{\infty} a_{df}(k\Delta x)\delta(x-k\Delta x) \qquad (3.3.4)$$

with a filter whose impulse response is equal to point spread function $\varphi_r(x)$ of the signal reconstruction device:

$$a_r(x) = \int_{-\infty}^{\infty} \tilde{a}(\xi)\varphi_r(x-\xi)d\xi \qquad (3.3.5)$$

as it is illustrated by the flow diagram in Fig. 3-11.

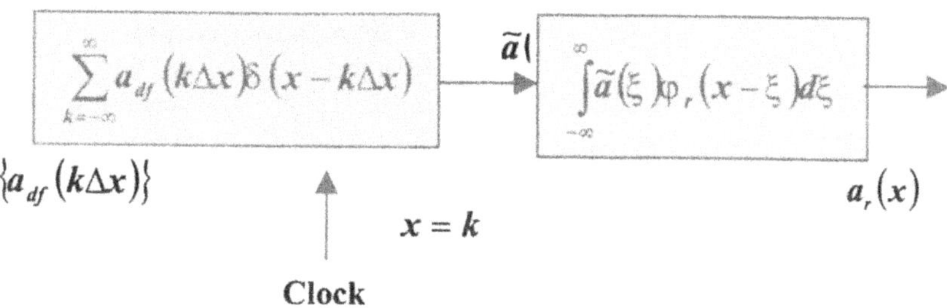

Figure 3-11. Mathematical model of the device for restoring continuous signals from their samples

Further analysis of signal sampling is much simplified in Fourier Transform domain. Compute Fourier Transforms of signals $a(x)$, $a_r(x), a_{df}(f)$ and $\tilde{a}(x)$

$$\alpha(f) \cong \alpha_r(f) = \left\{\sum_{k=-\infty}^{\infty} \alpha_k \exp(i2\pi kf\Delta x)\right\}\Phi_r(f) = \tilde{\alpha}_r(f)\Phi_r(f), \quad (3.3.6)$$

$$\tilde{\alpha}(f) = \int_{-\infty}^{\infty} \tilde{a}(x)\exp(i2\pi fx)dx = \sum_{k=-\infty}^{\infty} a_{df}(k\Delta x)\exp(i2\pi k\Delta xf), \quad (3.3.7)$$

$$\alpha_{df}(f) = \alpha(f)\Phi_d(f), \quad (3.3.8)$$

where

$$\alpha(f) = \int_{-\infty}^{\infty} a(x)exp(i2\pi fx)dx, \quad (3.3.9)$$

$$\alpha_r(f) = \int_{-\infty}^{\infty} a_r(x)exp(i2\pi fx)dx, \quad (3.3.10)$$

and

$$\tilde{\alpha}_r(f) = \sum_{k=-\infty}^{\infty} \alpha_k \exp(i2\pi k f \Delta x) = \sum_{k=-\infty}^{\infty} a_{df}(k\Delta x) \exp(i2\pi k f \Delta x), \qquad (3.3.11)$$

are Fourier spectra of signals $a(x)$, $a_r(x)$ and $\tilde{a}_r(x)$ and

$$\Phi_r(f) = \int_{-\infty}^{\infty} \varphi_r(x) \exp(i2\pi f x) dx \qquad (3.3.12)$$

and

$$\Phi_d(f) = \int_{-\infty}^{\infty} \varphi_d(x) \exp(i2\pi f x) dx \qquad (3.3.13)$$

are, respectively, frequency responses of the signal/image reconstruction/display and discretization devices.

Consider now Eq. 3.3.7. Samples $a_{df}(k\Delta x)$ of signal $a_{df}(x)$ can be found from its spectrum $\alpha_{df}(f)$ as:

$$a_{df}(k\Delta x) = \int_{-\infty}^{\infty} a_{df}(p) \exp(-i2\pi k \Delta x p) dp . \qquad (3.3.14)$$

Then obtain:

$$\tilde{\alpha}(f) = \sum_{k=-\infty}^{\infty} \int_{-\infty}^{\infty} a_{df}(p) \exp(-i2\pi k \Delta x p) dp \exp(i2\pi k \Delta x f) =$$

$$\int_{-\infty}^{\infty} a_{df}(p) dp \sum_{-\infty}^{\infty} \exp[i2\pi k \Delta x (f - p)] \qquad (3.3.15)$$

By Poisson's summation formula:

$$\sum_{k=-\infty}^{\infty} \exp[i2\pi k \Delta x (f - p)] = \Delta x \sum_{m=-\infty}^{\infty} \delta\left(f - p + \frac{m}{\Delta x}\right). \qquad (3.3.16)$$

Therefore,

$$\tilde{\alpha}(f) = \Delta x \sum_{m=-\infty}^{\infty} \alpha_{df}\left(f + \frac{m}{\Delta x}\right) \qquad (3.3.17)$$

which means that sampling signal $a_{df}(x)$ results, in Fourier domain, in periodical replication of its spectrum with a period inverse to sampling interval Δx. This phenomenon is illustrated in Fig. 3-12.

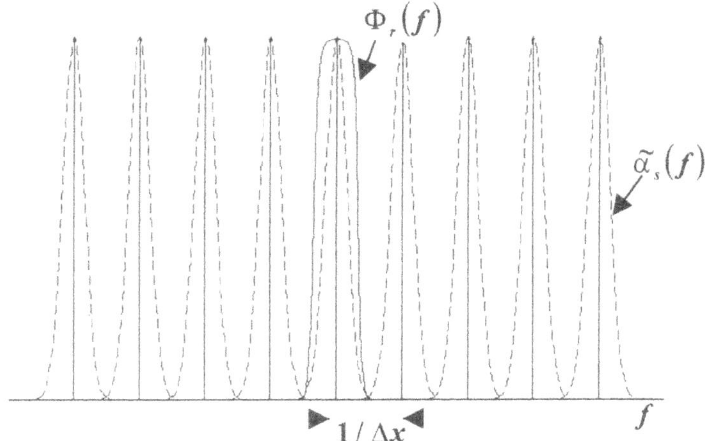

Figure 3-12. Illustration of Eq. (3.3.17) of periodical replication of signal spectrum in signal's sampled representation.

In this interpretation, reconstruction of signal $a(x)$ is reduced, according to Eq. 3.3.5, to removing, from the set of periodically replicated copies $\{\alpha_{df}(f + m/\Delta f)\}$ of spectrum of signal $a_{df}(x)$ all extra replicas but the central $\{\alpha_{df}(f)\}$ one by means of masking the set by frequency response $\Phi_r(f)$ of the signal/image reconstruction device:

$$\alpha(f) \cong \alpha_r(f) = \tilde{\alpha}_r(f)\Phi_r(f) = \left\{\sum_{m=-\infty}^{\infty} \alpha_{df}\left(f + \frac{m}{\Delta x}\right)\right\}\Phi_r(f) =$$

$$= \left\{\Delta x \sum_{m=-\infty}^{\infty} \alpha\left(f + \frac{m}{\Delta x}\right)\Phi_d\left(f + \frac{m}{\Delta x}\right)\right\}\Phi_r(f). \qquad (3.3.18)$$

From Eq. (3.3.18) one can easily see what type of signal distortions occur in the case of signal discretization by sampling. One can also see that perfect (distortion less) signal restoration from the result of its sampling is possible only if signals are band-limited:

$$\alpha(f) = \alpha(f)rect(f\Delta x + 1/2) \qquad (3.3.19)$$

and discretization and restoration devices act as ideal low-pass filters:

$$\Phi_d(f) \propto rect(f\Delta x + 1/2), \tag{3.3.20}$$

$$\Phi_r(f) = rect(f\Delta x + 1/2), \tag{3.3.21}$$

where

$$rect(x) = \begin{cases} 1, & 0 \le x \le 1 \\ 0, & otherwise \end{cases}. \tag{3.3.22}$$

Eqs. (3.3.20-21) imply that:

$$\varphi_d(x) = \frac{1}{\Delta x}\mathrm{sinc}(\pi x / \Delta x) \tag{3.3.23}$$

$$\varphi_r(x) = \mathrm{sinc}(\pi x / \Delta x) \tag{3.3.24}$$

where $\mathrm{sinc}(x) = \frac{\sin x}{x}$. In this ideal case

$$a(x) = \sum_{k=-\infty}^{\infty} \alpha_k \,\mathrm{sinc}[\pi (x - k\Delta x)/\Delta x] \tag{3.3.25}$$

provided

$$\alpha_k = \frac{1}{\Delta x}\int_{-\infty}^{\infty} a(x)\mathrm{sinc}[\pi (x - k\Delta x)/\Delta x]dx = \int_{-\infty}^{\infty} a(x)\delta(x - k\Delta x)dx = a(k\Delta x) \tag{3.3.26}$$

Eqs. (3.3.25 and 26) formulate ***the sampling theorem***: band-limited signals can exactly reconstructed from their samples taken at the inter-sample distance that is inversely proportional to the signal bandwidth.

The above analyses resulted in the sampling theorem establishes also a link between the volume of the signal discrete representation, i.e., the number of signal samples N, and signal length X and signal spectral bandwidth F specified by the selected discretization interval Δx :

$$N = X / \Delta x = XF \tag{3.3.27}$$

Therefore ***signal space-bandwidth product*** XF is a fundamental parameter that defines signal degrees of freedom N .

3.3.2 Sampling 2-D and multidimensional signals

For sampling 2-D signals, 2-D discretization and restoration basis functions should be used. The simplest way to generate and to implement in image discretization and display devices 2-D discretization and restoration basis functions is to use separable 2-D basis functions that are formed as a product of 1-D function and that work separately in two dimensions when they are used for signal/image discretization and reconstruction:

$$\varphi_d^{2D}(x_1, x_2) = \varphi_d^{(1)}(x_1)\varphi_d^{(2)}(x_2);$$

$$\varphi_r^{2D}(x_1, x_2) = \varphi_r^{(1)}(x_1)\varphi_r^{(2)}(x_2). \tag{3.3.28}$$

If 2-D separable basis functions are build from 1-D shift basis functions $\left\{\varphi_d^{(1)}(x_1 - k\Delta x_1); \varphi_d^{(2)}(x_2 - k\Delta x_2)\right\}$ and $\left\{\varphi_r^{(1)}(x_1 - k\Delta x_1); \varphi_r^{(2)}(x_2 - k\Delta x_2)\right\}$ 2-D signal sampling is carried out in sampling points arranged in a rectangular (Cartesian) raster which corresponds, in the domain of 2-D Fourier transform, to periodical replication of signal spectrum in Cartesian co-ordinates as it is shown in Fig. 3-12. Yet another example of a periodical replication of image spectra in Cartesian coordinates is shown in Fig. 3-13 (b).

One can see from these figures that the Cartesian periodical replication of 2-D signal spectrum does not necessarily ensures efficient spectra dense packing, and, correspondingly, least dense arrangement of signal sampling points for minimization of the volume of signal discrete representation. In general, tilted co-ordinate systems selected accordingly to the shape of the figure that encompasses signal 2-D spectrum will secure better spectra packing (Fig. 3-14 c,d). When sampling is carried out in a tilted co-ordinate system, sampling functions are not anymore separable functions of two Cartesian variables.

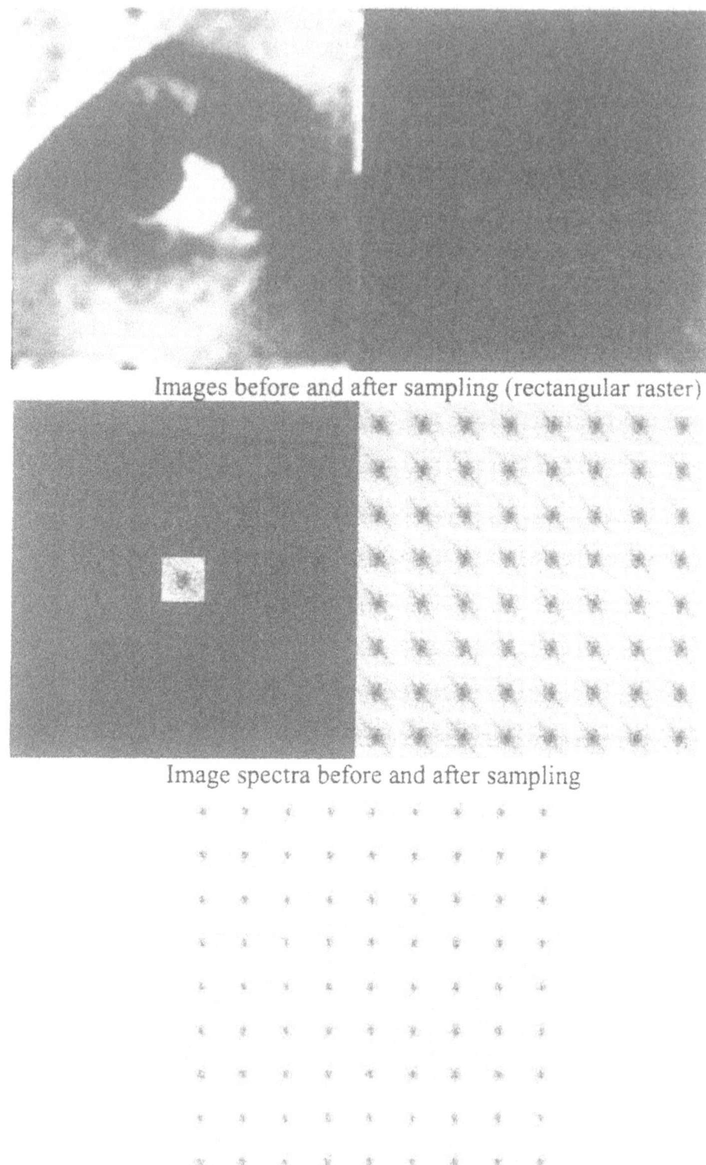

Figure 3-13. 2-D separable image sampling in a rectangular raster and periodical replication of its spectrum in a Cartesian co-ordinate system. Image at the bottom shows spectrum aliasing (seen as dark squares around main spectral lobes) that takes place when sampling interval is too large.

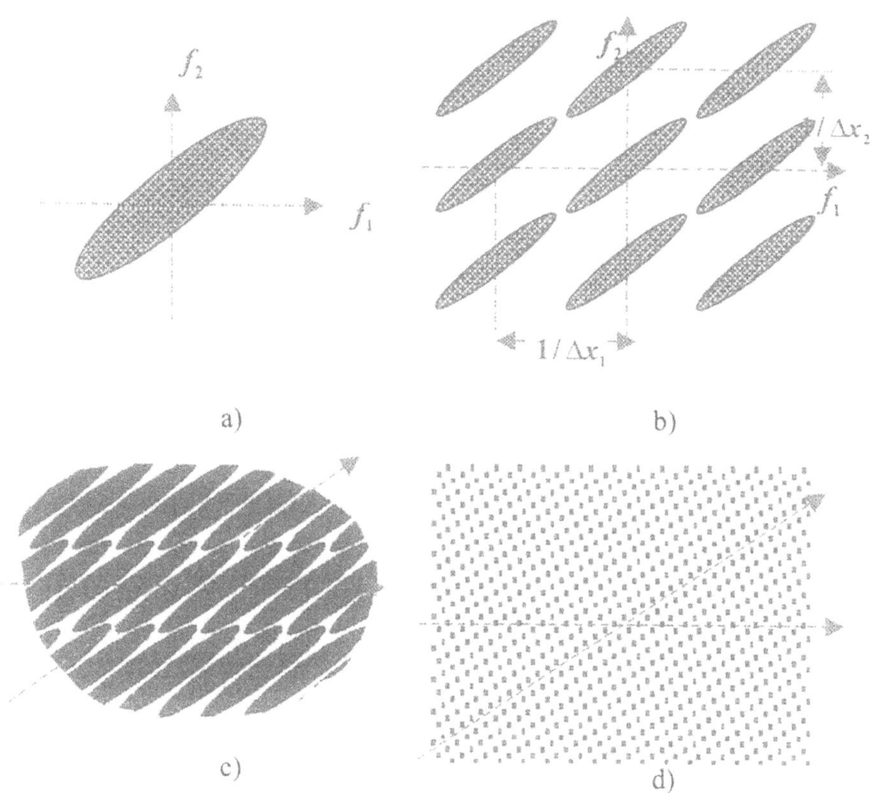

Figure 3-14. An example of an image spectrum (a), its periodical replication in rectangular (b) and tilted (c) coordinate systems and tilted discretization raster (d)

Two practical examples of sampling in tilted co-ordinate systems are illustrated in Figs. 3-15, 16. Fig. 3-15 illustrates that sampling in 45°-rotated raster coordinated with the sensitivity of human vision to spatial frequencies of different orientation (Fig. 3-15, a) is more efficient then that in Cartesian co-ordinates. This method of sampling is regularly used in print art (Fig. 3-15, f). Fig. 3-16 illustrates better efficiency of hexagonal raster when spectra of signals are isotropic and may be regarded bounded by a circle. This is the most natural assumption regarding spectra of images generated by optical imaging systems. Hexagonal sampling raster is frequently used in color displays and print art. Hexagonal arrangement of light sensitive cells can also be found in the retina of human eye (Fig. 3-16 (d)).

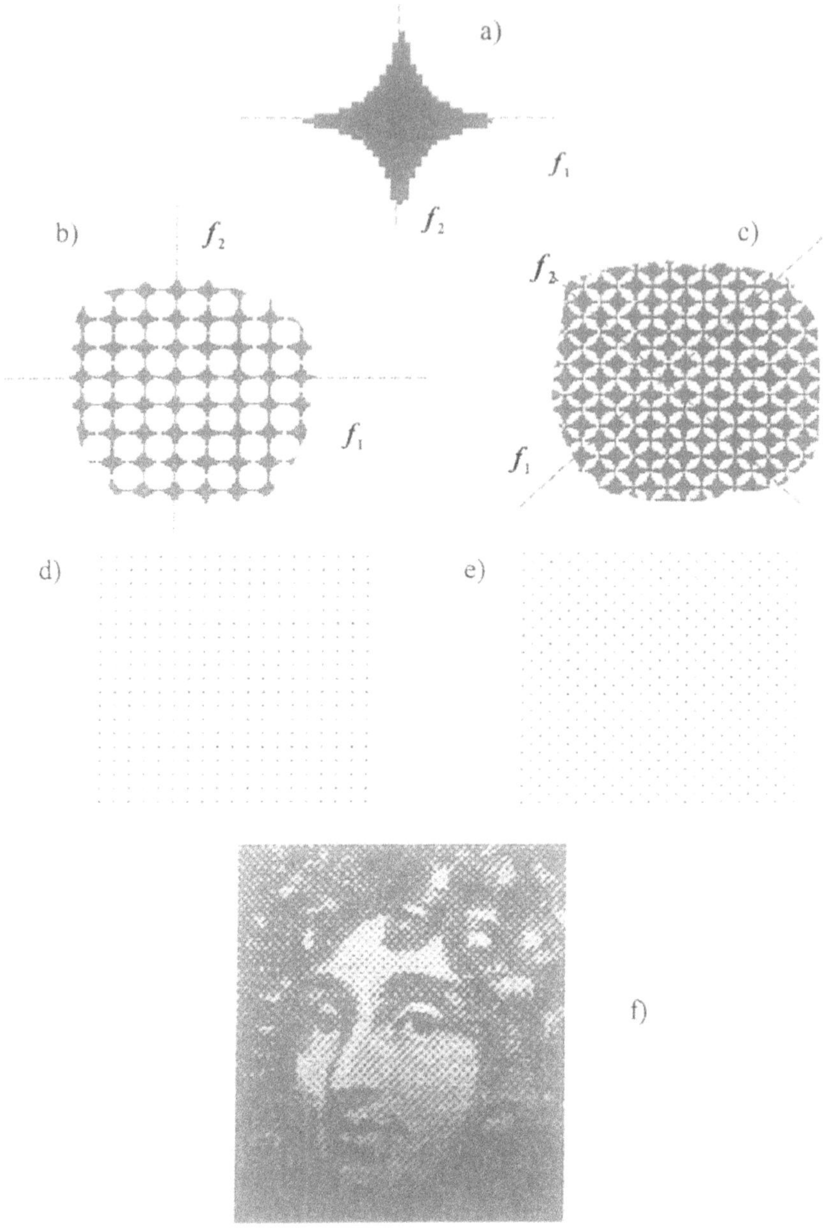

Figure 3-15. 2-D sampling: a) figure in frequency domain that illustrates different sensitivity of human visual system to different spatial frequencies: higher for horizontal and vertical and lower for diagonal ones; b) , c) – periodical replication of the figure a) in a conventional and 45°-rotated rectangle co-ordinate systems, respectively; d), e) – corresponding image sampling rasters; f) an example of an image taken from a conventional book printing

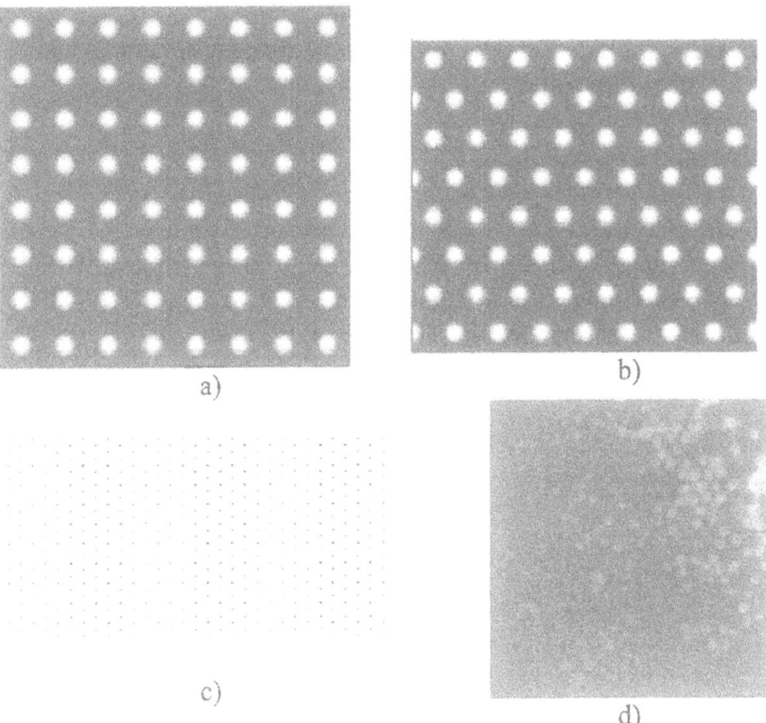

Figure 3-16. Image spectra replication for Cartesian (a) and hexagonal (b) sampling: the same number of circular spectra occupy less area when they are packed in a hexagonal co-ordinate system then in Cartesian co-ordinates; c) – hexagonal sampling raster; d) Trichromatic retinal mosaics: hexagonal arrangement of cones of human eye retina (adopted from [14]).

Rotating and tilting discretization rasters are not the only methods for the optimization of 2-D and multi dimensional sampling. Two other methods are "***discretization with gaps***" and ***discretization with sub band decomposition***.([15])

Method of the discretization with gaps is based on the use of signal spectra symmetry. It is illustrated in Fig. 3-17. Let signal spectrum is bounded as it is shown in Fig. 3-17 a). If one rotates the signal by 90° and superimposes the rotated copy onto the initial signal, a composite signal is obtained whose spectrum perfectly fills a square. Therefore, the composite signal can be optimally sampled in Cartesian co-ordinates. One can see that for all groups of four composite and initial signal samples centered around the rotation center, the following relationships hold:

$$z_1 = a_1 + a_4;\ z_2 = a_1 + a_2;\ z_3 = a_2 + a_3; z_4 = a_3 + a_4; \qquad (3.3.29)$$

from which it follows that, from four samples $\{z_1, z_2, z_3, z_4\}$ of the composite signal one sample, say, z_4, is redundant because it can be found from the other three ones:

$$z_4 = z_1 - z_2 + z_3 . \tag{3.3.30}$$

Therefore the use of signal spectrum symmetry by means of sampling the composite signal obtained by superimposing the signal and its 90° rotated copy enables saving 25% of signal samples. Note, however, that the initial redundancy of signal spectrum (Fig. 3.17, a) is 50%.

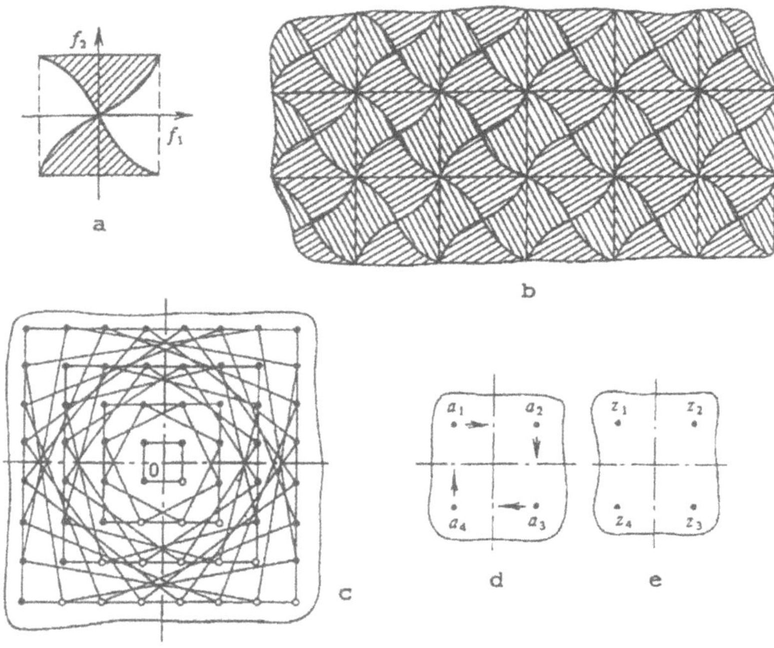

Figure 3-17. Discretization with gaps: a) – a figure bounding a signal spectrum (shaded); b) – the spectrum pattern obtained by superimposing the signal rotated by 90° and periodically extended in Cartesian co-ordinate system; c) the discretization raster corresponding to the spectrum's pattern (b); d) – group of four initial signal samples around the rotation center and directions of their translation after the rotation; e) – the corresponding group of four samples of the initial signal plus its 90° rotated copy. Empty circles in (c) denote redundant samples.

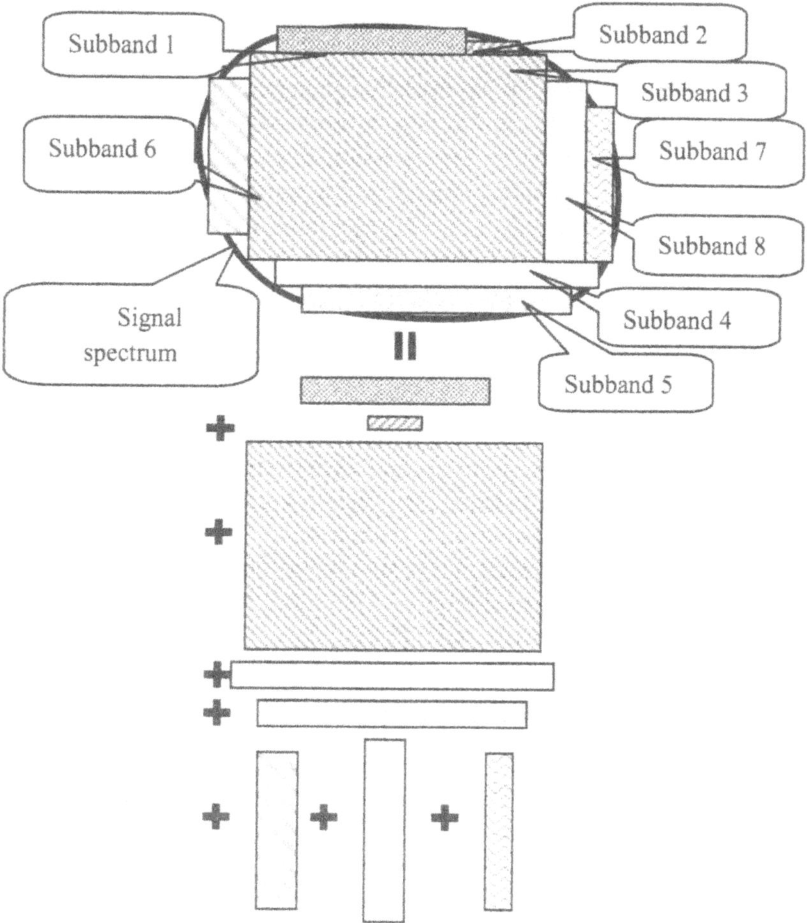

Figure 3-18. Explanation of discretization with sub-band decomposition

Method of discretization with sub-band decomposition is illustrated in Fig. 3-18. Area occupied by the signal spectrum is divided into sub-bands bounded by rectangles. Signals that correspond to sub-bands can then be optimally sampled in Cartesian rasters. For restoration of the initial signal, one should first restore, from their corresponding samples, sub-band signals and then add up them. Such a procedure allows, in principle, to achieve minimal volume of the signal discrete representation of $N = S_{x_1,x_2}\mathbf{S}_{f_1,f_2} = S_{x_1,x_2}\sum_k \mathbf{S}^{(k)}_{f_1,f}$, where S_{x_1,x_2} and $\mathbf{S}_{f_1,f_2}$ are areas occupied, respectively, signal and its Fourier spectrum and summation over k assumes summation over sub-bands.

3.3.3 Sampling artifacts: qualitative analysis

As it is stated by the sampling theorem, signal sampling should be carried out after signal ideal low pass filtering in the bandwidth that corresponds to the selected sampling interval. Signal reconstruction from its samples also requires using ideal low pass filtering in the same bandwidth. In this case distortions because of signal sampling are minimal. Their mean squared value is equal to the energy of signal spectral components left outside of the chosen bandwidth. Real signal sampling and reconstruction devices are incapable of implementing the ideal low pass filtering. This causes additional distortions in reconstructed signals commonly called ***aliasing effects***.

Two sources of aliasing effects should be distinguished: inappropriate signal pre-filtering for the sampling and inappropriate sampled signal low pass filtering for the reconstruction. Distortions caused by the former source are called ***strobe effects***. They exhibit themselves as a stroboscopic reduction of frequency of signal components outside of the chosen bandwidth. Fig. 3-19 and 20 illustrate this phenomenon for 1-D and 2-D signals.

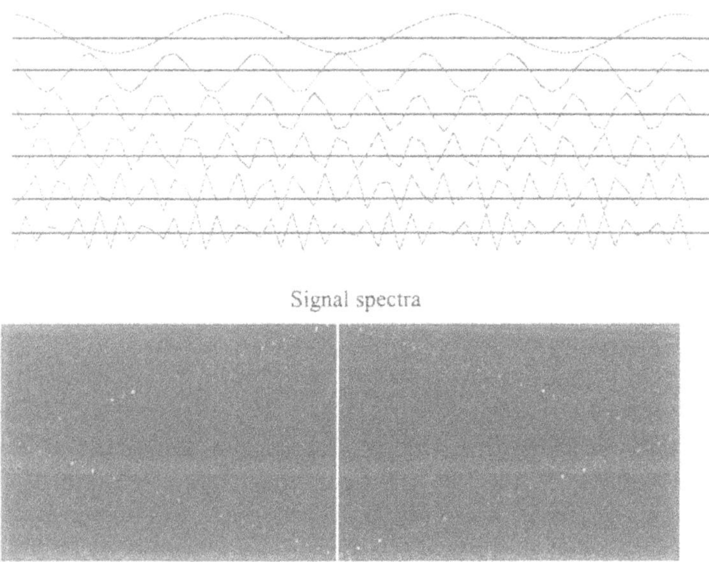

Figure 3-19. Strobe effects in sampling 1-D sinusoidal signals. Graphs on top show sinusoidal signals of different frequencies lower and higher the frequency that corresponds to the chosen discretization bandwidth. Image on bottom shows spectra (bright spots) of sampled sinusoidal signals of frequencies that increase from top to bottom. One can see that frequency of sampled sinusoids increases until it reaches the border of the bandwidth. After that the increase of signal frequency causes the decrease of the sampled signal frequency.

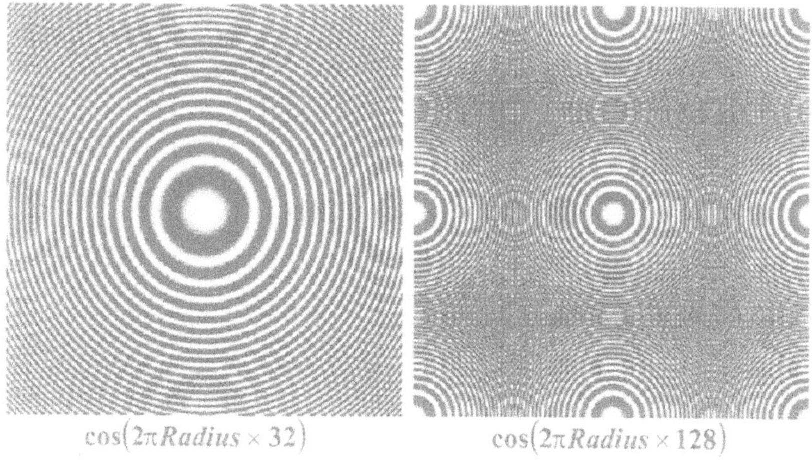

Figure 3-20. Strobe effects in sampling 2-D sinusoidal signals

Distortions caused by inappropriate sampled signal low pass filtering for the reconstruction exhibit themselves in appearance of signal high frequencies doubles from spectrum replicas not cleaned out by the reconstruction filter. For sinusoidal signals, their doubles cause low frequency beatings in the reconstructed signal that appear as moiré patterns. Owing to this reason, these aliasing effects are called ***moiré effects***. In images, moiré effects usually appear in form of "pixelation" illustrated in Fig. 3-21. Fig. 3-22 by Salvador Daly illustrates artistic use of the "pixelation" effect.

Figure 3-21. Image "pixelation" effects caused by inappropriate reconstruction filtering

Figure 3-22. Artistic use by Salvador Daly of the "pixelation" effect

Numerical evaluation of aliasing effects in terms of mean squared signal reconstruction error can be carried out using Fourier Transform domain representation of sampling and reconstruction discussed in Sects. 3.3.1 and 3.3.2. However, mean squared reconstruction error is not always an adequate measure of the aliasing. For instance, it does not directly evaluate such image quality feature as image readability. Fig. 3-23 illustrates importance of proper signal pre- and post-filtering in image sampling and reconstruction.

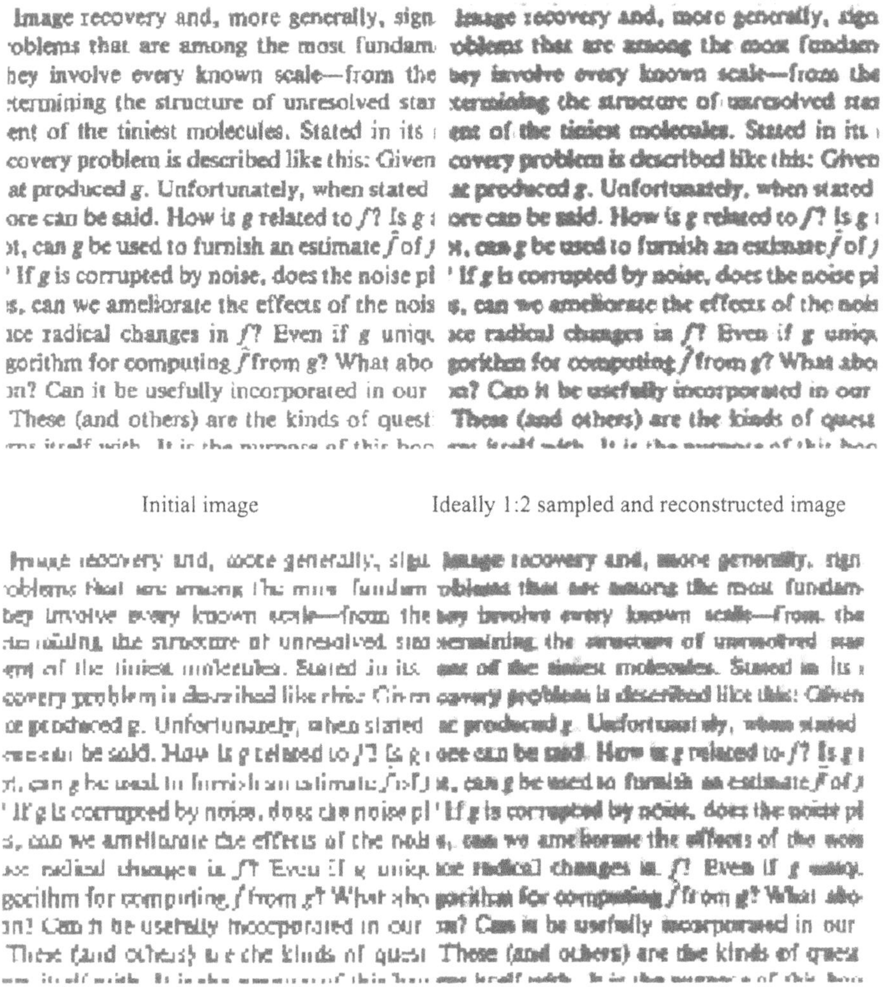

Initial image

Ideally 1:2 sampled and reconstructed image

1:2 sampled and reconstructed image with no low-pass pre-filtering and "box" post-filtering

1:2 sampled and reconstructed image with "box" pre-filtering and "box" post-filtering

Figure 3-23. Influence of sampling artifacts on image readability: image sampling and reconstruction with an ideal low pass filter and with a "box" filter that computes signal average over a square box of 3x3 samples around each pixel.

3.3.4 Sampling artifacts: quantitative analysis

In this section, we will derive formulas that can be used for numerical evaluation of discretization artifacts. For the sake of simplicity, 1-D case will be analyzed. For 2-D signals, one should regard signal and spectral coordinate and indices as two-component vectors.

For discretization aperture $\varphi_d(x)$, sampling interval Δx and signal reconstruction aperture $\varphi_r(x)$, signal samples at points $\{(k+u)\Delta x\}$, where $u\Delta x$ is a shift in positioning discretization aperture with respect to signal coordinate system x, are measured as (Eq. 3.3.1a):

$$a_k = \frac{1}{\Delta x}\int_{-\infty}^{\infty} a(x)\varphi_d[x-(k+u)\Delta x]dx =$$

$$\frac{1}{\Delta x}\int_{-\infty}^{\infty}\left\{\int_{-\infty}^{\infty} a(\xi)\varphi_d(x-\xi)d\xi\right\}\delta[x-(k+u)\Delta x]dx =$$

$$\frac{1}{\Delta x}\int_{-\infty}^{\infty} a_{df}(x)\delta[x-(k+u)\Delta x]dx, \tag{3.3.31}$$

where

$$a_{df}(x) = \int_{-\infty}^{\infty} a(\xi)\varphi_d(x-\xi)d\xi \qquad (3.3.32).$$

(see Fig. 3-10) and reconstructed signal is generated as

$$a_r(x) = \sum_{k=-N}^{N} a_k\varphi_r[x-(k+v)\Delta x] \tag{3.3.33}$$

where $2N+1$ is the number of samples used for the reconstruction and $v\Delta x$ is shift of the reconstruction aperture with respect to the signal coordinate system.

Find an explicit relationship between $a(x)$ and $a_r(x)$ that involves discretization and reconstruction device parameters:

$$a_r(x) = \sum_{k=-N}^{N} a_k\varphi_r[x-(k+v)\Delta x] =$$

$$\frac{1}{\Delta x}\sum_{k=-N}^{N}\left\{\int_{-\infty}^{\infty} a(\xi)\varphi_d[\xi-(k+u)\Delta x]d\xi\right\}\varphi_r[x-(k+v)\Delta x]=$$

$$\frac{1}{\Delta x}\int_{-\infty}^{\infty} a(\xi)d\xi\sum_{k=-N}^{N}\varphi_d[\xi-(k+u)\Delta x]\varphi_r[x-(k+v)\Delta x]=\int_{-\infty}^{\infty} a(\xi)h(x,\xi)d\xi\,, \tag{3.3.34}$$

where

$$h(x,\xi)=\frac{1}{\Delta x}\sum_{k=-N}^{N}\varphi_d[\xi-(k+u)\Delta x]\varphi_r[x-(k+v)\Delta x] \tag{3.3.35}$$

is an equivalent impulse response of the discretization/reconstruction procedure.

As we saw it in Sect. 3.3.1, it is easier to analyze sampling artifacts in signal Fourier domain. Find Fourier spectrum $\alpha_r(f)$ of the reconstructed signal and connect it with spectrum $\alpha(f)$ of the initial signal:

$$\alpha_r(f)=\int_{-\infty}^{\infty} a_r(x)\exp(i2\pi fx)dx=\int_{-\infty}^{\infty}\left\{\int_{-\infty}^{\infty} a(\xi)h(x,\xi)d\xi\right\}\exp(i2\pi fx)dx=$$

$$\int_{-\infty}^{\infty}\int_{-\infty}^{\infty}\left\{\int_{-\infty}^{\infty}\alpha(p)\exp(-i2\pi p\xi)dp\right\}h(x,\xi)\exp(i2\pi fx)dxd\xi=$$

$$\int_{-\infty}^{\infty}\alpha(p)dp\int_{-\infty}^{\infty}\int_{-\infty}^{\infty} h(x,\xi)\exp[i2\pi(fx-p\xi)]\exp(-i2\pi)dpdxd\xi=$$

$$\int_{-\infty}^{\infty}\alpha(p)H(f,p)dp\,, \tag{3.3.36}$$

where $H(f,p)$ is a frequency response of the sampling/reconstruction procedure:

$$H(f,p)=\int_{-\infty}^{\infty} h(x,\xi)\exp[i2\pi(fx-p\xi)]dxd\xi=$$

$$\frac{1}{\Delta x}\int_{-\infty}^{\infty}\int_{-\infty}^{\infty}\sum_{k=-N}^{N}\varphi_d[\xi-(k+u)\Delta x]\varphi_r[x-(k+v)\Delta x]\exp[i2\pi(fx-p\xi)]dxd\xi=$$

$$\frac{1}{\Delta x}\sum_{k=-N}^{N}\int_{-\infty}^{\infty}\int_{-\infty}^{\infty}\varphi_d[\xi-(k+u)\Delta x]\varphi_r[x-(k+v)\Delta x]\exp[i2\pi(fx-p\xi)]dxd\xi =$$

$$\frac{1}{\Delta x}\sum_{k=-N}^{N}\int_{-\infty}^{\infty}\varphi_d[\xi-(k+u)\Delta x]\exp(-i2\pi p\xi)d\xi\times$$

$$\int_{-\infty}^{\infty}\varphi_r[x-(k+v)\Delta x]\exp(i2\pi fx)dx=$$

$$\frac{1}{\Delta x}\sum_{k=-N}^{N}\int_{-\infty}^{\infty}\varphi_d(\tilde{\xi})\exp\{-i2\pi p[\tilde{\xi}+(k+u)\Delta x]\}d\tilde{\xi}\times$$

$$\int_{-\infty}^{\infty}\varphi_r(\tilde{x})\exp\{i2\pi f[\tilde{x}+(k+v)\Delta x]\}d\tilde{x}=$$

$$\frac{1}{\Delta x}\sum_{k=-\infty}^{\infty}\exp[-i2\pi p(k+u)\Delta x]\exp[i2\pi f(k+v)\Delta x]\times$$

$$\int_{-\infty}^{\infty}\varphi_d(\tilde{\xi})\exp(-i2\pi p\tilde{\xi})d\tilde{\xi}\int_{-\infty}^{\infty}\varphi_r(\tilde{x})\exp(i2\pi f\tilde{x})d\tilde{x}=$$

$$\frac{\cdot\Phi_d^*(p)\cdot\Phi_r(f)}{\Delta x}\exp[i2\pi(fv-pu)\Delta x]\cdot\sum_{k=-N}^{N}\exp[i2\pi(f-p)k\Delta x], \quad (3.3.37)$$

where $\Phi_d(\cdot)$ and $\Phi_r(\cdot)$ are frequency responses of signal discretization and reconstruction devices. Consider the summation term in Eq. (3.3.37):

$$\sum_{k=-N}^{N}\exp[i2\pi(f-p)k\Delta x]=$$
$$\frac{\exp[i2\pi(f-p)(N+1)\Delta x]-\exp[-i2\pi(f-p)N\Delta x]}{\exp[i2\pi(f-p)\Delta x]-1}=$$
$$\frac{\sin[\pi(f-p)(2N+1)\Delta x]}{\sin[\pi(f-p)\Delta x]}=(2N+1)\frac{\sin[\pi(f-p)(2N+1)\Delta x]}{(2N+1)\sin[\pi(f-p)\Delta x]}. \quad (3.3.38)$$

Using Eq. 3.3.38, overall frequency response of the signal discretization-reconstruction procedure can be written as

$$H(f,p)=\exp[i2\pi(fv-pu)\Delta x]\cdot\Phi_d^*(p)\cdot\Phi_r(f)\times$$
$$\frac{(2N+1)}{\Delta x}\frac{\sin[\pi(f-p)(2N+1)\Delta x]}{(2N+1)\sin[\pi(f-p)\Delta x]}. \quad (3.3.39)$$

This equation shows that the signal discretization/reconstruction procedure being implemented by shift basis function is nevertheless not shift invariant for shift invariant system with PSF $h(x,\xi)=h(x-\xi)$ frequency response has a form $H(f,p)=H(f)\delta(f-p)$. The reason of space variance is finite number of signal samples used for the reconstruction, or, in other words, boundary effects. When this number increases boundary effects can be neglected. Therefore consider asymptotic behavior of $H(f,p)$ when $N\to\infty$.

By Poisson's summation formula:

$$\lim_{N\to\infty}\sum_{k=-\infty}^{\infty}\exp[i2\pi k\Delta x(f-p)]=\Delta x\sum_{m=-\infty}^{\infty}\delta\left(f-p+\frac{m}{\Delta x}\right), \tag{3.3.40}$$

obtain from Eq. 3.3.35:

$$\lim_{N\to\infty}H(f,q)=\exp[i2\pi(fv-pu)\Delta x]\cdot\Phi_d^*(p)\cdot\Phi_r(f)\times$$
$$\Delta x\sum_{m=-\infty}^{\infty}\delta\left(f-p+\frac{m}{\Delta x}\right). \tag{3.3.41}$$

Substitution of Eq. 3.3.41 into Eq. 3.3.37 then gives:

$$\lim_{N\to\infty}\alpha_r(f)=$$
$$\int_{-\infty}^{\infty}\alpha(p)\exp[i2\pi(fv-pu)\Delta x]\cdot\Phi_d^*(p)\cdot\Phi_r(f)\sum_{m=-\infty}^{\infty}\delta\left(f-p+\frac{m}{\Delta x}\right)dp=$$
$$\Phi_r(f)\exp[i2\pi f(u-v)\Delta x]\sum_{m=-\infty}^{\infty}\alpha\left(f+\frac{m}{\Delta x}\right)\exp\left(-i2\pi\frac{mu}{\Delta x}\right)\cdot\Phi_d^*\left(f+\frac{m}{\Delta x}\right)=$$
$$\Phi_r(f)\Phi_d^*(f)\alpha(f)\exp[i2\pi(u-v)\Delta x]+\Phi_r(f)\exp[i2\pi f(u-v)\Delta x]\times$$
$$\sum_{m=1}^{\infty}\left[\alpha_{df}\left(f+\frac{m}{\Delta x}\right)\exp\left(-i2\pi\frac{mu}{\Delta x}\right)+\alpha_{df}\left(f-\frac{m}{\Delta x}\right)\exp\left(i2\pi\frac{mu}{\Delta x}\right)\right]. \tag{3.3.43}$$

One can interpret Eq. 3.3.43 as one showing that the signal $a_r(x)$ reconstructed from samples $\{a_k\}$ of signal $a(x)$ consists of 2 components. The first component is a copy of signal $a(x)$, shifted by $(u-v)\Delta x$ and modified by convolution with point spread functions of the discretization and reconstruction devices. Its spectrum is described by the first term of Eq. 3.3.41. The second component is aliasing one. Its spectrum is described by the second term of Eq. 3.3.41. It consists of periodical replicas $\left\{\alpha_{df}\left(f \pm \frac{m}{\Delta x}\right)\exp\left(\mp i2\pi \frac{mu}{\Delta x}\right)\right\}$ spectrum $\alpha_{df}(f)$ of signal $a_{df}(x)$ (Eq. 3.3.32).

In the numerical evaluation of signal distortions caused by the discretization/reconstruction, it is devisable to distinguish two types of distortions that are described by the above two reconstructed signal components: distortions owing to the signal blur by the discretization and reconstruction apertures and aliasing distortions. The former are completely specified by deviation of reconstructed signal spectrum from the initial one:

$$\varepsilon_{sp}(f) = \left[1 - \Phi_r(f)\Phi_d^*(f)\right]\alpha(f) \tag{3.3.44}$$

(note that shift factor $\exp[i2\pi(u - v)\Delta x]$ in the evaluation of this error can be neglected as irrelevant). For the evaluation of the second type of distortions, the most straightforward measure is that of its energy which, by Parceval theorem, is

$$\overline{\varepsilon_{el}^2} = \int_{-\infty}^{\infty} \varepsilon_{al}^2(f)df, \tag{3.3.45}$$

where

$$\varepsilon_{al}^2(f) = |\Phi_r(f)|^2 \times$$

$$\sum_{m=1}^{\infty}\left|\alpha\left(f + \frac{m}{\Delta x}\right)\right|^2\left|\Phi_d\left(f + \frac{m}{\Delta x}\right)\right|^2 + \sum_{m=1}^{\infty}\left|\alpha\left(f - \frac{m}{\Delta x}\right)\right|^2\left|\Phi_d\left(f - \frac{m}{\Delta x}\right)\right|^2 \tag{3.3.46}$$

For image spectra usually decay with frequency, the value

$$\overline{\varepsilon_{el}^2} = \int_{-\infty}^{\infty}|\Phi_r(f)|^2\sum_{m=1}^{\infty}\left\{\left|\Phi_d\left(f + \frac{m}{\Delta x}\right)\right|^2 + \left|\Phi_d\left(f - \frac{m}{\Delta x}\right)\right|^2\right\}df \quad (3.3.47)$$

can be used as an upper bound of aliasing error.

3.4 MULTI-RESOLUTION SAMPLING

The idea of optimal sampling with sub-band decomposition described in Sect. 3.3.2 has found its implementation in multi-resolution sampling that gained its popularity under the name of wavelet decomposition. Multi-resolution sampling can be most easily explained for 1-D signals. As it is illustrated in Fig. 3-24, signal is split into two components, low pass and high pass ones. The low pass component has bandwidth half of that of the initial signal and can be sampled with half of the sampling rate of the initial signal. The residual high pass component's spectrum bandwidth is also half of the initial bandwidth but its spectral components are situated in the upper half of the spectral range. Therefore, in order to sample it with half of the initial sampling rate one should shift its spectrum into the lower half of the range.

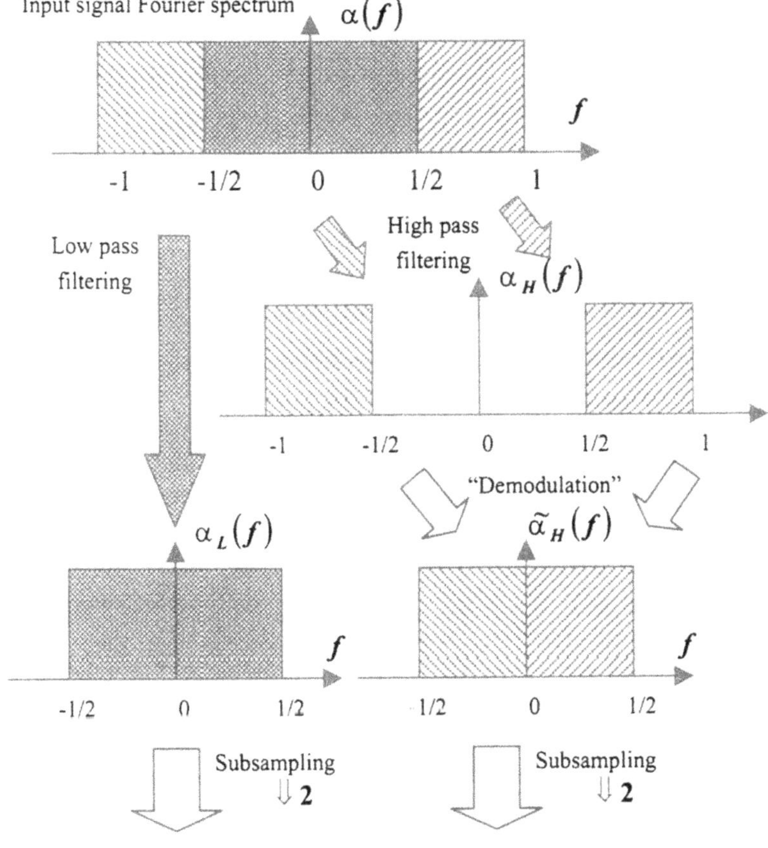

Figure 3-24. Multi-resolution sampling explained for a 1-D signal in Fourier domain

This is achieved by means of its "demodulation" which is implemented by multiplying the "high pass" signal component by a a sinusoidal signal whose frequency is equal to half of the initial signal highest frequency. In multi-resolution sampling, this procedure is repeated recursively to "low pass" components obtained on each previous step of the recursion as it is shown in Fig. 3-25.

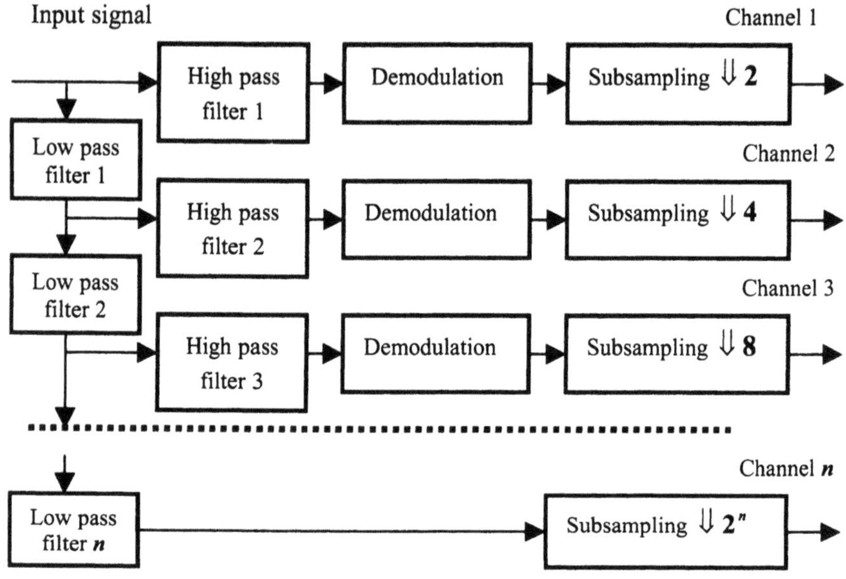

Figure 3-25. Multi resolution signal sampling flow diagram

On each step, bandwidths of low pass and high pass filters, and therefore signal sampling rate are half of those on the previous step. On the last n-th step, low pass component obtained on this stage has bandwidth equal to $1/2^n$-th fraction of the initial signal bandwidth and it is sampled accordingly with $1/2^n$-th of the rate that corresponds to this initial bandwidth. Signal reconstruction from its multi resolution sampled representation is also carried out component (channel) wise. Each sampled high pass component is reconstructed individually accordingly to its bandwidth and its spectrum is shifted correspondingly and than all reconstructed components are added up to form the entire reconstructed signal.

Obviously, the described multi resolution sampling does not reduce the number of signal samples with respect to that corresponding to the signal bandwidth. It however enables applying different quantization to different signal components and in this way to gain in the final volume of signal digital representation. The method has found its application in wavelet image coding ([9]).

3.5 UNCONVENTIONAL DIGITAL IMAGING METHODS

In many digital imaging devices, discretization basis functions and restoration basis functions belong to different families of functions. While restoration basis functions implemented in common display and printing devices are always shift (convolution) ones, in some unconventional imaging methods discretization basis functions are other then shift ones. This assumes that discrete image data collected in the discretization process in such devices should be transformed into a set of image samples for image display or printing. This process usually carried out in computers is called ***image reconstruction***. We will briefly review three unconventional digital imaging methods: method of coded apertures, computed tomography and magnetic resonance imaging.

Coded aperture imaging methods (see, for instance, [16, 17]) are suggested for non-optical imaging such as X-ray, gamma-ray and other nuclear radiation imaging when no optical focussing devices are available. They are an alternative to pinhole cameras that also build images directly without any focussing radiation. In coded aperture methods, radiation from objects to be imaged is sensed by means of binary (transparent/opaque) masks such as Fresnel zone plates, 2-D Walsh function masks, masks built from binary pseudo-random sequences (M-sequences) or by corresponding arrays of parallel sensors. Such binary masks collect radiation from half of the mask area, that is half of the image size, and thus they offer much better ratio of signal-to-quantum noise at sensor's output than pinhole cameras that collect radiation energy only within the area of the pinhole.

In ***computed tomography***, 2-D discretization is separable in polar coordinate system: linear array of sensors rotates around the body to be imaged and, for a discrete set of observation angles, samples projections of body slices. If $h_d(\cdot)$ is an aperture function of sensors in the array, (x,y) are coordinates in the slice plane (See Fig.2 17), (k,l) are basis function indices; $\Delta\xi$ and $\Delta\theta$ are projection and angle discretization intervals, discretization bases functions for parallel beam computed tomography is

$$\varphi_d(\theta,\xi) = h(x\cos l\Delta\theta - y\sin l\Delta\theta - k\Delta\xi), \qquad (3.4.1)$$

In ***magnetic resonance imaging*** ([18]), the object to be imaged is placed into a strong and spatially non-homogeneous magnetic field. When an electromagnetic excitation signal on a radio frequency is applied to the object, the object re-emits the signal but modulates its frequency. The intensity of the signal re-emitted by different elements of the body volume is proportional to the density of protons in these volume elements and its frequency is determined by the strength of the magnetic field in their coordinates. Because the magnetic field is spatially non-homogeneous, frequency of the re-emitted signal carries information about the spatial coordinates of the points. Therefore, for collecting data about distribution of proton density within the body, re-emitted signal is sampled in frequency domain and, therefore, the discretization bases functions are sinusoidal ones.

3.6 PRINCIPLES OF SIGNAL SCALAR QUANTIZATION

3.6.1 Optimal homogeneous non-uniform quantization

Scalar (element-wise) quantization is the second stage of signal digitization. It is applied to signal discrete representation obtained as the result of signal discretization. Scalar quantization implies that a dynamic range of a finite length is chosen in the entire range of the continuous signal representation coefficient values and is divided into ***quantization intervals*** (Fig. 3-24a). Any value falling within a particular interval is designated by a number (quantization interval's index) common to all values in the interval. In signal reconstruction, this number is replaced with a value selected as a representative of this interval (***quantized value***). Arrangement of quantization intervals and selection of representative values is governed by the accuracy criterion for representing the continuous signal by a digital one by probability distribution of the quantized values.

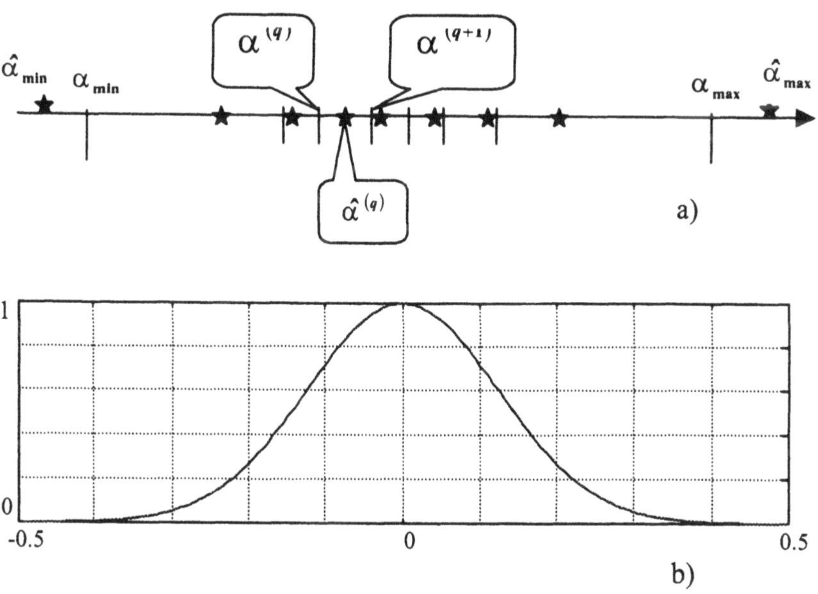

Figure 3-26. Non-uniform quantization. (a)-arrangement of quantization intervals and their representative values; (b) - probability distribution function of values to be quantized; (c) - non-uniform quantization by means of a nonlinear pre-distortion

Let α be value of signal's discrete representation coefficient to be quantized, $\hat{\alpha}_{min}$ and $\hat{\alpha}_{max}$ are the corresponding representatives of values outside the selected dynamic range, α_{min} and α_{max} are boundaries (minimum and maximum) of the dynamic range, $\alpha^{(q)}$, $\alpha^{(q+1)}$ and $\hat{\alpha}^{(q)}$, are, respectively, the left and right boundaries and representative value of the q-th quantization interval (Fig. 3-26a). One can characterize the quantization error at the q-th interval within the dynamic range as

$$\varepsilon^{(q)} = \alpha - \hat{\alpha}^{(q)}; \; \alpha \in \left[\alpha^{(q)}, \alpha^{(q+1)}\right] \tag{3.6.1}$$

and truncation errors of the dynamic range limitation as

$$\varepsilon^{(min)} = \alpha - \hat{\alpha}^{(min)}; \; \alpha < \alpha_{min}; \tag{3.6.2}$$

$$\varepsilon^{(max)} = \alpha - \hat{\alpha}^{(max)}; \; \alpha > \alpha_{max}. \tag{3.6.3}$$

The requirements for quantization accuracy are generally formulated in terms of the constraints imposed upon the quantization errors $\{\varepsilon^{(q)}\}$. The most common approach to formulating these constraints assumes that the quantization artifacts are the same for all signal representation coefficients and therefore all coefficients are quantized in the same way. Such quantization is called ***homogeneous***. For the design of homogeneous quantizer, quantization artifacts are evaluated on average over all possible coefficient values to be quantized. To this goal, loss functions $D_l(\varepsilon^{(min)})$, $D_{dr}(\varepsilon^{(q)})$, $D_r(\varepsilon^{(q)})$ that measure losses owing to the quantization errors within and outside the dynamic range and probability density $p(\alpha)$ of the values to be quantized are introduced, and quantization quality is evaluated, separately for quantization errors within and outside the dynamic range, as

$$\overline{D}_l = \int_{-\infty}^{\alpha_{min}} p(\alpha) D_l(\varepsilon_{min}) d\alpha \tag{3.6.4}$$

$$\overline{D}_r = \int_{\alpha_{max}}^{\infty} p(\alpha) D_r(\varepsilon_{max}) d\alpha \tag{3.6.5}$$

and

$$\overline{D}_{dr} = \sum_{q=1}^{Q-1} \int_{\alpha^{(q)}}^{\alpha^{(q+1)}} p(\alpha) D_{dr}(\varepsilon^{(q)}) d\alpha \; ; \; \alpha^{(1)} = \alpha_{min}; \; \alpha^{(Q)} = \alpha_{max}, \tag{3.6.6}$$

where Q is the number of quantization intervals. Separate evaluation of dynamic range limitation errors and quantization errors is advisable owing to the different nature of these errors: while quantization errors are limited in the range by the quantization interval size, dynamic range limitation errors may be substantially larger.

Eqs. 3.6.4 and 5 can be used for determining dynamic range boundaries $\{\alpha_{\min}, \alpha_{\max}\}$ given dynamic range limitation errors $\overline{D}_l$ and $\overline{D}_r$. Eq. (3.6.6) can be used for determining the set of quantization intervals boundaries $\{\alpha^{(q)}\}$ and representatives $\{\hat{\alpha}^{(q)}\}$ that minimize the number of quantization intervals Q given average quantization error $\overline{D}_{dr}$ or to minimize the average quantization error given the number of quantization intervals. Such set of quantization interval boundaries and quantization interval representatives determine ***optimal scalar quantizer***.

It is intuitively clear that optimal quantization intervals are, in general, non-uniform and are defined by two factors: by probability density $p(\alpha)$ and by loss function $D_{dr}(\varepsilon)$. The lower is the probability density in a certain sub-range of the dynamic range of α the smaller is contribution of the quantization error in this sub-range into the average quantization error and therefore the larger can be quantization intervals in this sub-range. The loss function affects size of the quantization intervals similarly.

There are two approaches to the design of the optimal quantizer, a numerical optimization approach and a ***compressor-expander (compander)*** one. Numerical optimization approach assumes analytical or numerical solving the optimization equation:

$$\{\alpha^{(q)}, \hat{\alpha}^{(q)}\} = \underset{\{\alpha^{(q)}, \hat{\alpha}^{(q)}\}}{\arg\min} \left\{ \overline{D}_{dr} = \sum_{q=1}^{Q-1} \int_{\alpha^{(q)}}^{\alpha^{(q+1)}} p(\alpha) D_{dr}(\varepsilon^{(q)}) d\alpha \right\} \tag{3.6.7}$$

The optimization is much simplified if loss function $D_{dr}(\varepsilon^{(q)})$ is an even function: $D_{dr}(\varepsilon) = D_{dr}(-\varepsilon)$ such as, for instance, a quadratic loss function $D_{dr}(\varepsilon) = \varepsilon^2$ is. In this case, from $\frac{\partial}{\partial \alpha^{(q)}} D_{dr} = 0$, it follows that optimal boundaries of the discretization intervals should be places just halfway from the corresponding quantized values:

$$\alpha_{opt}^{(q)} = \left(\hat{\alpha}_{opt}^{(q-1)} + \hat{\alpha}_{opt}^{(q)}\right)/2. \tag{3.6.8}$$

As for the representatives, they still should be found by a numerical or, when it is possible, by analytical optimization. For instance, for the quadratic loss function, from $\frac{\partial}{\partial \hat{\alpha}^{(q)}} \boldsymbol{D}_{dr} = \mathbf{0}$ it follows, in addition to Eq. (3.6.8), that optimal representatives are centers of mass of the probability density within the discretization intervals:

$$\hat{\alpha}_{opt}^{(q)} = \int_{\alpha^{(q)}}^{\alpha^{(q+1)}} \alpha p(\alpha) d\alpha \Bigg/ \int_{\alpha^{(q)}}^{\alpha^{(q+1)}} p(\alpha) d\alpha \qquad (3.6.9)$$

Such a solution of the quantization optimization problem for the quadratic loss function is called ***Max-Lloyd quantization***. In Ref. [19] one can find tables of $\{\alpha_{opt}^{(q)}, \hat{\alpha}_{opt}^{(q)}\}$ for optimal mean square quantizers for normal and Laplacian probability densities.

Compressor-expander quantization was initially suggested as a method for hardware implementation of non-uniform optimal quantization using readily available uniform quantizers in which signal dynamic range is split into equal quantization intervals and centers of the intervals are used as quantization levels. In order to implement a non-uniform quantization with uniform quantizers, signal, before being sent to uniform quantizer is subjected to a compressive nonlinear point wise transformation. Correspondingly, at the signal reconstruction stage, quantized values are to be subjected, in digital-to-analog converters, to the expanding nonlinear transformation inverse to the compressing one. (Fig. 3-27).

Optimization of the compressor-expander quantization is achieved by an appropriate selection of the compressive nonlinear point wise transformation. For analytical optimization, quantization optimization equation (3.6.7) should be reformulated in terms of the compression transformation function. Let $w(.)$ is a compression transformation function, Δ_u is interval of uniform quantization.

Then, for a particular value α to be quantized, quantization interval Δ that corresponds to uniform quantization of values of function $w(\alpha)$ will be equal to (Fig. 3-27, a)

$$\Delta = \frac{\Delta_u}{dw(\alpha)/d\alpha} \qquad (3.6.10)$$

and one can obtain optimal function $w(\alpha)$ as a solution of equation:

$$w(\alpha) = \arg\min_{w(\alpha)} \left\{ \overline{\overline{D}}_{dr} = \int_{\alpha_{min}}^{\alpha_{max}} p(\alpha) \overline{D}_{dr} \left(\frac{\Delta_u}{dw(\alpha)/d\alpha} \right) d\alpha \right\}, \qquad (3.6.11)$$

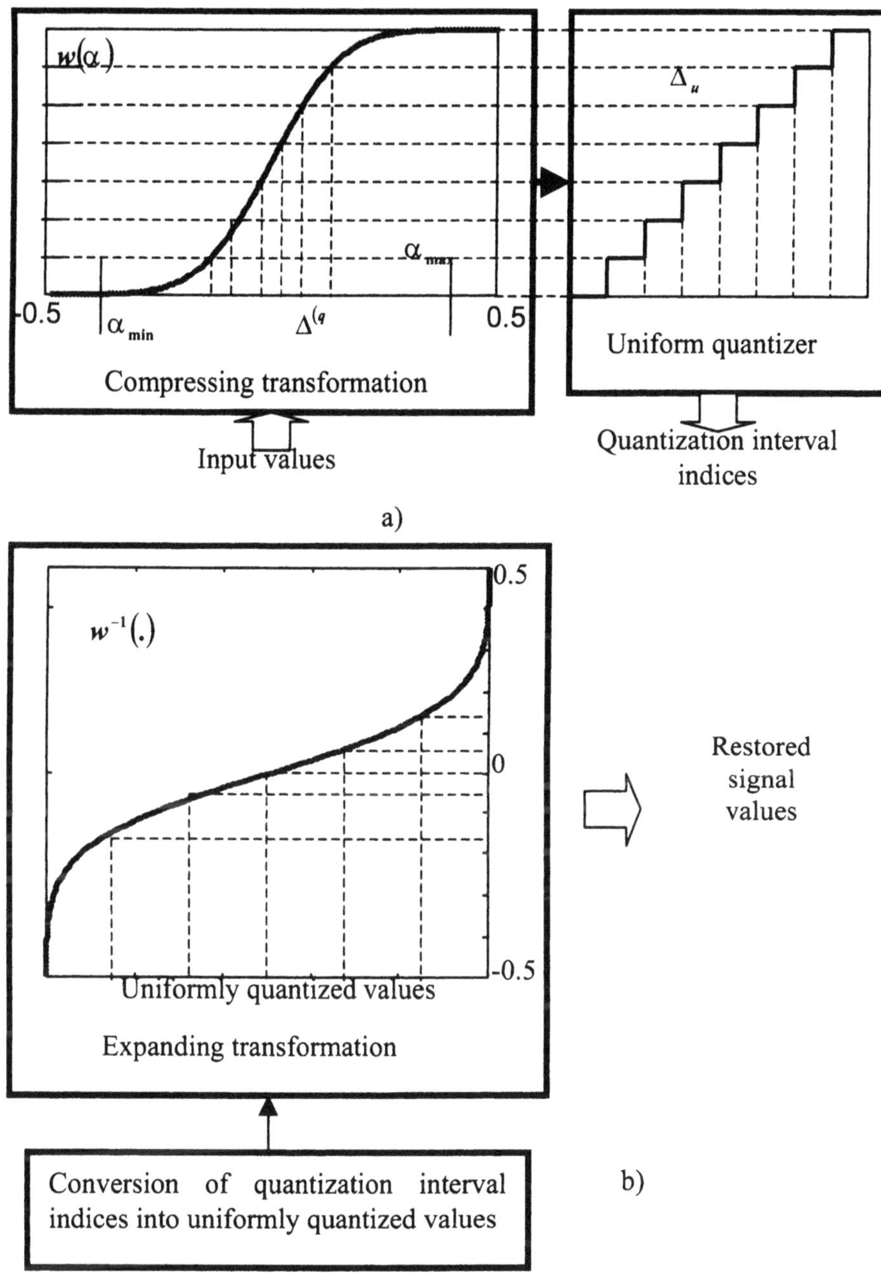

Figure 3-27. Schematic diagram of the companding quantization (a) and expanding restoration (b)

where $\overline{D}_{dr}(\cdot)$ is a quantization loss function formulated in terms of the range of quantization errors for each particular value α. The solution is provided by the Euler-Lagrange equation, which in this case is written as:

$$\frac{\partial}{\partial \dot{w}}\left\{p(\alpha)\overline{D}_{dr}\left(\Delta_u / \dot{w}\right)\right\} = const, \tag{3.6.12}$$

where $\dot{w} = dw(\alpha)/d\alpha$.

The following special cases will provide an insight into how probability distribution $p(\alpha)$ and the type of loss function $\overline{D}_{dr}(\cdot)$ affect optimal arrangement of quantization intervals within the quantized value dynamic range. We will begin with threshold quantization quality criteria that assume that quantization error is nil if it does not exceed a certain threshold. For threshold criteria, optimal arrangement of quantization intervals does not depend on probability distribution $p(\alpha)$ because, according to the criteria, losses owing to quantization should be kept zero for all quantization intervals.

Example 1. Uniform threshold criterion for Absolute value of Quantization Error:

$$\overline{D}_{dr}\left(\Delta_u / \dot{w}\right) = \begin{cases} 0, & |\Delta| \le \Delta_{thr} \\ 1, & otherwise \end{cases}. \tag{3.6.13}$$

From Eq. (3.6.13) it follows that, for monotonic functions $w(\alpha)$,

$$\dot{w} = 2\Delta_u / \Delta_{thr}, \tag{3.6.14}$$

and, therefore, uniform quantization is optimal:

$$\frac{w(\alpha) - w(\alpha_{min})}{w(\alpha_{max}) - w(\alpha_{min})} = \frac{\alpha - \alpha_{min}}{\alpha_{max} - \alpha_{min}}. \tag{3.6.15}$$

Example 2. Uniform threshold criterion of Relative value of Quantization Error:

$$\overline{D}_{dr}\left(\Delta_u / \dot{w}\right) = \begin{cases} 0, & |\Delta| \le \Delta_{thr} = \delta_{thr}\alpha \\ 1, & otherwise \end{cases}. \tag{3.6.16}$$

In this case,

$$\dot{w} = 2\Delta_u / \delta_{thr}\alpha\ , \tag{3.6.17}$$

and, therefore, uniform quantization in a logarithmic scale is optimal:

$$\frac{w(\alpha) - w(\alpha_{min})}{w(\alpha_{max}) - w(\alpha_{min})} = \frac{\ln(\alpha / \alpha_{min})}{\ln(\alpha_{max} / \alpha_{min})}. \tag{3.6.18}$$

This particular case is of a special interest in image processing. Signal quantization results in piece wise constant signals. In quantized images, boundaries between these pieces visually appear as "***false contours***" (see Fig. 3-28). A natural primary requirement for image quantization is that the number and arrangement of quantization levels should be selected so as to secure invisibility of false contours in displayed digital images. Visibility of patches of constant gray level depends on their contrast and of size of patches the constant gray level. According to known ***Weber-Fechner's law***, the ratio of brightness contrast ΔB_{thr} of a stimulus with brightness $B + \Delta B_{thr}$ on the threshold of its visibility to background brightness B is constant for a wide range of the brightness:

$$\frac{\Delta B_{thr}}{B} = \delta_{thr} = const\ . \tag{3.6.19}$$

The constant decreases with the stimulus size. Its lowest value of 1% - 3% corresponds to stimuli of large angular size. Therefore it follows from Weber-Fechner's law that optimal, for digital image display, quantization of image samples should be uniform in the logarithmic scale. The number Q of required quantization levels for logarithmic quantization can be found from equation

$$Q = \frac{w(\alpha_{max}) - w(\alpha_{min})}{\Delta_u} = \frac{\ln(\alpha_{max} / \alpha_{min})}{\delta_{thr}} \tag{3.6.20}$$

For good quality photographic, TV and computer displays, dynamic range $\alpha_{max} / \alpha_{min}$ of displayed image brightness is about 100. Using this estimation of the dynamic range and taking visual sensitivity threshold equal to 2%, one can obtain from Eq. 3.6.20 that the required number of quantization levels is 243. Similar figures are characteristic also for human audio sensibility. This was the main reason why 256 quantization levels (8 bits) were chosen as a standard for image, sound and other analog signal representation and why 8 bits (byte) had become a basic unit in the computer industry.

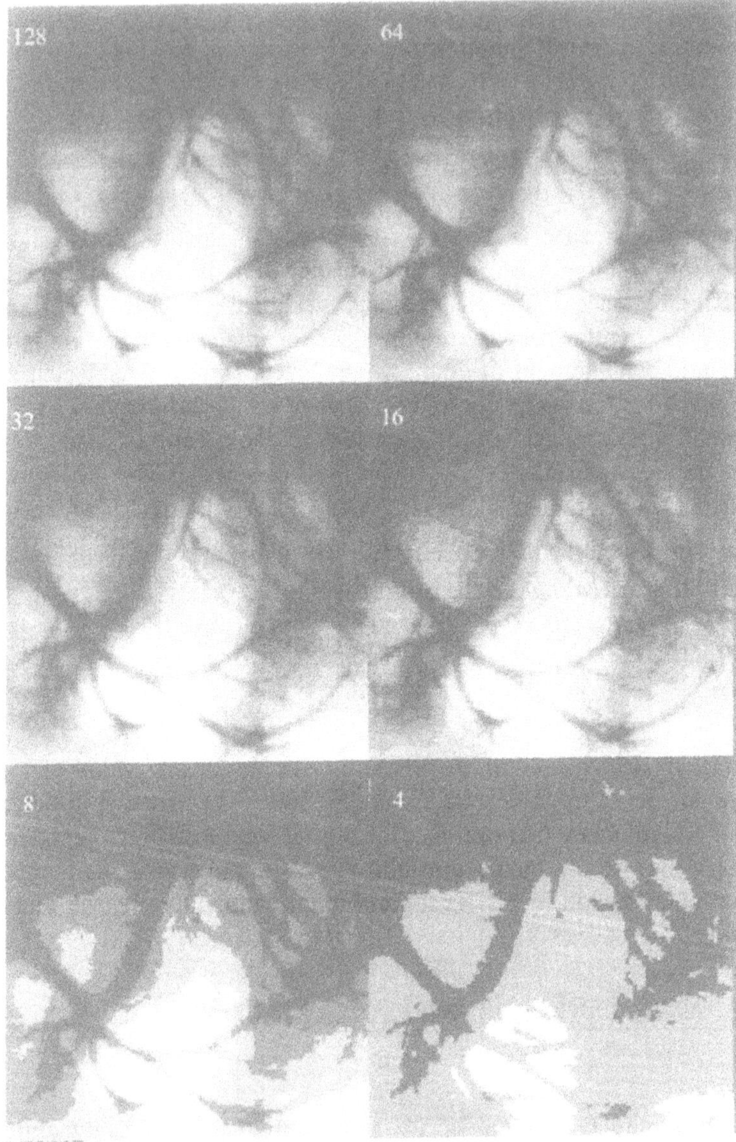

Figure 3-28. False contours in image quantization (128; 64; 32; 16; 8; 4 quantization levels)

Example 3. *Exponential Criterion of Absolute Quantization Error:*

$$\overline{D}_{dr}(\Delta)=|\Delta|^{2P}=\left|\Delta_u \big/ \dot{w}\right|^{2q} \qquad (3.6.21)$$

Substituting (3.6.21) into (3.6.12) and solving the resulting differential equation, obtain

$$\frac{w(\alpha)-w(\alpha_{min})}{w(\alpha_{max})-w(\alpha_{min})}=\frac{\int\limits_{\alpha_{min}}^{\alpha}(p(\alpha))^{1/(2q+1)}d\alpha}{\int\limits_{\alpha_{min}}^{\alpha_{max}}(p(\alpha))^{1/(2q+1)}d\alpha}. \qquad (3.6.22)$$

Thus, the required degree of nonlinear pre-distortion depends solely on the probability distribution of the quantized values. The meaning of this relationship becomes evident from the expression:

$$\Delta(\alpha)=\Delta_u / \dot{w}(\alpha)\propto(p(\alpha))^{-1/(2q+1)} \qquad (3.6.23)$$

which implies that the size of quantization intervals for the various values of α are inversely proportional to their probability densities raised to the corresponding power. For the widely used mean squared quantization error criterion ($q=1$),

$$\frac{w(\alpha)-w(\alpha_{min})}{w(\alpha_{max})-w(\alpha_{min})}=\frac{\int\limits_{\alpha_{min}}^{\alpha}(p(\alpha))^{1/3}d\alpha}{\int\limits_{\alpha_{min}}^{\alpha_{max}}(p(\alpha))^{1/3}d\alpha}. \qquad (3.6.24)$$

Note that an interesting special case

$$\frac{w(\alpha)-w(\alpha_{min})}{w(\alpha_{max})-w(\alpha_{min})}=\int\limits_{\alpha_{min}}^{\alpha}p(\alpha)d\alpha \Big/ \int\limits_{\alpha_{min}}^{\alpha_{max}}p(\alpha)d\alpha \qquad (3.6.25)$$

takes place when $q=0$. Such a transformation is known as ***histogram equalization*** and is one of popular image enhancement transformations. For image quantization, this transformation assumes that quantization errors are equally important whatever they values are (see Eq. (3.6.21)) and makes quantization intervals to be inversely proportional to the probability density for the level to be quantized (see Eq. 3.6.23).

Sometimes, as in the case quantizing spectral coefficients of signals in Fourier, Walsh and other bases, one may regard the quantized coefficients of the discrete signal representation as distributed according to a truncated Gaussian probability density distribution on the interval $[\alpha_{min}, \alpha_{max}]$:

$$p(\alpha) \propto \exp\left[-(\alpha - \bar{\alpha})^2 / 2\sigma_\alpha^2\right]. \tag{3.6.26}$$

Then, for mean squared error criterion ($P = 1$),

$$\frac{w(\alpha) - w(\alpha_{\min})}{w(\alpha_{\max}) - w(\alpha_{\min})} = \frac{\Phi\left[(\alpha - \bar{\alpha})/\sqrt{3}\sigma_\alpha\right] - \Phi\left[(\alpha_{\min} - \bar{\alpha})/\sqrt{3}\sigma_\alpha\right]}{\Phi\left[(\alpha_{\max} - \bar{\alpha})/\sqrt{3}\sigma_\alpha\right] - \Phi\left[(\alpha_{\min} - \bar{\alpha})/\sqrt{3}\sigma_\alpha\right]} \tag{3.6.27}$$

where

$$\Phi(x) = \int_{-\infty}^{x} \exp\left(-\xi^2/2\right) d\xi \tag{3.6.28}$$

One can find that, for this case, the reduction in the required number of quantization levels, as compared , for the same averaged quantization error, to the case of uniform quantization, is equal to

$$g = \frac{\alpha_{\max/\min}}{\sqrt[4]{27}\sqrt{2\pi}} \frac{\left[\Phi(\alpha_{\max/\min}/2) - \Phi(\alpha_{\max/\min}/2)\right]^{1/2}}{\left[\Phi\left(\alpha_{\max/\min}/2\sqrt{3}\right) - \Phi\left(\alpha_{\max/\min}/2\sqrt{3}\right)\right]^{3/2}}, \tag{3.6.29}$$

where $\alpha_{max/min} = (\alpha_{max} - \alpha_{min})/\sigma_\alpha$. For large $\alpha_{max/min}$, the gain tends to $\alpha_{max/min}/5.7$.

<u>Example 4</u>. *Exponential Criterion of Relative Quantization Error:*

$$\overline{D}_{dr}(\Delta) = |\Delta/\alpha|^{2P} = \left|\Delta_u \Big/ \overset{\bullet}{w\alpha}\right|^{2q} \tag{3.6.30}$$

In this case, solution of the Euler-Lagrange equation (3.6.12) yields:

$$\frac{w(\alpha)-w(\alpha_{\min})}{w(\alpha_{\max})-w(\alpha_{\min})}=\frac{\int\limits_{\alpha_{\min}}^{\alpha}\left(p(\alpha)/\alpha^{2q}\right)^{1/(2q+1)}d\alpha}{\int\limits_{\alpha_{\min}}^{\alpha_{\max}}\left(p(\alpha)/\alpha^{2q}\right)^{1/(2q+1)}d\alpha}. \quad (3.6.31)$$

If quantized values α are distributed uniformly in the dynamic range, optimal compression function is

$$\frac{w(\alpha)-w(\alpha_{\min})}{w(\alpha_{\max})-w(\alpha_{\min})}=\frac{\alpha^{1/(2q+1)}-\alpha_{\min}^{1/(2q+1)}}{\alpha_{\max}^{1/(2q+1)}-\alpha_{\min}^{1/(2q+1)}}. \quad (3.6.32)$$

For the mean squared relative quantization error criterion (Eq. (3.6.30), $q=1$):

$$\frac{w(\alpha)-w(\alpha_{\min})}{w(\alpha_{\max})-w(\alpha_{\min})}=\frac{\alpha^{1/3}-\alpha_{\min}^{1/3}}{\alpha_{\max}^{1/3}-\alpha_{\min}^{1/3}}. \quad (3.6.33)$$

We will refer to such type of nonlinear transformations of Eq. (3.6.32) as to ***"P-th law quantization"***:

$$\frac{w(\alpha)-w(\alpha_{\min})}{w(\alpha_{\max})-w(\alpha_{\min})}=\frac{\alpha^{P}-\alpha_{\min}^{P}}{\alpha_{\max}^{P}-\alpha_{\min}^{P}}. \quad (3.6.34)$$

P-th law quantization proved to be very useful for quantizing Fourier and DCT transform coefficients of images. The optimal value of the nonlinearity index ***P*** for which standard deviation of errors in reconstructed images because of quantization of their spectra is minimal is usually about 0.3 (see, for instance, Fig. 3-29). Remarkable enough, this corresponds to the optimum for the mean squared relative quantization error criterion defined by Eq. 3.6.3 3. Another option in quantization image spectral coefficients is ***inhomogeneous quantization*** in which different coefficients are quantized in different way. Inhomogeneous quantization in for of ***zonal quantization*** has found its application in image transform coding (see Sect. 3.7.2)

3.6.2 Quantization in digital holography

Quantization in digital holography has its peculiarities. First of all, as it follows from Sect. 2.3.2 (see also Sect. 13.1), this is quantization in Fourier

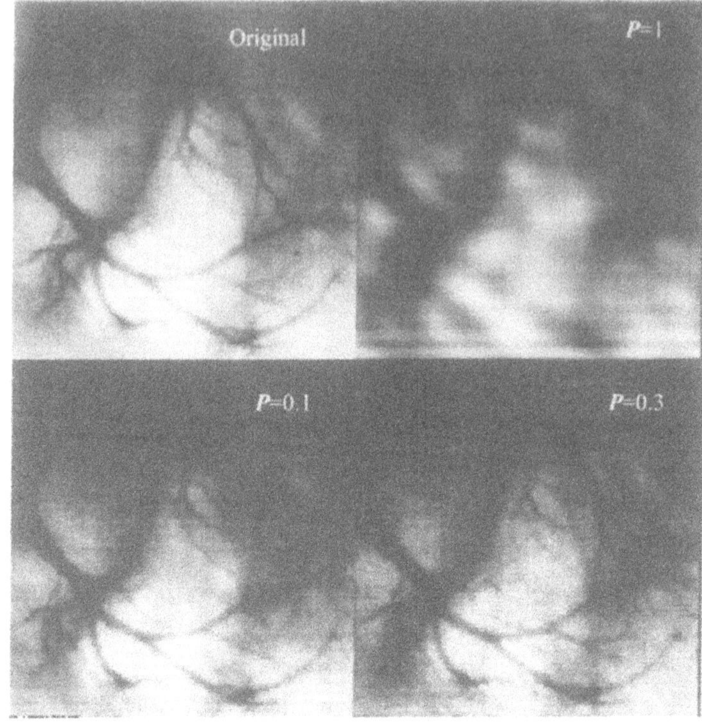

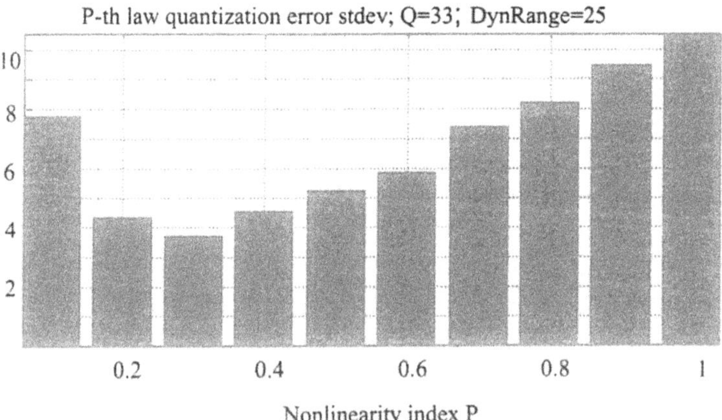

Figure 3-29. Optimization of ***P***-th law quantization of image spectrum (33 quantization levels of spectrum real and imaginary parts in the dynamic range of ±25 of their standard deviation): Initial image and images reconstructed from uniformly (P=1) and ***P***-th (***P***=0.3 and 0.1) law quantized of spectral coefficients. Bar diagram at the bottom shows how standard deviation of quantization error depends on the nonlinearity index ***P***. Note that while not much difference between images for P=0.1 and P=0.3 is visible, standard deviations of the quantization error differ substantially for these two cases.

or Fresnel transform domains. Therefore, in quantization of holograms for digital reconstruction of images and for recording computer generated holograms, ***P***-th law quantization is advisable. For digital reconstruction of holograms, corresponding correcting expanding transformations should be implemented in hologram preprocessing (see Sect. 8. 6.3). For recording computer generated holograms, correcting expanding transformation should be implemented in hologram encoding for recording on optical medium.

For encoding computer generated display holograms, yet another quantization method, the ***pseudo-random diffuser*** one, can be employed. The method makes use of the fact that reproduction of object wave front amplitude is only required for computer generated display holograms. Therefore the phase component which is not visually observed can be selected arbitrarily so as to minimize hologram quantization artifacts.

The method is mathematically described as follows. Let $\{\bar{A}^2_{k,l}\}$ be samples of the object brightness with (k,l) being sample indices. Then samples of the object wave front $\tilde{\bar{A}}_{k,l}$ are defined as:

$$\tilde{\bar{A}}_{k,l} = \bar{A}_{k,l} \exp(i\theta_{k,l}), \tag{3.6.35}$$

where $\{\theta_{k,l}\}$ is an array of pseudo-random numbers taken from the range $[0,2\pi]$. The simplest solution is to use for $\{\theta_{k,l}\}$ binary statistically independent numbers that assume values 0 and π with equal probabilities.

Such a pseudo-random phase modulation of the object wave front redistributes wave front Fourier and Fresnel spectrum energy uniformly (statistically) between all spectral coefficients so that all coefficients have the same dynamic range. Fig. 3.30 illustrates this phenomenon of spectrum "equalization" by image pseudo-random phase modulation.

Initial image

Spectrum of the initial image

Spectrum of pseudo-random phase modulated image

Figure 3-30. Image Fourier spectrum equalization by means of pseudo-random phase modulation

Image pseudo-random phase modulation eliminates the need of nonuniform or inhomogeneous quantization and simplifies the hologram encoding for recording on optical media.

The pseudo-random phase component can be further optimized by an iterative optimization procedure aimed at reducing image reconstruction error and similar to that described in Sect. 7.2.4. The optimization is especially required for recording ***kinoforms***, computer generated holograms in which hologram amplitude is quantized to only one level (see Sect. 7.2.4). Fig. 3-31 illustrates effectiveness of ***P***-th law and pseudo-random diffuser quantizations for hologram encoding.

Figure 3-31. ***P***-th law and pseudo-random diffuser quantization for encoding computer generated display Fourier holograms: a) , b) - simulated images reconstructed from a Fourier hologram of the image of Fig. 13-30, real and imaginary parts of which are, correspondingly, uniformly and ***P***-th law (P=0.3) quantized to 257 quantization levels (2x128 for positive and negative values plus one level for zero value); c) image reconstructed from a Fourier hologram of the same image computed with the use of pseudo-random binary (0 and π) phase component assigned to the image.

3.7 BASICS OF SIGNAL CODING AND DATA COMPRESSION

3.7.1 Signal rate distortion function, entropy and statistical coding

Purpose of signal coding and data compression is encoding signals from a signal ensemble into a stream of binary digits, or bits (binary code) of the least possible length as evaluated on average over all possible signals. For continuous signals, the lower bound of the code length is given by rate distortion function $\boldsymbol{H}_{\varepsilon}$ of the signal ensemble $\boldsymbol{A}$ given accuracy of signal reconstruction from the obtained digital signals:

$$\boldsymbol{H}_{\varepsilon}=\min \boldsymbol{I}\left(\hat{\boldsymbol{A}}, \boldsymbol{A} / \boldsymbol{d}\left(\boldsymbol{A}, \hat{\boldsymbol{A}}\right) \leq \varepsilon\right), \tag{3.7.1}$$

where $\hat{\boldsymbol{A}}$ is a signal ensemble reconstructed from digital signals obtained by the encoding process from signal ensemble $\boldsymbol{A}$ and $\boldsymbol{I}\left(\hat{\boldsymbol{A}}, \boldsymbol{A} / \boldsymbol{D}\left(\hat{\boldsymbol{A}}, \boldsymbol{A}\right) \leq \varepsilon\right)$ is mutual information between the ensembles provided error $\boldsymbol{d}\left(\hat{\boldsymbol{A}}, \boldsymbol{A}\right)$ of representation of signals from ensemble $\boldsymbol{A}$ by corresponding reconstructed signals form ensemble $\hat{\boldsymbol{A}}$ does not exceed certain value ε ([20]). The meaning of this measure can be illustrated by rate distortion function of an ensemble of signals that can be modeled as Gaussian random processes with uniform spectral density within finite bandwidth $\boldsymbol{F}$ and that are required to be reconstructed with root mean squared error ε :

$$\boldsymbol{H}_{\varepsilon}=\boldsymbol{F} \log_2 \frac{\overline{|\boldsymbol{A}|^2}}{\varepsilon^2}, \tag{3.7.2}$$

where $\overline{|\boldsymbol{A}|^2}$ is variance of the signal ensemble. The lower bound defined by Eq. (3.7.1) can, in principle, be achieved as a result of ***general digitization*** process discussed in Sect. 2.1.2.

In practice, data compression is applied for digital signals obtained for continuous signals by their discretization and element-wise quantization discussed in this chapter. For digital signals, lower bound for the code length is determined by the entropy of digital signal ensemble ([20]). If $\{\boldsymbol{A}_n\}$ is a set of digital signals to be encoded, and $\boldsymbol{P}(\boldsymbol{A}_n)$ is probability with which signal $\boldsymbol{A}_k$ may occur, entropy

$$H(A) = -\sum_{n} P(A_n)\log_2 P(A_n) \tag{3.7.3}$$

of the signal ensemble defined by the signal set $\{A_n\}$ and their probabilities $\{P(A_n)\}$ determines minimal number of bits per signal that are necessary to specify any particular digital signal from the ensemble. When all signals are equally probable, entropy, and therefore number of bits required, is maximal and equal to

$$H_{\max}(A) = \log_2(\textit{Number of possible signals}). \tag{3.7.4}$$

If, however, signal probability distribution is substantially nonuniform, the number of required bits is much less than the upper bound defined by Eq. 3.7.4. The stream of bits obtained in data coding according to Eq. 3.7.4 is in this case redundant. Ratio

$$R = H(A)/H_{\max}(A) \tag{3.7.5}$$

represents a measure of the digital ***data redundancy***. It determines the limit for possible data compression.

If, as we assume, digital signals are represented by their quantized representation coefficients $\{\{\alpha_r\}_n\}$ over discretization basis functions, signal ensemble entropy can also be computed as

$$H(A) = -\sum_{\{\alpha_n\}} p(\{\alpha_n\})\log_2 p(\{\alpha_n\}), \tag{3.7.6}$$

where $p(\{\alpha_n\})$ are mutual probabilities of the sets $\{\alpha_n\}$ of coefficients for signals $\{A_n\}$. In this case it is more convenient to evaluate entropy per coefficient (e.g., per sample, per pixel):

$$H(\mathrm{A}) = -\frac{1}{N}\sum_{\{\alpha_n\}} p(\{\alpha_n\})\log_2 p(\{\alpha_n\}), \tag{3.7.7}$$

where N is number of the representation coefficients per signal. If representation coefficients happen to be mutually statistically independent such that

$$p(\{\alpha_n\}) = \prod_{n=0}^{N-1} p(\alpha_n), \tag{3.7.8}$$

then

$$H_{ind}(A) = -\frac{1}{N}\sum_{n}\sum_{q=0}^{Q-1} p\left(\hat{\alpha}_n^{(q)}\right)\log_2 p\left(\hat{\alpha}_n^{(q)}\right) \geq H(A), \quad (3.7.9)$$

where $p\left(\hat{\alpha}_n^{(q)}\right)$ is probability of q-th quantization level of coefficient α_n and Q is number of coefficient quantization levels. Furthermore, if probability distribution of quantized coefficients does not depend on their index n, entropy per coefficient will further increase to

$$H_{iid}(A) = -\sum_{q=0}^{Q-1} p\left(\hat{\alpha}^{(q)}\right)\log_2 p\left(\hat{\alpha}^{(q)}\right) \geq H_{iid}(A) \geq H(A). \quad (3.7.10)$$

The upper limit to the per-coefficient entropy is given by entropy of uniformly distributed quantization levels:

$$H_{uiid}(A) = \log_2 Q \geq H_{iid}(A) \geq H_{iid}(A) \geq H(A). \quad (3.7.11)$$

In practice of digital holography and image processing as well as in all other digital signal processing applications, digital data obtained as a result of discretization and scalar quantization are very redundant because signal representation coefficients are usually highly correlated and distributed non-uniformly, and therefore their entropy is substantially lower then the upper limit defined by Eq. 3.7.11.

From the above reasoning, one can see that data compression which removes their redundancy requires, in general, assigning binary codes to signals as a whole and that the length of the code, as it follows from Eq. 3.7.2, should be equal to binary logarithm of the inverse to its probability. This, however is not practically feasible for the same reason why general digitization is not practically workable and, in addition, because determination of signal probabilities is not practically feasible. This why binary encoding is applied, with rare exceptions, to individual quantized signal representation coefficients. For this reason, data compression is implemented in either two or three steps. The two step procedure:

- Decorrelating signal representation coefficients. If successful, this produces data that are less correlated but instead distributed highly non-uniformly.
- Encoding decorrelated data by allocating to each of the decorrelated coefficients a binary code with the number of bits as close as possible to binary logarithm of inverse to its probability.

is called ***statistical (entropy) coding***. It has also obtained a jargon name ***loss-less coding*** because it does not assume introducing any irreversible changes (such as quantization) into the data and enables perfect restoration of initial signal representation coefficients. Because decorrelating signal representation coefficients is most frequently far to be complete, it is usually supplemented with a secondary quantization of decorrelated data. This three step procedure:

- decorrelating signal representation coefficients
- secondary quantization of decorrelated data
- binary statistical encoding re-quantized data

is called ***lossy compression*** to emphasize the fact that original signal representation coefficients can not be perfectly restored from their binary codes because of the secondary quantization.

For more then 60 years since the first works on digital data coding, numerous data compression methods have been developed and implemented. They will be briefly reviewed in the next section. They, however, differ only in the implementation of the first two steps of the above three step compression procedure. The last one, binary statistical coding is common to all of them.

Two mutually complement methods of binary statistical coding are known and overwhelmingly used: ***variable length coding*** (***VL-coding***) and ***coding of rare symbols***. In data coding jargon, objects of encoding are called symbols. VL-coding is aimed at generating, for each symbol to be coded, a binary code word of the length as close to logarithm of the inverse to its probability. Originally suggested by Shannon ([20]) and Fano ([21]), it was later improved by Huffman ([22]), and the Huffman's coding procedure had become the preferred implementation of VL-coding. VL-Huffman coding is an iterative process, in which, at each iteration, to codes of two symbols with the least probabilities bits 0 and 1, correspondingly, are added and then the symbols are united in a new auxiliary symbol whose probability is sum of the united symbols. The iteration process is repeated with respect to the modified at the previous iteration set of symbols until all symbols are encoded. If probabilities of symbols are integer power of ½, such a procedure generates binary codes with number of bits per symbol exactly equal to binary logarithm of the inverse to its probability, and therefore average number of bits per symbol is equal to the entropy of the symbol ensemble. An example of Huffman-coding is illustrated in Table 3.2 for 8 symbols (A to H).

As one can see, when one of the symbols of the ensemble has probability that is much higher than ½ while common probability of others is much less than ½, Huffman coding is inefficient because it does not allow to allocate to symbols less than one bit. In such cases, coding of rare symbols is applied.

Table 3-2. An example of ShFH-coding of 8 symbols

Iteration	**A** *P*(A)= 0.49	**B** *P*(B)= 0.27	**C** *P*(C)= 0.12	**D** *P*(D)= 0.065	**E** *P*(E)= 0.031	**F** *P*(F)= 0.014	**G** *P*(G)= 0.006	**H** *P*(H)= 0.004
1-st	-	-	-	-	-	-	0	1
							P(GH)=0.01	
2-nd	-	-	-	-	-	0	1	
						P(FGH)=0.024		
3-d	-	-	-	-	0	1		
					P(EFGH)=0.055			
4-th	-	-	-	0	1			
				P(DEFGH)=0.12				
5-th	-	-	0	1				
			P(CDEFGH)=0.24					
6-th	-	0	1					
		P(BCDEFGH)=0.51						
7-th	0	1						
Binary code	**0**	**10**	**110**	**1110**	**11110**	**111110**	**1111110**	**1111111**
Entropy *H*=**1.9554**								
Average number of bits per symbol: **1.959**								

There are two most popular varieties of methods for coding rare symbols: ***run length coding*** and ***coding co-ordinates of rare symbols***. In run length coding, it is assumed that symbols to be coded form a sequence in which runs of the most frequent symbol occur. The encoder computes length of the runs and generates a new sequence of symbols in which the runs of the most frequent initial symbol are replaced with new auxiliary symbols that designate the run lengths. This new sequence can then be subjected to VL-coding if necessary. Run length coding is an essentially 1-D procedure. For coding 2-D data, it is applied after 2-D data are converted into 1-D data by means of a zigzag scanning. Principle of zigzag scanning is illustrated below in Fig. 3-34.

Coding coordinates of rare symbols is an alternative to the run length coding. In this method, positions of rare symbols (others than the most frequent one) are found. Then a new sequence of symbols is generated from the initial one in which all occurrences of the most frequent symbol are removed and auxiliary symbols are added to each occurrence of rare symbols that designate their position in the initial sequence. Coding coordinates of rare symbols is less sensitive to errors in transmitting the binary code than run length coding. For the former, transmission errors cause only localized errors in the decoded symbol sequence while they result in shifts of entire decoded symbol sequence after the erroneous one in the latter. This property is of especial importance in image coding for transmission where transmission channel errors, in case of run length coding, may cause substantial deformation of object boundaries.

3.7.2 Image compression methods: a review

Digital data compression methods have been developed for and found their major applications in audio and video data compression. In this section, we will briefly review methods for image and video compression. Classification diagram of image coding methods is presented in Fig. 3-32. Coding methods are classified in it into two groups: methods in which data decorrelation and quantization are carried out separately one after another and methods in which one can not separate decorrelation from quantization. The latter attempt to implement, though in a reduced form, the idea of encoding of signals as a whole to achieve the theoretical minimum of bits defined by Eq. 3.7.3.

In decorrelation-then-quantization methods, data decorrelation is carried out with linear transforms. Two large families of decorrelating linear transforms are in use: "predictive" and orthogonal transforms. In predictive transforms, for each pixel is compared its "predicted" value is generated in a certain way from other pixels and a prediction error is computed as a difference between the pixel value and its "predicted" value. The prediction error is then sent to scalar quantizer. The simplest method that implements decorrelation by prediction is known as ***DPCM*** (from "Differential Pulse Code Modulation). Block diagram of DPCM coding and decoding are shown in Fig. 3-33. DPCM assumes that images are scanned row-wise/column-wise and, for each current pixel $a_{k,l}$, prediction error is computed as (Fig. 3-34)

$$\varepsilon_{k,l} = a_{k,l} - \bar{a}_{k,l} = a_{k,l} - c_{-1,0}a_{k-1,l} - c_{0,1}a_{k,l+1} - c_{-1,-1}a_{k-1,l-1} + c_{-1,1}a_{k-1,l+1} \,. \quad (3.7.12)$$

where $\bar{a}_{k,l}$ is a prediction value. Prediction coefficients $\{c\}$ are usually chosen so as to minimize prediction error standard deviation as computed for an ensemble of pixels subjected to coding. For instance, in a natural assumption that images are statistically isotropic in all directions,

$$c_{-1,0} = c_{0,1} = c_1;\ c_{-1,-1} = c_{-1,1} = c_2 \,, \quad (3.7.13)$$

and c_1 and c_2 are found from the equation:

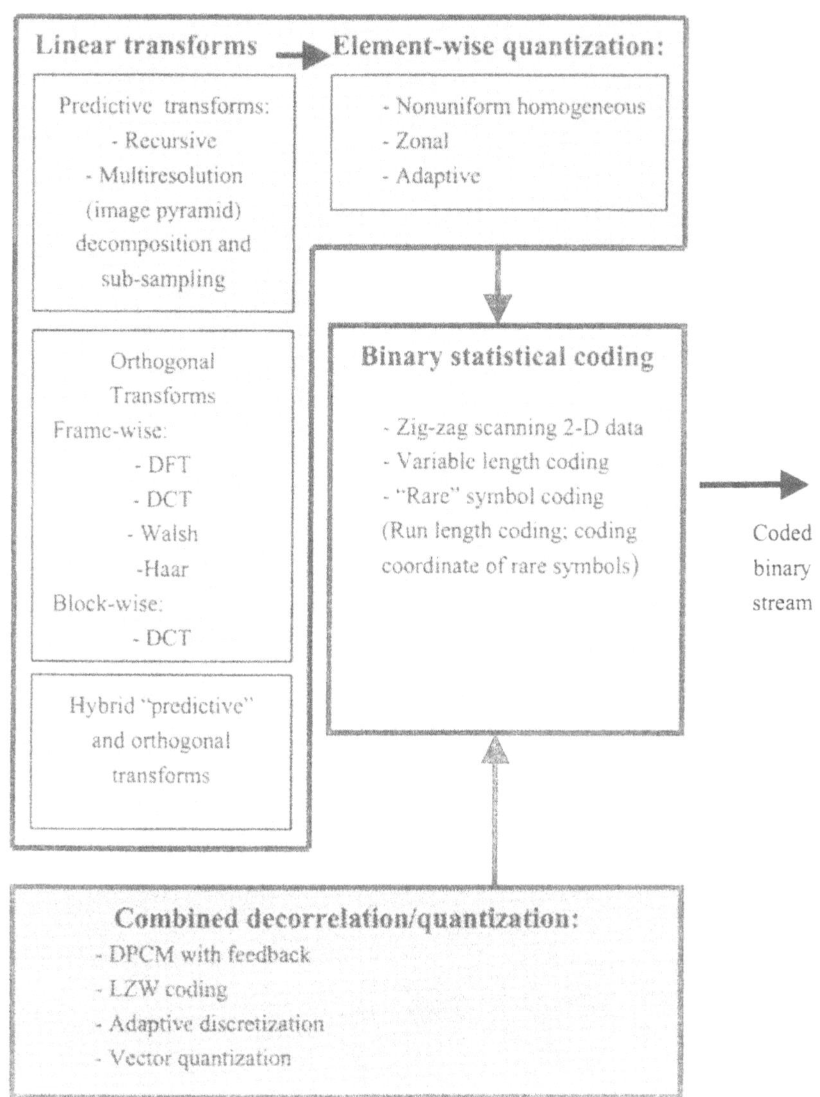

Figure 3-32. Classification of digital data compression methods

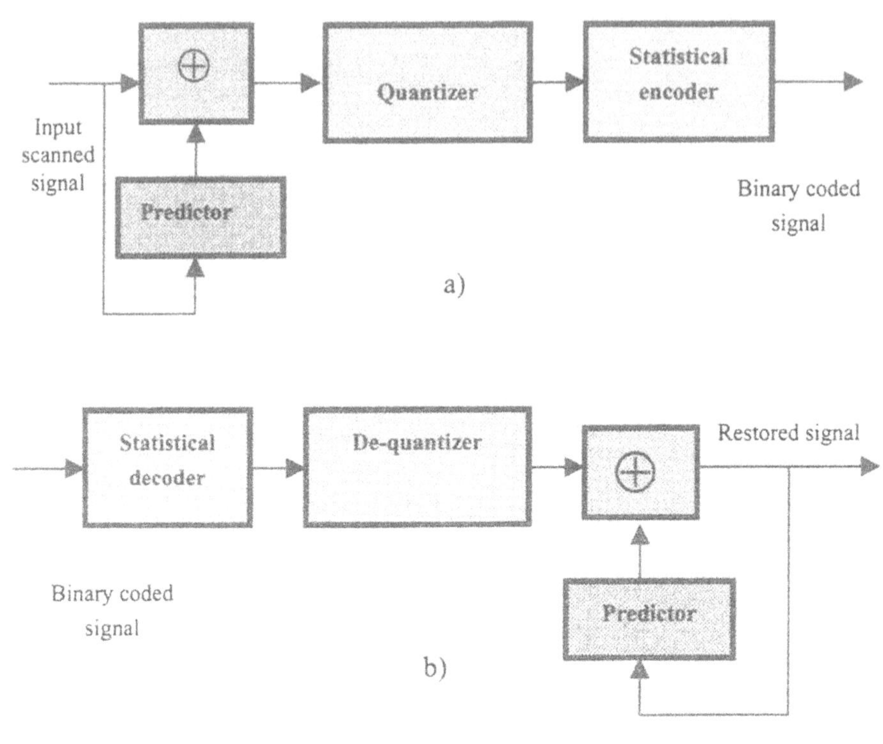

Figure 3-33. Block digram of DPCM coding (a) and decoding (b)

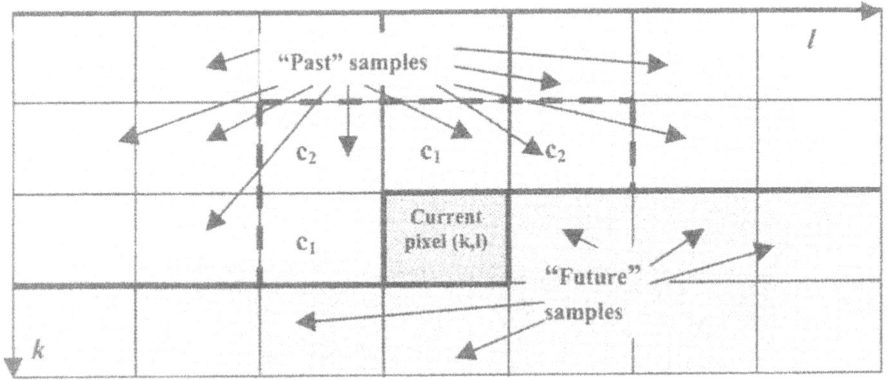

Figure 3-34. 2-D prediction for row-column image scanning method

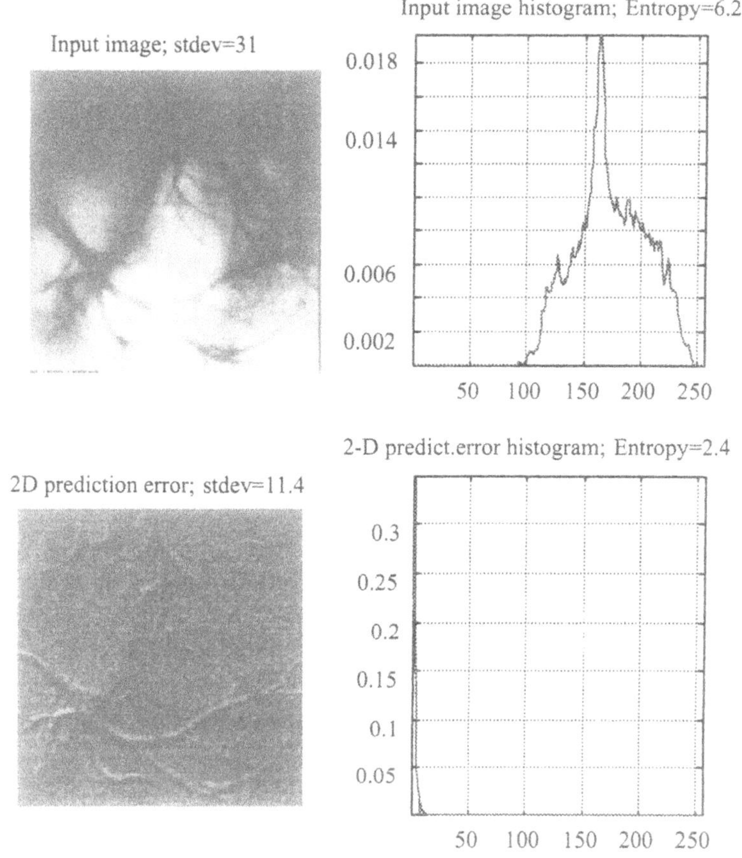

Figure 3-35. Dynamic range, standard deviation and entropy reduction by predictive decorrelation in DPCM

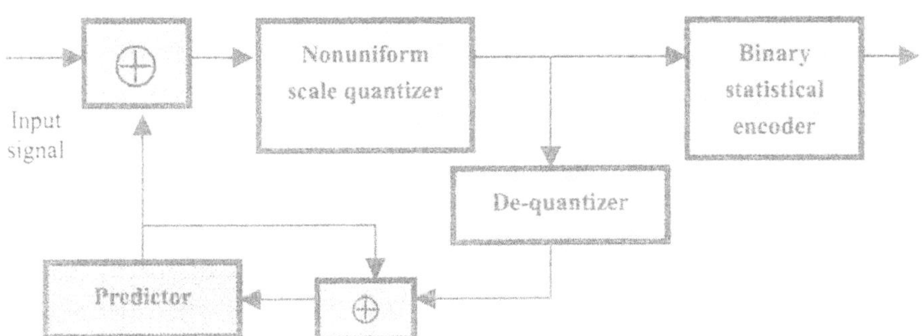

Figure 3-36. Block diagram of DPCM with feedback.

$$\begin{cases} c_1\left(1+\rho_d\right)+c_2\dfrac{3\rho_{hv}+\rho_{2hv}}{2}=\rho_{hv} \\ c_1\dfrac{3\rho_{hv}+\rho_{2hv}}{2}+c_2\left(1+\rho_{2h}\right)=\rho_d \end{cases}, \tag{3.7.14}$$

where

$$\rho_{hv}=\frac{AV\left(a_{k-1,l}a_{k,l}+a_{k,l-1}a_{k,l}\right)}{2AV\left(\left(a_{k,l}\right)^2\right)};\quad \rho_d=\frac{AV\left(a_{k-1,l-1}a_{k,l}\right)}{AV\left(\left(a_{k,l}\right)^2\right)}; \tag{3.7.15}$$

$$\rho_{2h}=\frac{AV\left(a_{k-1,l-1}a_{k-1,l+1}\right)}{AV\left(\left(a_{k,l}\right)^2\right)};\qquad \rho_{2hv}=\frac{AV\left(a_{k,l-1}a_{k-1,l+1}\right)}{AV\left(\left(a_{k,l}\right)^2\right)}; \tag{3.7.16}$$

are correlation coefficients of pixels with their immediate neighbors in vertical or horizontal direction, of pixels with their immediate neighbor in diagonal direction, of pixels with their next to immediate horizontal neighbors and of pixels that are distant one from another one sampling interval in vertical and two intervals in horizontal directions, respectively. Fig. 3-34 illustrates how predictive decorrelation reduces dynamic range and entropy of signal-to-be-quantized.

DPCM with no quantization is used for loss-less image coding. With quantization, substantially larger compression is possible although images are reconstructed with losses. Quantization artifacts can be reduced if prediction error is computed from already quantized data as it is done in ***DPCM with feedback*** (Fig. 3-35). From classification point of view, DPCM with feedback can be regarded as a combined discretization/quantization method.

In DPCM, the prediction is practically made over only immediate vicinity of pixels. More efficient decorrelation is achieved when larger spatial neighborhood of pixels are involved in the prediction as it is implemented in multi-resolution image expansion. This method, depending on the implementation, is known under different names: ***pyramid coding***, ***sub-band decomposition coding***, ***wavelet coding***. In this method, image transform coefficients obtained at each resolution (Fig. 3-6) are optimally quantized and then statistically encoded. Fig. 3.36 demonstrates an example of such an image multi-resolution decomposition.

Decorrelation with orthogonal transforms is an alternative to predictive decorrelation. Redundancy of the image signals exhibits itself in transform domain mostly in compaction of signal energy in small number of transform coefficients and in contraction of the dynamic range of higher order coefficients. While Karhunen-Loeve transform is the statistically the best one in terms of energy compaction property, it is difficult to generate and is computationally expensive. In practice, transforms such as DFT, DCT, Walsh transforms that can be computed with fast transform algorithms (see Chapt. 5) are used.

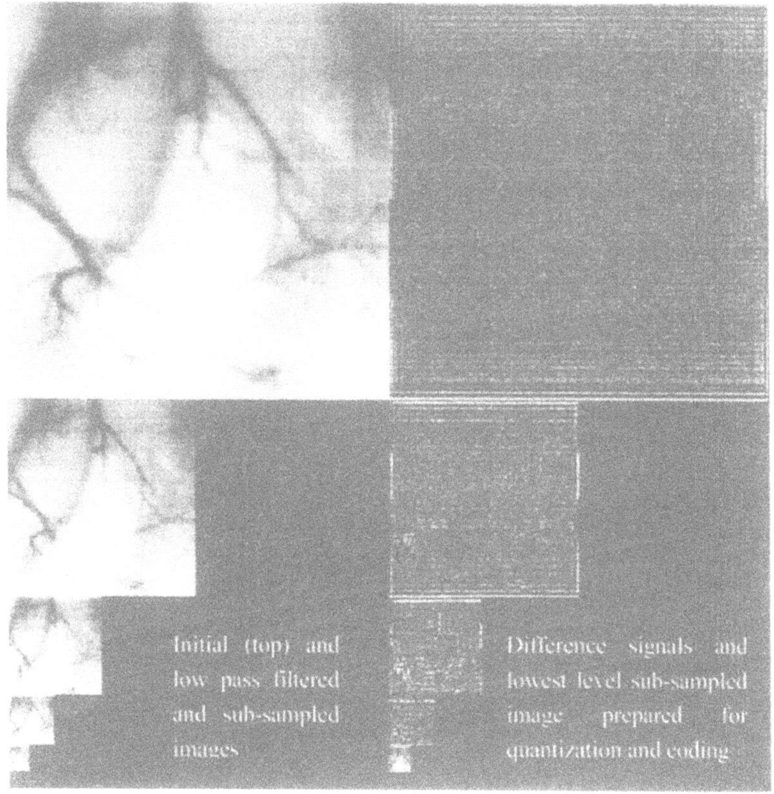

Figure 3-37. Image pyramid

Two versions of transform coding are known: frame-wise and block-wise ones. In frame-wise coding, image frame as a whole is transformed, transform coefficients with very low energy are truncated and the others are quantized. In block-wise coding, image is split into blocks and individual blocks are transformed separately. For every block, low energy block transform coefficients are truncated and the others are quantized. Truncation and quantization of transform coefficients is either predefined by a special quantization table (***zonal quantization***) or adaptively changed for every block.

In terms of energy compaction, frame-wise transforms are, in principle, more efficient then block-wise ones. However, block-wise transform requires much less computations and can even be implemented in inexpensive hardware. Moreover, it is well suited to spatial image inhomogeneity and allows using adaptive quantization in which way truncating and quantizing coefficients is optimized for each individual block.

Among the block transforms, DCT had proved to be the best one and it is put in the base of image compression standards H.261, H.262, H.263, H.320, JPEG and MPEG ([22]). The principle of image DCT block coding is illustrated in Fig. 3-38.

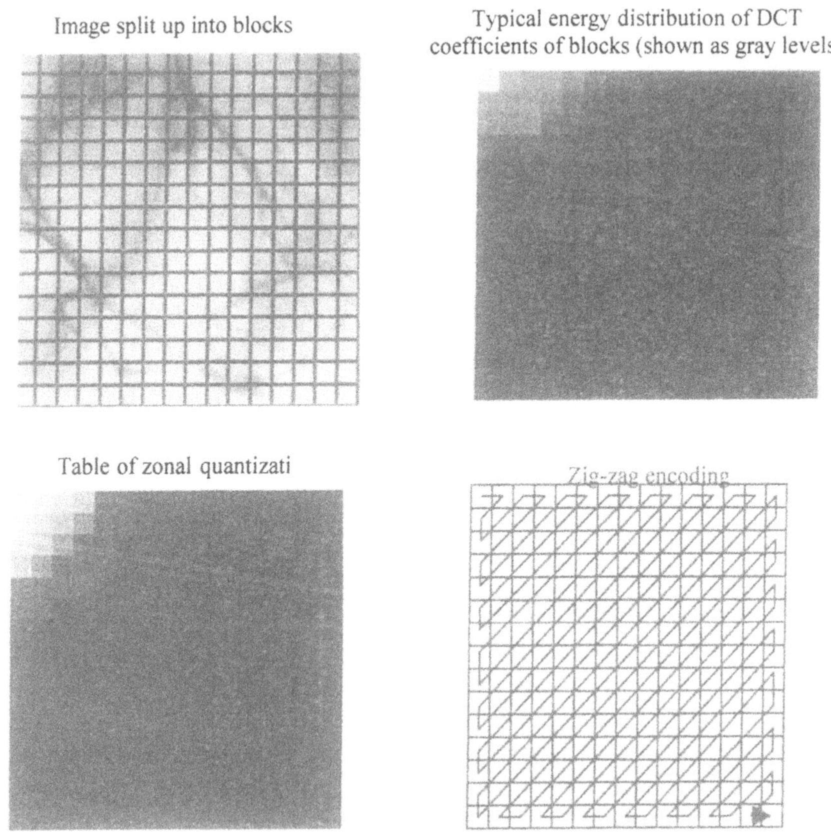

Figure 3-38. Principle of DCT block transform coding

In the image block transform coding, image is split up into blocks. DCT of each block is then computed and spectral coefficients are quantized individually according to a quantization table of zonal quantization that specifies the number of quantization levels allocated to each coefficients as a function of the coefficient indices. The table is built according to typical energy distribution of the transform coefficients: the higher is variance of the coefficient, the larger number of quantization levels is allocated for quantization of this coefficient. Obtained 2-D arrays of quantized numbers for each block are then converted, for transmitting and storing, into 1-D sequences by zigzag scanning of the arrays.

Quantization of DCT coefficients causes certain distortions in reconstructed images. A typical artifact caused by quantization in image block transform coding is the so called blocking effect: visible discontinuities at the borders of the blocks. These discontinuities can be substantially mitigated if blocks are overlapping. A most known implementation of this idea is lapped transform coding ([23]).

While in coding still images predictive or orthogonal transform decorrelation are mostly used, hybrid combination of predictive and transform coding has also proved its usefulness. For instance, JPEG standard for image coding assumes using DPCM for coding dc components of image blocks and in video coding intra-frame redundancy is removed with block DCT coding whereas for removing inter-frame redundancy the ideas of predictive decorrelations are used in a form of object motion compensation.

To conclude this brief review, mention other then DPCM with feedback combined decorrelation/quantization methods. For loss-less coding, LZW (Lempel-Ziv-Welch) coding is widely used in which "a code book" is constructed for individual symbols as well as their combinations in the transmitted sequence of symbols. Adaptive discretization assumes taking samples of the signal only if difference in the signal values from the previous sample exceeds a quantization interval. For images, such coding results in generating an image "contour map". Vector quantization is an exact implementation of the general coding signals as a whole. However, owing to computer memory and computational capacity limitations it is applied to groups of signal/image samples small enough to match these limitations.

References

1. E.C. Titchmarsh, Introduction to the theory of Fourier integrals, Oxford, New York, 1948
2. H. F. Harmuth, Transmission of Information by Orthogonal Functions, Springer Verlag, Berlin, 1970
3. S. Mallat, A Wavelet Tour of Signal Processing, 2nd edition, Academic Press, 1999
4. N. Ahmed, K.R. Rao, Orthogonal Transforms for Digital Signal Processing, Springer Verlag, Berlin, Heidelberg,New York, 1975
5. L. Yaroslavsky, Digital Picture Processing, An Introduction, Springer Verlag, Berlin, Heidelberg, New York, Tokyo, 1985
6. K. G. Beauchamp, Transforms for Engineers, A Guide to Signal Processing, Clarendon Press, Oxford, 1987.
7. O.J. Burt, E.A. Adelson, The Laplacian Pyramid as a compact Image Code, IEEE Trans. on Com., Vol. 31, No. 4, pp. 5320540, Apr., 1983
8. Y. Meyer, Wavelets, Algorithms and Applications, SIAM, Philadelphia, 1993
9. M. Vetterli, J. Kovaĉević, Wavelets and Subband Coding, Prentice Hall PTR, Englewwod Cliffs, New Jersey, 199
10. H. Karhunen, Über Lineare Methoden in der Wahscheinlichkeitsrechnung", Ann. Acad. Science Finn., Ser. A.I.37, Helsinki, 1947
11. M. Loeve, Fonctions Aleatoires de Seconde Ordre, in.: P. Levy, Processes Stochastiques et Movement Brownien, Paris, France: Hermann, 1948
12. H. Hottelling, Analysis of a Complex of Statistical Variables into Principal Components, J. Educ. Psychology 24 (1933), 417:441, 498-520
13. R. Dawkins, Climbing Mount Unprobable, W.W.Norton Co, N.Y.,1998
14. H. Hofer, D. R. Willams, The Eye's Mechanisms for Autocalibration, Optics & Photonics News, Jan. 2002
15. L. Yaroslavsky, Digital Picture Processing, An Introduction, Springer Verlag, Berlin, Heidelberg, New York, Tokyo, 1985
16. E. Caroli, J. B. Stephen, G. Di Cocco, L. Natalucci, and A. Spizzichino, "Coded Aperture Imaging in X- and Gamma-ray astronomy", *Space Science Reviews*, vol. 45, 1987, pp. 349-403
17. G. K. Skinner, "Imaging with coded aperture masks", Nuclear Instruments and Methods in Physics Research, vol. 221, 1984, pp. 33-40
18. W.S. Hinshaw, A.H. Lent, An introduction to NMR imaging, Proc. IEEE, Special issue on Computerized Tomography, 71, No. 3, March 1983
19. A. K. Jain, Fundamentals of Digital Image Processing, Prentice Hall, Englewood Cliffs, 1989
20. C. Shannon, A Mathematical Theory of Communication, Bell System Techn. J., vol. 27, No. 3, 379-423; No. 4, 623-656, 194
21. R.M. Fano, Technical Report No. 65, The Research Laboratory of Electronics, MIT, March 17, 1949
22. D. A. Huffman, A Method for the Construction of Minimum Redundancy Codes, Prioc. IRE, vol. 40, No. 10, pp. 1098-1101
22. R. Gonzalez, R. E. Woods, Digital Image Processing, Prentice Hall, Upper Saddle River, N.J., 2002
23. H. S. Malvar, Signal Processing with Lapped Transforms, Artech House, Norwood, MA, 1992

Chapter 4

DIGITAL REPRESENTATION OF SIGNAL TRANSFORMATIONS

4.1 THE PRINCIPLES

Two principles lie in the base of digital representation of continuous signal transformations:

- Consistency principle with digital representation of signals
- Mutual correspondence principle between continuous and discrete transformations

The ***consistency principle*** requires that digital representation of signal transformations should parallel that of signals. The ***mutual correspondence principle*** between continuous and digital transformations is said to hold if both act to transform identical input signals into identical output signals. Thus, methods for representing, specifying and characterizing continuous transforms digitally should account for procedures of digitizing continuous signals and reconstructing continuous signals from their digital representatives (Fig.4.1).

In Sect. 3.1 we admitted that signal digitization is carried out in a two step procedure: discretization and element-wise quantization. Let signal $a(x)$ be represented in a digital form by quantized coefficients $\{\hat{\alpha}_k^{(q)}\}$ of its discrete representation $\mathbf{a}=\{\alpha_k\}$ over a chosen discretization basis. In Sect.2.2.1, two classes of signal transformations were specified, point-wise nonlinear and linear ones. The digital representation of point-wise transformations is based on the definition of such transformations as functions $\left\{F_k\left(\hat{\alpha}_k^{(q)}\cdot\right)\right\}$ of the quantized representation coefficients.

$$\mathbf{PWTa} = \left\{F_k\left(\hat{\alpha}_k^{(q)}\right)\right\} \tag{4.1.1}$$

In general, the transformation function may depend on the coefficient index k, and transformation is ***inhomogeneous***. Most frequently, ***homogeneous transformations*** are regarded whose function does not depend on coefficient indices.

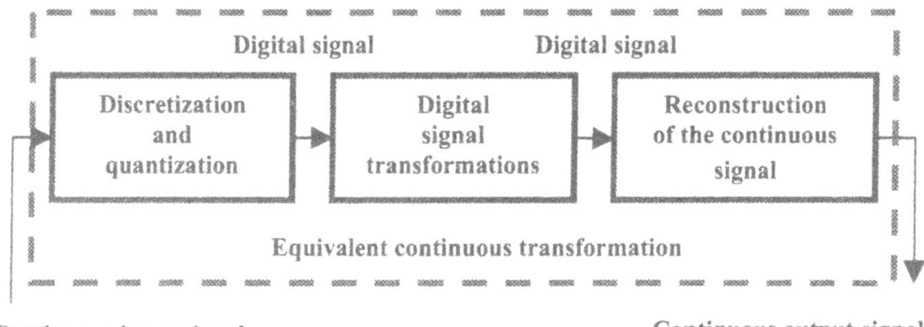

Figure 4-1. Mutual correspondence principle between continuous and digital signal transformations

Because the coefficient values are quantized functions $\{F_k(\cdot)\}$ may be most easily specified in a tabular form. The volume of the table depends upon the number of the signal quantization levels. If this number is too large and especially when numbers in computers are represented in floating point format, other ways of specifying the function may be used. The most commonly used way is to specify point-wise transformation functions in the form of a polynomial:

$$F_k\left(\alpha_k = \sum_{r=0}^{N-1} c_r \alpha_k^r\right). \tag{4.1.2}$$

This settles the problem of digital representation of point-wise signal transformations.

Representing linear transforms in digital form is, similarly to signal digitization, also split into two steps: discretization, and subsequent quantization of the discrete representation. The discrete representation of continuous linear transforms proceeds by analogy to the discrete representation of continuous signals.

Let $\{\varphi_n^{(d)}(x)\}$, $\{\varphi_n^{(r)}(x)\}$, $\{\phi_k^{(d)}(x)\}$, and $\{\phi_k^{(r)}(x)\}$ be, respectively, the discretization and reciprocal reconstruction bases for input $a(\xi)$ and output $b(x)$ signals of a linear transformation $\mathbf{L}$ and let $\{\alpha_n\}$ and $\{\beta_k\}$ be the corresponding sets of signal representing coefficients on these bases:

$$a(\xi) = \sum_n \alpha_n \varphi_n^{(r)}(\xi); \; \alpha_n = \int_X a(\xi)\varphi_n^{(d)}(\xi)d\xi; \tag{4.1.3}$$

$$b(x) = \mathbf{L}a(\xi) = \sum_k \beta_k \phi_k^{(r)}(x); \; \beta_k = \int_X b(x)\varphi_k^{(d)}(x)dx. \tag{4.1.4}$$

Find what numerical parameters of the linear transformation specified by its point spread function $h(x,\xi)$ are needed to compute the set of representation coefficients $\{\beta_k\}$ of the linear transformation output signal from those $\{\alpha_n\}$ of its input signal:

$$\begin{aligned}
\beta_k &= \int_X b(x)\varphi_k^{(d)}(x)dx = \int_X \mathbf{L}a(\xi)\phi_k^{(d)}(x)dx = \\
&\int_X \mathbf{L}\sum_n \alpha_n \varphi_n^{(r)}(\xi)\phi_k^{(d)}(x)dx = \sum_n \alpha_n \int_X \mathbf{L}\left(\varphi_n^{(r)}(\xi)\right)\phi_k^{(d)}(x)dx = \\
&\sum_n \alpha_n \int_X \left[\int_\Xi h(x,\xi)\varphi_n^{(r)}(\xi)d\xi\right]\phi_k^{(d)}(x)dx = \sum_n \alpha_n h_{k,n}
\end{aligned} \tag{4.1.5}$$

where

$$\left\{h_{k,n} = \int_X \left[\int_\Xi h(x,\xi)\varphi_n^{(r)}(\xi)d\xi\right]\phi_k^{(d)}(x)dx\right\}. \tag{4.1.6}$$

Therefore $\{h_{k,n}\}$ are coefficients that form discrete representation of the linear transformation with PSF $h(x,\xi)$ over discretization basis functions of output signals $\{\phi_k^d(x)\}$ and reconstructing basis functions $\{\varphi_n^{(r)}(x)\}$ of input signals. They completely specify linear transformations for the chosen sets of basis functions. We will refer to them as the ***discrete point spread function*** (DPSF) or ***discrete impulse response*** of the linear transformation. Equation 4.1.5 is the equation of discrete linear transformation discussed in Sect. 2.2.1. We refer to the transformation defined by Eq. 4.1.5 to as the ***digital filter***. In the context of digital processing of continuous signals, it represents a discrete analog of integral linear transformation of Eq. 2.2.7.

Eq. 4.1.5 can be rewritten in a vector-matrix form as

$$\mathbf{B} = \mathbf{H}\mathbf{A}^t, \tag{4.1.7}$$

where $\mathbf{B} = \{\beta_k\}$and is $\mathbf{A} = \{\alpha_n\}$ are vector-columns of the representative coefficients of output and input signals, superscript t denotes matrix transposition and H is matrix of coefficients $\{h_{k,n}\}$.

In a special case, when the bases of the input and output signals are identical, orthogonal and are built from the "eigen" functions of the transformation:

$$\{\varphi_k^{(r)}(\cdot) = \phi_k^{(r)}(\cdot)\}; \; \mathbf{L}(\phi_k^{(r)}(\cdot)) = \eta_k \phi_k^{(r)}(\cdot), \tag{4.1.8}$$

matrix H degenerates into a diagonal matrix

$$h_{k,n} = \eta_k \delta_{k,n}, \tag{4.1.9}$$

where $\delta_{k,n}$ is the Kronecker delta (Eq. 2.1.10), and the relationship of Eq. 4.1.5 between input and output representation coefficients is reduced to a simple element-wise multiplication:

$$\beta_k = \eta_k \alpha_k \tag{4.1.10}$$

A discrete linear transform described by a diagonal matrix will be referred to as the ***scalar filter***. A general transformation described by Eq 4.1.5 will be referred to as the ***vector filter***.

Eq. 4.1.6 provides a way to determine a discrete linear transformation that corresponds to a particular continuous linear transformation for the chosen bases of input and output signals of the transformation. One can also formulate an inverse problem: given discrete transformation with DPSF $\{h_{k,n}\}$ what will be the point spread function of the corresponding continuous transformation? Substituting Eqs. 4.1.3 and 5 and into Eq. 4.1.4 obtain:

$$b(x)=\sum_k \beta_k \phi_k^{(r)}(x)=\sum_k\left(\sum_n h_{n,k}\alpha_n\right)\phi_k^{(r)}(x)=$$
$$\sum_k\left[\sum_n h_{n,k}\int_\Xi a(\xi)\varphi_n^{(d)}(\xi)d\xi\right]\phi_k^{(r)}(x)= \qquad (4.1.11)$$
$$\int_\Xi a(\xi)\left[\sum_k\sum_n h_{k,n}\varphi_n^{(d)}(\xi)\phi_k^{(r)}(x)\right]d\xi,$$

from which it follows that PSF $h_{eq}(x,\xi)$ of the continuous linear transformation equivalent, according to the mutual correspondence principle, to the discrete transformation with DPSF $\{h_{k,n}\}$ is

$$h_{cont}(x,\xi)=\sum_k\sum_n h_{k,n}\varphi_n^{(d)}(\xi)\phi_k^{(r)}(x). \qquad (4.1.12)$$

In conclusion note that in digital processing DPSF coefficients of linear transformations are quantized. However, the number of quantization levels is usually much higher then that of signals and therefore effects of DPSF quantization will generally be disregarded except for special cases such as that of Sect. 6.5.

As it was stated in Sect. 3.3.1, the most widely used method of signal discretization is that of signal sampling that uses shifted, or convolution bases. In what follows, we derive discrete representation, for shifted bases, of convolution integrals, Fourier, Fresnel and wavelet transforms.

4.2 DISCRETE REPRESENTATION OF CONVOLUTION INTEGRAL. DIGITAL FILTERS

The convolution integral of a signal $a(x)$ with shift invariant kernel $h(x)$ (point spread function) is defined as (Eqs. 2.3.3):

$$b(x) = \int_{-\infty}^{\infty} a(\xi) h(x-\xi) d\xi . \qquad (4.2.1)$$

Let signal discretization and reconstruction bases be

$$\varphi_k^{(d)}(x) = \varphi^{(d)}(x - k\Delta x) \text{ and } \varphi_k^{(r)}(x) = \varphi^{(r)}(x - k\Delta x). \qquad (4.2.2)$$

Then, substituting Eqs. 4.2.1 and 2 in Eq. 4.1.6, obtain:

$$h_{k,n} = \int_{-\infty}^{\infty} \left[\int_{-\infty}^{\infty} h(x-\xi) \varphi^{(r)}(\xi - n\Delta x) d\xi \right] \phi^{(d)}(x - k\Delta x) dx = \\ \int_{-\infty}^{\infty} \int_{-\infty}^{\infty} h[x - \xi - (k-n)\Delta x] \varphi^{(r)}(\xi) \phi^{(d)}(x) dx d\xi = h_{k-n} \qquad (4.2.3)$$

and Eq. 4.1.5 becomes

$$b_k = \sum_n a_n h_{k-n} \qquad (4.2.4)$$

where summation is carried out over all available signal samples. Because the spatial extent of the convolution kernel $h(x)$ is usually much less then that of signals and, consequently, the number of samples $\{h_k\}$ of the filter DPSF is much less then that of signals, it is convenient to rewrite Eq. 4.2.4 in an alternative form as

$$b_k = \sum_{n=0}^{N_h - 1} h_n a_{k-n} \qquad (4.2.5a)$$

assuming that N_h is the number of samples of $\{h_k\}$, and that the number of signal samples is, in accordance with infinite limits in the convolution integral, infinite. For two-dimensional signals, one can, in a similar way, obtain the following discrete representation of 2-D convolution integral:

$$b_{k_1,k_2} = \sum_{n_1=0}^{N_{1,h}-1} \sum_{n_2=0}^{N_{2,h}-1} h_{n_1,n_2} a_{k_1-n_1,k_2-n_2} \tag{4.2.5b}$$

Signal transformation defined by Eqs. 4.2.5 is commonly referred to as ***digital filtering*** and these equations are regarded as basic equations of signal digital filtering by digital filters defined by their ***discrete PSF*** $\{h_n\}$.

According to the correspondence principle between continuous and digital signal transformations, let us find continuous and discrete impulse responses and frequency responses of digital filters. In the derivation, we repeat that of Eq. 4.1.12 for digital filters redefined in Eqs. 4.2.5 and take into account that output signal of a digital filter $\{b_k\}$ has finite number, say N_b, of samples. For a digital filter defined by its discrete PSF $\{h_n\}$ and discretization and reconstruction basis functions as given by Eq. 4.2.2 obtain that

$$b(x) = \sum_{k=0}^{N_b-1} b_k \varphi^{(r)}(x - k\Delta x) = \sum_{k=0}^{N_b-1}\left(\sum_{n=0}^{N_h-1} h_n a_{k-n}\right)\varphi^{(r)}(x - k\Delta x) =$$
$$\sum_{k=0}^{N_b-1}\left[\sum_{n=0}^{N_h-1} h_n \int_{-\infty}^{\infty} a(\xi)\varphi^{(d)}(\xi - (k-n)\Delta x)d\xi\right]\varphi^{(r)}(x - k\Delta x) =$$
$$\int_{-\infty}^{\infty} a(\xi)\left[\sum_{k=0}^{N_b-1}\sum_{n=0}^{N_h-1} h_n \varphi^{(r)}(x - k\Delta x)\varphi^{(d)}(\xi - (k-n)\Delta x)\right]d\xi . \tag{4.2.6}$$

Therefore, PSF of the equivalent continuous filter is

$$h_{eq}(x,\xi) = \sum_{k=0}^{N_b-1}\sum_{n=0}^{N_h-1} h_n \varphi^{(r)}(x - n\Delta x)\varphi^{(d)}[\xi - (k-n)\Delta x] \tag{4.2.7}$$

It is more convenient to characterize equivalent continuous filter by its frequency response found as Fourier Transform of its impulse response:

$$H_{eq}(f,p) = \int_{-\infty}^{\infty}\int_{-\infty}^{\infty} h_{eq}(x,\xi)\exp[i2\pi(fx - p\xi)]dxd\xi =$$
$$\int_{-\infty}^{\infty}\int_{-\infty}^{\infty}\sum_{k=0}^{N_b-1}\sum_{n=0}^{N_h-1} h_n \varphi^{(r)}(x - n\Delta x)\varphi^{(d)}[\xi - (k-n)\Delta x]\times$$
$$\exp[i2\pi(fx - p\xi)]dxd\xi = \sum_{k=0}^{N_b-1}\sum_{n=0}^{N_h-1} h_k \int_{-\infty}^{\infty}\int_{-\infty}^{\infty}\varphi^{(d)}(\xi)\varphi^{(r)}(x)\times$$

$$\exp\{i2\pi[f(x+k\Delta x)-p(\xi+(k-n)\Delta x)]\}=$$
$$\left[\sum_{n=0}^{N_h-1} h_n \exp(i2\pi pn\Delta x)\right]\left[\sum_{k=0}^{N_b-1}\exp[i2\pi(f-p)k\Delta x]\right]\times$$
$$\int_{-\infty}^{\infty}\varphi^{(r)}(x)\exp(i2\pi fx)dx\int_{-\infty}^{\infty}\varphi^{(d)}(\xi)\exp(-i2\pi p\xi)d\xi . \qquad (4.2.8)$$

This expression contains 4 terms:

$$H_{eq}(f,p)=DFR(p)\cdot\Phi^{(r)}(f)\cdot\Phi^{(r)}(-p)\cdot SV(f,p), \qquad (4.2.9)$$

where

$$DFR(p)=\sum_{n=0}^{N_h-1} h_n \exp(i2\pi pn\Delta x) \qquad (4.2.10)$$

is determined by digital filter discrete point spread function $\{h_n\}$,

$$\Phi^{(r)}(f)=\int_{-\infty}^{\infty}\varphi^{(r)}(x)\exp(i2\pi fx)dx \qquad (4.2.11)$$

and

$$\Phi^{(d)}(-p)=\int_{-\infty}^{\infty}\varphi^{(d)}(x)\exp(-i2\pi px)dx \qquad (4.2.12)$$

are frequency responses of the signal reconstruction and discretization devices, respectively, and term

$$SV(f,p)=\sum_{k=0}^{N_b-1}\exp[i2\pi(f-p)k\Delta x]=\frac{\exp[i2\pi(f-p)N_b\Delta x]-1}{\exp[i2\pi(f-p)\Delta x]-1}=$$

$$\frac{\sin[\pi(f-p)N_b\Delta x]}{\sin[\pi(f-p)\Delta x]}\exp[i\pi(f-p)(N_b-1)\Delta x]. \qquad (4.2.13)$$

that depends only on the number of output signal samples.

The term $DFR(p)$ is referred to as ***discrete frequency response*** of the digital filter. It is a periodical function in frequency domain. If signal reconstruction and discretization devices are, as it is required by the sampling theorem, ideal low pass filters, terms $\Phi^{(r)}(f)$ and $\Phi^{(d)}(-p)$ remove all but one periods of the discrete frequency response (compare with signal reconstruction from its samples). Because this is not the case for all real devices, one should anticipate aliasing effects similar to those in signal reconstruction.

Term $SV(f,p)$ reflects the fact that digital filter defined by Eqs. 4.2.5 and obtained as a discrete representation of the convolution integral is nevertheless not spatially invariant owing to the finite number N_b samples $\{b_k\}$ of the filter output signal involved in the reconstruction continuous output signal $b(x)$. When this number increases, contribution of boundary effects into the reconstructed signal decreases and quality of the digital filter approximation to the convolution integral improves. In the limit, when $N_b \to \infty$, the filter is spatially homogeneous:

$$\lim_{N_b \to \infty} SV(f,p) =$$

$$\lim_{N_b \to \infty} N_b \frac{\sin[\pi(f-p)N_b \Delta x]}{N_b \sin[\pi(f-p)\Delta x]} \exp[i\pi(f-p)(N_b-1)\Delta x] = \delta(f-p)$$

(4.2.14)

Boundary effects in digital filtering require especial treatment of signal borders. From Eq. 4.2.5 it follows that $N_b \le N_a$, where N_a is the number of filter input signal. The equality takes place only if input signal samples $\{a_n\}$ are extrapolated outside signal borders at least by the number of samples N_h of the filter discrete PSF. Methods of extrapolating and border processing in digital filtering will be discussed in Sect. 5.1.4.

4.3 DISCRETE REPRESENTATION OF FOURIER INTEGRAL TRANSFORM.

4.3.1 Discrete Fourier Transforms

Transformation kernel of integral Fourier transform $\exp(i2\pi fx)$ is shift variant. Therefore, in the derivation of its discrete representation one should, in distinction to the case of the convolution integral, account for possible arbitrary shifts of signal and its spectrum samples positions with respect to signal and its spectrum coordinate systems as it is shown in Fig. 4.2.

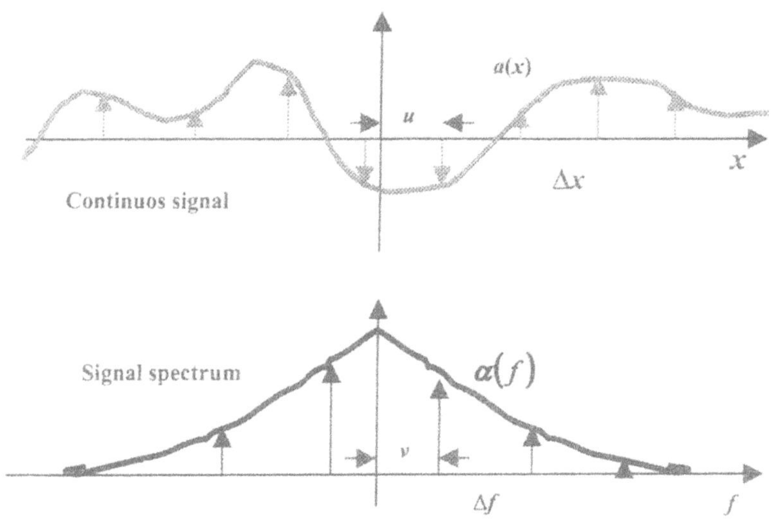

Figure 4-2. Geometry of dispositions of signal and its Fourier spectrum samples

Let signal reconstructing and spectrum discretization basis functions be $\left\{\varphi_k^{(r)}(x) = \phi^{(r)}\left[x-(k+u)\Delta x\right]\right\}$, and $\left\{\varphi_r^{(d)}(x) = \phi^{(d)}\left[f-(r+v)\Delta f\right]\right\}$, where k and r are integer indices of signal and its spectrum samples and u and v are, respectively, shifts, in fractions of the discretization intervals Δx and Δf, of signal and its spectrum samples with respect to the corresponding co-ordinate system. Then discrete representation of the Fourier transform kernel is, according to Eq. 4.1.6,

$$h_{k,r} = \int_{-\infty}^{\infty}\int_{-\infty}^{\infty} \exp(i2\pi fx)\varphi^{(r)}[x-(k+u)]\varphi^{(d)}[f-(r+v)\Delta f]dxdf =$$

$$\int_{-\infty}^{\infty}\varphi^{(d)}(f)df \int_{-\infty}^{\infty}\exp\{i2\pi[(f+(r+v)\Delta f)(x+(k+u)\Delta x)]\}\varphi^{(r)}(x)dx =$$

$$\exp[i2\pi(k+u)(r+v)\Delta x\Delta f]\int_{-\infty}^{\infty}\varphi^{(d)}(f)\exp\{i2\pi f(k+u)\Delta x\}df \times$$

$$\int_{-\infty}^{\infty}\exp\{i2\pi[f+(r+v)\Delta f]x\}\varphi^{(r)}(x)dx = \exp[i2\pi(k+u)(r+v)\Delta x\Delta f]\times$$

$$\int_{-\infty}^{\infty}\Phi^{(r)}[f+(r+v)\Delta f]\varphi^{(d)}(f)\exp\{i2\pi f(k+u)\Delta x\}df \; . \qquad (4.3.1a)$$

where $\Phi^{(r)}(f)$ is frequency response of signal reconstruction device.

For discrete representation of Fourier Transform, only term

$$\tilde{h}_{k,r} = \exp[i2\pi(k+u)(r+v)\Delta x\Delta f] \qquad (4.3.1b)$$

is usually used and term

$$\int_{-\infty}^{\infty}\Phi^{(r)}[f+(r+v)\Delta f]\varphi^{(d)}(f)\exp\{i2\pi f(k+u)\Delta x\}df \qquad (4.3.1c)$$

is disregarded, though, in principle, it also depends on sample indices k and r. This can be approved by the following reasoning. Spectrum discretization kernel $\varphi^{(d)}(f)$ is a function compactly concentrated in an interval of about Δf around point $f=0$. Within this interval, frequency response of the signal reconstruction device $\Phi^{(r)}[f+(r+v)\Delta f]$ can be approximated by a constant because it is a Fourier transform of the reconstruction device PSF $\varphi^{(r)}(x)$ that is compactly concentrated around point $x=0$ in an interval of about Δx. Then Eq. 4.3.1 c can be approximated as

$$\int_{-\infty}^{\infty}\Phi^{(r)}[f+(r+v)\Delta f]\varphi^{(d)}(f)\exp[i2\pi f(k+u)\Delta x]df \cong$$

$$\int_{-\infty}^{\infty} \varphi^{(d)}(f) \exp\{i2\pi f(k+u)\Delta x\} df = \Phi^{(d)}((k+u)\Delta x) \tag{4.3.1d}$$

where $\Phi^{(d)}(\bullet)$ is a frequency response of the spectrum discretization kernel $\varphi^{(d)}(f)$ that, in its turn and for the same reason of the compactness of $\varphi^{(d)}(f)$, can also be approximated by a constant for argument values $(k+u)\Delta x$.

Let X be an interval occupied by the signal $a(x)$. Then, assuming that spectrum discretization is carried out, according to this interval, with $\Delta f = 1/X$, obtain that

$$\Delta x \Delta f = \frac{\Delta x}{X} = \frac{1}{N}, \tag{4.3.2}$$

where N is the number of signal samples within interval X , and Eq. 4.3.1b can be rewritten as

$$\tilde{h}_{k,r} = \exp\left[i2\pi \frac{(k+u)(r+v)}{N} \right]. \tag{4.3.3}$$

Then basic equation 4.1.5 of digital filtering takes, for the representation of Fourier transform, the form:

$$\alpha_r^{(u,v)} \propto \sum_{k=0}^{N-1} a_k \exp\left[i2\pi \frac{(k+u)(r+v)}{N} \right]. \tag{4.3.4}$$

where superscripts (u,v) are kept to denote that signal and its Fourier spectrum samples are assumed to be taken with displacements u and v with repect to signal and spectrum coordinate systems.

Discrete transformation (4.3.4) is orthogonal. In order to make it orthonormal, introduce a normalizing multiplier $1/\sqrt{N}$ and obtain:

$$\alpha_r^{(u,v)} = \frac{1}{\sqrt{N}} \sum_{k=0}^{N-1} a_k \exp\left[i2\pi \frac{(k+u)(r+v)}{N} \right] \tag{4.3.5}$$

Similarly, for inverse Fourier Transform one can obtain:

$$a_k^{(u,v)} = \frac{1}{\sqrt{N}} \sum_{k=0}^{N-1} \alpha_r^{(u,v)} \exp\left[-i2\pi \frac{(k+u)(r+v)}{N}\right] \tag{4.3.6}$$

Eqs. 4.3.5 and 6 contain multipliers $\exp\left(i2\pi \frac{uv}{N}\right)$ and $\exp\left(-i2\pi \frac{uv}{N}\right)$ that do not depend on sample indices and can be disregarded. In this way we arrive at the following transforms

$$\alpha_r^{(u,v)} = \frac{1}{\sqrt{N}} \left\{ \sum_{k=0}^{N-1} a_k \exp\left[i2\pi \frac{k(r+v)}{N}\right] \right\} \exp\left(i2\pi \frac{ru}{N}\right) \tag{4.3.7}$$

$$a_k^{(u,v)} = \frac{1}{\sqrt{N}} \left\{ \sum_{r=0}^{N-1} \alpha_r \exp\left[-i2\pi \frac{r(k+u)}{N}\right] \right\} \exp\left(-i2\pi \frac{kv}{N}\right). \tag{4.3.8}$$

as discrete representation of direct and inverse integral Fourier Transforms. A special case of these transforms for signal and its spectrum sample shifts $u = 0; v = 0$,

$$\alpha_r = \frac{1}{\sqrt{N}} \sum_{k=0}^{N-1} a_k \exp\left(i2\pi \frac{kr}{N}\right) \tag{4.3.9}$$

and

$$a_k = \frac{1}{\sqrt{N}} \sum_{r=0}^{N-1} \alpha_r \exp\left(-i2\pi \frac{kr}{N}\right) \tag{4.3.10}$$

are known as direct and inverse ***Discrete Fourier Transforms*** (DFT). In order to distinguish the general case from this special case, we will refer to transforms defined by Eqs. 4.3.7 and 8 as ***Shifted Discrete Fourier Transforms*** SDFT(u,v).

The presence of shift parameters makes the SDFT to be more flexible in simulating integral Fourier Transform then the conventional DFT. As we will see in Chapt. 9, it enables

- performing continuous spectrum analysis;
- computing convolution and correlation with sub-pixel resolution;
- flexibly implementation sinc-interpolation of discrete data required by the discrete sampling theorem.

In all applications, the DFT is the basic transform that can be computed with Fast Fourier Transform algorithm (see Chapt. 6). Shifted DFTs can be, for any shift parameters, computed via the DFT:

$$\alpha_r^{(u,v)} = \frac{1}{\sqrt{N}} \exp\left(i2\pi \frac{ru}{N}\right) \sum_{k=0}^{N-1} \left[a_k \exp\left(i2\pi \frac{kv}{N}\right) \right] \exp\left(i2\pi \frac{kr}{N}\right); \tag{4.3.11}$$

$$a_k^{(u,v)} = \frac{1}{\sqrt{N}} \exp\left(-i2\pi \frac{kv}{N}\right) \sum_{r=0}^{N-1} \left[\alpha_r \exp\left(-i2\pi \frac{ru}{N}\right) \right] \exp\left(-i2\pi \frac{kr}{N}\right). \tag{4.3.12}$$

Two-dimensional SDFT($u,v;p,q$) is defined for 2-D signals of $N_1 \times N_2$, samples as, correspondingly,

$$\alpha_{r,s}^{(u,v;p,q)} = \frac{1}{\sqrt{N_1 N_2}} \left(\sum_{k=0}^{N_1-1} \sum_{l=0}^{N_2-1} a_{k,l} \exp\left\{ i2\pi \left[\frac{k(r+v)}{N_1} + \frac{l(s+q)}{N_2} \right] \right\} \right) \times \exp\left[i2\pi \left(\frac{ru}{N_1} + \frac{sp}{N_2} \right) \right]; \tag{4.3.13}$$

$$a_{k,l}^{(u,v;p,q)} = \frac{1}{\sqrt{N_1 N_2}} \left(\sum_{r=0}^{N_1-1} \sum_{s=0}^{N_2-1} \alpha_{r,s} \exp\left\{ -i2\pi \left[\frac{r(k+u)}{N_1} + \frac{s(l+p)}{N_2} \right] \right\} \right) \times \exp\left[-i2\pi \left(\frac{kv}{N_1} + \frac{lq}{N_2} \right) \right]; \tag{4.3.14}$$

and 2-D DFT as:

$$\alpha_{r,s} = \frac{1}{\sqrt{N_1 N_2}} \sum_{k=0}^{N_1-1} \sum_{l=0}^{N_2-1} a_{k,l} \exp\left[i2\pi \left(\frac{kr}{N_1} + \frac{ls}{N_2} \right) \right]; \tag{4.3.15}$$

$$a_{k,l} = \frac{1}{\sqrt{N_1 N_2}} \sum_{r=0}^{N_1-1} \sum_{s=0}^{N_2-1} \alpha_{r,s} \exp\left[-i2\pi \left(\frac{kr}{N_1} + \frac{ls}{N_2} \right) \right]. \tag{4.3.16}$$

4.3.2 Properties of Discrete Fourier Transforms

In this section, we will review basic properties of Discrete Fourier Transforms. It is instructive to compare them with the corresponding properties of the integral Fourier transform.

Invertibility and sincd-function

Let $\left\{\alpha_r^{u,v}\right\}$ be SDFT(u,v) coefficients of a discrete signal $\left\{a_k\right\}$, $k = 0,1,...,N-1$. Compute the inverse SDFT with shift parameters (p,q) from a subset of K spectral samples $\left\{\alpha_r^{u,v}\right\}$:

$$\widetilde{a}_n^{u/p,v/q} = \frac{1}{\sqrt{N}}\left[\sum_{r=0}^{K-1}\alpha_r^{u,v}\exp\left(-i2\pi\frac{rp}{N}\right)\right]\exp\left[-i2\pi\frac{n(r+q)}{N}\right] =$$

$$\frac{1}{\sqrt{N}}\sum_{r=0}^{K-1}\left\{\frac{1}{\sqrt{N}}\sum_{k=0}^{N-1}a_k\exp\left(i2\pi\frac{kv}{N}\right)\exp\left[i2\pi\frac{(k+u)r}{N}\right]\right\}\times$$

$$\exp\left(-i2\pi\frac{rp}{N}\right)\exp\left[-i2\pi\frac{n(r+q)}{N}\right] =$$

$$\frac{1}{N}\sum_{k=0}^{N-1}a_k\exp\left(i2\pi\frac{kv-nq}{N}\right)\sum_{r=0}^{K-1}\exp\left[i2\pi\frac{(k-n+u-p)r}{N}\right] =$$

$$\frac{1}{N}\sum_{k=0}^{N-1}a_k\exp\left(i2\pi\frac{kv-nq}{N}\right)\frac{\exp[i2\pi(k-n+u-p)K/N]-1}{\exp[i2\pi(k-n+u-p)/N]-1} =$$

$$\frac{1}{N}\sum_{k=0}^{N}a_k\frac{\exp\left\{\dfrac{i\pi K[(k+u)-(n+p)]}{N}\right\}-\exp\left\{-\dfrac{i\pi K[(k+u)-(n+p)]}{N}\right\}}{\exp\left[i\pi\dfrac{(k+u)-(n+p)}{N}\right]-\exp\left[-i\pi\dfrac{(k+u)-(n+p)}{N}\right]}\times$$

$$\exp\left\{i\pi\frac{K-1}{N}[(k+u)-(n+p)]\right\}\exp\left(i2\pi\frac{kv-nq}{N}\right) =$$

$$\sum_{k=0}^{N}a_k\frac{\sin\left\{\pi\dfrac{K}{N}[(k+u)-(n+p)]\right\}}{N\sin\{\pi[(k+u)-(n+p)]\}}$$

$$\exp\left[i\pi\frac{(K-1)(k+u-n-p)}{N}\right]\exp\left(i2\pi\frac{kv-nq}{N}\right) =$$

$$\left\{\sum_{k=0}^{N-1} a_k \exp\left[i\pi k\left(\frac{K-1}{N}+2v\right)\right]\mathrm{sincd}\left[K;N;(k-n+u-p)\right]\right\}\times$$
$$\exp\left[-i\pi\left(\frac{K-1}{N}+2q\right)n\right]\exp\left[i\pi\frac{K-1}{N}(u-p)\right], \qquad (4.3.17)$$

where

$$\mathrm{sincd}(K;N;x)=\frac{\sin\left(\pi\frac{K}{N}x\right)}{N\sin\left(\pi\frac{x}{N}\right)} \qquad (4.3.18)$$

is the ***discrete sinc-function***, discrete analog of the continuous sinc-function defined by Eq. 2.4.6. For integer $x = k-n$ function $\mathrm{sincd}(N;N;k-n)$ is identical to Kronecker delta-function:

$$\mathrm{sincd}(N;N;k-n)=\frac{\sin(\pi(k-n))}{N\sin\left(\pi\frac{k-n}{N}\right)}=\begin{cases}1, & k=n\\ 0, & k\neq n\end{cases}. \qquad (4.3.19)$$

Therefore, when $u=p$, $v=q$ and $K=N$,

$$\tilde{a}_n^{u/p,v/q}=\sum_{k=0}^{N-1} a_n\,\mathrm{sincd}(N;N;k-n)=a_n \qquad (4.3.20)$$

which confirms the invertibility of the SDFT(u,v). However, the substance of Eq. 4.3.17 is richer. It shows that the SDFT(u,v) is invertible in a more general sense. Provided appropriate modulation and demodulation of signal and its ISDFT spectrum by exponential multipliers in Eq. 4.2.17, one can generate sincd-interpolated signal copies arbitrarily shifted with respect to the original signal samples by $(u-p)$-th fraction of the discretization interval. This property is closely connected with the ***discrete sampling theorem*** proved below.

Discrete sinc-function is a discrete analog of the continuous sinc-function defined by Eq. 2.4.6. They are almost identical within interval of N samples for the discrete sinc-function and its corresponding interval $N\Delta x$ for the continuous delta-function as one see from Fig. 4-3. But outside this interval they differ dramatically: while sinc-function continues to decay, sincd-function is a periodical function and the type of its periodicity depends on whether N is odd or even number: $\mathbf{sincd}(N;N;k+gN)=(-1)^{g(N-1)}$.

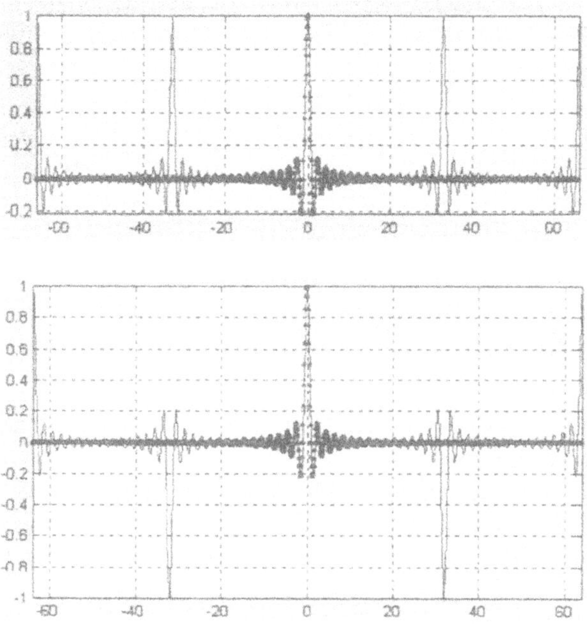

Figure 4-3. Discrete sincd- (solid line) and continuous sinc- (dotted line) functions for $N=33$ (top plot) and $N=34$ (bottom plot)

Separability of 2-D transforms

Separability of 2-D and, generally, multi-dimensional SDFTs easily seen from their definitions (Eqs. 4.3.13 - 16) follows from the separability of 2-D (multi-dimensional) integral Fourier transform and the separability of sampling basis functions.

Periodicity

From Eq. 4.3.8 it follows that, for integer number g, discrete signal $\{a_k^{u,v}\}$ reconstructed by the inverse SDFT(u,v) from SDFT(u,v) spectrum of signal $\{a_k\}$is defined outside the index interval $(0, N-1)$ as

$$a_{k+gN}^{(u,v)} = a_k^{(u,v)} \exp\left(-i2\pi \frac{gv}{N}\right) \qquad (4.3.21)$$

In particular, when $\boldsymbol{v} = \mathbf{0}$, it is a periodical function of $\boldsymbol{k}$:

$$a_{k+gN}^{(u,0)} = a_k^{(u,0)} = a_{(k)\bmod N}^{(u,0)}, \tag{4.3.22}$$

where $(k)\bmod N$ is residual of k/N. Similar periodicity property holds for SDFT($\boldsymbol{u}$,$\boldsymbol{v}$) spectra:

$$\alpha_{r+gN}^{(u,v)} = \alpha_r^{(u,v)} \exp\left(i2\pi\frac{gu}{N}\right); \qquad \alpha_{r+gN}^{(0,v)} = \alpha_r^{(0,v)} = \alpha_{(r)\bmod N}^{(0,v)} \tag{4.3.23}$$

For the DFT, both signal and its spectrum are regarded as periodical:

$$a_k^{(0,0)} = a_{(k)\bmod N}^{(0,0)}; \qquad \alpha_r^{(0,0)} = \alpha_{(r)\bmod N}^{(0,0)}. \tag{4.3.24}$$

Owing to the separability of the DFTs, the same holds for 2-D and multi-dimensional DFTs for all indices.

The periodicity property of DFTs is their main distinction from the integral Fourier transform whose discrete representation they are. It originates from image sampling with shift basis functions and from the fact that the number of samples in signal definition is finite.

Shift theorem

Shift theorem for the DFTs is completely analogous to that for integral Fourier transform:

$$\text{If } \{a_k\} \xrightarrow{\text{SDFT}(u,v)} \{\alpha_r^{u,v}\} \text{ and } \{\alpha_r^{u,v}\} \xrightarrow{\text{ISDFT}(u,v)} \{a_k^{u,v}\}$$

$$\text{then } \{a_{k+k_0}\} \xrightarrow{\text{SDFT}(u,v)} \left\{\alpha_r^{u,v} \exp\left(-i2\pi k_0 \frac{r+v}{N}\right)\right\} \tag{4.3.25a}$$

$$\text{and } \{\alpha_{r+r_0}^{u,v}\} \xrightarrow{\text{ISDFT}(u,v)} \left\{a_k^{u,v} \exp\left(i2\pi r_0 \frac{k+u}{N}\right)\right\}. \tag{4.3.25b}$$

Permutation theorem

The permutation theorem is an analog of the integral Fourier transform scaling theorem. For the DFT, if P is a number that has no divisors in common with N, and Q is an integer inverse to P such that their modulo N product $(PQ)_{\mathrm{mod}\,N} = 1$,

$$\left\{a_{(Pk)_{\mathrm{mod}\,N}}\right\} \xrightarrow{DFT} \left\{\alpha_{(Qr)_{\mathrm{mod}\,N}}\right\}. \tag{4.3.26}$$

As one can see from Eq. 4.3.26, multiplying signal sample index results, in distinction to the continuous case, not in the corresponding scaling of signal and its spectrum but rather in permutations of signal and its spectrum samples provided the scaling factor and the number of signal samples have no divisors in common. The distinctive feature of the finite dimensional discrete signal space in which the DFT operates is that sample indices are counted modulo N and no other indices are defined.

Convolution theorem

Let $\left\{\alpha_r^{u_a,v_a}\right\}$ be SDFT(u_a,v_a) of a signal $\{a_k\}$ and $\left\{\beta_r^{u_b,v_b}\right\}$ be SDFT(u_b,v_b) of a signal $\{b_k\}$, $k,r = 0,1,...,N-1$. Compute the ISDFT(u_c,v_c) of their product:

$$c_k^{u_c,v_c} = \frac{1}{\sqrt{N}}\exp\left(-i2\pi\frac{kv_c}{N}\right)\sum_{r=0}^{N-1}\alpha_r^{u_a,v_a}\beta_r^{u_b,v_b}\exp\left[-i2\pi\frac{r(k+u_c)}{N}\right] =$$

$$\frac{1}{\sqrt{N}}\exp\left(-i2\pi\frac{kv_c}{N}\right)\sum_{r=0}^{N-1}\left\{\frac{1}{\sqrt{N}}\exp\left(i2\pi\frac{ru_a}{N}\right)\sum_{n=0}^{N-1}a_n\exp\left[i2\pi\frac{n(r+v_a)}{N}\right]\times\right.$$

$$\left.\frac{1}{\sqrt{N}}\exp\left(i2\pi\frac{ru_b}{N}\right)\sum_{m=0}^{N-1}b_m\exp\left[i2\pi\frac{m(r+v_b)}{N}\right]\exp\left[-i2\pi\frac{r(k+u_c)}{N}\right]\right\} =$$

$$\frac{1}{\sqrt{N^3}}\exp\left(-i2\pi\frac{kv_c}{N}\right)\sum_{n=0}^{N-1}a_n\exp\left(2\pi\frac{nv_a}{N}\right)\times$$

$$\left\{\sum_{m=0}^{N-1}b_m\exp\left(i2\pi\frac{mv_b}{N}\right)\sum_{r=0}^{N-1}\exp\left[i2\pi\frac{r(n+m-k+u_a+u_b-u_c)}{N}\right]\right\} =$$

$$\frac{1}{\sqrt{N^3}}\exp\left(-i2\pi\frac{kv_c}{N}\right)\sum_{n=0}^{N-1}a_n\exp\left(2\pi\frac{nv_a}{N}\right)\times$$

$$\left\{\sum_{m=0}^{N-1}b_m\exp\left(i2\pi\frac{mv_b}{N}\right)\frac{\sin[\pi(n+m-k+u_a+u_b-u_c)]}{\sin\pi(n+m-k+u_a+u_b-u_c)/N]}\times\right.$$

$$\left.\exp\left[i\pi\frac{N-1}{N}(n+m-k+u_a+u_b-u_c)\right]\right\} =$$

$$\frac{1}{\sqrt{N}}\exp\left\{-i2\pi\left[\frac{k}{N}\left(v_c-\frac{N-1}{2N}\right)-\frac{N-1}{2N}(u_a+u_b-u_c)\right]\right\}\times$$
$$\sum_{n=0}^{N-1}a_n\exp\left[i2\pi\frac{n}{N}\left(v_a+\frac{N-1}{2N}\right)\right]\times$$
$$\sum_{m=0}^{N-1}b_m\exp\left[i2\pi\frac{m}{N}\left(v_b+\frac{N-1}{2N}\right)\right]\mathrm{sincd}[N;N;\pi(n+m-k+u_a+u_b-u_c)]$$
(4.3.27)

Select $v_a=v_b=-v_c=-\frac{N-1}{2N}$ to obtain:

$$c_k^{u_c v_c}=\frac{\exp\left\{i2\pi\left[\frac{N-1}{N}(u_a+u_b-u_c)\right]\right\}}{\sqrt{N}}\sum_{n=0}^{N-1}a_n\tilde{b}_{k-n}, \quad (4.3.28)$$

where

$$\tilde{b}_n=\sum_{m=0}^{N-1}b_m\,\mathrm{sincd}[N;N;\pi(m-n+u_a+u_b-u_c)] \quad (4.3.29)$$

is sincd-interpolated copy of signal $\{b_k\}$. Eq. 4.3.27 is a discrete convolution of signals $\{b_k\}$ and $\{\tilde{b}_n\}$. On signal borders, it assumes that the signals are extended outside index interval $[0,N-1]$ according to the shift theorem (Eq. 4.3.25). In a special case of the DFT ($u_a=u_b=u_c=v_a=v_b=-v_c=0$), it is a ***cyclic convolution***:

$$c_{(k)_{\mathrm{mod}\,N}}=\frac{1}{\sqrt{N}}\sum_{n=0}^{N-1}a_{(n)_{\mathrm{mod}\,N}}b_{(k-n)_{\mathrm{mod}\,N}} \quad (4.3.30)$$

In order to distinguish it from the general case of Eq. 4.3.28, we will refer to Eqs. 4.3.27 and 28 as to ***shifted cyclic convolution***

Symmetry properties

Finite number of samples affect also symmetry properties of DFTs as compared to those of integral Fourier transform:

$$\left\{a_{N-k}^{u,v}\exp\left[-i2\pi\frac{2v(k-1/2)}{N}\right]\right\}\xrightarrow{SDFT(u,v)}\left\{\alpha_{N-r}^{u,v}\exp\left[i2\pi\frac{2u(r-1/2)}{N}\right]\right\};\quad(4.3.31)$$

$$\left\{a_k^{u,v}=\pm a_{N-k}^{u,v}\exp\left[-i2\pi\frac{2v(k-1/2)}{N}\right]\right\}\xrightarrow{SDFT(u,v)}\left\{\alpha_r^{u,v}=\pm\alpha_{N-r}^{u,v}\exp\left[i2\pi\frac{2u(r-1/2)}{N}\right]\right\};\quad(4.3.32)$$

$$\left\{\left(a_k^{u,v}\right)^*\exp\left(-i2\pi\frac{2kv}{N}\right)\right\}\xrightarrow{SDFT(u,v)}\left\{\left(\alpha_{N-r}^{u,v}\right)^*\exp(i2\pi u)\right\};\quad(4.3.33)$$

$$\left\{\left(a_{N-k}^{u,v}\right)^*\exp(i2\pi v)\right\}\xrightarrow{SDFT(u,v)}\left\{\left(\alpha_r^{u,v}\right)^*\exp\left(i2\pi\frac{2ru}{N}\right)\right\};\quad(4.3.34)$$

If $2u$ and $2v$ are integer,

$$\left\{(-1)^{2v}a_{N-2u-k}\right\}\xrightarrow{SDFT(u,v)}\left\{(-1)^{2u}\alpha_{N-2v-r}\right\};\quad(4.3.35)$$

$$\left\{a_k^{u,v}=\pm(-1)^{2v}a_{N-2u-k}^{u,v}\right\}\xrightarrow{SDFT(u,v)}\left\{\alpha_r^{u,v}=\pm(-1)^{2u}\alpha_{N-2v-k}^{u,v}\right\};\quad(4.3.36)$$

If $2v$ is integer,

$$\left\{\left(a_k^{u,v}\right)^*\exp\left(i2\pi\frac{2uv}{N}\right)\right\}\xrightarrow{SDFT(u,v)}\left\{\left(\alpha_{N-2v-r}^{u,v}\right)^*\exp(i2\pi u)\right\};\quad(4.3.37)$$

If $2v$ is integer,

$$\left\{\left(a^{u,v}_{N-2u-k}\right)^{*}\exp(i2\pi v)\right\} \xrightarrow{SDFT(u,v)} \left\{\left(\alpha^{u,v}_{r}\right)^{*}\exp\left(i2\pi\frac{2uv}{N}\right)\right\}; \quad (4.3.38)$$

Eqs. 4.3.34 - 37 show that semi-integer shift parameters play an especial role in symmetry properties of SDFTs.

For the DFT, above equations are reduced to

$$\{a_{N-k}\} \xrightarrow{SDFT(u,v)} \{\alpha_{N-r}\}; \quad (4.3.39)$$

$$\{a_k = \pm a_{N-k}\} \xrightarrow{SDFT(u,v)} \{\alpha_r = \pm\alpha_{N-r}\}; \quad (4.3.40)$$

$$\left\{(a_k)^{*}\right\} \xrightarrow{SDFT(u,v)} \left\{(\alpha_{N-r})^{*}\right\}; \quad (4.3.41)$$

$$\left\{a_k = \pm(a_k)^{*}\right\} \xrightarrow{SDFT(u,v)} \left\{\begin{array}{l}\alpha_r = \pm(\alpha_{N-r})^{*}; \\ \alpha_0 = (\alpha_0)^{*}; \text{For even N}, \alpha_{N/2} = (\alpha_{N/2})^{*}\end{array}\right\}; \quad (4.3.42)$$

$$\left\{a_k = (a_k)^{*} = \pm a_{N-k}\right\} \xrightarrow{SDFT(u,v)} \left\{\alpha_r = \pm(\alpha_{N-r})^{*} = \pm\alpha_{N-r}\right\}. \quad (4.3.43)$$

Symmetry properties of 2-D SDFTs may be obtained from the separability of the transforms. For DFT, they are illustrated in Fig. 4-4.

Term α_0 plays a role of the zero frequency component of the integral Fourier transform. It is proportional to the signal ***dc-component***:

$$\alpha_0 = \sqrt{N}\left(\frac{1}{N}\sum_{k=0}^{N-1} a_k\right). \quad (4.3.44)$$

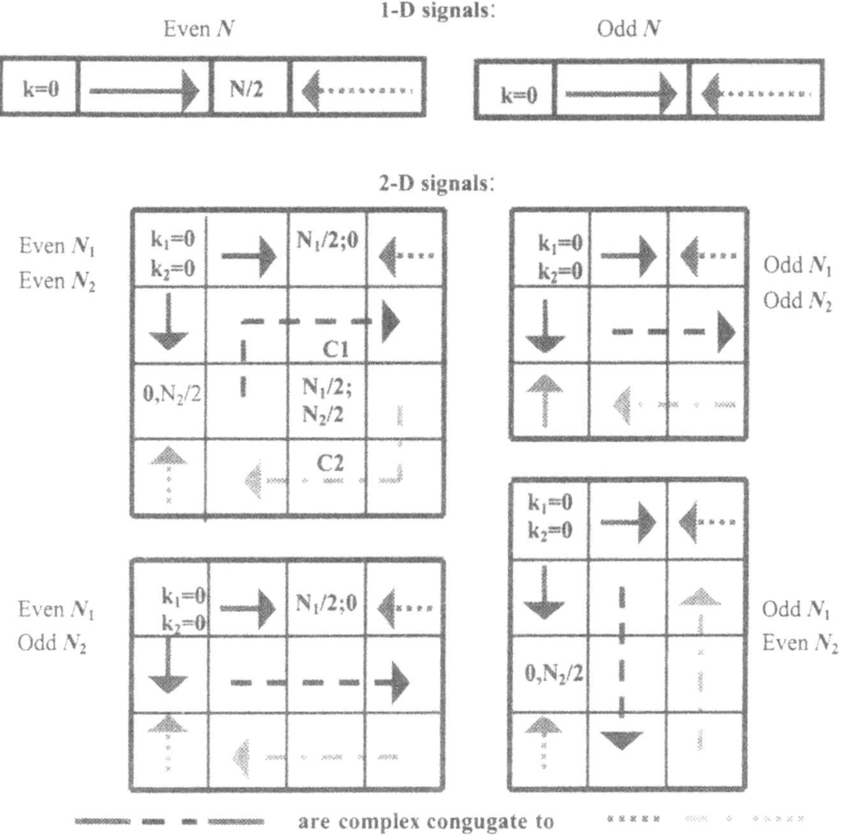

Figure 4-4. Types of DFT spectra symmetries for real valued signals

For even numbered N, coefficient $\alpha_{N/2}$ plays a role of the highest signal frequency $f = \frac{N}{2}\Delta f$:

$$\alpha_{N/2} = \frac{1}{\sqrt{N}} \sum_{k=0}^{N-1} (-1)^k a_k . \tag{4.3.45}$$

For odd numbered N, coefficients $\alpha_{(N-1)/2} = \left(\alpha_{(N+1)/2}\right)^*$ (for signals with real values) plays a role of the highest signal frequency $f = \frac{N-1}{2}\Delta f$.

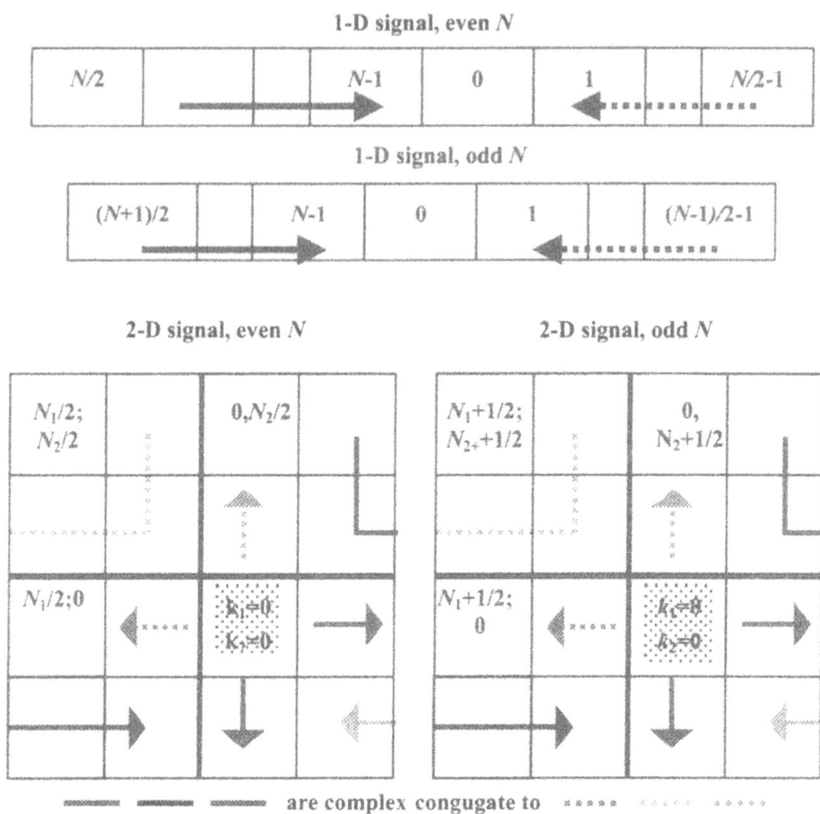

Figure 4-5. Centering DFT spectrum (numbers in boxes indicate spectral indices)

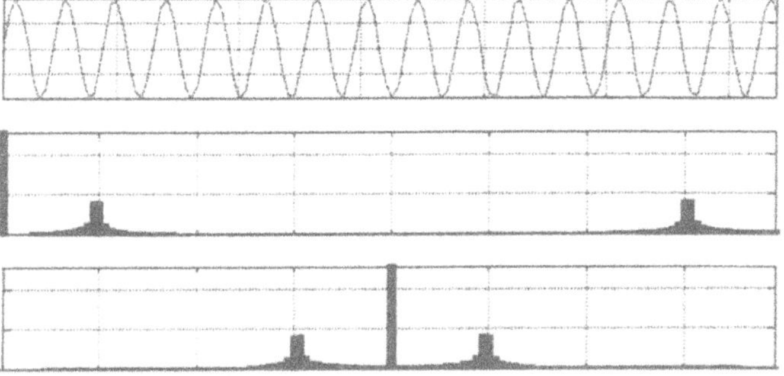

Figure 4-6. Spectrum of a sinusoidal signal plotted without (central plot) and with (bottom plot) spectrum centering

In order to maintain similarity with DFT spectrum $\{\alpha_r\}$ of discrete signal $\{a_k\}$ and Fourier spectrum $\alpha(f)$, $f \in [-\infty, \infty]$ of the corresponding continuous signal $a(x)$, it is convenient to rearrange sequence $\{\alpha_r\}$ as it is shown in Fig. 4-5. Fig. 4-6 illustrates spectrum of a sinusoidal signal without and with spectrum centering.

Discrete sampling theorem

Let discrete signal $\{\tilde{a}_k\}$ of LN samples be obtained from signal $\{a_{k_N}\}$ of N samples ($k_N = 0,1,...N-1$) by placing between its samples (L-1) zero samples:

$$\tilde{a}_k = a_{k_N} \delta(k_L), \; k_L = 0,1,..., L-1, \tag{4.3.46}$$

where $\delta(\cdot)$ is Kronecker delta (see Fig. 4-7 a,, b). Let us compute DFT of this signal:

$$\begin{aligned} \tilde{\alpha}_r &= \frac{1}{\sqrt{LN}} \sum_{k=0}^{LN-1} \tilde{a}_k \exp\left(i2\pi \frac{kr}{LN}\right) = \\ & \frac{1}{\sqrt{LN}} \sum_{k_2=0}^{L-1} \sum_{k_1=0}^{N-1} a_{k_1} \delta(k_2) \exp\left[i2\pi \frac{(k_1 L + k_2)}{LN} r\right] = \\ & \frac{1}{\sqrt{LN}} \sum_{k_1=0}^{N-1} a_{k_1} \exp\left(i2\pi \frac{k_1}{N} r\right) = \frac{1}{\sqrt{L}} \alpha_{(r) \bmod N} \end{aligned} \tag{4.3.47}$$

where $\{\alpha_r\}, r = 0,1,..., LN-1$ is DFT of signal $\{a_k\}$. Eq. (4.3.47) shows that placing zeros between its samples results in a periodical replication of its DFT spectrum (Fig. 4-7, c) with the number of replicas equal to the number of zeros plus one (Fig. 4-7, d).

Let now multiply spectrum $\{\tilde{\alpha}_r\}$ by a mask $\{\mu_r\}$ that zeroes all periods but one (Fig. 4-7, e). According to the convolution theorem, multiplying signal DFT spectrum by a mask function results in signal cyclic convolution with IDFT of the mask. In order to secure that the convolution kernel will be a real valued function, the mask should maintain spectrum symmetry for real valued signals (Eq. 4.3.42). Therefore, for odd N, the mask should be:

$$\mu_r = 1 - rect \frac{r - (N+1)/2}{LN - N - 1}, \tag{4.3.48}$$

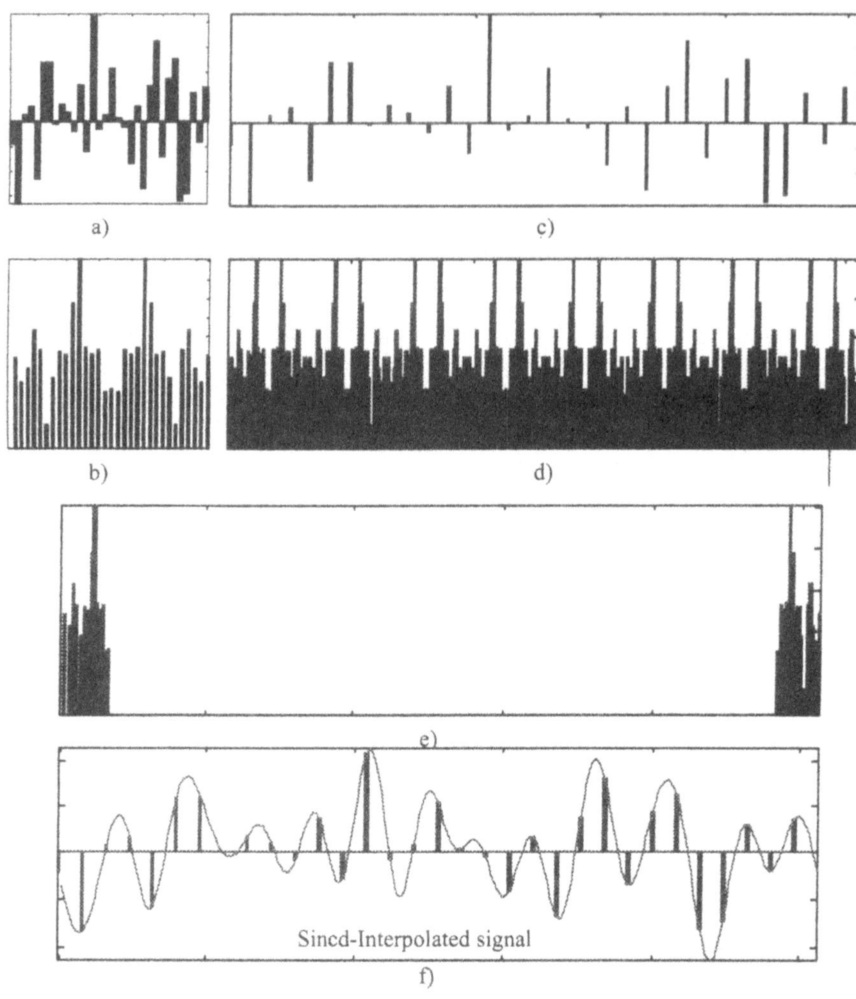

Figure 4-7. Graphical illustration of the discrete sampling theorem: a) initial signal; b) DFT spectrum of signal (a); c) initial signal with zeros placed between its samples; d) its DFT spectrum: periodical replication of the initial signal spectrum; e) removing all spectrum replicas but one by the ideal discrete low pass filter; f) discrete sinc-interpolated signal resulted from inverse DFT of spectrum (e)

where $rect(x) = \begin{Bmatrix} 1, & 0 < x < 1 \\ 0, & otherwise \end{Bmatrix}$.

In this case, inverse DFT $IDFT\left\{\left[1 - rect\frac{r-(N+1)/2}{LN-N-1}\right]\alpha_{(r)\bmod N}\right\}$ produces a signal

$$\tilde{\tilde{a}}_k = \frac{1}{\sqrt{LN}}\sum_{r=0}^{LN-1}\left[1 - rect\frac{r-(N+1)/2}{LN-N-1}\right]\alpha_{(r)\bmod N}\exp\left(-i2\pi\frac{kr}{LN}\right) =$$

$$\frac{1}{\sqrt{LN}}\left[\sum_{r=0}^{(N-1)/2}\alpha_r\exp\left(-i2\pi\frac{kr}{LN}\right) + \right.$$

$$\left.\sum_{r=LN-(N-1)/2}^{LN-1}\alpha_{(r)\bmod N}\exp\left(-i2\pi\frac{kr}{LN}\right)\right] =$$

$$\frac{1}{\sqrt{LN}}\left\{\sum_{r=0}^{(N-1)/2}\left[\frac{1}{\sqrt{N}}\sum_{n=0}^{N-1}a_n\exp\left(i2\pi\frac{nr}{N}\right)\right]\exp\left(-i2\pi\frac{kr}{LN}\right) + \right.$$

$$\left.\sum_{r=r=LN-(N-1)/2}^{LN-1}\left[\frac{1}{\sqrt{N}}\sum_{n=0}^{N-1}a_n\exp\left(i2\pi\frac{nr}{N}\right)\right]\exp\left(-i2\pi\frac{kr}{LN}\right)\right\} =$$

$$\frac{1}{N\sqrt{L}}\left\{\sum_{n=0}^{N-1}a_n\left[\sum_{n=0}^{(N-1)/2}\exp\left(-i2\pi\frac{(k-Ln)r}{LN}\right) + \right.\right.$$

$$\left.\left.\sum_{r=LN-(N-1)/2}^{LN-1}\exp\left(-i2\pi\frac{(k-Ln)r}{LN}\right)\right]\right\} =$$

$$\frac{1}{N\sqrt{L}}\left(\sum_{n=0}^{N-1}a_n\left\{\frac{\exp\left[-i2\pi\frac{(k-Ln)(N+1)}{2LN}\right]-1}{\exp\left[-i2\pi\frac{(k-Ln)}{LN}\right]-1} + \right.\right.$$

$$\left.\left.\frac{\exp\left[-i2\pi\frac{LN(k-Ln)}{LN}\right]-\exp\left[-i2\pi\frac{(k-Ln)(LN-(N-1)/2)}{LN}\right]}{\exp\left[-i2\pi\frac{(k-Ln)}{LN}\right]-1}\right\}\right) =$$

$$\frac{1}{N\sqrt{L}}\sum_{n=0}^{N-1}a_n\frac{\exp\left[-i2\pi\frac{(k-Ln)(N+1)}{2LN}\right]-\exp\left[-i2\pi\frac{(k-Ln)(N-1)}{2LN}\right]}{\exp\left[-i2\pi\frac{(k-Ln)}{LN}\right]-1} =$$

$$\frac{1}{\sqrt{L}}\sum_{n=0}^{N-1} a_n \frac{\sin\left[\pi \frac{N(k-Ln)}{LN}\right]}{N\sin\left[2\pi\frac{(k-Ln)}{LN}\right]} = \frac{1}{\sqrt{L}}\sum_{n=0}^{N-1} a_n \operatorname{sincd}[N;N;(k-nL)/L],$$

(4.3.49)

Using for k two indices $k = Lk_N + k_L$, we obtain finally:

$$\tilde{\tilde{a}}_{Lk_N+k_L} = \frac{1}{\sqrt{L}}\sum_{n=0}^{N-1} a_n \operatorname{sincd}\left(N;N;k_N - n + \frac{1}{L}k_L\right). \tag{4.3.50}$$

This equation shows that, disregarding constant multiplier $1/\sqrt{L}$, nonzero samples of signal $\tilde{a}_k$ remain, in signal $\tilde{\tilde{a}}_k$, unchanged and equal to original samples of signal $\{a_{k_N}\}$

$$\tilde{\tilde{a}}_{Lk_N} = \frac{1}{\sqrt{L}} a_{k_N} \tag{4.3.51}$$

while its additional zero samples are interpolated with the discrete sinc-function from the initial nonzero ones (Fig. 4-7, f).

For even N, term with index $(N+1)/2$ required by Eq. (4.3.46) does not exist and maintaining the spectrum symmetry is possible either with the mask

$$\mu_r^{(0)} = 1 - rect\frac{r - N/2}{LN - N} \tag{4.3.52}$$

that zeros $N/2$-th spectrum component or with the mask

$$\mu_r^{(2)} = 1 - rect\frac{r - N/2 - 1}{LN - N - 2} \tag{4.3.53}$$

that leaves two copies of this component in the extended spectrum. In these two cases one can, similarly to the derivation of Eq. 4.3.49, obtain that inverse DFT of the masked extended spectrum results in signals:

$$\tilde{\tilde{a}}_k^{(0)} = \mathbf{IDFT}\left\{\mu_r^{(0)}\alpha_{(r)\bmod N}\right\} =$$
$$\frac{1}{\sqrt{L}}\sum_{n_1=0}^{N-1} a_n \mathrm{sincd}\left(N-1; N; k_N - n + \frac{1}{L}k_L\right) \tag{4.3.54}$$

or, correspondingly,

$$\tilde{\tilde{a}}_k = \mathbf{IDFT}\left\{\mu_r^{(2)}\alpha_{(r)\bmod N}\right\} =$$
$$\frac{1}{\sqrt{L}}\sum_{n_1=0}^{N-1} a_n \mathrm{sincd}\left(N+1; N; k_N - n + \frac{1}{L}k_L\right) \tag{4.3.55}$$

Eqs. (4.3.50, 54 and 55) are formulations of the discrete sampling theorem. It is a discrete analog of the sampling theorem for continuous signals formulated in Sect. 3.3.1. Their dissimilarities are associated with the finite number of signal and its spectrum samples. Masking functions $\{\mu_r\}$ (Eqs. 4.3.48, 52 and 53) describe discrete ***ideal low pass filters***. They are different for odd and even N. The role of the sinc-function is played, for discrete signals, by the discrete sincd-function defined in Eq. 4.3.18. If N is an even number, original signal samples are not retained in the interpolated signal because of a special treatment required for the spectral coefficient with index $N/2$. In Sect. 9.2 we will show how the discrete sampling theorem can be used for digital resampling signals specified by their samples.

4.3.3 Discrete cosine and sine transforms

There is a number of special cases of SDFTs worth of a special discussion. All of them are associated with representation of signal and/or spectra that exhibit certain symmetry. In digital image processing the number of image samples is most commonly even. If signals are symmetrical with respect to a certain point, their corresponding discrete signal with even number of samples will retain signal symmetry if the samples are shifted with respect to the signal symmetry center by half of the discretization interval. Therefore semi-integer shift parameters of SDFT represent especial interest in this respect.

The most important special case is that of ***Discrete Cosine Transform*** (***DCT***). Let, for a signal $\{a_k\}$, $k = 0,1,\ldots, N-1$, form an auxiliary signal

$$\tilde{a}_k = \begin{cases} a_k, & k = 0,1,...,N-1 \\ a_{2N-k-1}, & k = N,...,2N-1 \end{cases} \tag{4.3.56}$$

and compute its SDFT(1/2,0):

$$\mathbf{SDFT}_{1/2,0}\{\tilde{a}_k\} = \frac{1}{\sqrt{2N}} \sum_{k=0}^{2N-1} \tilde{a}_k \exp\left[i2\pi \frac{(k+1/2)r}{2N}\right] =$$
$$\frac{1}{\sqrt{2N}}\left\{\sum_{k=0}^{N-1} \tilde{a}_k \exp\left[i2\pi \frac{(k+1/2)r}{2N}\right] + \sum_{k=N}^{2N-1} \tilde{a}_k \exp\left[i2\pi \frac{(k+1/2)r}{2N}\right]\right\} =$$
$$\frac{1}{\sqrt{2N}}\left\{\sum_{k=0}^{N-1} a_k \exp\left[i2\pi \frac{(k+1/2)r}{2N}\right] + \sum_{k=N}^{2N-1} a_{2N-k-1} \exp\left[i2\pi \frac{(k+1/2)r}{2N}\right]\right\} =$$
$$\frac{1}{\sqrt{2N}}\left\{\sum_{k=0}^{N-1} a_k \exp\left[i2\pi \frac{(k+1/2)r}{2N}\right] + \sum_{k=0}^{N-1} a_k \exp\left[i2\pi \frac{(2N-k-1/2)r}{2N}\right]\right\} =$$
$$\frac{1}{\sqrt{2N}}\left\{\sum_{k=0}^{N-1} a_k \exp\left[i2\pi \frac{(k+1/2)r}{2N}\right] + \sum_{k=0}^{N-1} a_k \exp\left[-i2\pi \frac{(k+1/2)r}{2N}\right]\right\} =$$
$$\frac{2}{\sqrt{2N}} \sum_{k=0}^{N-1} a_k \cos\left[\pi \frac{(k+1/2)r}{N}\right]. \tag{4.3.57}$$

In this way we arrive at the Discrete Cosine Transform defined as:

$$\alpha_r^{DCT} = \frac{2}{\sqrt{2N}} \sum_{k=0}^{N-1} a_k \cos\left[\pi \frac{(k+1/2)r}{N}\right]. \tag{4.3.58}$$

DCT signal spectrum is, as one can see from Eq.(4.3.58), always an odd (anti-symmetric) sequence if regarded outside its base interval $[0, N-1]$:

$$\alpha_r^{DCT} = -\alpha_{2N-r}^{DCT}; \alpha_N = 0 \tag{4.3.59}$$

while signal is, as it follows from Eq. 4.3.56, even (symmetric):

$$a_k = a_{2N-k-1}. \tag{4.3.60}$$

From the above derivation of the DCT it also follows that, for the DCT, signals are regarded as periodical with period $2N$:

$$a_k = a_{(k)\mathrm{mod}\,2N}\,. \tag{4.3.61}$$

The inverse DCT can easily be found as the inverse SDFT(1/2,0) of anti-symmetric spectrum of Eq.(4.3.59):

$$\begin{aligned}
\tilde{a}_k = {} & \frac{1}{\sqrt{2N}} \sum_{r=0}^{2N-1} \alpha_r^{DCT} \exp\left[-i2\pi\frac{(k+1/2)r}{2N}\right] = \\
& \frac{1}{\sqrt{2N}} \left\{ \alpha_0^{DCT} + \sum_{r=1}^{N-1} \alpha_r^{DCT} \exp\left[-i2\pi\frac{(k+1/2)r}{2N}\right] + \right. \\
& \sum_{r=N+1}^{2N-1} \alpha_r^{DCT} \exp\left[-i2\pi\frac{(k+1/2)r}{2N}\right] = \\
& \frac{1}{\sqrt{2N}} \left\{ \alpha_0^{DCT} + \sum_{r=1}^{N-1} \alpha_r^{DCT} \exp\left[-i2\pi\frac{(k+1/2)r}{2N}\right] + \right. \\
& \sum_{r=1}^{N-1} \alpha_r^{DCT} \exp\left[-i2\pi\frac{(k+1/2)(2N-r)}{2N}\right] = \\
& \frac{1}{\sqrt{2N}} \left\{ \alpha_0^{DCT} + \sum_{r=1}^{N-1} \alpha_r^{DCT} \exp\left[-i2\pi\frac{(k+1/2)r}{2N}\right] + \right. \\
& \sum_{r=1}^{N-1} \alpha_r^{DCT} \exp\left[i2\pi\frac{(k+1/2)r}{2N}\right] = \\
& \frac{1}{\sqrt{2N}} \left(\alpha_0^{DCT} + 2\sum_{r=1}^{N-1} \alpha_r^{DCT} \cos\left[\pi\frac{(k+1/2)r}{N}\right] \right)
\end{aligned} \tag{4.3.62}$$

Coefficient α_0^{DCT} is, similarly to the case of DFT, proportional to the signal dc-component:

$$\alpha_0^{DCT} = \sqrt{2N}\left(\frac{1}{N} \sum_{k=0}^{N-1} a_k \right). \tag{4.3.63}$$

Coefficient α_{N-1}^{DCT}

$$\alpha_{N-1}^{DCT} = \frac{2}{\sqrt{2N}} \sum_{k=0}^{N-1} a_k (-1)^k \sin\left(\pi\frac{k+1/2}{N}\right) \tag{4.3.64}$$

represents signal's highest frequency (compare with Eq. 4.3.45 for DFT).

The DCT can be regarded as a discrete analog of integral cosine transform (Eq.2.4.24). It was introduced by Ahmed and Rao ([1]) as a transform for image coding. It has a very good energy compaction property and is very frequently considered to be an approximation to Karhunen-Loeve transform ([2]). From the above derivation, it becomes clear that the energy compaction property of the DCT has a very simple and fundamental explanation. It should be attributed to the fact that the DCT is a shifted DFT of a signal that is extended outside the base interval $[0, N-1]$ in a way (4.3.56) that removes the signal discontinuity at the signal border, the main cause of poor convergence property of DFT and other bases defined on a finite interval. The absence of discontinuity at signal borders makes the DCT to be attractive not only as a good transform for signal discretization but also as a substitute for DFT for the implementation of signal digital convolution in transform domain with the use of Fast Transforms. We will discuss this application in Sect. 5.2.

The DCT has its complement sine transform, the ***DcST***:

$$\alpha_0^{DcST} = \frac{2}{\sqrt{2N}} \sum_{k=0}^{N-1} a_k \sin\left[\pi \frac{(k+1/2)r}{N}\right]. \quad (4.3.65)$$

The DcST is an imaginary part of the SDFT(1/2,0) of a signal

$$\tilde{a}_k = \begin{cases} a_k, & k = 0,1,\ldots, N-1 \\ -a_{2N-k-1}, & k = N, N+1,\ldots,2N-1 \end{cases} \quad (4.3.66)$$

extended to interval $[0,2N-1]$ in an odd symmetric way:

$$\frac{1}{i\sqrt{2N}} \sum_{k=0}^{2N-1} \tilde{a}_k \exp\left[i2\pi \frac{(k+1/2)r}{2N}\right] =$$

$$\frac{1}{i\sqrt{2N}} \left\{ \sum_{k=0}^{N-1} \tilde{a}_k \exp\left[i2\pi \frac{(k+1/2)r}{2N}\right] + \sum_{k=N}^{2N-1} \tilde{a}_k \exp\left[i2\pi \frac{(k+1/2)r}{2N}\right] \right\} =$$

$$\frac{1}{i\sqrt{2N}} \left\{ \sum_{k=0}^{N-1} a_k \exp\left[i2\pi \frac{(k+1/2)r}{2N}\right] - \sum_{k=N}^{2N-1} a_{2N-k-1} \exp\left[i2\pi \frac{(k+1/2)r}{2N}\right] \right\} =$$

$$\frac{1}{i\sqrt{2N}} \left\{ \sum_{k=0}^{N-1} a_k \exp\left[i2\pi \frac{(k+1/2)r}{2N}\right] - \sum_{k=0}^{N-1} a_k \exp\left[i2\pi \frac{(2N-k-1/2)r}{2N}\right] \right\} =$$

$$\frac{1}{i\sqrt{2N}} \left\{ \sum_{k=0}^{N-1} a_k \exp\left[i2\pi \frac{(k+1/2)r}{2N}\right] - \sum_{k=0}^{N-1} a_k \exp\left[-i2\pi \frac{(k+1/2)r}{2N}\right] \right\} =$$

$$\frac{2}{\sqrt{2N}}\sum_{k=0}^{N-1}a_k \sin\left[\pi\frac{(k+1/2)r}{N}\right]. \quad (4.3.67)$$

DcST spectrum exhibits even symmetry when regarded in interval $[0,2N-1]$:

$$\alpha_r^{DcST}=\alpha_{2N-r}^{DcST}. \quad (4.3.68)$$

and assumes periodical replication, with a period $2N$, of the signal complemented with its odd symmetrical copy:

$$\tilde{a}_k=\tilde{a}_{(k)\bmod 2N}. \quad (4.3.69)$$

In distinction from the DFT and the DCT, the DcST does not contain signal's dc-component:

$$\alpha_0^{DcST}=0 \quad (4.3.70)$$

and the inverse DcST :

$$a_k=\frac{1}{\sqrt{2N}}\left(\alpha_N^{DcST}+2\sum_{r=1}^{N-1}\alpha_r^{DcST}\sin\left[\pi\frac{(k+1/2)r}{N}\right]\right) \quad (4.3.71)$$

involves spectral coefficients with indices $\{1,2,...,N\}$ rather then $\{0,1,...,N-1\}$ for the DCT.

In a similar way one can introduce different modifications of the SDFT for different shifts and different types of signal symmetry. For instance, SDFT(1/2,1/2) of a signal extended in an anti-symmetric way (Eq. 4.3.66) is a modification of the DCT

$$\alpha_r^{DCTIV}=\sqrt{\frac{2}{N}}\sum_{k=0}^{N-1}a_k\cos\left[\pi\frac{(k+1/2)(r+1/2)}{N}\right] \quad (4.3.72)$$

that obtained a name DCT-IV ([3]). One can easily verify that the DCT-IV spectrum also exhibits an odd symmetry:

$$\left\{\alpha_r^{DCTIV}=-\alpha_{2N-r-1}^{DCTIV}\right\}. \quad (4.3.73)$$

Therefore the inverse DCT-IV is identical to the direct one

$$a_k^{DCTIV} = \sqrt{\frac{2}{N}} \sum_{r=0}^{N-1} \alpha_r \cos\left[\pi \frac{(k+1/2)(r+1/2)}{N}\right]. \tag{4.3.74}$$

Because of 1/2-shift in frequency domain, the DCT-IV of a signal assumes periodical replication, with alternating sign, of the auxiliary signal of Eq. 4.3.66.

Similarly to the DCT-IV, one can define the DST-IV ([3]):

$$\alpha_r^{DSTIV} = \sqrt{\frac{2}{N}} \sum_{k=0}^{N-1} a_k \sin\left[\pi \frac{(k+1/2)(r+1/2)}{N}\right] \tag{4.3.75}$$

that is obviously an imaginary part of the SDFT(1/2,1/2) of a symmetrized auxiliary signal of Eq. 4.3.56.

In the same way one can show that the SDFT(1,1) of an auxiliary signal $\{\tilde{a}_k\}$ formed from a signal $\{a_k, k = 0,1,..., N-1\}$ by extending it in an odd way to $2N+2$ samples:

$$\tilde{a}_k = \begin{cases} a_k, & 0 \le k \le N-1 \\ 0, & k = N \\ -a_{2N-k}, & N \le k \le 2N \\ 0, & k = 2N+1 \end{cases} \tag{4.3.76}$$

reduces to the transform

$$\alpha_r^{DST} = \frac{1}{\sqrt{2N}} \sum_{k=0}^{N-1} a_k \sin\left[\pi \frac{(k+1)(r+1)}{N}\right] \tag{4.3.77}$$

known as ***Discrete Sine Transform*** (***DST***). DST spectrum is also and in the same way anti-symmetric:

$$\{\alpha_r^{DST} = -\alpha_{2N-r}^{DST}\};\ \alpha_N^{DST} = \alpha_{2N+1}^{DST} = 0 \tag{4.3.78}$$

and, therefore, inverse DST is also identical to the direct one:

$$a_k = \frac{1}{\sqrt{2N}} \sum_{r=0}^{N-1} \alpha_r^{DST} \sin\left[\pi \frac{(k+1)(r+1)}{N}\right]. \quad (4.3.79)$$

DST of a signal assumes periodical replication, with a period $2N+2$ of the auxiliary signal of Eq. 4.3.76.

2-D and multi-dimensional DCTs and DSTs are usually defined as separable to 1-D transforms on each of the coordinates. For instance, 2-D DCT of a signal $\{a_{k,l}\}$, $k = 0,1,...,N_1 - 1$, $l = 0,1,...,N_2 - 1$ is defined as

$$\alpha_{r,s}^{DCT} = \frac{2}{\sqrt{N_1 N_2}} \sum_{k=0}^{N_1-1} \sum_{l=0}^{N_2-1} a_{k,l} \cos\left[\pi \frac{(k+1/2)r}{N_1}\right] \cos\left[\pi \frac{(l+1/2)s}{N_2}\right] \quad (4.3.80)$$

that corresponds to 4-fold image symmetry illustrated in Fig. 2-15.

4.3.4 Lapped Transforms

Discrete orthogonal transforms such as DFT, DCT and DST can be applied to signals either globally, i.e. to all available signal samples, or block wise or in a sliding window. If they are applied block wise, signal is subdivided into blocks and the corresponding transform is applied to individual blocks. In block transform signal processing, independent treatment of individual blocks may cause so called blocking effects or artifacts on the borders of blocks. Radical solution of this problem is processing in transform domain in a sliding window. We will discuss application of sliding window transform domain processing in Chs. 8 and 9. Another option that allows to substantially reduce blocking effects is signal processing in hopping overlapping windows. Most popular and computationally efficient is half window overlapping. Transforms that are used in this processing are called "Lapped" transforms ([4]).

One of the most popular lapped transforms is ***Modified DCT (MDCT)*** or ***Modulated Lapped Transform*** (***MLT***). The MDCT of a signal sequence $\{a_k\}$, $k = 0,1,...,2N-1$ is defined as

$$\alpha_r^{MDCT} = \sqrt{\frac{2}{N}} \sum_{k=0}^{2N-1} a_k h_k \cos\left[\pi\left(k + \frac{N+1}{2}\right)\left(r + \frac{1}{2}\right)\Big/N\right] \quad (4.3.81)$$

where $\{h_k\}$ is a window function:

$$h_k = \sin\left(\pi \frac{k+1/2}{N}\right). \quad (4.3.82)$$

One can show that MLT is equivalent to the SDFT ($\frac{N+1}{2}, \frac{1}{2}$) of a signal $\{\tilde{a}_k = a_k h_k\}$ modified in a certain way:

$$\alpha_r^{MDCT} = \sqrt{\frac{2}{N}} \sum_{k=0}^{2N-1} \tilde{a}_k \cos\left[\pi\left(k + \frac{N+1}{2}\right)\left(r + \frac{1}{2}\right)\Big/N\right] =$$

$$\sqrt{\frac{1}{2N}} \sum_{k=0}^{2N-1} \tilde{a}_k \left\{\exp\left[i\pi\left(k + \frac{N+1}{2}\right)\left(r + \frac{1}{2}\right)\Big/N\right] + \right.$$

$$\left.\exp\left[-i\pi\left(k + \frac{N+1}{2}\right)\left(r + \frac{1}{2}\right)\Big/N\right]\right\} =$$

$$\sqrt{\frac{1}{2N}} \sum_{k=0}^{N-1} \tilde{a}_k \exp\left[i\pi\left(k + \frac{N+1}{2}\right)\left(r + \frac{1}{2}\right)\Big/N\right] +$$

$$\sqrt{\frac{1}{2N}} \sum_{k=0}^{N-1} \tilde{a}_k \exp\left[-i\pi\left(k + \frac{N+1}{2}\right)\left(r + \frac{1}{2}\right)\Big/N\right] +$$

$$\sqrt{\frac{1}{2N}} \sum_{k=N}^{2N-1} \tilde{a}_k \exp\left[i\pi\left(k + \frac{N+1}{2}\right)\left(r + \frac{1}{2}\right)\Big/N\right] +$$

$$\sqrt{\frac{1}{2N}} \sum_{k=N}^{2N-1} \tilde{a}_k \exp\left[-i\pi\left(k + \frac{N+1}{2}\right)\left(r + \frac{1}{2}\right)\Big/N\right]. \qquad (4.3.83)$$

Replacement of summation index k in the second and fourth terms of Eq. 4.3.83 by $(N-1-k)$ and $(3N-1-k)$, respectively, yields:

$$\alpha_r^{MDCT} = \sqrt{\frac{1}{2N}} \sum_{k=0}^{N-1} \tilde{a}_k \exp\left[i\pi\left(k + \frac{N+1}{2}\right)\left(r + \frac{1}{2}\right)\Big/N\right] +$$

$$\sqrt{\frac{1}{2N}} \sum_{k=0}^{N-1} \tilde{a}_{N-1-k} \exp\left\{-i\pi\left[2N - \left(k + \frac{N+1}{2}\right)\right]\left(r + \frac{1}{2}\right)\Big/N\right\} +$$

$$\sqrt{\frac{1}{2N}} \sum_{k=N}^{2N-1} \tilde{a}_k \exp\left[i\pi\left(k + \frac{N+1}{2}\right)\left(r + \frac{1}{2}\right)\Big/N\right] +$$

$$\sqrt{\frac{1}{2N}} \sum_{k=N}^{2N-1} \tilde{a}_{3N-1-k} \exp\left\{-i\pi\left[4N - \left(k + \frac{N+1}{2}\right)\right]\left(r + \frac{1}{2}\right)\Big/N\right\} =$$

$$\sqrt{\frac{1}{2N}} \sum_{k=0}^{N-1} \tilde{a}_k \exp\left[i\pi\left(k + \frac{N+1}{2}\right)\left(r + \frac{1}{2}\right)\Big/N\right] +$$

$$\sqrt{\frac{1}{2N}}\sum_{k=0}^{N-1}\tilde{a}_{N-1-k}\exp\left[-i2\pi\left(r+\frac{1}{2}\right)\right]\exp\left[i\pi\left(k+\frac{N+1}{2}\right)\left(r+\frac{1}{2}\right)\Big/N\right]+$$
$$\sqrt{\frac{1}{2N}}\sum_{k=N}^{2N-1}\tilde{a}_{k}\exp\left[i\pi\left(k+\frac{N+1}{2}\right)\left(r+\frac{1}{2}\right)\Big/N\right]+$$
$$\sqrt{\frac{1}{2N}}\sum_{k=N}^{2N-1}\tilde{a}_{3N-1-k}\exp\left[-i4\pi\left(r+\frac{1}{2}\right)\right]\exp\left[i\pi\left(k+\frac{N+1}{2}\right)\left(r+\frac{1}{2}\right)\Big/N\right]. \tag{4.3.84}$$

With substitution $\exp\left[-i2\pi\left(r+\frac{1}{2}\right)\right]=-1$, $\exp\left[-i4\pi\left(r+\frac{1}{2}\right)\right]=1$ and

$$\ddot{a}_k=\begin{cases}\tilde{a}_k-\tilde{a}_{N-1-k}, & k=0,1,...,N-1\\ \tilde{a}_k+\tilde{a}_{3N-1-k}, & k=N,N+1,...,2N-1\end{cases} \tag{4.3.85}$$

obtain finally:

$$\alpha_r^{MDCT}=\frac{1}{\sqrt{2N}}\sum_{k=0}^{2N-1}\ddot{a}_k h_k\exp\left[i\pi\left(k+\frac{N+1}{2}\right)\left(r+\frac{1}{2}\right)\Big/N\right], \tag{4.3.86}$$

the SDFT ($\frac{N+1}{2},\frac{1}{2}$) of a signal $\{\ddot{a}_k\}$ formed from the initial windowed signal $\{\tilde{a}_k\}$ according to Eq. 4.3.85. Physical interpretation of Eq. 4.3.85 is straightforward. According to it, first half of the signal samples is mirrored and overlaid onto the signal in negative and second half of the signal samples is mirrored as well but overlaid onto the signal in positive. In signal processing with MLT, successive window positions overlap by half of the window of $2N$ samples. With this overlapping, same mirrored halves of the signal within window overlaid, in two successive window positions, with opposite signs and compensate each other. This secures perfect reconstruction of the signal.

It follows from Eqs 4.3.81 and 4.3.86 that spectral coefficients of the MDCT exhibit two-fold redundancy:

$$\alpha_r=(-1)^{N+1}\alpha_{2N-1-r}. \tag{4.3.87}$$

From Eq. 4.3.86 it follows that the inverse MDCT coincides with the direct one:

$$\tilde{\ddot{a}}_k = \frac{1}{\sqrt{2N}} \sum_{k=0}^{N-1} \alpha_r^{MDCT} \cos\left[\pi\left(k + \frac{N+1}{2}\right)\left(r + \frac{1}{2}\right)\Big/N\right] \qquad (4.3.88)$$

and that the MDCT of a signal $\{a_k\}$ assumes sign alternating periodical replication with period $2N$ of signal $\tilde{\ddot{a}}_k$ of Eq. 4.3.85:

$$\tilde{\ddot{a}}_{k+2gN} = (-1)^g \tilde{\ddot{a}}_k . \qquad (4.3.89)$$

4.4 DISCRETE REPRESENTATION OF FRESNEL INTEGRAL TRANSFORM

4.4.1 Discrete Fresnel Transform

Transformation kernel of integral Fresnel transform $\exp\left[-i\pi(x-f)^2\right]$ is shift invariant. Nevertheless, in the transform discrete representation, it is useful to account for possible arbitrary shifts of signal and its Fresnel spectrum samples positions with respect to the signal and spectrum coordinate systems in a similar way as it was done for discrete representation of Fourier Transform in Sect. 4.3.1.

Let signal reconstructing and spectrum discretization basis functions be $\varphi_k^{(r)}(x)=\varphi^{(r)}\left[x-(k+u)\Delta x\right]$, and $\varphi_r^{(d)}(x)=\varphi^{(d)}\left[f-(r+v)\Delta f\right]$, where k and r are integer indices of signal and its spectrum samples and u and v are, respectively, shifts, in fractions of the discretization intervals Δx and Δf, of signal and its spectrum samples with respect to the corresponding coordinate system. Then discrete representation of the Fresnel transform kernel is, according to Eq. 4.1.6,

$$h_{k,r}=\int_{-\infty}^{\infty}\int_{-\infty}^{\infty}\exp\left[-i\pi(x-f)^2\right]\varphi^{(r)}\left[x-(k+u)\right]\varphi^{(d)}\left[f-(r+v)\Delta f\right]dx\,df=$$

$$\int_{-\infty}^{\infty}\varphi^{(d)}(f)df\int_{-\infty}^{\infty}\exp\left\{-i\pi\left[x-f+(k+u)\Delta x-(r+v)\Delta f\right]^2\right\}\varphi^{(r)}(x)dx$$

(4.4.1)

Introduce, temporarily, a variable

$$s_{kr}=(k+u)\Delta x-(r+v)\Delta f \tag{4.4.2}$$

Then

$$h_{k,r}=\int_{-\infty}^{\infty}\varphi^{(d)}(f)df\int_{-\infty}^{\infty}\exp\left[-i\pi(x-f+s_{rk})^2\right]\varphi^{(r)}(x)dx=$$

$$\exp\left(-i\pi s_{rk}^2\right)\int_{-\infty}^{\infty}\varphi^{(d)}(f)\exp\left(-i\pi f^2\right)\exp\left(i2\pi f s_{rk}\right)df\times$$

$$\int_{-\infty}^{\infty} \varphi^{(r)}(x)\exp\left(-i\pi x^2\right)\exp\left[i2\pi x\left(f - s_{rk}\right)\right]dx =$$

$$\exp\left(-i\pi s_{rk}^2\right)\int_{-\infty}^{\infty} \varphi^{(d)}(f)\exp\left(-i\pi f^2\right)\tilde{\Phi}\left(f - s_{rk}\right)\exp\left(i2\pi f s_{rk}\right)df \text{ ,} \quad (4.4.3)$$

where

$$\tilde{\Phi}(f) = \int_{-\infty}^{\infty} \varphi^{(r)}(x)\exp\left(-i\pi x^2\right)\exp\left(i2\pi x f\right)dx \quad (4.4.4)$$

is Fourier Transform of signal reconstruction basis function $\varphi^{(r)}(x)$ modulated by the chirp-function. For $\varphi^{(r)}(x)$ is a function compactly concentrated around point $x = 0$ in an interval of about Δx, $\exp\left(-i\pi x^2\right) \cong 1$ within this interval and one can regard $\tilde{\Phi}(f)$ as an approximation to frequency response of a hypothetical signal reconstruction device assumed in the signal discrete representation. With similar reservations as those made for discrete representation of Fourier integral, for discrete representation of Fresnel integral only term

$$\tilde{h}_{k,r} = \exp\left(-i\pi s_{rk}^2\right) = \exp\left\{-i\pi\left[(k+u)\Delta x - (r+v)\Delta f\right]^2\right\} \quad (4.4.5)$$

of Eq. 4.4.3 is used.

To complete the derivation, introduce dimensionless variables:

$$\mu = \left(\Delta x / \Delta f\right)^{1/2} \quad (4.4.6)$$

and

$$w = u\mu - v/\mu \text{ .} \quad (4.4.7)$$

Then, using relationship $N = 1/\Delta x\Delta f$ between the number of signal samples and discretization intervals (Eq. 4.3.2), obtain for $\tilde{h}_{k,r}$ finally:

$$\tilde{h}_{k,r} = \exp\left\{-i\pi\left[(k+u)\Delta x - (r+v)\Delta f\right]^2\right\} =$$

$$\exp\left\{-i\pi\frac{[(k+u)\Delta x-(r+v)\Delta f]^2}{N\Delta x\Delta f}\right\}=\exp\left[-i\pi\frac{(k\mu-r/\mu+w)^2}{N}\right] \tag{4.4.8}$$

With this discrete PSF, basic equation Eq. 4.1.5 of digital filtering takes, for the representation of Fresnel Transform, the form:

$$\alpha_r^{(\mu,w)}=\frac{1}{\sqrt{N}}\sum_{k=0}^{N-1}a_k\exp\left[-i\pi\frac{(k\mu-r/\mu+w)^2}{N}\right] \tag{4.4.9}$$

where multiplier $1/\sqrt{N}$ is introduced, similarly to Discrete Fourier Transforms, for normalization purposes. Eq. 4.4.9 can be regarded as a discrete representation of the Fresnel integral transform ([5]). We will refer to it as to ***Shifted Discrete Fresnel Transform*** (***SDFrT***). Its parameter μ plays a role of the distance parameter λD in the Fresnel approximation of Kirchhof's integral (Eq. 2.3.10). Parameter w is a combined shift parameter of discretization raster shifts in signal and Fresnel Transform domains. It is frequently considered equal to zero (see for instance, [6,7]) and ***Discrete Fresnel Transform*** (***DFrT***) is commonly defined as

$$\alpha_r^{(\mu,w)}=\frac{1}{\sqrt{N}}\sum_{k=0}^{N-1}a_k\exp\left[-i\pi\frac{(k\mu-r/\mu)^2}{N}\right]. \tag{4.4.10}$$

SDFrT (μ, w) is quite obviously connected with SDFT($0,-w\mu$):

$$\alpha_r^{(\mu,w)}=\frac{1}{\sqrt{N}}\left\{\sum_{k=0}^{N-1}a_k\exp\left(-i\pi\frac{k^2\mu^2}{N}\right)\exp\left[i2\pi\frac{k(r-w\mu)}{N}\right]\right\}\times$$
$$\exp\left[-i\pi\frac{(r-w\mu)^2}{N\mu^2}\right] \tag{4.4.11}$$

From this relationship one can conclude that SDFrT is invertible and that inverse SDFrT is defined as

$$a_k^{(\mu,w)}=\frac{1}{\sqrt{N}}\sum_{r=0}^{N-1}\alpha_r^{(\mu,w)}\exp\left[i\pi\frac{(k\mu-r/\mu+w)^2}{N}\right]. \tag{4.4.12}$$

Because shift parameter w is a combination of shifts in signal and spectral domains, shift in signal domain causes a corresponding shift in transform domain which, however depends on the focal parameter μ according to Eq. 4.4.7. One can break this interdependence if, in the definition of the discrete representation of integral Fresnel transform, impose a symmetry condition:

$$\alpha_r^{(\mu,w)} = \alpha_{N-r}^{(\mu,w)} \tag{4.4.13}$$

of the transform of a point source $\delta(k)$

$$\alpha_r^{(\mu,w)} = \exp\left[-i\pi\frac{(r/\mu - w)^2}{N}\right]. \tag{4.4.14}$$

This symmetry condition is satisfied when

$$w = \frac{N}{2\mu} \tag{4.4.15}$$

and SDFrT for such shift parameter takes form:

$$\alpha_r^{\left(\mu,\frac{N}{2\mu}\right)} = \frac{1}{\sqrt{N}}\sum_{k=0}^{N-1} a_k \exp\left\{-i\pi\frac{[k\mu - (r - N/2)/\mu]^2}{N}\right\}. \tag{4.4.16}$$

Transform defined by Eq. (4.4.16) allows to keep position of objects reconstructed from Fresnel holograms invariant to focus distance. We will cal this version of the discrete Fresnel transform ***Focal Plane Invariant Discrete Fresnel Transform***.

Discrete representations of the partial Fresnel Transform are obtained by removing, from Eqs. 4.4.9 and 16, exponential phase terms that do not depend on k. Therefore, ***Partial Discrete Shifted Fresnel Transform*** is defined as

$$\tilde{\alpha}_r^{(\mu,w)} = \frac{1}{\sqrt{N}}\sum_{k=0}^{N-1} a_k \exp\left(-i\pi\frac{k^2\mu^2}{N}\right)\exp\left[i2\pi\frac{k(r - w\mu)}{N}\right]. \tag{4.4.17}$$

with its inverse transform as:

$$a_k^{(\mu,w)} = \frac{1}{\sqrt{N}}\left\{\sum_{r=0}^{N-1} \hat{\alpha}_r^{(\mu,w)} \exp\left[-i2\pi\frac{(r-w\mu)k}{N}\right]\right\}\exp\left(i\pi\frac{k^2\mu^2}{N}\right). \quad (4.4.18)$$

and ***Focal Plane Invariant Partial Discrete Fresnel Transform*** as

$$\hat{\hat{\alpha}}_r^{(\mu,w)} = \frac{1}{\sqrt{N}}\sum_{k=0}^{N-1} a_k \exp\left(-i\pi\frac{k^2\mu^2}{N}\right)\exp\left[i2\pi\frac{k(r-N/2)}{N}\right] =$$

$$\frac{1}{\sqrt{N}}\sum_{k=0}^{N-1} a_k (-1)^k \exp\left(-i\pi\frac{k^2\mu^2}{N}\right)\exp\left(i2\pi\frac{kr}{N}\right) \quad (4.4.19)$$

with its inverse transform as

$$a_k = \frac{1}{\sqrt{N}}(-1)^k \exp\left(-i\pi\frac{k^2\mu^2}{N}\right)\sum_{k=0}^{N-1} \hat{\hat{\alpha}}_r^{\mu} \exp\left(-i2\pi\frac{kr}{N}\right). \quad (4.4.20)$$

Fig. 4-8 illustrates using Discrete Fresnel and Focal Plane invariant Discrete Fresnel Transforms for reconstruction of optical holograms.

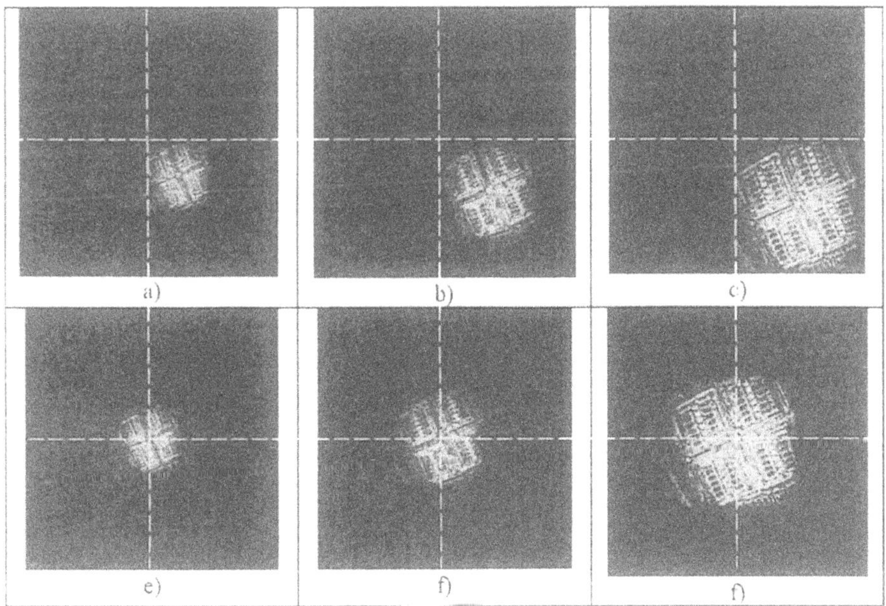

Figure 4-8. Reconstruction of a hologram recorded in a form of amplitude and phase. a) - c): reconstruction on different depth using Discrete Fresnel Transform (Eq. 4.4.10); d) - f) - reconstruction using Focal Plane Invariant Discrete Fresnel Transform (Eq. 4.4.16). (The hologram was provided by Dr. G. Pedrini[3],[8])

Fig. 4-9 illustrates optical reconstruction of a computer generated hologram that combines holograms computed with the use of Discrete Fourier and Discrete Fresnel Transforms. As one can see, the hologram is capable of reconstructing different object in different planes. One can also see in the images zero order diffraction spot that appears in reconstruction of holograms recorded on amplitude-only medium (See Sect. 13.3).

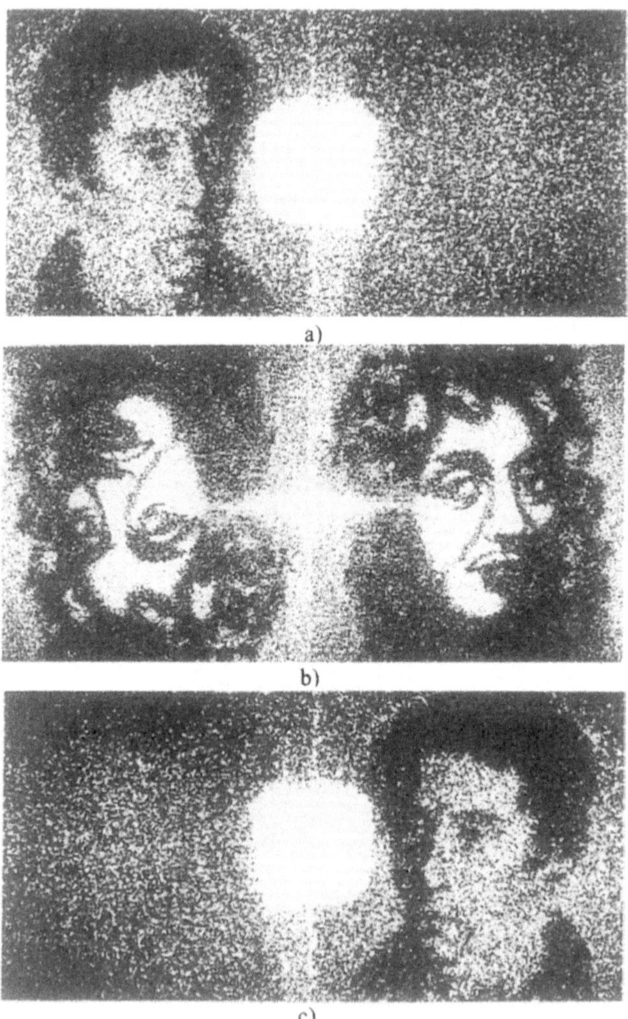

Figure 4-9. Optical reconstruction of a computer generated hologram computed using Discrete Fourier Transform and Discrete Fresnel Transform for visualizing objects in different planes: a), c) - front and rear focal planes of the Fresnel transform; b) - Fourier plane (adopted from [6]).

4.4.2 Properties of Discrete Fresnel Transforms

Invertibility and frincd-function

For a discrete signal $\{a_k\}$, $k = 0,1,...,N-1$, let us compute inverse SDFrT with shift parameters (μ_-, w_-) of its SDFrT spectrum computed with parameters (μ_+, w_+):

$$a_k^{(\mu^\pm, w^\pm)} = \frac{1}{\sqrt{N}} \sum_{r=0}^{N-1} \alpha_r^{(\mu^+, w^+)} \exp\left\{ i\pi \frac{[k\mu_- - r/\mu_- + w_-]^2}{N} \right\} =$$
$$\frac{1}{\sqrt{N}} \sum_{r=0}^{N-1} \left\{ \frac{1}{\sqrt{N}} \sum_{n=0}^{N-1} a_n \exp\left[-i\pi \frac{(n\mu_+ - r/\mu_+ + w_+)^2}{N} \right] \right\} \times$$
$$\exp\left[i\pi \frac{(k\mu_- - r/\mu_- + w_-)^2}{N} \right] =$$
$$\frac{1}{N} \sum_{r=0}^{N-1} \sum_{n=0}^{N-1} a_n \exp\left\{ i\pi \left[\frac{(n\mu_+ - r/\mu_+ + w_+)^2 - (k\mu_- - r/\mu_- + w_-)^2}{N} \right] \right\} =$$
$$\frac{1}{N} \exp\left[-i\pi \frac{(k\mu_- + w_-)^2}{N} \right] \sum_{n=0}^{N-1} a_n \exp\left[i\pi \frac{(n\mu_+ + w_+)^2}{N} \right] \times$$
$$\sum_{r=0}^{N-1} \exp\left[i\pi \frac{r^2 \left(1/\mu_+^2 - 1/\mu_-^2\right)}{N} \right] \exp\left(-i2\pi r \frac{n - k + w_+/\mu_+ - w_-/\mu_-}{N} \right). \tag{4.4.21}$$

Introduce the following substitutions:

$$w_+ = \frac{N}{2\mu_+} + \overline{w}_+; \qquad w_- = \frac{N}{2\mu_-} + \overline{w}_-; \qquad q = 1/\mu_+^2 - 1/\mu_-^2;$$
$$\overline{w}_\pm = \overline{w}_+/\mu_+ - \overline{w}_-/\mu_- . \tag{4.4.22}$$

and obtain finally:

$$a_k^{(\mu^\pm, w^\pm)} = \frac{1}{N} \exp\left[-i\pi \frac{(k\mu_- + w_-)^2}{N} \right] \sum_{n=0}^{N-1} a_n \exp\left[i\pi \frac{(n\mu_+ + w_+)^2}{N} \right] \times$$

$$\sum_{r=0}^{N-1}\exp\left[i\pi\frac{rq(r-N)}{N}\right]\exp\left(-i2\pi r\frac{n-k+\overline{w}_+/\mu_+-\overline{w}_-/\mu_-}{N}\right)=$$

$$\frac{1}{N}\exp\left[-i\pi\frac{(k\mu_-+w_-)^2}{N}\right]\times$$

$$\sum_{n=0}^{N-1}a_n\exp\left[i\pi\frac{(n\mu_++w_+)^2}{N}\right]\text{frincd}(N;q;n-k+\overline{w}_\pm),\tag{4.4.23}$$

where

$$\text{frincd}(N;q;x)=\frac{1}{N}\sum_{r=0}^{N-1}\exp\left[i\pi\frac{qr(r-N)}{N}\right]\exp\left(-i2\pi\frac{xr}{N}\right)\tag{4.4.24}$$

is a ***frincd-function***, a discrete analog of the frinc-function of integral Fresnel Transform (Eq. 2.4.51). It is also an analog of sincd-function of the DFT and is identical to it when $q=0$:

$$\text{frincd}(N;0,x)=\text{sincd}(N;N;x)\tag{4.4.25}$$

In particular, when $q=0$, from Eq. 4.4.23 obtain:

$$a_k^{(\mu,w^\pm)}\exp\left[-i\pi\frac{(k\mu_-+w_-)^2}{N}\right]\times$$

$$\sum_{n=0}^{N-1}a_n\exp\left[i\pi\frac{(n\mu_++w_+)^2}{N}\right]\text{sincd}(N;N;n-k+\overline{w}_\pm)\tag{4.4.26}$$

Thus, the inconsistency only in direct and inverse transforms shift parameters results in reconstruction of a chirp-demodulated sincd-interpolated, $\overline{w}_\pm=(w^+-w^-)/\mu$ - shifted initial signal modulated by a chirp-function.

Parameter q of frincd-function plays a role of a focussing parameter associated with distance parameter λD in near zone diffraction equation (Eq. 2.3.11) modeled by the integral Fresnel Transform. From Eq. 4.4.24 one can see that frincd-function is a periodical function with period of N samples:

$$\text{frincd}(N;q,x)=\text{frincd}(N;q,(x)_{\text{mod}\,N})\tag{4.4.27}$$

and that

$$\left(\mathbf{frincd}\left(N;q,x\right)\right)^{*}=\mathbf{frincd}\left(N;-q,N-x\right); \tag{4.4.28}$$

$$\mathbf{frincd}\left(N;q,0\right)=\frac{1}{N}\sum_{r=0}^{N-1}\exp\left(i\pi\frac{qr\left(r-N\right)}{N}\right). \tag{4.4.29}$$

Fig. 4-10 illustrates the behavior of frincd-function for different values of focussing parameter q.

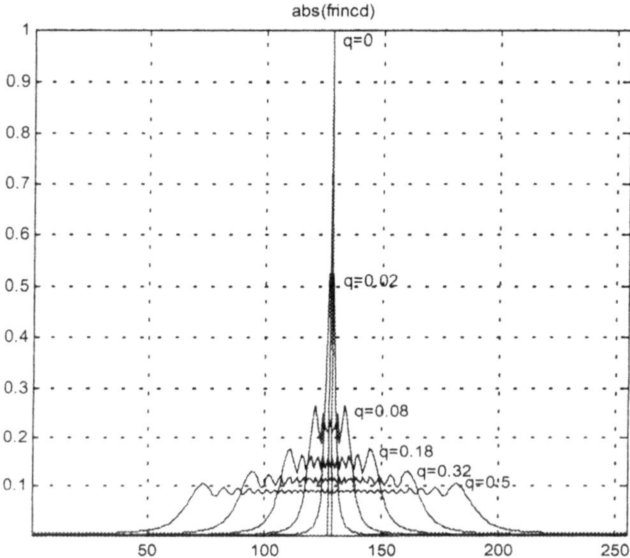

Figure 4-10. Absolute values of function ***frincd (N;q;x)*** for N=256 and different values of "focussing" parameter q. (The function is shown shifted to be centered around the middle point of the range of its argument)

Periodicity:

For integer h and g:

$$a_{k+Nh}=a_k\exp\left[i2\pi h\mu^2\left(k+\frac{Nh}{2}+\frac{w}{\mu}\right)\right]. \tag{4.4.31}$$

$$\alpha_{r+Ng}=\alpha_r\exp\left[-i2\pi\frac{g}{\mu^2}\left(r+\frac{Ng}{2}+w\mu\right)\right]. \tag{4.4.32}$$

Signal and Fresnel spectrum sinusoidal modulation and shift:

$$a_{k-k_0}\exp\left[-i\pi\frac{(k-k_0)^2\mu^2}{N}\right] \xleftrightarrow{SDFrT_{\mu,w}} \alpha_r \exp\left[i2\pi\frac{k_0(r+w\mu)}{N}\right]. \tag{4.4.33}$$

$$a_r\exp\left[i\pi\frac{(r-r_0)^2}{\mu^2 N}\right] \xleftrightarrow{SDFrT_{\mu,w}} a_k\exp\left[-i2\pi\frac{r_0(k+w/\mu)}{N}\right]. \tag{4.4.34}$$

SDFrTs of selected elementary signals:

$$\{a_k=\delta_{k-k_0}\} \xleftrightarrow{SDFrT_{\mu,w}} \left\{\frac{1}{\sqrt{N}}\exp\left[-i\pi\frac{(k_0\mu-r/\mu+w)^2}{N}\right]\right\}; \tag{4.4.35}$$

$$\left\{\frac{1}{\sqrt{N}}\exp\left[i\pi\frac{(k\mu-r_0/\mu+w)^2}{N}\right]\right\} \xleftrightarrow{SDFrT_{\mu,w}} \{\alpha_r=\delta_{r-r_0}\}; \tag{4.4.36}$$

$$\frac{1}{\sqrt{N}} \xleftrightarrow{ISDFrT_{\mu,w}} \mathrm{frincd}\left(N;\frac{1}{\mu^2};k+\frac{w}{\mu}-\frac{N}{2\mu^2}\right)\exp\left[i\pi\frac{(k+N/2\mu^2)^2}{N}\right] \tag{4.4.37}$$

References

1. N. Ahmed, K.R. Rao, Orthogonal Transforms for Digital Signal Processing, Springer Verlag, Berlin, Heidelberg, New York, 1975
2. A. K. Jain, Fundamentals of Digital Image Processing, Prentice Hall, Englewood Cliffs, 1989
3. Wang, Z., Fast Algorithms for the discrete W transform and for the discrete Fourier Transform", IEEE Trans. Acoust., Speech, Signal Processing, vol. ASSP-32, Aug. 1984, pp. 803-816
4. H. S. Malvar, Signal Processing with Lapped Transforms, Artech House Inc., Norwood, MA, 1992
5. L. Yaroslavsky, M. Eden, Fundamentals of Digital Optics, Birkhauser, Boston, 1996
6. L. Yaroslavsky, N. Merzlyakov, Methods of Digital Holography, Consultant Bureau, N.Y., 1980
7. T. Kreis, Holographic Interferometry, Principles and Methods, Akademie Verlag, Berlin, 1996
8. G. Pedrini, H. J. Tiziani, "Short-Coherence Digital Microscopy by Use of a Lensless Holographic Imaging System", Applied Optics-OT, 41, (22), 4489-4496, (2002).

Chapter 5

METHODS OF DIGITAL FILTERING

5.1 FILTERING IN SIGNAL DOMAIN

5.1.1 Transversal and recursive filtering

As it was stated in Sect. 4.2, basic formula of digital filtering a signal defined by its samples $\{a_k, k = 0,1,...,N_a - 1\}$ by a filter defined by its discrete PSF $\{h_n, n = 0,1,...,N_h - 1\}$ is

$$b_k = \sum_{n=0}^{N_h-1} h_n a_{k-n} \qquad (5.1.1)$$

Direct computational implementation of this formula is called ***filtering in signal domain***.

An important issue in digital filtering is its computational complexity. In conventional serial computers filtering in signal domain is carried out by signal weighted summation according to Eq. 5.1.1 in sliding window of N_h signal samples as it is illustrated in Fig. 5-1. This filtering technique is referred to as ***transversal filtering***. The computational complexity of transversal filtering according to Eq. 5.1.1 is defined by the extent of digital filter PSF: N_h addition and $N_h - 1$ multiplication operations per output signal sample. This complexity may be reduced if to use, in course of filtering in each filter window position, filtering results obtained on previous positions. Such filtering technique referred to as ***recursive filtering*** is described by equation:

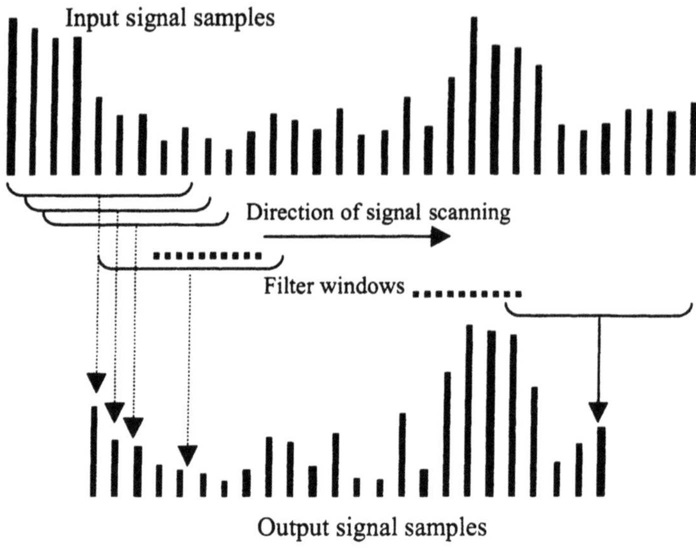

Figure 5-1. Signal filtering in sliding window

$$b_k = \sum_{n=0}^{N_h^{(t)}-1} h_n^{(t)} a_{k-n} + \sum_{s=1}^{N_h^{(r)}} h_s^{(r)} b_{k-s}, \tag{5.1.2}$$

where $N_h^{(t)}$ and $N_h^{(r)}$ are the numbers of addends in the transversal (first sum) and recursive (second sum) parts, respectively, of the filter. In hardware implementation, recursive filters correspond to feedback circuitry.

Eq. 5.1.2 assumes that, for computing each output signal sample, $N_h^{(t)}$ input signal samples and $N_h^{(r)}$ previously computed output signal samples are used. To illustrate computational saving with recursive filtering consider the simplest case, when $N_h^{(t)} = 1$ and $N_h^{(r)} = 1$:

$$b_k = h_0^{(t)} a_k + h_1^{(r)} b_{k-1}. \tag{5.1.3}$$

Computations according Eq. 5.1.3 require two multiplications and one addition operations, while the same filter in the transversal form:

$$b_k = h_0^{(t)} a_k + h_0^{(t)} h_1^{(r)} a_{k-1} + h_0^{(t)} \left(h_1^{(r)}\right)^2 a_{k-2} + h_0^{(t)} \left(h_1^{(r)}\right)^3 a_{k-3} + \ldots.. \tag{5.1.4}$$

may contain unlimited number of terms.

A good practical example of the recursive filter is the filter that computes signal ***local mean*** in a window of $2N+1$ samples:

$$b_k = \frac{1}{2N+1}\sum_{-N}^{N} a_{k-n} \,. \tag{5.1.5}$$

Eq. 5.1.5 can be represented in a recursive form

$$b_k = b_{k-1} + \left(a_{k+N} - a_{k-N-1}\right)/\left(2N+1\right) \tag{5.1.6}$$

that requires two addition and one multiplication operation independently on window size while, in its transversal form, filter requires $2N+1$ additions and one multiplication.

Computational saving possible with recursive filtering suggests that whenever possible, the desired filter should be approximated by a recursive one.

In filtering 2-D signals, there is no natural sequential ordering of signal samples. The definition of "past" and "future" samples in sequential signal processing is a matter of a convention. Usually, 2-D discrete signals are treated as 2-D arrays of numbers and sequential filtering is carried out row wise from top to bottom rows of the array. Arrangement of "past" (bygone) and "future" samples in such a signal scanning is illustrated in Fig. 5-2.

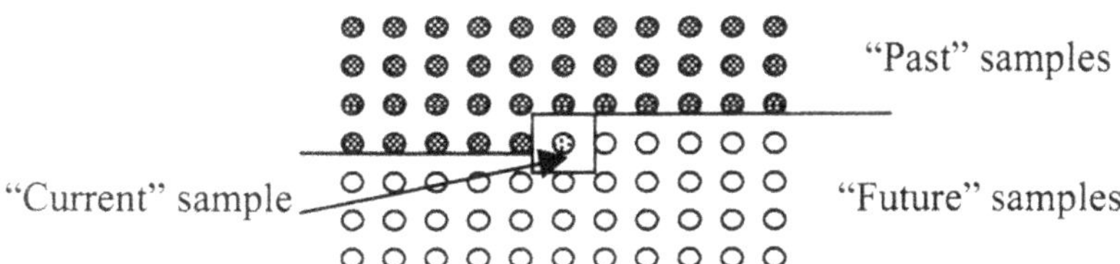

Figure 5-2. "Past" and "future" samples on a rectangular sampling raster at 2-D row wise filtering

5.1.2 Separable, cascade, parallel and parallel recursive filtering

General digital filtering of a 2-D signal defined by its samples $\{a_{k,l}\}$ is described by the equation:

$$b_{k,l} = \sum_{n=0}^{N-1}\sum_{m=0}^{M-1} h_{n,m} a_{k-n,l-m} \,. \tag{5.1.7}$$

Direct computations according to this formula require NM multiplication and summation operations per output sample. This number can be substantially reduced if filter PSF $\{h_{n,m}\}$ is a separable function of its indices:

$$h_{n,m} = h_n^{(1)} h_m^{(2)}. \tag{5.1.8}$$

In this case filtering can be carried out separately row-wise and column-wise with corresponding 1-D filters

$$b_{k,l} = \sum_{m=0}^{M-1} h_m^{(2)} \sum_{n=0}^{N-1} h_n^{(1)} a_{k-n,l-m} \tag{5.1.9}$$

which requires $N+M$ multiplication and summation operations rather then MN ops. In Chapt. 6 we will see that representation of transform in a separable form is a key point for synthesis of fast transform algorithms.

Separable representation of digital filters is an example of decomposition of complex filters into a set of more simple and fast ones. Two other methods for such a decomposition are ***cascade*** and ***parallel*** implementation of filters. In cascade implementation, desired filtering is achieved by sequential application of a number of sub-filters as it is illustrated in the flow diagram in Fig. 5-3.

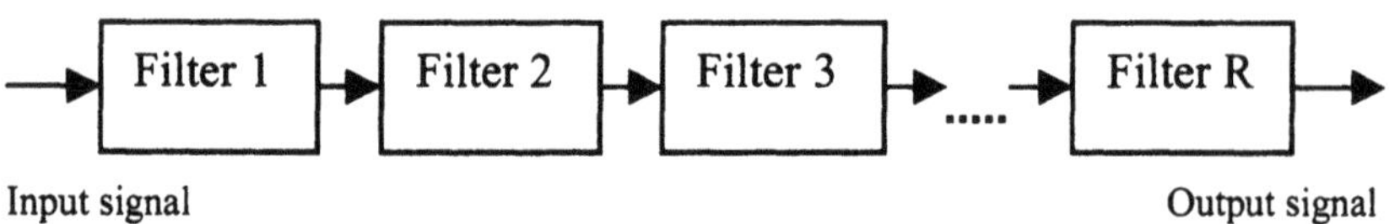

Figure 5-3. The principle of cascade filtering

PSF of a filter implemented as a cascade of R sub-filters with PSFs $\{h_n^{(r)}, r = 1,2,..., R\}$ is a discrete convolution of PSFs of those filters:

$$h_n = \sum_{n_R} h_{n_R - n_{R-1}}^{(R)} \sum h_{n_{R-1} - n_{R-2}}^{(R-1)} \cdots \sum_{n_1} h_{n_2 - n_1}^{(2)} h_{n_1}^{(1)}. \tag{5.1.10}$$

As it follows from the convolution theorem of the DFT, discrete frequency response of the cascade of R sub-filters is a product

$$DFR(p) = \prod_{r=1}^{R} DFR_r(p). \tag{5.1.11}$$

of discrete frequency responses $\{DFR_r(p)\}$ of the sub-filters defined by Eq. 4.2.10.

Factorizing DFR of complex filters into a product of DFRs of more simple sub-filters is a way to design cascade implementation of complex filters. A good example of the cascade implementation of filters are ***splines***. In spline theory terminology ([1,2]), signal moving average filter (Eq. 5.1.5) is a zero-order spline. First order spline is obtained as the convolution of two zero-order splines. Its PSF is obviously a triangle. Second order spline is the convolution of the first and zero-order splines. Higher order splines are corresponding higher order convolutions.

Let us evaluate potential computational gain that can be achieved by cascade filtering with respect to direct filtering. Consider first 1-D filters. Let N is the number of samples in PSFs of sub-filters. Each sub-filter in the cascade after the first one adds to this number $N-1$ samples. Then, for R sub-filters in cascade, equivalent PSF will have $R(N-1)+1$ samples. This figure defines the computational complexity of the direct filtering while the computational complexity of the cascade filtering is defined by the sum NR of the number of samples in PSFs of sub-filters which is higher then $R(N-1)+1$. Therefore, the computational complexity of the cascade implementation of 1-D filters may even be higher then that of direct filtering unless sub-filters are recursive ones. For 2-D filtering, the situation is completely different. If NM is the number of samples in PSFs of R 2-D sub-filters in cascade, the equivalent number of samples in PSF of the entire cascade that defines its computational complexity in the direct filtering is $[R(N-1)+1]\times[R(M-1)+1]$. The complexity of the cascade filtering is defined by the sum RNM of PSF samples of R sub-filters. Therefore the computational complexity of 2-D cascade filtering is approximately R times less then that of the direct filtering.

Parallel implementation of filters assumes filter decomposition into a number of sub-filters that act in parallel (see Fig. 5-4).

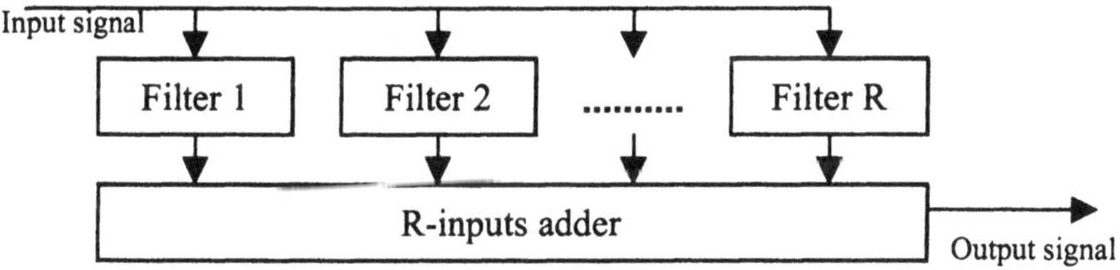

Figure 5-4. The principle of parallel implementation of filtering

For such an implementation, filter PSF $\{h_n\}$ has to be decomposed into a sum

$$h_n = \sum_{r=1}^{R} h_n^{(r)} \tag{5.1.12}$$

of PSFs $\{h_n^{(r)}, r = 1,2,\dots,R\}$ of sub-filters.

Parallel filtering is well suited for parallel computing. In serial computers, it can be carried out by sequential signal filtering with sub-filters and subsequent addition of the filtering results. Computational saving can in this case be achieved in this case only if sub-filters are recursive.

Let us show how an arbitrary digital filter can be represented in such a parallel and recursive form. Consider a digital filter:

$$b_k = \sum_{-N_1^{(h)}}^{N_2^{(h)}} h_n a_{k-n} \tag{5.1.13}$$

and assume that filter PSF $\{h_n\}$ can be approximated, with the desired accuracy, by expanding it into a series over a system of basis functions $\{\varphi_r(n), r = 1,2,\dots,R;\ R \le N_1^{(h)} + N_2^{(h)} + 1\}$:

$$h_n \cong \sum_{r=1}^{R} \eta_r \varphi_r(n). \tag{5.1.14}$$

Then filter output signal can then be approximated as

$$b_k \cong \sum_{-N_1^{(h)}}^{N_2^{(h)}} a_{k-n} \sum_{r=1}^{R} \eta_r \varphi_r(n) = \sum_{r=1}^{R} \eta_r \sum_{-N_1^{(h)}}^{N_2^{(h)}} \varphi_r(n) a_{k-n} = \sum_{r=1}^{R} \eta_r \beta_r(k), \tag{5.1.15}$$

where

$$\beta_r(k) = \sum_{-N_1^{(h)}}^{N_2^{(h)}} \varphi_r(n) a_{k-n} \,. \tag{5.1.16}$$

Eq. 5.1.15 describes signal filtering with sub-filters with PSFs $\{\varphi_r(n)\}$ in R parallel branches and adding up the sub-filtering results with weights $\{\eta_r\}$.

Let us now find conditions under which coefficients $\beta_r(k)$ in branch filters can be computed recursively. In the simplest first order recursive filter, every given output signal sample is computed from the output sample determined at the preceding filter window position. Let us establish a link between $\{\beta_r(k)\}$ and $\{\beta_r(k-1)\}$:

$$\beta_r(k)=\sum_{n=-N_1^{(r)}}^{N_2^{(r)}}\varphi_r(n)a_{k-n}=\sum_{n=-N_1^{(r)}}^{N_2^{(r)}}\varphi_r(n)a_{k-1-(n-1)}=\sum_{m=-N_1^{(r)}-1}^{N_2^{(r)}-1}\varphi_r(m+1)a_{k-1-m}=$$
$$\sum_{m=-N_1^{(r)}}^{N_2^{(r)}-1}\varphi_r(m+1)a_{k-1-m}+\varphi_r\left(-N_1^{(r)}\right)a_{k+N_1^{(r)}}. \tag{5.1.17}$$

and choose $\varphi_r(m)$ such that

$$\varphi_r(m+1)=\varphi_r^{(0)}\varphi_r(m). \tag{5.1.18}$$

Then obtain:

$$\beta_r(k)=\varphi_r^{(0)}\sum_{m=-N_1^{(r)}}^{N_2^{(r)}-1}\varphi_r(m)a_{k-1-m}+\varphi_r\left(-N_1^{(r)}\right)a_{k+N_1^{(r)}}=$$
$$\varphi_r^{(0)}\sum_{m=-N_1^{(r)}}^{N_2^{(r)}}\varphi_r(m)a_{k-1-m}+\varphi_r\left(-N_1^{(r)}\right)a_{k+N_1^{(r)}}-\varphi_r^{(0)}\varphi_r\left(N_2^{(r)}\right)a_{k-N_1^{(r)}}=$$
$$\beta_r(k-1)+\varphi_r\left(-N_1^{(r)}\right)a_{k+N_1^{(r)}}-\varphi_r^{(0)}\varphi_r\left(N_2^{(r)}\right)a_{k-N_1^{(r)}}. \tag{5.1.19}$$

Eq. 5.1.19 signifies that when basis functions $\{\varphi_r(n)\}$ satisfy Eq. 5.1.18 coefficients $\beta_r(k)$ can be computed recursively with the recursiveness on one preceding step.

It follows from Eq. (5.1.18) that the class of the recursive basis functions is that of exponential functions:

$$\varphi_r(m)=\left[\varphi_r^{(0)}\right]^m. \tag{5.1.20}$$

Two most important special cases are the basis of discrete exponential functions:

$$\varphi_r(m) = \exp\left(i2\pi p_0 \frac{rm}{M}\right); \quad M = N_1^{(h)} + N_2^{(h)} + 1\,; \tag{5.1.21}$$

where p_0 is a free parameter and the linearly independent basis of rectangular functions

$$\varphi_r(m) = \mathrm{rect}\left(\frac{m + N_1^{(r)}}{N_1^{(r)} + N_2^{(r)} + 1}\right) \tag{5.1.22}$$

which can be obtained from Eq. (5.1.21) when $p_0 = 0$. The latter corresponds to recursive computation of signal local mean:

$$\bar{a}_k = \frac{1}{2N^{(w)} + 1} \sum_{n=-N^{(w)}}^{N^{(w)}} a_{k-n} = \frac{1}{2N^{(w)} + 1}\left[\bar{a}_{k-1} + \frac{a_{k+N^{(w)}} - a_{k-N^{(w)}-1}}{2N^{(w)} + 1}\right]. \tag{5.1.23}$$

When the basis of Eq. 5.1.21 with $p = 1$ is used each of the R parallel recursive filters (5.1.16) performs signal discrete Fourier analysis in the window of $M = N_1^{(h)} + N_2^{(h)} + 1$ samples at the frequency r. The resultant spectral coefficients are multiplied element-wise by the spectral coefficients of the filter pulse response, i.e. by samples of its discrete frequency response, and summed up in order to obtain the filtering result. The basis of Eq. 5.1.21 with $p_0 = 1/2$ corresponds to local DCT/DcST analysis. Corresponding recursive algorithms of sliding window DFT and DCT/DcST are discussed in Sect. 5.1.3

When the basis (5.1.22) is used, the digital filter is composed of a set of ***local mean*** filters with different window size. The impulse response of the filter is a sum of R rectangular functions of various extents. Synthesis of such a representation for a particular filter is therefore reduced to quantization of values of its PSF to R quantization levels (Fig. 5-5). For minimization of filter PSF the approximation error caused its quantization, optimal quantization methods discussed in Sect. 3.5 may be used.

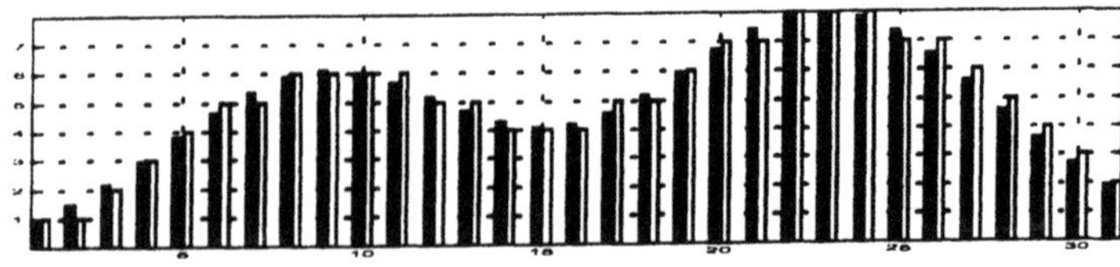

Figure 5-5. An example of a filter PSF (bold bars) and its quantized approximation with 8 quantization levels (clear bars)

For two-dimensional digital filters,

$$b_{k,l} = \sum_{n=-N_1^{(h)}}^{N_2^{(h)}} \sum_{m=-M_1^{(h)}}^{M_2^{(h)}} h_{m,n} a_{k-n,l-m} \tag{5.1.24}$$

basis functions $\{\varphi_{r,s}(m,n)\}$ in the expansion of the filter PSF

$$h_{m,n} \cong \sum_{r=0}^{R-1} \sum_{s=0}^{S-1} \eta_{r,s} \varphi_{r,s}(m,n) \tag{5.1.25}$$

are represented in the simplest possible way as separable functions $\varphi_{r,s}(m,n) = \varphi_r^{(1)}(m)\varphi_s^{(2)}(n)$. The associated parallel filters

$$\beta_{r,s}(k,l) = \sum_{n=-N_1^{(h)}}^{N_2^{(h)}} \varphi_s^{(2)} \sum_{m=-M_1^{(h)}}^{M_2^{(h)}} \varphi_r^{(1)}(m) a_{k-n,l-m} \tag{5.1.26}$$

would then be the simplest possible. The above results are obviously applicable to the one-dimensional functions $\varphi_r^{(1)}(m)$ and $\varphi_s^{(2)}(n)$. Hence in the two-dimensional case, "pure" bases of exponential and rectangular functions may be supplemented by "mixed" bases, rectangular in one coordinate and exponential in the other coordinate.

There are also inseparable recursive two-dimensional bases. One of them is a basis of rectangular functions within an arbitrary shaped area on a rectangular raster. An analogous separable basis is bounded by a rectangle. The aperture support of an inseparable two-dimensional filter is shown in Fig.5-6, b) and its separable analog is shown in Fig.5-6, a) in which the boundary elements of the aperture are cross hatched.

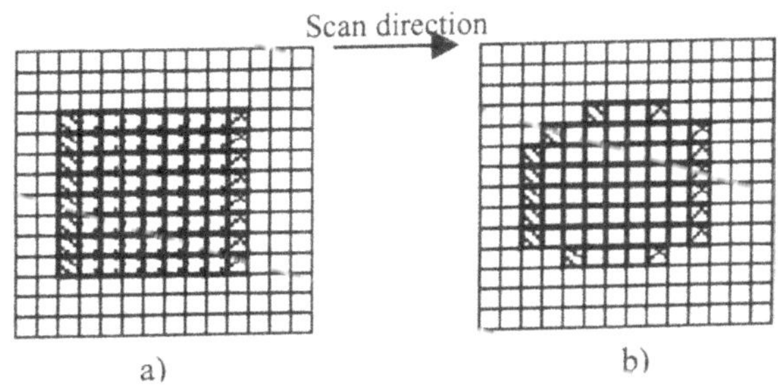

Figure 5-6. Pattern of support of separable (a) and inseparable (b) 2-D moving average bases (pixels that are drop out the filter aperture are single line hatched, those that enter it are double line hatched).

5.1.3 Recursive algorithms of sliding window DFT and DCT/DcST

Let us consider DFT of two adjacent $(k-1)$-th and k-th fragments $\{a_{k-1-fix((N_w-1)/2)},\dots a_{k-1+fix(N_w/2)}\}$ and $\{a_{k-fix((N_w-1)/2)},\dots a_{k+fix(N_w/2)}\}$ in sliding window of N_w samples of a signal $\{a_k\}$

$$\alpha_r^{(k-1)} = \frac{1}{\sqrt{N_w}} \sum_{n=0}^{N_w-1} a_{k-1-fix((N_w-1)/2)+n} \exp\left(i2\pi \frac{nr}{N_w}\right) \tag{5.1.27}$$

and

$$\alpha_r^{(k)} = \frac{1}{\sqrt{N_w}} \sum_{n=0}^{N_w-1} a_{k-fix((N_w-1)/2)+n} \exp\left(i2\pi \frac{nr}{N_w}\right) \tag{5.1.28}$$

where $fix(x)$ is integer part of x. Spectrum $\{\alpha_r^{(k)}\}$ in k-th position of the sliding window can be computed from spectrum $\{\alpha_r^{(k-1)}\}$ in $(k-1)$-th window position in the following way:

$$\alpha_r^{(k)} = \frac{1}{\sqrt{N_w}} \sum_{n=0}^{N_w-1} a_{k-fix((N_w-1)/2)+n} \exp\left(i2\pi \frac{nr}{N_w}\right) =$$

$$\frac{1}{\sqrt{N_w}} \sum_{n=1}^{N_w} a_{k-1-fix((N_w-1)/2)+n} \exp\left(i2\pi \frac{(n-1)r}{N_w}\right) =$$

$$\left[\frac{1}{\sqrt{N_w}} \sum_{n=0}^{N_w-1} a_{k-1-fix((N_w-1)/2)+n} \exp\left(i2\pi \frac{nr}{N_w}\right) + \right.$$

$$\left. \frac{a_{k-1-fix((N_w-1)/2)+N_w} - a_{k-1-fix((N_w-1)/2)}}{\sqrt{N_w}} \right] \exp\left(-i2\pi \frac{r}{N_w}\right) =$$

$$\left(\alpha_r^{(k-1)} + \frac{a_{k-1-fix((N_w-1)/2)+N} - a_{k-1-fix((N_w-1)/2)}}{\sqrt{N_w}}\right) \exp\left(-i2\pi \frac{r}{N_w}\right). \tag{5.1.29}$$

Eq. 5.1.29 describes the recursive algorithm for computing DFT signal spectrum in sliding window of N_w samples that requires, for signals whose samples are real numbers, $2(N_w-1)$ multiplication operations and (N_w+1) addition operations.

The DCT and DcST in a running window can be computed in a similar way ([3]). For adjacent $(k-1)$-th and k -th signal fragments, their DCT and DcST spectra $\{\alpha_r^{(k-1)}\}$and $\{\alpha_r^{(k)}\}$are, correspondingly, real and imaginary parts of

$$\alpha_r^{(k-1)} = \frac{1}{\sqrt{N_w}} \sum_{n=0}^{N-1} a_{k-1-fix((N_w-1)/2)+n} \exp\left(i\pi \frac{(n+1/2)r}{N_w} \right) \qquad (5.1.30)$$

and

$$\alpha_r^{(k)} = \frac{1}{\sqrt{N_w}} \sum_{n=0}^{N-1} a_{k-fix((N_w-1)/2)+n} \exp\left(i\pi \frac{(n+1/2)r}{N_w} \right) \qquad (5.1.31)$$

Spectrum $\{\tilde{\alpha}_r^{(k)}\}$ can be represented through spectrum $\{\tilde{\alpha}_r^{(k-1)}\}$ as following:

$$\tilde{\alpha}_r^{(k)} = \tilde{\alpha}_r^{(k-1)} \exp\left(-i\pi \frac{r}{N_w} \right) +$$

$$\left[(-1)^r a_{k-1-fix((N_w-1)/2)+N_w} - a_{k-1-fix((N_w-1)/2)} \right] \exp\left(-i\pi \frac{r}{2N_w} \right). \qquad (5.1.32)$$

Recursive computations according to Eq. 5.1.32 require $4(N_w-1)$ multiplication and $6N_w-3$ addition operations for both DCT and DcST transform in sliding window of N samples.

Extension of these algorithms to the recursive computation of 2-D DFT, DCT and DcST is straightforward owing to the separability of the transforms. For instance, for DCT/DcST in the window of $N_w \times M_w$ samples obtain:

$$\tilde{\alpha}_{r,s}^{(k,l)} = \tilde{\alpha}_{r,s}^{(k-1,l)} \exp(-i\pi r / N_w) + \left[a_{\tilde{k}+N,\tilde{l}} (-1)^r - a_{\tilde{k},\tilde{l}} \right] \exp\left(-i\pi \frac{r}{2N_w} \right) \qquad (5.1.33)$$

where $\tilde{k} = k-1-fix((N_w-1)/2)$, $\tilde{l} = l-1-fix((M_w-1)/2)$.

As one can see, in the recursive algorithm for computing DCT and DcST in sliding window described by Eq. 5.1.31, DCT and DcST are inseparable:

$$\alpha_r^{(k)DCT} = \text{Real}\{\tilde{\alpha}_r^{(k)}\} = \alpha_r^{(k-1)DCT} C_1 + \alpha_r^{(k-1)DST} S_1 + \Delta a_k C_2; \qquad (5.1.34)$$

and

$$\alpha_r^{(k)DcST} = \mathrm{Imag}\left\{\tilde{\alpha}_r^{(k)}\right\} = \alpha_r^{(k-1)DcST} C_1 - \alpha_r^{(k-1)DCT} S_1 - \Delta a_k S_2, \quad (5.1.35)$$

where

$$C_1 = \cos\left(\pi \frac{r}{N_w}\right);\ C_2 = \cos\left(\pi \frac{r}{2N_w}\right);\ S_1 = \sin\left(\pi \frac{r}{N_w}\right);\ S_2 = \sin\left(\pi \frac{r}{2N_w}\right);$$

$$\Delta a_k = (-1)^r a_{k-1-fix((N_w-1)/2)+N_w} - a_{k-1-fix((N_w-1)/2)}. \quad (5.1.36)$$

Many applications such as those discussed in Chapts 8 and 9 require computing only the DCT in sliding window. In these cases, computations of both DCT and DcST according to Eqs. 5.1.32 and 33 are redundant. This redundancy associated with the need to compute the DcST for recursive computing the DCT can be eliminated in an algorithm that uses DCT spectrum found on two previous window positions ([4]). To derive this algorithm, compute DFTs A_s^{DCT} and A_s^{DcST} of $\left\{\alpha_r^{(k)DCT}\right\}$ and $\left\{\alpha_r^{(k)DcST}\right\}$, respectively, over index k assuming that the number of signal samples is N_k. According to the shift theorem of the DFT, obtain for A_s^{DCT} and A_s^{DcST} from Eqs. 5.1.34 and 35:

$$A_s^{DCT} = A_s^{DCT} C_1 Z + A_s^{DcST} S_1 Z + \Delta A_s C_2 \quad (5.1.37)$$

$$A_s^{DcST} = A_s^{DcST} C_1 Z - A_s^{DCT} S_1 Z - \Delta A_s S_2, \quad (5.1.38)$$

where s - index in the DFT domain, ΔA_s - DFT of Δa_k and $Z = \exp\left(i2\pi \frac{s}{N_k}\right)$ - the spectral shift factor that corresponds to signal shift one sample back. Express, using Eq. 5.1.38, A_s^{DcST} through A_s^{DCT} and substitute it into Eq. 5.1.37. Then obtain:

$$A_s^{DCT}[1 - C_1 Z]^2 = -A_s^{DCT} S_1^2 Z^2 + \Delta A_s S_2 S_1 Z + \Delta A_s C_2 (1 - C_1 Z), \quad (5.1.39)$$

from which it follows that

$$A_s^{DCT} = 2C_1 A_s^{DCT} Z - \left(C_1^2 + S_1^2\right) A_s^{DCT} Z^2 - \left(C_2 C_1 - S_2 S_1\right) \Delta A_s Z + C_2 \Delta A_s . \tag{5.1.40}$$

or, as $C_1^2 + S_1^2 = 1$ and $C_2 C_1 - S_2 S_1 = C_2$,

$$A_s^{DCT} = 2C_1 A_s^{DCT} Z - A_s^{DCT} Z^2 + C_2\left(\Delta A_s - \Delta A_s Z\right). \tag{5.1.41}$$

Finally, taking inverse DFT of both part of Eq. 5.1.42 and using the fact that spectral multiplicands Z and Z^2 correspond to signal shifts one and two samples back, respectively, obtain for the sliding window DCT:

$$\alpha_r^{(k)DCT} = 2\alpha_r^{(k-1)DCT} \cos\left(\pi \frac{r}{N_w}\right) - \alpha_r^{(k-1)DCT} + \left(\Delta a_k - \Delta a_{k-1}\right)\cos\left(\pi \frac{r}{2N_w}\right). \tag{5.1.42}$$

Similar recursive relationship can be derived for the sliding window DcST:

$$\alpha_r^{(k)DcST} = 2\alpha_r^{(k-1)DcST} \cos\left(\pi \frac{r}{N_w}\right) - \alpha_r^{(k-1)DcST} - \left(\Delta a_k + \Delta a_{k-1}\right)\sin\left(\pi \frac{r}{2N_w}\right) \tag{5.1.43}$$

Computations according to Eqs. 5.1.42 and 43 require only $2N_w$ multiplication and $2N_w + 4$ addition operations.

5.1.4 Border processing in digital filtering

Eq. 5.1.1 defines digital filter output signal samples $\{b_k\}$ only for $k = N_h, N_h + 1, N_a - 1$ because input signal samples $\{a_k\}$ for $k < 0$ are not defined. Samples for $k < N_h$ and $k > N_a - 1$ are not defined. Therefore in digital filtering the number of output signal samples is shrinking away unless unavailable input signal samples are not additionally defined by means of a certain procedure of signal ***border processing***.

Border processing is approximating samples of initial continuous signal lost in the result of signal discretization by extrapolating available signal samples onto unavailable ones. Two approaches are possible to border processing, statistical and deterministic ones.

In the statistical approach, available signal samples are regarded as a realization of a random process with certain statistical characteristics such as mean value, standard deviation, correlation function and alike. As a first approximation, one can regard assigning to unavailable signal samples the mean value of the random process, for instance, zero if the process is zero mean. This is the rationale behind assigning zeros to unavailable signal samples which is very frequently practiced in digital signal processing. The zero mean assumption is however quite unnatural in image processing and digital holography unless it is a priori known that unavailable signal samples are zero. This is for instance the case in antenna array processing when it is a priori known that that there is no radiation outside antenna's aperture. Such situation in fact assumes deterministic approach to border processing.

In the deterministic approach, border processing is designed for every particular signal individually and implements certain requirements regarding signal behavior. One of the most natural requirements is that of the absence of discontinuities at signal borders. This requirement can be met by the

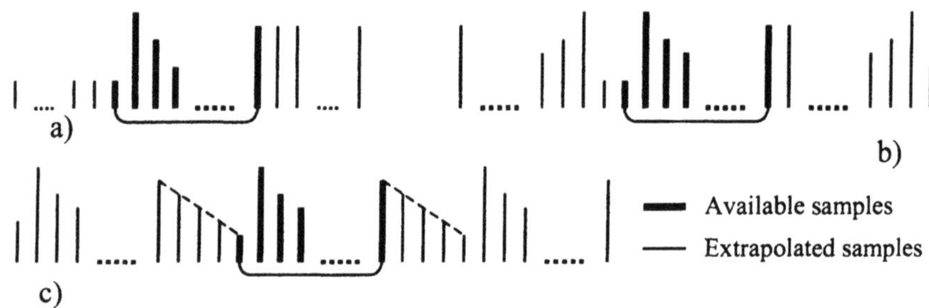

Figure 5-7. Three signal extrapolation methods that secure the absence of signal discontinuity on the borders: a) - repeating border samples; b) - signal mirroring; c) - interpolation between border samples in the assumption of signal periodical replication

the following three extrapolation methods illustrated in Fig.5-7:

- Replicating border samples of signal $\{a_k\}$of N_a samples when extended signal copy is obtained as:

$$\tilde{a}_k = \begin{cases} a_0, & k < 0 \\ a_k, & 0 \le k \le N_a - 1; \\ a_{N_a-1}, & k > N_a - 1 \end{cases} \tag{5.1.44}$$

- mirroring signal border samples:

$$\tilde{a}_k = \begin{cases} a_{k+1}, & k \le -1 \\ a_k, & 0 \le k \le N_a - 1 \; ; \\ a_{2N_a-k-1}, & k > N_a - 1 \end{cases} \tag{5.1.45}$$

- interpolating, by, for instance, linear interpolation, signal samples between left/right border samples assuming signal periodical replication with a certain period $N_p > N_a$:

$$\tilde{a}_k = \begin{cases} \dfrac{(N_p - N_a + k)a_0 - k a_{N-1}}{N_p - N_a}, & -(N_p - N_a) < k < 0 \\ a_k, & 0 \le k \le N_a - 1 \\ \dfrac{(N-1-k)a_0 + (k - N + 1 + N_p - N_a)a_{N-1}}{N_p - N_a}, & -1 < k - N < N_p - N_a - 1 \end{cases} \quad (5.1.46)$$

These methods should be regarded as just good practical stratagems. Two latter methods, signal mirroring and signal interpolating between border samples have a direct correspondence to filtering methods in the domain of DFT and DCT discussed in Sect. 5.2. Fig. 5-8 illustrates these three extrapolation methods applied to an image.

Original image

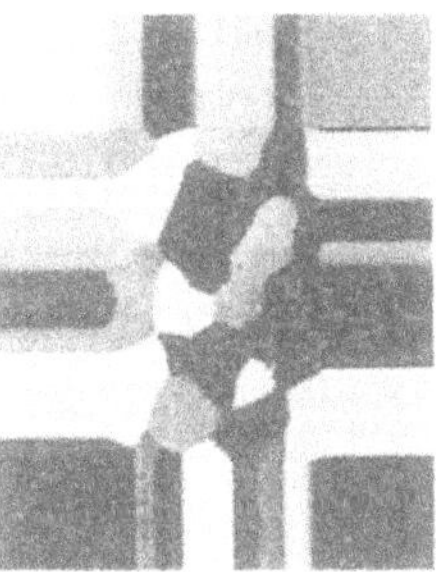

Extrapolation by replicating border samples

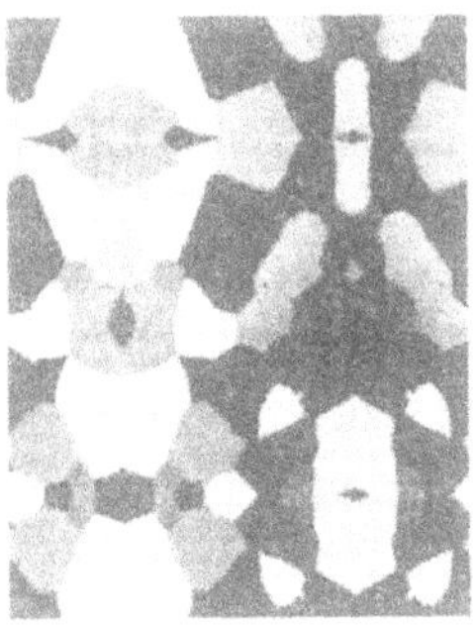

Extrapolation by mirroring

Linear interpolation between borders

Figure 5-8. Three methods for extrapolating images without discontinuities on its borders

5.2 FILTERING IN TRANSFORM DOMAIN

5.2.1 DFT and cyclic convolution. Border processing

According to the convolution theorem for Discrete Fourier Transform (Eq.4.3.30), inverse DFT of the product $\{\alpha_r \eta_r\}$ of DFT spectra $\{\alpha_r\}$ and $\{\eta_r\}$ of signals $\{a_k\}$ and $\{h_k\}$ of N samples is proportional to the cyclic convolution of the signals:

$$c_{(k)\bmod N} = \frac{1}{\sqrt{N}} \sum_{n=0}^{N-1} a_{(n)\bmod N} h_{(k-n)\bmod N}, \tag{5.2.1}$$

This property can be used for computing signal cyclic convolution as

$$\mathbf{c} = \mathbf{IDFT}\{\mathbf{DFT}\{\mathbf{a}\} \bullet \mathbf{DFT}\{\mathbf{h}\}\}, \tag{5.2.2}$$

where $\mathbf{a}$, $\mathbf{b}$ and $\mathbf{c}$ are vectors of signals and their cyclic convolution, $\mathbf{DFT}\{\ \}$ and $\mathbf{IDFT}\{\ \}$ denote operators of direct and inverse Discrete Fourier Transforms and $\bullet$ is component-wise multiplication (***Hadamard product***) of vectors. We will refer to this method of signal filtering as to ***filtering in the DFT domain***.

Filtering in DFT domain is a special case of ***filtering in transform domain***:

$$\mathbf{c} = \mathbf{T}^{-1}\{\mathbf{T}\{\mathbf{a}\} \bullet \mathbf{T}\{\mathbf{b}\}\}, \tag{5.2.3}$$

where T and T^{-1} are direct and inverse transform.

Because of the existence of Fast Fourier Transform (FFT) algorithms for computing direct and inverse DFTs, such an implementation of the convolution can be advantageous to direct convolution in signal domain described in the previous section. As it will be shown in Chapt. 6, the number of arithmetic operations for computing DFT with the use of FFTs grows with N as $N \log_2 N$, or $\log_2 N$ per signal sample. Therefore computational complexity of signal cyclic convolution via FFT is $O(\log_2 N)$ per signal sample while direct signal convolution requires $O(N_h)$ operations for a digital filter PSF of N_h samples unless recursive, cascade or parallel recursive implementation are used. This comparison shows that for large enough N_h the cyclic convolution via FFT may become faster then the direct convolution. This is especially true for 2-D and multi-dimensional signal for which the computational complexity of the direct convolution grows as N_h^m, where m is the signal dimensionality and N_h is 1-D size of the filter PSF, while FFT complexity grows as $m\log_2 N$. The same comparison holds for general filtering in transform domain (Eq. 5.2.3) provided that fast transform algorithms exist. They will be reviewed in Chapt. 6.

The cyclic periodicity of digital convolution via FFT is rather unnatural for real life signals. It may cause substantial filtering artifacts in the vicinity of signal borders. With cyclic signal replication that is assumed when the DFT is used, signal samples situated near opposite signal borders are regarded as near neighbors though, in fact, they are separated by the entire signal extent. In particular, strong discontinuities on signal borders may cause substantial Gibbs effects in filtering. These cyclic convolution artifacts can be reduced by means of signal extension by linear or other interpolation between opposite signal borders described in the previous section (Eq. 5.1.46 and Fig. 5-8). A more radical solution of the problem of border artifacts is signal mirroring which leads to the signal convolution in the DCT domain.

5.2.2 Signal convolution in the DCT domain

Let signal $\{\tilde{a}_k\}$ is obtained from signal $\{a_k\}$ of N samples by its extension by mirror reflection and periodical replication of the result with a period of $2N$ samples:

$$\tilde{a}_{(k)\bmod 2N} = \begin{cases} a_k, & k = 0,1,\dots,N-1; \\ a_{2N-k-1}, & k = N, N=1,\dots,2N-1 \end{cases}. \tag{5.2.4}$$

and let $\{h_n\}$ is a convolution kernel of N samples ($n = 0,1,\dots,N-1$). Then

$$\tilde{c}_k = \sum_{n=0}^{N-1} h_n \tilde{a}_{(k-n+[N/2])\bmod 2N} \tag{5.2.5}$$

where

$$\left[\frac{N}{2}\right] = \begin{cases} N/2, & \text{for even } N \\ (N-1)/2, & \text{for odd } N \end{cases} \tag{5.2.6}$$

is a convolution of extended, by mirror reflection, signal $\{a_k\}$ with kernel $\{h_n\}$. It will coincide with the cyclic convolution of period $2N$:

$$c_{(k)\bmod 2N} = \sum_{n=0}^{N-1} \tilde{h}_{(n)\bmod 2N} \tilde{a}_{(k-n+[N/2])\bmod 2N} \tag{5.2.7}$$

for kernel

$$\tilde{h}_{(n)\bmod 2N} = \begin{cases} 0 \quad , & n = 0,...,[N/2]-1 \\ h_{n-[N/2]}, & n = [N/2],...,[N/2]+N-1 \\ 0 \quad , & [N/2]+N,...,2N-1 \end{cases} . \tag{5.2.8}$$

The cyclic convolution described by Eqs. 5.2.7 and 8 is illustrated in Fig. 5-9.

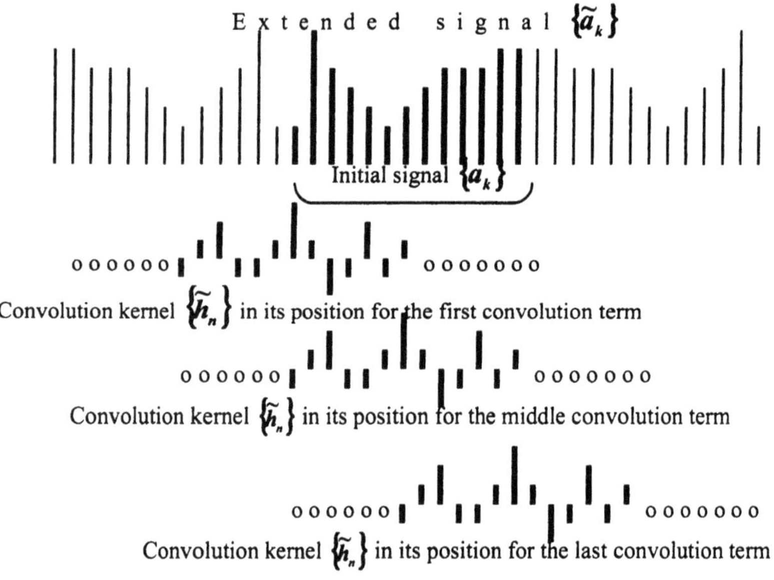

Figure 5-9. Cyclic convolution of a signal extended by its mirror reflection from its borders

As it follows from Eq. 4.2.27 for shifted cyclic convolution, for such signals of double length $2N$, one can retain advantages of computing convolution with the use of a FFT if one DFT in the Fourier domain convolution algorithm described by Eq. 5.2.2 is replaced by SDFT(u,v) with shift parameters $u = 1/2$ and $v = 0$:

$$\mathbf{c} = \mathbf{ISDFT}_{1/2,0}\left\{\mathbf{SDFT}_{1/2,0}\{\tilde{\mathbf{a}}\} \bullet \mathbf{DFT}\{\tilde{\mathbf{h}}\}\right\} \tag{5.2.9}$$

It was shown in Sect. 4.3.3, that for signals extended to their double length by mirror reflection, SDFT$(1/2,0)$ is reduced to the Discrete Cosine Transform. Therefore Eq. 5.2.9 is equivalent to the equation:

$$\mathbf{c} = \mathbf{ISDFT}_{1/2,0}\left\{\mathbf{DCT}\{\mathbf{a}\} \bullet \mathbf{DFT}\{\tilde{\mathbf{h}}\}\right\}, \tag{5.2.10}$$

where vector $\{\alpha_r^{DCT}\} = \mathbf{DCT}\{\mathbf{a}\}$ of DCT transform coefficients is assumed to be extended to its double length according to its symmetry property (Eq. 4.3.59):

$$\alpha_r^{DCT} = -\alpha_{2N-r}^{DCT};\ \alpha_N^{DCT} = 0. \tag{5.2.11}$$

In virtue of this symmetry inverse SDFT $ISDFT_{1/2,0}$ for computing convolution according to Eq. 5.2.7 is reduced to inverse DCT and DCsT transforms:

$$\begin{aligned}
c_k &= \frac{1}{\sqrt{2N}} \sum_{r=0}^{2N-1} \alpha_r^{DCT} \tilde{\eta}_r \exp\left[-i2\pi \frac{(k+1/2)r}{2N}\right] = \\
&\frac{1}{\sqrt{2N}} \left\{ \sum_{r=0}^{N-1} \alpha_r^{DCT} \tilde{\eta}_r \exp\left[-i2\pi \frac{(k+1/2)r}{2N}\right] + \right. \\
&\left. \sum_{r=N}^{2N-1} \alpha_r^{DCT} \tilde{\eta}_r \exp\left[-i2\pi \frac{(k+1/2)r}{2N}\right] \right\} = \\
&\frac{1}{\sqrt{2N}} \left\{ \alpha_0^{DCT} \eta_0 + \sum_{r=1}^{N-1} \alpha_r^{DCT} \left[\eta_r \exp\left(-i\pi \frac{(k+1/2)}{N} r \right) \right] + \right. \\
&\left. \sum_{r=1}^{N-1} \alpha_r^{DCT} \eta_r^* \exp\left[i\pi \frac{(k+1/2)}{N} r \right] \right\} = \\
&\frac{1}{\sqrt{2N}} \left\{ \alpha_0^{DCT} \eta_0 + 2 \sum_{r=1}^{N-1} \alpha_r^{DCT} \eta_r^{re} \cos\left[\pi \frac{(k+1/2)}{N} r \right] \right. \\
&\left. -2 \sum_{r=1}^{N-1} \alpha_r^{DCT} \eta_r^{im} \sin\left[\pi \frac{(k+1/2)}{N} r \right] \right\},
\end{aligned} \tag{5.2.12}$$

where $\{\eta_r^{re}\}$ and $\{\eta_r^{im}\}$ are real and imaginary parts of DFT coefficients $\{\eta_r\}$ of interpolation kernel $\{\tilde{h}_k\}$:

$$\eta_r = \frac{1}{\sqrt{N}} \sum_{k=0}^{2N-1} \tilde{h}_k \exp\left(i2\pi \frac{kr}{2N} \right), \tag{5.2.13}$$

and $\{\eta_r^*\}$ are complex conjugates to $\{\eta_r\}$.

The computational complexity of computing signal convolution in the DCT domain by Eq. 5.2.10 is, by the order of magnitude, the same as that of the above DFT domain convolution algorithm. However fast algorithms for DCT, IDCT and IDST exist [5-12] that require, in fact, even less computations than FFT and IFFT involved in the DFT domain algorithm.

5.3 COMBINED ALGORITHMS FOR COMPUTING DFT AND DCT OF REAL VALUED SIGNALS

As it was already mentioned, for computing the DFT and SDFT, FFT algorithms are used that substantially reduce the computational burden of digital filtering. The use of FFT assumes that data are complex numbers (see Sect. 6.1). However in many image processing and digital holography applications data are represented by real numbers. For direct application of FFT algorithms to such data one should treat them as complex numbers with zero imaginary part. This leads to additional computation and memory expenditures. The DCT deals with real numbers and there exist special fast DCT algorithms that are described in Sect. 6.3.6. However the DCT can also be computed with FFT algorithms in virtue of the link between the DCT and SDFT(1/2,0). Direct use of a FFT for computing the DCT exhibits four-fold redundancy: two-fold redundancy owing to the fact that data are real numbers and additional two-fold redundancy owing to the signal symmetrization. Combined algorithms described in this section enable avoiding these unnecessary expenditures.

5.3.1 Combined algorithms for computing DFT of signals with real samples

Let $\{a_k = a_k^*\}$ is signal with N real samples and $\{\alpha_r\}$ is a set of its complex DFT coefficients. According to Eq. 4.3.42, $\{\alpha_r\}$ exhibit complex conjugate symmetry

$$\{\alpha_r = \alpha_{N-r}^*\} \tag{5.3.1}$$

This means that one needs to compute only half of the coefficients $\{\alpha_r\}$, the other half can then be found from Eq. 5.3.1. However FFT algorithms are capable of efficient computing of only all transform coefficient vector as a whole unless more complex and less widespread "pruned" transforms (See Sect. 6.4) are available. One can avoid this limitation and efficiently exploit the spectrum symmetry for reducing the number of operations needed for computing signal DFT spectrum with conventional FFT algorithms. Two ways are available to this goal: (i) by simultaneous computing spectra of two signals with real samples and (ii) by splitting one signal with real samples into two signals, computing their spectra and then combining them into the spectrum of the initial signal.

Consider the first method. Let $\{a_k = a_k^*\}$ and $\{b_k = b_k^*\}$ are two signals with N real samples and $\{\alpha_r = \alpha_{N-r}^*\}$ and $\{\beta_r = \beta_{N-r}^*\}$ are their respective DFT spectra. For an auxiliary signal with complex samples:

$$c_k = a_k + ib_k \tag{5.3.2}$$

and compute its DFT spectrum:

$$\{\gamma_r\} = \mathbf{DFT}\{c_k\} = \mathbf{DFT}\{a_k\} + i\mathbf{DFT}\{b_k\} = \{\alpha_r + i\beta_r\} \tag{5.3.3}$$

Consider now

$$\{\gamma_{N-r}^*\} = \{\alpha_{N-r}^* - i\beta_{N-r}^*\} = \{\alpha_r - i\beta_r\} \tag{5.3.4}$$

and obtain that spectra $\{\alpha_r\}$ and $\{\beta_r\}$ can be found as:

$$\{\alpha_r\} = \{(\gamma_r + \gamma_{N-r}^*)/2\}; \tag{5.3.5}$$

$$\{\beta_r\} = \{(\gamma_r - \gamma_{N-r}^*)/2i\}. \tag{5.3.6}$$

Flow chart of this combined algorithm is shown in Fig. 5-10 a.

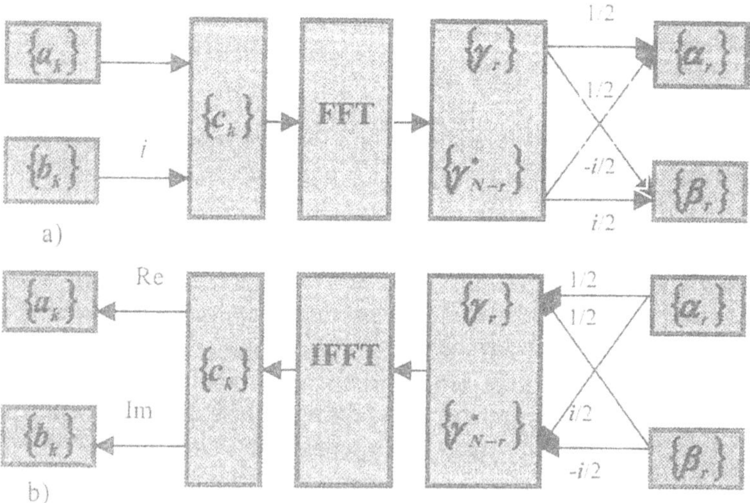

Figure 5-10. Combined algorithms for simultaneous computing direct and inverse DFT of two signals via FFT

Reversion of this algorithm as it is shown in Fig. 5-10 b) gives a combined algorithm for simultaneous inverse DFT of spectra of two signals with real samples.

Consider now the second method. Let $\{a_k\}$ is a signal with $2N$ real samples. Its DFT can be computed as

$$
\begin{aligned}
\alpha_r &= \frac{1}{\sqrt{2N}}\sum_{k=0}^{2N-1} a_k \exp\left(i2\pi\frac{kr}{2N}\right) = \\
&\frac{1}{\sqrt{2N}}\left[\sum_{k=0}^{N-1} a_{2k} \exp\left(i2\pi\frac{2kr}{2N}\right) + \sum_{k=0}^{N-1} a_{2k+1} \exp\left(i2\pi\frac{2k+1}{2N}r\right)\right] = \\
&\frac{1}{\sqrt{2}}\left\{\frac{1}{\sqrt{N}}\sum_{k=0}^{N-1} a_{2k} \exp\left(i2\pi\frac{kr}{N}\right) + \right. \\
&\left.\left[\frac{1}{\sqrt{N}}\sum_{k=0}^{N-1} a_{2k+1} \exp\left(i2\pi\frac{kr}{N}\right)\right]\exp\left(i\pi\frac{r}{2N}\right)\right\} = \\
&\frac{1}{\sqrt{2}}\left[\mathrm{DFT}\{a_{2k}\} + \exp\left(i\pi\frac{r}{N}\right)\cdot \mathrm{DFT}\{a_{2k+1}\}\right].
\end{aligned}
\tag{5.3.7}
$$

Therefore computing the DFT of the signal can be reduced to computing two DFTs of two auxiliary signals formed from its even and odd samples. The latter can be carried out by the above-described combined algorithm.

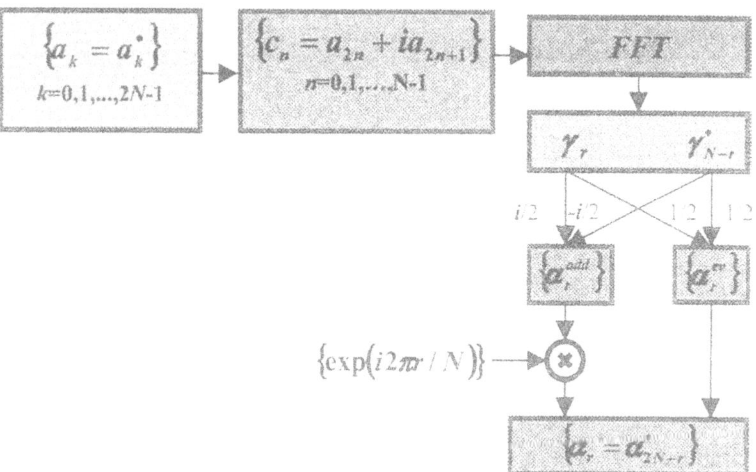

Figure 5-11. Flow diagram of the combined algorithm for computing DFT of a signal with even number of real samples

2-D DFT is usually computed as two successive passes as raw wise and column wise 1-D DFTs. The above combined algorithms can be used at the first pass, and the first algorithm is well suited for simultaneous computing 1-D DFT of pairs of image rows. At the second (column-wise) pass, data are complex. However, in virtue of their complex conjugate symmetry, additional saving is possible as schematic diagram of Fig. 5-12 shows. Because 1st and (*N*/2+1)-th columns of 2-D array of 1-D row spectra obtained at the first pass, where *N* is the number of image columns, are real numbers, their column-wise spectra can be computed simultaneously with the combined algorithm. Then one needs to additionally compute 1-D DFTs of *N*/2-1 columns beginning from the second one.

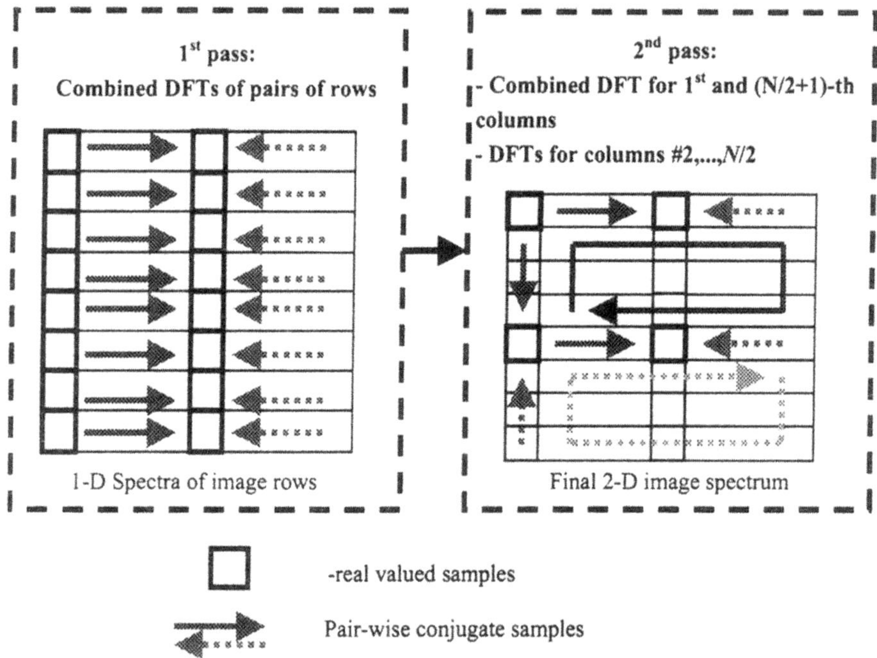

Figure 5-12. Flow diagram of two pass 2-D DFT with the use of combined algorithms

5.3.2 Combined algorithm for computing the DCT via FFT

Though special fast algorithms for computing the DCT are available (Sec. 6.4), it can also be computed using FFT packages that are more widespread. This option makes use of the fact that the DCT is SDFT(1/2,0) of signals with even symmetry:

$$\{a_k = a_{2N-k-1}\},\ k = 0,1,...,N-1 \tag{5.3.8}$$

From symmetry properties of SDFTs (Sect. 4.3.2) it follows that, for SDFT(1/2,0)

$$\{a_k = a_{2N-k-1}\} \leftrightarrow SDFT(1/2,0) \leftrightarrow \{\alpha_r^{1/2,0} = -\alpha_{2N-r}^{1/2,0} = -\alpha_{2N+r}^{1/2,0}\} \tag{5.3.9}$$

and

$$\left\{a_k \exp\left[i2\pi \frac{N(k+1/2)}{2N}\right] = i(-1)^k a_k\right\} \leftrightarrow SDFT(1/2,0) \leftrightarrow \{\alpha_{r+N}^{1/2,0}\} \tag{5.3.10}$$

These relationships suggest the following method for simultaneous computing the DCT of four signals. Let $\{a_{1,k}\}$, $\{a_{2,k}\}$, $\{a_{3,k}\}$, $\{a_{4,k}\}$ are four signals with N real samples. Form auxiliary signals

$$\begin{aligned} &b_k = a_{1,k} + ia_{2,k}; \qquad c_k = a_{3,k} + ia_{4,k}; \\ &\tilde{b}_k = \begin{cases} b_k, & k = 0,1,...,N-1 \\ b_{2N-k-1}, & k = N, N+1,...,2N-1 \end{cases}; \\ &\tilde{c}_k = \begin{cases} c_k, & k = 0,1,...,N-1 \\ c_{2N-1-k}, & k = N, N+1,...,2N-1 \end{cases} \end{aligned} \tag{5.3.11}$$

and

$$z_k = \tilde{b}_k + i(-1)^k \tilde{c}_k \tag{5.3.12}$$

and compute SDFT(1/2,0) of the composite signal $\{z_k\}$. Then obtain, with an account for Eq. 5.2.10:

$$\begin{aligned} &\zeta_r^{(1/2,0)} = \mathbf{SDFT}_{1/2,0}\{z_k\} = \mathbf{SDFT}_{1/2,0}\{\tilde{b}_k\} + \mathbf{SDFT}_{1/2,0}\{i(-1)^k \tilde{c}_k\} = \\ &\tilde{\beta}_r^{(1/2,0)} + \tilde{\gamma}_{N+r}^{(1/2,0)} \end{aligned} \tag{5.3.13}$$

From this and the symmetry property (Eq. 5.2.9) two equations follow:

$$\begin{aligned} &\zeta_r^{(1/2,0)} = \tilde{\beta}_r^{(1/2,0)} - \tilde{\gamma}_{N-r}^{(1/2,0)} \\ &\zeta_{2N-r}^{(1/2,0)} = \tilde{\beta}_{2N-r}^{(1/2,0)} + \tilde{\gamma}_{3N-r}^{(1/2,0)} = -\tilde{\beta}_r^{(1/2,0)} - \gamma_{N-r}^{(1/2,0)}. \end{aligned} \tag{5.3.14}$$

Therefore

$$\tilde{\beta}_r^{(1/2,0)} = \left(\varsigma_r^{(1/2,0)} - \varsigma_{2N-r}^{(1/2,0)}\right)/2; \qquad (5.3.15)$$

$$\tilde{\beta}_{N-r}^{(1/2,0)} = \left(\varsigma_r^{(1/2,0)} + \varsigma_{2N-r}^{(1/2,0)}\right)/2. \qquad (5.3.16)$$

The latter equation can be written alternatively as

$$\tilde{\beta}_r^{(1/2,0)} = \left(\varsigma_{N+r}^{(1/2,0)} + \varsigma_{N-r}^{(1/2,0)}\right)/2. \qquad (5.3.17)$$

Finally we obtain from Eqs. 5.3.15, 16 and 11 that DCT spectra of signals $\{a_{1,k}\}$, $\{a_{2,k}\}$, $\{a_{3,k}\}$, $\{a_{4,k}\}$ can be found as:

$$\begin{aligned}
\alpha_{1,r}^{DCT} &= \mathbf{DCT}\{a_{1,k}\} = \mathrm{Re}\left(\varsigma_r^{(1/2,0)} - \varsigma_{2N-r}^{(1/2,0)}\right)/2; \\
\alpha_{2,r}^{DCT} &= \mathbf{DCT}\{a_{2,k}\} = \mathrm{Im}\left(\varsigma_r^{(1/2,0)} - \varsigma_{2N-r}^{(1/2,0)}\right)/2; \\
\alpha_{3,r}^{DCT} &= \mathbf{DCT}\{a_{3,k}\} = \mathrm{Re}\left(\varsigma_{N+r}^{(1/2,0)} + \varsigma_{N-r}^{(1/2,0)}\right)/2; \\
\alpha_{4,r}^{DCT} &= \mathbf{DCT}\{a_{4,k}\} = \mathrm{Im}\left(\varsigma_{N+r}^{(1/2,0)} + \varsigma_{N-r}^{(1/2,0)}\right)/2.
\end{aligned} \qquad (5.3.18)$$

Block diagram of this combined algorithm is shown in Fig. 5-13.

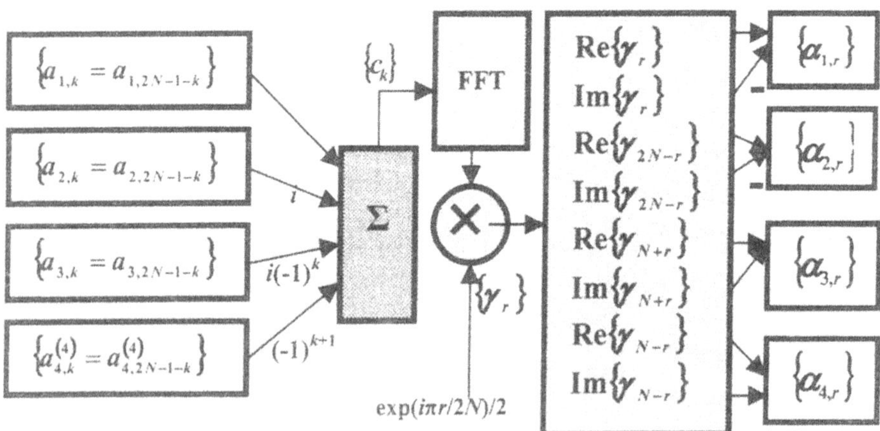

Figure 5-13. Block diagram of the combined algorithm for computing DCT of four signals

References

1. C. de Boor, A Practical Guide to Splines, Springer Verlag, N.Y., 1978
2. M. Unser, A. Aldroubi, "Polynomial Splines and Wavelets - A Signal Processing Perspective, In: Wavelets - A Tutorial in Theory and Applications, C. K. Chui, Ed., San Diego, CA, Academic, 1992, pp. 91-122
3. R. Y. Vitkus and L.P. Yaroslavsky, "Recursive Algorithms for Local Adaptive Linear Filtration", in *Mathematical Research.*, L.P. Yaroslavskii, A. Rosenfeld and W. Wilhelmi, Ed., Berlin: Academy Verlag, pp. 34-39, 1987.
4. Jiangtao Xi, J. F. Chicharo, Computing Running DCT's and DST's based on Their Second Order Shift Properties, IEEE Trans. On Circuits and Systems –1: Fundamental Theory and Applications, vol. 47, No. 5, May 2000, pp. 779-783
5. Z. Wang, "Fast algorithms for the discrete W transform and for the Discrete Fourier Transform", IEEE Trans. ASSP, vol. ASSP-32, pp. 803-816, Aug. 1984
6. H.S. Hou, "A fast recursive algorithm for computing the Discrete Cosine Transform", IEEE Trans. on ASSP, vol. ASSP-35, No. 10, Oct. 1987, p. 1455-1461
7. Z. Wang, "A simple structured algorithm for the DCT, in *Proc 3rd Ann. Conf. Signal Process. Xi'an*, China, Nov 1988, pp.28-31 (in Chinese).
8. A. Gupta, K. R. Rao, "A fast recursive algorithm for the discrete sine transform", IEEE Trans. on ASSP, vol. ASSP-38, No. 3, March 1990, pp. 553-557.
9. Z. Wang, "Pruning the Fast Discrete Cosine Transform", IEEE Transactions on communications, Vol. 39, No. 5, May 1991.
10. Z. Cvetkovic, M. V. Popovic, "New fast recursive algorithms for the computation of Discrete Cosine and Sine Transforms", IEEE Trans. Signal Processing, vol. 40, No. 8, August 1992, pp. 2083-2086.
11. J. Ch. Yao, Ch-Y. Hsu, "Further results on " New fast recursive algorithms for the discrete cosine and sine transforms", IEEE Trans. on Signal Processing, vol. 42, No. 11, Nov. 1994, pp. 3254-3255
12. L. Yaroslavsky, A. Happonen, Y. Katyi, "Signal discrete sinc-interpolation in DCT domain: fast algorithms", *SMMSP 2002, Second International Workshop on Spectral Methods and Multirate Signal Processing*, Toulouse (France), 07.09.2002 - 08.09.2002.

Chapter 6

FAST TRANSFORMS

6.1 THE PRINCIPLE OF FAST FOURIER TRANSFORMS

Fast Fourier Transforms (FFT) are algorithms for fast computation of the DFT. The principle of the FFT can easily be understood if one compares 1-D and 2-D DFT. Let $\{a_n^{(1)}\}$ and $\{a_{k,l}^{(2)}\}$ be 1-D and 2-D arrays with the same number $N = N_1 N_2$ of samples: $n = 0,1,...,N_1 N_2 - 1$; $k = 0,1,...,N_1 - 1$; $l = 0,1,...,N_2 - 1$. Direct computing of the 1-D DFT of array $\{a_n^{(1)}\}$

$$\alpha_r = \frac{1}{\sqrt{N}} \sum_{n=0}^{N-1} a_n^{(1)} \exp\left(i2\pi \frac{nr}{N} \right) \qquad (6.1.1)$$

requires $N^2 = N_1^2 N_2^2$ operations with complex numbers while computing the 2-D DFT of array $\{a_{k,l}^{(2)}\}$

$$\alpha_{r,s} = \frac{1}{\sqrt{N}} \sum_{k=0}^{N_1-1} \sum_{l=0}^{N_2-1} a_{k,l}^{(2)} \exp\left[i2\pi \left(\frac{kr}{N_1} + \frac{ls}{N_2} \right) \right] \qquad (6.1.2)$$

requires only $N_1^2 N_2 + N_1 N_2^2 = N_1 N_2 (N_1 + N_2)$ operations as the 2-D DFT is separable into two 1-D DFTs:

$$\alpha_{r,s} = \frac{1}{\sqrt{N}} \sum_{k=0}^{N_1-1} \exp\left[i2\pi \frac{kr}{N_1}\right] \sum_{l=0}^{N_2-1} a_n \exp\left[i2\pi \frac{ls}{N_2}\right] \qquad (6.1.3)$$

Therefore one can accelerate computing the DFT by representing it in a separable multi-dimensional form. One can do it if the size of the array is a composite number. Let, as in the above example, $N = N_1 N_2$. Represent indices k and r of signal and its transform samples as two-dimensional ones:

$$\begin{aligned} &k = k_2 N_1 + k_1; \quad k_1 = 0,1,...,N_1 - 1; \quad k_2 = 0,1,...,N_2 - 1; \\ &r = r_1 N_2 + r_2; \quad r_1 = 0,1,...,N_1 - 1; \quad r_2 = 0,1,...,N_2 - 1; \end{aligned} \qquad (6.1.4)$$

Then the 2-D DFT in Eq. 6.1.2 can be split into two successive 1-D DFTs:

$$\begin{aligned} \alpha_{r_1,r_2} &= \frac{1}{\sqrt{N_1 N_2}} \sum_{k_2=0}^{N_2-1} \sum_{k_1=0}^{N_1-1} a_{k_1,k_2} \exp\left[i2\pi \frac{(k_2 N_1 + k_1)(r_1 N_2 + r_2)}{N_2 N_1}\right] = \\ &\frac{1}{\sqrt{N}} \sum_{k_1=0}^{N_1-1} \exp\left(i2\pi \frac{k_1 r_1}{N_1}\right) \left[\exp\left(i2\pi \frac{k_1 r_2}{N}\right) \sum_{k_2=0}^{N_2-1} a_{k_1,k_2} \exp\left(i2\pi \frac{k_2 r_2}{N_2}\right)\right] = \\ &\mathbf{DFT}_{N_1}\left\{\exp\left(i2\pi \frac{k_1 r_2}{N}\right) \bullet \mathbf{DFT}_{N_2}\left\{a_{k_1,k_2}\right\}\right\}. \end{aligned} \qquad (6.1.5)$$

This computation scheme is illustrated in flow diagram of Fig. 6-1.

Obviously, the larger is the number of factors of N the higher is the dimensionality to which 1-D DFT can be decomposed in this way, and the higher is the computational complexity reduction. For n factors of $N = N_0 \cdot N_1 \cdot,...,\cdot N_{n-1}$ the DFT can be computed with only $N \cdot (N_0 + N_1 +,...,+ N_{n-1})$ operations with complex numbers instead of $N^2 = N(N_0 \cdot N_1 \cdot,...,\cdot N_{n-1})$ operations required for the direct computation. For $N = 2^n$ computational complexity of the DFT is $O(N \log_2 N)$. Fast Fourier Transforms for signal size that is a power of 2 are called ***radix-2 FFT***. They are most widespread in the software packages for signal processing.

FFT algorithms are usually illustrated in the form of graphs. A typical graph of an 8 points radix 2 FFT is shown in Fig. 6-2. Signs "-" at graph nodes denote that values at the correspondent graph edges are subtracted. Circles at the nodes signify multiplication by exponential "twiddle" factors (see Eq. 6.1.5).

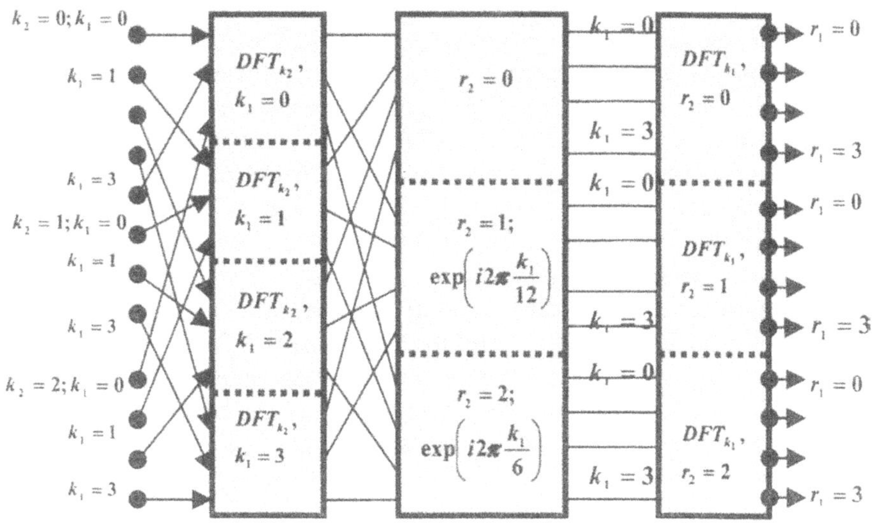

Figure 6-1. Flow diagram of FFT for $N = 12 = 3 \times 4$

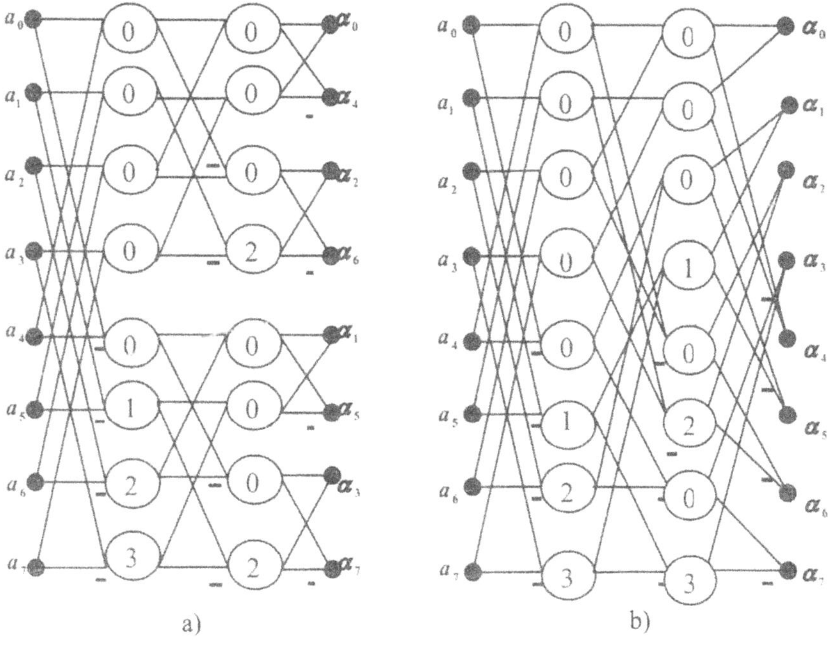

Figure 6-2. Flow charts of an "in place" FFT (lcfl) and of FFT without permutations; numbers in circles are values of index m in exponential twiddle factors $\exp(i2\pi m/8)$

One can see on this graph that the computations are split into $\log_2 8 = 3$ stages and are carried out in each stage "in-place", without a need to allocate separate computer memory storage for intermediate arrays of values. One can also note that output spectral data are arranged in this algorithm in a bit-reversed order (see also inverse order of two-component representation of signal and its spectrum indices in Eq. 6.1.4). The bit-reversed order is the order sequential for binary codes of indices written in reversed order of binary digits. Such an arrangement of output data may require an additional computational stage for "***bit reversal***" data reshuffling into their natural order. Algorithms with bit-reversed order of spectral coefficients are, in a signal processing jargon, called ***"decimation-in-time" FFT*** algorithms. By means of left-to-right mirroring the graph a ***"decimation-in-frequency" FFT*** graph with bit reversed ordering of input data is obtained. The data or spectra reshuffling can be avoided by incorporating them into the graph as it is illustrated in the graph of Fig. 6-2, b). The FFT algorithms with natural ordering of input and output data do not enable "in-place" computation.

FFT algorithms are a special case of a broad class of fast transform algorithms that have similar structure and whose computational complexity is $O(N \log N)$ or lower.

6.2 MATRIX TECHNIQUES IN FAST TRANSFORMS

Graph representation of fast algorithms illustrated in Fig. 6.2 enables visualizing the algorithms for any particular transform size and helps to understand the algorithm structure. Matrix techniques are compact and general means for describing transformation and deriving fast transform algorithms in a way that permits automatic conversion of the algorithm formal description into a program code and facilitates evaluation of its computational complexity ([1]).

In matrix interpretation, signal transformations are treated as multiplication of a vector-column composed of signal samples by a transform matrix and fast transform algorithms are interpreted as factorization of transform matrix into a product of sparse matrices.

We will describe the matrix techniques using the following denotations.

Matrices:

- Matrix $\mathbf{M}_{V,H}$ is a rectangle array of $V \times H$ numbers $\boldsymbol{\mu}_{v,h}$:

$$\mathbf{M}_{V,H} = \begin{bmatrix} \mu_{0,0} & \mu_{0,1} & \cdots & \mu_{0,H-1} \\ \mu_{1,0} & \mu_{1,1} & \cdots & \cdots \\ \cdots & \cdots & \cdots & \cdots \\ \mu_{V-1,0} & \cdots & \cdots & \mu_{V-1,H-1} \end{bmatrix};$$

- ***Zero - matrix*** $\mathbf{0}_{V,H}$ is a matrix of $V \times H$ zeros;
- ***Ones - matrix*** $\mathbf{1}_{V,H}$ is a matrix of $V \times H$ ones;
- "***Bit***"-***matrix*** $[\mathbf{10}]$ is a vector-row $[1 \quad 0]$;
- Reverse "***Bit***"-***matrix*** $[\mathbf{01}]$ is a vector-row $[0 \quad 1]$;
- ***Delta-matrix*** is a vector-row

$$\mathbf{\Delta}_S^{(s_0)} = \{\delta(s - S_0)\},$$

where $\delta(\cdot)$ is Kronecker delta and $s = 0,1,...,S-1$.

- $[\mathbf{11}] = [\mathbf{10}] + [\mathbf{01}] = [1 \quad 1]$;
- $[\mathbf{1}\bar{\mathbf{1}}] = [\mathbf{10}] - [\mathbf{01}] = [1 \quad -1]$;
- ***Identity matrix*** $\mathbf{I}_S$ is a square $S \times S$ matrix

$$\mathbf{I}_s = \begin{bmatrix} 1 & 0 & \cdots & 0 \\ 0 & 1 & \cdots & \cdots \\ \cdots & \cdots & \cdots & 0 \\ 0 & \cdots & 0 & 1 \end{bmatrix};$$

- ***Conjugate identity matrix*** $\mathbf{I}^*$ is an operator of converting vectors to their complex conjugate: $\mathbf{I}^* \cdot \mathbf{a} = \mathbf{a}^*$
- ***Reverse identity matrix*** $\bar{\mathbf{I}}_S$ is a square $S \times S$ matrix:

$$\bar{\mathbf{I}}_S = \begin{bmatrix} 0 & \cdots & 0 & 1 \\ \cdots & \cdots & 1 & 0 \\ 0 & \cdots & \cdots & \cdots \\ 1 & 0 & \cdots & 0 \end{bmatrix};$$

Multiplied by a vector, it changes the order of vector elements.

- ***Hadamard matrix of order 2***:

$$\mathbf{h}_2 = \begin{bmatrix} 1 & 1 \\ 1 & -1 \end{bmatrix}.$$

Matrix operations:

- ***Matrix sum of matrices***:

$$\sum_{k=0}^{N-1} \mathbf{M}_{V,H}^{(k)} = \begin{bmatrix} \sum_{k=0}^{N-1} \mu_{0,0}^{(k)} & \sum_{k=0}^{N-1} \mu_{0,1}^{(k)} & \cdots & \sum_{k=0}^{N-1} \mu_{0,H-1}^{(k)} \\ \sum_{k=0}^{N-1} \mu_{1,0}^{(k)} & \sum_{k=0}^{N-1} \mu_{1,1}^{(k)} & \cdots & \cdots \\ \cdots & \cdots & \cdots & \cdots \\ \sum_{k=0}^{N-1} \mu_{V-1,0}^{(k)} & \cdots & \cdots & \sum_{k=0}^{N-1} \mu_{V-1,H-1}^{(k)} \end{bmatrix}.$$

- ***Matrix product of two of matrices***:

$$\mathbf{M}_{R,S} = \mathbf{M}_{R,Q}^{(1)} \cdot \mathbf{M}_{Q,S}^{(2)} = \begin{bmatrix} \sum_{q=0}^{Q-1} \mu_{0,q}^{(1)} \mu_{q,0}^{(2)} & \sum_{q=0}^{Q-1} \mu_{0,q}^{(1)} \mu_{q,1}^{(2)} & \cdots & \sum_{q=0}^{Q-1} \mu_{0,q}^{(1)} \mu_{q,S-1}^{(2)} \\ \sum_{q=0}^{Q-1} \mu_{1,q}^{(1)} \mu_{q,0}^{(2)} & \sum_{q=0}^{Q-1} \mu_{1,q}^{(1)} \mu_{q,1}^{(2)} & \cdots & \cdots \\ \cdots & \cdots & \cdots & \cdots \\ \sum_{q=0}^{Q-1} \mu_{V-1,q}^{(1)} \mu_{q,0}^{(2)} & \cdots & \cdots & \sum_{q=0}^{Q-1} \mu_{V-1,q}^{(1)} \mu_{q,S-1}^{(2)} \end{bmatrix};$$

- ***Inverse matrix*** $\mathbf{M}^{-1}$ to matrix $\mathbf{M}$ is defined by the equation

$$\mathbf{M}^{-1} \cdot \mathbf{M} = \mathbf{I};$$

- ***Matrix transposition***:

$$
\left(\mathbf{M}_{V,H}\right)^{tr} = \begin{bmatrix} \mu_{0,0} & \mu_{0,1} & \cdots & \mu_{0,H-1} \\ \mu_{1,0} & \mu_{1,1} & \cdots & \cdots \\ \cdots & \cdots & \cdots & \cdots \\ \mu_{V-1,0} & \cdots & \cdots & \mu_{V-1,H-1} \end{bmatrix}^{tr} = \begin{bmatrix} \mu_{0,0} & \mu_{1,0} & \cdots & \mu_{V-1,0} \\ \mu_{0,1} & \mu_{1,1} & \cdots & \cdots \\ \cdots & \cdots & \cdots & \cdots \\ \mu_{0,H-1} & \cdots & \cdots & \mu_{V-1,H-1} \end{bmatrix};
$$

- ***Direct matrix sum*** of two matrices:

$$
\mathbf{M}^{(1)}_{V_1,H_1} \oplus \mathbf{M}^{(2)}_{V_2,H_2} = \begin{bmatrix} \mathbf{M}^{(1)}_{V_1,H_1} & \mathbf{0}_{V_1,H_2} \\ \mathbf{0}_{V_2,H_1} & \mathbf{M}^{(2)}_{V_2,H_2} \end{bmatrix};
$$

- ***Direct matrix sum*** of N matrices:

$$
\mathop{\oplus}_{k=0}^{N-1} \mathbf{M}^{(k)}_{V_k,H_k} = \begin{bmatrix} \mathbf{M}^{(0)}_{V_0,H_0} & \mathbf{0}_{V_0,H_1} & \cdots & \mathbf{0}_{V_0,H_{N-1}} \\ \mathbf{0}_{V_1,H_0} & \mathbf{M}^{(1)}_{V_1,H_1} & \cdots & \cdots \\ \cdots & \cdots & \cdots & \cdots \\ \mathbf{0}_{V_{N-1},H_0} & \cdots & \cdots & \mathbf{M}^{(1)}_{V_{N-1},H_{N-1}} \end{bmatrix};
$$

- ***Vertical*** (row-wise) ***matrix sum*** of two matrices

$$
\mathbf{M}^{(1)}_{V_1,H} \,\Xi\, \mathbf{M}^{(2)}_{V_2,H} = \begin{bmatrix} \mathbf{M}^{(1)}_{V_1,H} \\ \mathbf{M}^{(2)}_{V_2,H} \end{bmatrix};
$$

- ***Vertical*** (row-wise) ***matrix sum*** of N matrices:

$$
\mathop{\Xi}_{k=0}^{N-1} \mathbf{M}^{(k)}_{V_2,H} = \begin{bmatrix} \mathbf{M}^{(0)}_{V_0,H} \\ \mathbf{M}^{(1)}_{V_1,H} \\ \cdots \\ \mathbf{M}^{(N-1)}_{V_{N-1},H} \end{bmatrix};
$$

- ***Horizontal*** (column-wise) ***matrix sum*** of two matrices:

$$
\mathbf{M}^{(1)}_{V,H_1} \,[|]\, \mathbf{M}^{(2)}_{V,H_2} = \begin{bmatrix} \mathbf{M}^{(1)}_{V,H_1} & \mathbf{M}^{(2)}_{V,H_2} \end{bmatrix};
$$

- ***Horizontal*** (column-wise) ***matrix sum*** of N matrices:

$$\prod_{k=0}^{N-1} \mathbf{M}^{(k)}_{V,H_k} = \left[\mathbf{M}^{(0)}_{V,H_0} \quad \mathbf{M}^{(1)}_{V,H_1} \ldots \quad \mathbf{M}^{(N-1)}_{V,H_{N-1}}\right];$$

- ***Kronecker product*** of two matrices:

$$\mathbf{M}^{(1)}_{V_1,H_1} \otimes \mathbf{M}^{(2)}_{V_2,H_2} = \begin{bmatrix} \mu^{(1)}_{0,0}\mathbf{M}^{(2)}_{V_2,H_2} & \mu^{(1)}_{0,1}\mathbf{M}^{(2)}_{V_2,G_2} & \ldots & \mu^{(1)}_{0,G_1-1}\mathbf{M}^{(2)}_{V_2,G_2} \\ \mu^{(1)}_{1,0}\mathbf{M}^{(2)}_{V_2,G_2} & \mu^{(1)}_{1,1}\mathbf{M}^{(2)}_{V_2,G_2} & \ldots & \ldots \\ \ldots & \ldots & \ldots & \ldots \\ \mu^{(1)}_{V_1-1,0}\mathbf{M}^{(2)}_{V_2,G_2} & \ldots & \ldots & \mu^{(1)}_{V_1-1,G_1-1}\mathbf{M}^{(2)}_{V_2,G_2} \end{bmatrix};$$

- ***Kronecker product*** of N matrices:

$$\bigotimes_{k=0}^{N-1} \mathbf{M}^{(k)}_{V_k,H_k} = \mathbf{M}^{(k)}_{V_0,H_0} \otimes \mathbf{M}^{(k)}_{V_1,H_1} \otimes \ldots \otimes \mathbf{M}^{(1)}_{V_{N-2},H_{N-2}} \otimes \mathbf{M}^{(0)}_{V_{N-1},H_{N-1}};$$

- ***N-th Kronecker power of a matrix*** (Kronecker product of N identical matrices):

$$\mathbf{M}^{\otimes N} = \bigotimes_{k=0}^{N-1} \mathbf{M}$$

- ***Kronecker matrix*** is matrix that is representable as a Kronecker product of other matrices.
- ***Layered Kronecker matrix*** is matrix that is representable as a vertical sum of Kronecker matrices.

For facilitating algebraic manipulations with matrices, the following matrix identities are useful ([1,2]):

$$\left(\bigoplus_{k=0}^{N-1} \mathbf{M}^{(1)}_{V_k^{(1)},H_k^{(1)}}\right) \cdot \left(\bigoplus_{k=0}^{N-1} \mathbf{M}^{(2)}_{V_k^{(2)},H_k^{(2)}}\right) = \bigoplus_{k=0}^{N-1} \mathbf{M}^{(1)}_{V_k^{(1)},H_k^{(1)}} \cdot \mathbf{M}^{(2)}_{V_k^{(2)},H_k^{(2)}} \tag{6.2.1}$$

$$\bigoplus_{k=0}^{N-1} \mathbf{M} = \mathbf{I}_N \otimes \mathbf{M}; \tag{6.2.2}$$

$$\left(\prod_{k=0}^{N-1} \mathbf{M}^{(1,k)}_S\right) \otimes \left(\prod_{k=0}^{N-1} \mathbf{M}^{(2,k)}_S\right) = \prod_{k=0}^{N-1} \left(\mathbf{M}^{(1,k)}_S \otimes \mathbf{M}^{(2,k)}_S\right) \tag{6.2.3}$$

$$\left(\bigotimes_{k=0}^{N-1}\mathbf{M}_{S_k}^{(1,k)}\right)\cdot\left(\bigotimes_{k=0}^{N-1}\mathbf{M}_{S_k}^{(2,k)}\right)=\bigotimes_{k=0}^{N-1}\mathbf{M}_{S_k}^{(1,k)}\cdot\mathbf{M}_{S_k}^{(2,k)}. \tag{6.2.4}$$

$$\mathop{\Xi}_{k=0}^{N-1}\mathbf{M}_{S_k}^{(1,k)}\otimes\mathbf{M}_S=\left(\mathop{\Xi}_{k=0}^{N-1}\mathbf{M}_{S_k}^{(1,k)}\right)\otimes\mathbf{M}_S. \tag{6.2.5}$$

$$\mathop{\Xi}_{k=0}^{N-1}\mathbf{R}_{1,H}\otimes\mathbf{M}_S^{(k)}=\mathbf{R}_{1,H}\otimes\left(\mathop{\Xi}_{k=0}^{N-1}\mathbf{M}_S^{(k)}\right). \tag{6.2.6}$$

$$\mathop{\Xi}_{k=0}^{N-1}\left(\mathbf{M}_{R_k,Q}^{(1,k)}\cdot\mathbf{M}_{Q,P}^{(2)}\right)=\left(\mathop{\Xi}_{k=0}^{N-1}\mathbf{M}_{R_k,Q}^{(1,k)}\right)\cdot\mathbf{M}_{Q,P}^{(2)}. \tag{6.2.7}$$

$$\mathop{\Xi}_{k=0}^{N-1}\left(\mathbf{M}_{R,Q}^{(1)}\cdot\mathbf{M}_{Q,P}^{(2,k)}\right)=\left(\mathbf{I}_N\otimes\mathbf{M}_{R,Q}^{(1)}\right)\left(\mathop{\Xi}_{k=0}^{N-1}\mathbf{M}_{Q,P}^{(2,k)}\right). \tag{6.2.8}$$

$$\left(\mathbf{M}^{(1)}\cdot\mathbf{M}^{(2)}\right)^{-1}=\left(\mathbf{M}^{(2)}\right)^{-1}\cdot\left(\mathbf{M}^{(1)}\right)^{-1}. \tag{6.2.9}$$

$$\left(\bigoplus_{k=0}^{N-1}\mathbf{M}^{(k)}\right)^{-1}=\vec{\bigoplus_{k=0}^{N-1}}\left(\mathbf{M}^{(k)}\right)^{-1}. \tag{6.2.10}$$

$$\left(\bigotimes_{k=0}^{N-1}\mathbf{M}^{(k)}\right)^{-1}=\bigotimes_{k=0}^{N-1}\left(\mathbf{M}^{(k)}\right)^{-1}. \tag{6.2.11}$$

Matrix factorization identities:

$$\mathop{\Xi}_{k=0}^{N-1}\left(\mathbf{M}_{R_k,Q_k}^{(1,k)}\cdot\mathbf{M}_{Q_k,P_k}^{(2,k)}\right)=\left(\bigoplus_{k=0}^{N-1}\mathbf{M}_{R_k,Q_k}^{(1,k)}\right)\cdot\left(\mathop{\Xi}_{k=0}^{N-1}\mathbf{M}_{Q_k,P_k}^{(2,k)}\right). \tag{6.2.12}$$

$$\mathop{\Xi}_{k=0}^{N-1}\mathbf{M}_{V_k,H_k}^{(k)}=\left(\bigoplus_{k=0}^{N-1}\mathbf{M}_{V_k,H_k}^{(k)}\right)\cdot\left(\mathop{\Xi}_{k=0}^{N-1}\mathbf{I}_{H_k}^{(k)}\right). \tag{6.2.13}$$

$$\mathop{\Xi}_{k=0}^{N-1} \mathbf{M}^{(1,k)}_{R_k,Q} \cdot \mathbf{M}^{(2)}_{Q,P} = \left(\mathop{\Xi}_{k=0}^{N-1} \mathbf{M}^{(1,k)}_{R_k,Q} \right) \cdot \mathbf{M}^{(2)}_{Q,P} . \qquad (6.2.14)$$

$$\mathbf{M}^{(1)}_{V_1,H_1} \otimes \mathbf{M}^{(2)}_{V_2,H_2} = \left(\mathbf{M}^{(1)}_{V_1,H_1} \otimes \mathbf{I}_{V_2} \right) \cdot \left(\mathbf{I}_{H_1} \otimes \mathbf{M}^{(2)}_{V_2,H_2} \right). \qquad (6.2.15)$$

$$\mathop{\Xi}_{k=0}^{N-1} \left(\mathbf{M}^{(1,k)}_{V_{1,k},H_{1,k}} \otimes \mathbf{M}^{(2,k)}_{V_{2,k},H_{2,k}} \right) = \left[\bigoplus_{k=0}^{N-1} \left(\mathbf{M}^{(1,k)}_{V_{1,k},G_{1,k}} \otimes \mathbf{I}_{V_{2,k}} \right) \right] \cdot \left[\mathop{\Xi}_{k=0}^{N-1} \left(\mathbf{I}_{G_{1,k}} \otimes \mathbf{M}^{(2,k)}_{V_{2,k},H_{2,k}} \right) \right]. \qquad (6.2.16)$$

$$\mathop{\Xi}_{k=0}^{N-1} \left(\mathbf{M}^{(k)}_{V_{1,k},G_{1,k}} \otimes \mathbf{R}^{(k)}_{1,G_{2,k}} \right) = \left(\bigoplus_{k=0}^{N-1} \mathbf{M}^{(k)}_{V_{1,k},G_{1,k}} \right) \cdot \left[\mathop{\Xi}_{k=0}^{N-1} \left(\mathbf{I}_{G_{1,k}} \otimes \mathbf{R}^{(k)}_{1,G_{2,k}} \right) \right]. \qquad (6.2.17)$$

$$\mathop{\Xi}_{k=0}^{N-1} \left(\mathbf{R}^{(k)}_{1,G_{2,k}} \otimes \mathbf{M}^{(k)}_{V_{1,k},G_{1,k}} \right) = \left(\bigoplus_{k=0}^{N-1} \mathbf{M}^{(k)}_{V_{1,k},G_{1,k}} \right) \cdot \left[\mathop{\Xi}_{k=0}^{N-1} \left(\mathbf{R}^{(k)}_{1,G_{2,k}} \otimes \mathbf{I}_{G_{1,k}} \right) \right]. \qquad (6.2.18)$$

$$\left(\mathop{\Xi}_{k=0}^{N_1-1} \mathbf{R}^{(1,k)}_{1,H_1} \right) \otimes \left(\mathop{\Xi}_{k=0}^{N_2-1} \mathbf{R}^{(1,k)}_{1,H_1} \right) = \left[\mathop{\Xi}_{k=0}^{N_1-1} \left(\mathbf{I}_{N_2} \otimes \mathbf{R}^{(1,k)}_{1,H_1} \right) \right] \cdot \left[\mathop{\Xi}_{k=0}^{N_2-1} \left(\mathbf{I}_{G_1} \otimes \mathbf{R}^{(2,k)}_{1,H_2} \right) \right]. \qquad (6.2.19)$$

Matrix permutation identities:

$$\mathbf{M}^{(1)}_{V_1,G_1} \otimes \mathbf{M}^{(2)}_{V_2,G_2} = \left[\mathop{\Xi}_{k=0}^{V_1-1} \left(\mathbf{I}_{V_2} \otimes \Delta^{(k)}_{V_1} \right) \right] \cdot \left(\mathbf{M}^{(2)}_{V_2,G_2} \otimes \mathbf{M}^{(1)}_{V_1,G_1} \right) \cdot \left[\mathop{\Xi}_{k=0}^{V_1-1} \left(\mathbf{I}_{G_1} \otimes \mathbf{\Delta}^{(k)}_{G_2} \right) \right]; \qquad (6.2.20)$$

$$\mathop{\Xi}_{k=0}^{N-1} \mathbf{M}^{(k)}_{V_k,H} \otimes \mathbf{R}^{(k)}_{1,P} = \left[\mathop{\Xi}_{k=0}^{N-1} \left(\mathbf{R}^{(k)}_{1,P} \otimes \mathbf{M}^{(k)}_{V_k,H} \right) \right] \cdot \left[\mathop{\Xi}_{k=0}^{P-1} \left(\mathbf{I}_H \otimes \mathbf{\Delta}^{(k)}_{P} \right) \right]; \qquad (6.2.21)$$

$$\begin{aligned} &\left[\mathbf{M}^{(1)}_{V_1,G} \otimes \mathbf{R}^{(1)}_{2} \right] \Xi \left[\mathbf{M}^{(2)}_{V_2,G} \otimes \mathbf{R}^{(2)}_{2} \right] = \\ &\left[\left(\mathbf{R}^{(1)}_{2} \otimes \mathbf{M}^{(1)}_{V_1,G} \right) \Xi \left(\mathbf{R}^{(2)}_{2} \otimes \mathbf{M}^{(2)}_{V_2,G} \right) \right] \cdot \left[\left(\mathbf{I}_G \otimes [\mathbf{10}] \right) \Xi \left(\mathbf{I}_G \otimes [\mathbf{01}] \right) \right] \end{aligned} \qquad (6.2.22)$$

6.3 TRANSFORMS AND THEIR FAST ALGORITHMS IN MATRIX REPRESENTATION

In this section we will provide matrix formulas for orthogonal transforms listed in Table 3.1 and some other useful transforms and for their corresponding fast algorithms beginning with simplest "binary" transforms.

6.3.1 Walsh-Hadamard transform

Walsh-Hadamard transform matrix of order $\mathbf{2}^n$ is representable as a n-th Kronecker power of order 2 Hadamard matrix $\mathbf{h}_2$:

$$\mathbf{HAD}_{2^n} = (\mathbf{h}_2)^{\otimes n} = \bigotimes_{r=0}^{n-1} \mathbf{h}_2 . \qquad (6.3.1)$$

It is obviously representable also in a recursive form as:

$$\mathbf{HAD}_{2^n} = \mathbf{h}_2 \otimes \mathbf{HAD}_{2^{n-1}} = ([\mathbf{1}\mathbf{1}] \otimes \mathbf{HAD}_{2^{n-1}}) \Xi ([\mathbf{1}\bar{\mathbf{1}}] \otimes \mathbf{HAD}_{2^{n-1}}) \qquad (6.3.2)$$

and in a layered Kronecker matrix form as:

$$\mathbf{HAD}_{2^n} = [\mathbf{1}\mathbf{1}]^{\otimes n} \Xi \left[\Xi_{r=1}^{n-1} \left[[\mathbf{1}\mathbf{1}]^{\otimes(n-r)} \otimes [\mathbf{1}\bar{\mathbf{1}}] \otimes \mathbf{h}_2^{\otimes(r-1)} \right] \right] =$$

$$[\mathbf{1}\mathbf{1}]^{\otimes n} \Xi \left[\Xi_{r=1}^{n-1} \left[[\mathbf{1}\mathbf{1}]^{\otimes(n-r)} \otimes [\mathbf{1}\bar{\mathbf{1}}] \otimes \mathbf{HAD}_{2^{r-1}} \right] \right]. \qquad (6.3.3)$$

Using matrix factorization identity of Eq. 6.2.15, one can straightforwardly factorize Hadamard transform matrix into a product of two sparse matrices:

$$\mathbf{HAD}_{2^n} = \mathbf{HAD}_{2^{n-1}} \otimes \mathbf{h}_2 = [\mathbf{HAD}_{2^{n-1}} \otimes \mathbf{I}_2] \cdot [\mathbf{I}_{2^{n-1}} \otimes \mathbf{h}_2]. \qquad (6.3.4)$$

Continuing the factorization recursively, obtain finally:

$$\mathbf{HAD}_{2^n} = \prod_{r=0}^{n-1} \left[\mathbf{I}_{2^{n-r-1}} \otimes \mathbf{h}_2 \otimes \mathbf{I}_{2^r} \right] =$$

$$\prod_{r=0}^{n-1} \left\{ \mathbf{I}_{2^{n-r-1}} \otimes \left[([\mathbf{1}\mathbf{1}] \otimes \mathbf{I}_{2^r}) \Xi ([\mathbf{1}\bar{\mathbf{1}}] \otimes \mathbf{I}_{2^r}) \right] \right\}. \qquad (6.3.5)$$

Eq. 6.3.5 is a matrix formula for ***fast Hadamard transform***. It can be directly translated into a program code. Graph of 8 point fast Walsh-Hadamard transform is shown in Fig. 6-3, a).

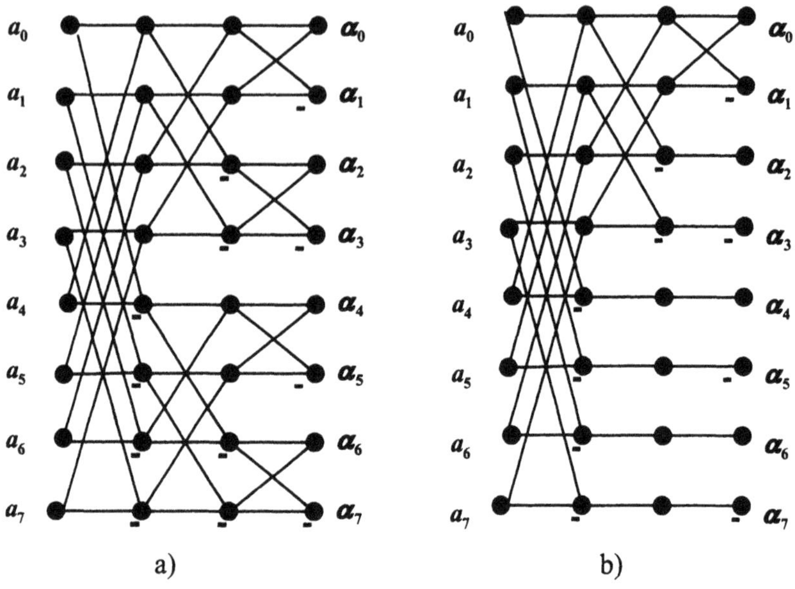

Figure 6-3. Flow diagrams of Fast Hadamard (a) and Fast Haar (b) Transform algorithms.

One can observe the isomorphism between Eq. 6.3.5 and the graph representation of the fast Walsh-Hadamard transform shown in Fig. 6-3, (a). Term $r = 0$ in Eq. 6.3.5 corresponds to the right most stage in the graph. Term $r = n-1$ corresponds to the left most stage in the graph. Term described by matrix $\mathbf{h}_2$ is represented in the graph by the "butterfly" unit that computes sum and difference of two inputs.

Eq. 6.3.5 explicitly shows also that Walsh-Hadamard transform of a vector of $N = 2^n$ samples can be computed in $n = \log_2 N$ steps with N addition/subtraction operations in each step, or, totally, with $N \log_2 N$ additions/subtractions.

6.3.2 Walsh-Paley Transform and Bit Reversal Permutation

Walsh-Paley Transform matrix is obtained from Walsh-Hadamard Transform matrix by means of ***bit reversal permutation*** of its rows ([3]):

$$\mathbf{PAL}_{2^n} = \mathbf{BRP}_{2^n} \cdot \mathbf{HAD}_{2^n}, \tag{6.3.6}$$

where $\mathbf{BRM}_{2^n}$ is a bit-reversal permutation matrix. Walsh-Paley Transform matrix can be represented in a recursive form as

$$\mathbf{PAL}_{2^n} = \left(\mathbf{PAL}_{2^{n-1}} \otimes [\mathbf{11}]\right) \Xi \left(\mathbf{PAL}_{2^{n-1}} \otimes [\mathbf{1}\bar{\mathbf{1}}]\right). \tag{6.3.7}$$

Using Eq. 6.3.7 recursively yields representation of Walsh-Paley transform matrix in the form of a layered Kronecker matrix as

$$\mathbf{PAL}_{2^n} = [\mathbf{11}]^{\otimes n} \Xi \left[\mathop{\Xi}_{r=1}^{n} \left(\mathbf{PAL}_{2^{r-1}} \otimes [\mathbf{1}\bar{\mathbf{1}}]\right) \otimes [\mathbf{11}]^{\otimes(n-r)} \right]. \tag{6.3.8}$$

One can compare Eqs. 6.3.7 and 6.3.8 and Eqs. 6.3.2 and 6.3.3 to see that bit reversal permutation that converts Walsh-Hadamard transform matrix into Walsh-Paley matrix can be attributed to permutation of their corresponding constituent elementary matrices.

Bit reversal permutation matrix, in its turn, can be represented in layered Kronecker matrix and recursive forms as

$$\mathbf{BRP}_{2^n} = [\mathbf{10}]^{\otimes n} \Xi \left[\mathop{\Xi}_{r=1}^{n} \left(\mathbf{BRP}_{2^{r-1}} \otimes [\mathbf{01}]\right) \otimes [\mathbf{10}]^{\otimes(n-r)} \right]; \tag{6.3.9, a}$$

$$\mathbf{BRP}_{2^n} = \left(\mathbf{BRP}_{2^{n-1}} \otimes [\mathbf{10}]\right) \Xi \left(\mathbf{BRP}_{2^{n-1}} \otimes [\mathbf{01}]\right). \tag{6.3.9, b}$$

As one can see, it is isomorphic to the Walsh-Paley matrix with substitutions $[\mathbf{10}] \leftrightarrow [\mathbf{11}]$ and $[\mathbf{01}] \leftrightarrow [\mathbf{1}\bar{\mathbf{1}}]$.

Walsh-Paley transform matrix can be factorized using factorization identities (6.2.14-16) to obtain the following matrix formulas for ***fast Walsh-Paley transform***:

$$\mathbf{PAL}_{2^n} = \prod_{r=0}^{n-1} \left\{ \mathbf{I}_{2^{n-r-1}} \otimes \left[\left(\mathbf{I}_{2^r} \otimes [\mathbf{11}]\right) \Xi \left(\mathbf{I}_{2^r} \otimes [\mathbf{1}\bar{\mathbf{1}}]\right) \right] \right\}. \tag{6.3.10, a}$$

and, from Eqs. 6.3.5 and 6.3.6,

$$\mathbf{PAL}_{2^n} = \mathbf{BRP}_{2^n} \cdot \prod_{r=0}^{n-1} \left\{ \mathbf{I}_{2^{n-r-1}} \otimes \mathbf{h}_2 \otimes \mathbf{I}_{2^r} \right\}. \qquad (6.3.10, b)$$

In virtue of the isomorphism between Walsh-Paley and bit reversal permutation matrices, one can also obtain a matrix formula for fast bit reversal permutation:

$$\mathbf{BRP}_{2^n} = \prod_{r=0}^{n-1} \left\{ \mathbf{I}_{2^{n-r-1}} \otimes \left[\left(\mathbf{I}_{2^r} \otimes [\mathbf{10}] \right) \boxminus \left(\mathbf{I}_{2^r} \otimes [\mathbf{01}] \right) \right] \right\}. \qquad (6.3.11)$$

from which it follows that bit reversal permutation of a vector of $N = 2^n$ samples can be carried out with $N \log_2 N$ permutations of two samples (n steps with N permutations in each).

Note that bit reversal permutation is a special case of matrix transposition. Matrix transposition can, in a similar way, be implemented through transpositions of small sub-matrices and subsequent transposition of the order of sub-matrices ([1]).

6.3.3 Walsh Transform and Binary-to-Gray Code and Hadamard Permutations

To introduce Walsh transform matrix, we need a notion of Gray code for binary representation of numbers. m-th binary digit τ_m^{GC} of Gray code of a number written in its direct binary code representation as $\{\tau_m^{DBC}\}$, $m = 0,1,...,n-1$ is computed as

$$\tau_m^{GC} = \tau_m^{DBC} \underset{bin}{+} \tau_{m+1}^{DBC} \qquad (6.3.12)$$

where $m = 0$ corresponds to the least significant digit and $\left(\underset{bin}{+}\right)$ denotes modulo 2 addition. An illustrative example of binary code to Gray code conversion is given in Tab. 6-1. For comparison, the table contains also bit reversal binary codes.

Table 6-1. Binary and Gray code representation of 8 first natural numbers

Decimal index	0	1	2	3	4	5	6	7
Direct binary code	000	001	010	011	100	101	110	111
Bit reversal code	000	100	010	110	001	101	011	111
Gray code	000	001	011	010	110	111	101	100

Walsh transform matrix is obtained from Walsh-Paley matrix by means of permutation of its rows according to Gray codes of their index ([3]):

$$\mathbf{WAL}_{2^n} = \mathbf{BGCP}_{2^n} \cdot \mathbf{PAL}_{2^n} = \mathbf{BGCP}_{2^n} \cdot \mathbf{BRP}_{2^n} \cdot \mathbf{HAD}_{2^n}. \qquad (6.3.13)$$

where $\mathbf{BGCP}_{2^n}$ is a ***direct binary-to-Gray code permutation*** matrix. The combination $\mathbf{BGCP}_{2^n} \cdot \mathbf{BRP}_{2^n}$ of direct binary-to-Gray code permutation and bit reversal permutation converts Hadamard transform matrix into Walsh transform matrix of which rows are arranged according to the number of zero crossing in the sequence of elements of their rows (sequency, see Sect. 3.2.2.2)

Walsh transform matrix may be represented in both recursive form:

$$\mathbf{WAL}_{2^n} = \left(\mathbf{WAL}_{2^{n-1}} \otimes [\mathbf{11}]\right) \Xi \left[\left(\bar{\mathbf{I}}_{2^{n-1}} \cdot \mathbf{WAL}_{2^{n-1}}\right) \otimes [\mathbf{1\bar{1}}]\right] \qquad (6.3.14)$$

and a layered Kronecker matrix form as:

$$\mathbf{WAL}_{2^n} = [\mathbf{11}]^{\otimes n} \Xi \left\{ \mathop{\Xi}_{r=1}^{n} \left[\left(\bar{\mathbf{I}}_{2^{r-1}} \cdot \mathbf{WAL}_{2^{r-1}}\right) \otimes [\mathbf{1\bar{1}}] \otimes [\mathbf{11}]^{\otimes(n-r)}\right] \right\}. \qquad (6.3.15)$$

Comparison of Eqs. 6.3.14 and 6.3.15 with corresponding Eqs. 6.3.7 and 6.3.8 shows that direct binary-to-Gray code permutation is associated with reverse identity matrix $\bar{\mathbf{I}}_2$.

Matrix $\mathbf{HADP}_{2^n} = \mathbf{BGCP}_{2^n} \cdot \mathbf{BRP}_{2^n}$ that connects Walsh transform and Walsh-Hadamard Transform is called ***Hadamard permutation matrix*** ([4]). It is isomorphic to the Walsh transform matrix with substitutions $[\mathbf{11}] \leftrightarrow [\mathbf{10}]$ and $[\mathbf{1\bar{1}}] \leftrightarrow [\mathbf{10}]$. With these substitutions obtain from Eq. 6.3.14:

$$\mathbf{HADP}_{2^n} = \left(\mathbf{HADP}_{2^{n-1}} \otimes [\mathbf{10}]\right) \Xi \bar{\mathbf{I}}_{2^{n-1}} \cdot \left(\mathbf{HADP}_{2^{n-1}} \otimes [\mathbf{01}]\right) \qquad (6.3.16, a)$$

and from Eq. 6.3.15:

$$\mathbf{HADP}_{2^n} = [\mathbf{10}]^{\otimes n} \Xi \left\{ \left\{ \mathop{\Xi}_{r=1}^{n} \left[\left(\bar{\mathbf{I}}_{2^{r-1}} \cdot \mathbf{HADP}_{2^{r-1}}\right) \otimes [\mathbf{1\bar{1}}] \otimes [\mathbf{11}]^{\otimes(n-r)}\right] \right\} \right\}.$$

(6.3.16, b)

By inversion of the order of Kronecker products in Eq. 6.3.16, b), one can separate from the product $\mathbf{HADP}_{2^n} = \mathbf{BGCP}_{2^n} \cdot \mathbf{BRP}_{2^n}$ the direct binary-to-Gray code permutation matrix:

$$\mathbf{BGCP}_{2^n} = [10]^{\otimes n} \Xi \left\{ \left\{ \overset{n}{\underset{r=1}{\Xi}} [11]^{\otimes(n-r)} \otimes [1\bar{1}] \otimes \left(\bar{\mathbf{I}}_{2^{r-1}} \cdot \mathbf{BGCP}_{2^{r-1}} \cdot \mathbf{BRP}_{2^{r-1}} \right) \right\} \right\} \tag{6.3.17}$$

Applying recursively factorization identities (Eqs. 6.2.11-18), one can obtain a matrix formula for ***Fast Walsh Transform***:

$$\begin{aligned} \mathbf{WAL}_{2^n} = & \left\{ \prod_{r=0}^{n-2} \left\{ \mathbf{I}_{2^{n-2-r}} \otimes \left[\left(\mathbf{I}_{2^{r1}} \otimes [11] \right) \Xi \left(\mathbf{I}_{2^r} \otimes [1\bar{1}] \right) \right] \oplus \right. \right. \\ & \left. \left. \bar{\mathbf{I}}_{2^{r+1}} \cdot \left[\left(\mathbf{I}_{2^{r1}} \otimes [11] \right) \Xi \left(\mathbf{I}_{2^r} \otimes [1\bar{1}] \right) \right] \cdot \bar{\mathbf{I}}_{2^{r+1}} \right\} \right\} \times \\ & \left[\left(\mathbf{I}_{2^{n-1}} \otimes [11] \right) \Xi \bar{\mathbf{I}}_{2^{n-1}} \left(\mathbf{I}_{2^{r-1}} \otimes [1\bar{1}] \right) \right] \end{aligned} \tag{6.3.18}$$

Substitutions $[11] \leftrightarrow [10]$ and $[1\bar{1}] \leftrightarrow [10]$ convert this formula into the formula for fast Hadamard permutation:

$$\begin{aligned} \mathbf{HADP}_{2^n} = & \left\{ \prod_{r=0}^{n-2} \left\{ \mathbf{I}_{2^{n-2-r}} \otimes \left[\left(\left[\left(\mathbf{I}_{2^{r1}} \otimes [10] \right) \Xi \left(\mathbf{I}_{2^r} \otimes [01] \right) \right] \right) \right] \oplus \right. \right. \\ & \left. \left. \bar{\mathbf{I}}_{2^{r+1}} \cdot \left[\left(\mathbf{I}_{2^{r1}} \otimes [10] \right) \Xi \left(\mathbf{I}_{2^r} \otimes [01] \right) \right] \cdot \bar{\mathbf{I}}_{2^{r+1}} \right\} \right\} \times \\ & \left[\left[\left(\mathbf{I}_{2^{n-1}} \otimes [10] \right) \Xi \bar{\mathbf{I}}_{2^{n-1}} \left(\mathbf{I}_{2^{r-1}} \otimes [01] \right) \right] \right] \end{aligned} \tag{6.3.19}$$

Matrix formula for fast direct binary-to-Gray code permutation can be obtained from the relationship of Eq. 6.3.13 between **WAL** and **PAL** matrices as ([5]):

$$\mathbf{BGCP}_{2^n} = \prod_{r=0}^{n-1} \left[\mathbf{I}_{2^r} \otimes \left(\mathbf{I}_{2^{n-r-1}} \oplus \bar{\mathbf{I}}_{2^{n-r-1}} \right) \right]. \tag{6.3.20}$$

6.3.4 Haar Transform

Haar transform matrix can also be represented both in a recursive form:

$$\mathbf{HAR}_{2^n} = \left(\mathbf{HAR}_{2^{n-1}} \otimes [11] \right) \Xi \left(\mathbf{I}_{2^{n-1}} \otimes [1\bar{1}] \right) \tag{6.3.21}$$

and as a layered Kronecker matrix:

$$\mathbf{HAR}_{2^n} = [11]^{\otimes n} \Xi \mathop{\Xi}_{r=0}^{n-1}\left(\mathbf{I}_{2^{r-1}} \otimes [1\bar{1}]\right) \otimes [11]^{\otimes(n-r)}. \tag{6.3.21}$$

Similarity between Haar transform matrix as described by Eq. 6.3.21 and layered Kronecker matrix representation of Walsh-Paley transform (Eq. 6.3.8) is obvious. Recursive factorization of Eq. 6.3.20 yields the following matrix formula for ***Fast Haar Transform***:

$$\mathbf{HAR}_{2^n} = \prod_{r=0}^{n-1}\left\{\left[\left(\mathbf{I}_{2^r} \otimes [11]\right) \Xi \left(\mathbf{I}_{2^r} \otimes [1\bar{1}]\right)\right] \oplus \mathbf{I}_{2^n - 2^{r+1}}\right\}. \tag{6.3.22}$$

Basis functions of Haar transform in its matrix representations of Eqs. 6.3.20-23 are not normalized to be orthonormal. With the normalization, fast Haar transform can be written as ([5]):

$$\mathbf{HAR}_{2^n} = \left(\mathbf{1} \oplus \bigotimes_{r=0}^{n-1} 2^{(r-n)/2}\mathbf{I}_{2^r}\right)\prod_{r=0}^{n-1}\left\{\left[\left(\mathbf{I}_{2^r} \otimes [11]\right) \Xi \left(\mathbf{I}_{2^r} \otimes [1\bar{1}]\right)\right] \oplus \mathbf{I}_{2^n - 2^{r+1}}\right\}. \tag{6.3.23}$$

6.3.5 Discrete Fourier Transform

Matrix of the DFT of order $N = 2^n$ is, to the accuracy of the normalization factor $1/\sqrt{N}$ the matrix

$$\mathbf{FOUR}_{2^n} = \left\{ w_n^{kr} = \exp\left(i2\pi k \frac{r}{2^n} \right) \right\}. \tag{6.3.24}$$

Represent index r of row elements of this matrix via binary digits $\{r_s\}$, $s = 0,1,\ldots,n-1$ of their binary code:

$$r = \sum_{s=0}^{n-1} r_s 2^s \tag{6.3.25}$$

Then matrix $\mathbf{FOUR}_{2^n}$ can be written in a layered Kronecker matrix form as:

$$\mathbf{FOUR}_{2^n} = \left\{ w_n^{k\sum_{s=0}^{n-1} r_s 2^s} \right\} = \left\{ \prod_{s=0}^{n-1} w_n^{2^s k r_s} \right\} = \mathop{\Xi}_{k=0}^{2^n-1}\left(\bigotimes_{s=0}^{n-1} \left[1 \quad w_n^{2^s k}\right] \right). \tag{6.3.26}$$

In this representation, matrix $\mathbf{FOUR}_{2^n}$ contains $N = 2^n$ layers. One can convert it into a recursive layered Kronecker representation with n layers by means of the following algebraic matrix manipulations. First, represent index k in the vertical matrix sum in Eq. 6.3.26 through binary digits $\{k_m\}$, $m = 0,1,...,n-1$ of its binary code representation:

$$k = \sum_{m=0}^{n-1} k_m 2^m \tag{6.3.27}$$

and rearrange vertical sum in Eq. 6.3.26 into n sums grouped according $\{k_m\}$:

$$\mathbf{FOUR}_{2^n} = \mathop{\Xi}_{k=0}^{2^n-1}\left(\bigotimes_{s=0}^{n-1}\begin{bmatrix}1 & w_n^{2^s k r_s}\end{bmatrix}\right) = \mathop{\Xi}_{k_{n-1}=0}^{1}\left(\mathop{\Xi}_{k_{n-2}=0}^{1}\cdots\left(\mathop{\Xi}_{k_0=0}^{1}\left(\bigotimes_{s=0}^{n-1}\begin{bmatrix}1 & w_n^{2^s \sum_{m=0}^{n-1} k_m 2^m}\end{bmatrix}\right)\right)\right). \tag{6.3.28}$$

Change now the order of summation over binary code digits $\{k_m\}$ of k in Eq.6.3.27 to the reverse one while keeping indexing of $\{k_m\}$ that corresponds to Eq. 6.3.27. This will require introducing into Eq. 6.3.28 a bit reversal matrix:

$$\mathbf{FOUR}_{2^n} = \mathbf{BRP}_{2^n} \cdot \left\{\mathop{\Xi}_{k_0=0}^{1}\left(\mathop{\Xi}_{k_1=0}^{1}\cdots\left(\mathop{\Xi}_{k_{n-1}=0}^{1}\left(\bigotimes_{s=0}^{n-1}\begin{bmatrix}1 & w_n^{2^s \sum_{m=0}^{n-1} k_m 2^m}\end{bmatrix}\right)\right)\right)\right\}. \tag{6.3.29}$$

Introduce, for the convenience, the bit reversed DFT matrix $\overline{\mathbf{FOUR}}_{2^n} = \mathbf{BRP}_{2^n} \cdot \mathbf{FOUR}_{2^n}$:

$$\overline{\mathbf{FOUR}}_{2^n} = \left\{\mathop{\Xi}_{k_0=0}^{1}\left(\mathop{\Xi}_{k_1=0}^{1}\cdots\left(\mathop{\Xi}_{k_{n-1}=0}^{1}\left(\bigotimes_{s=0}^{n-1}\begin{bmatrix}1 & w_n^{2^s \sum_{m=0}^{n-1} k_m 2^m}\end{bmatrix}\right)\right)\right)\right\}. \tag{6.3.30}$$

Rewrite Eq. 6.3.30 as

$$\overline{\mathbf{FOUR}}_{2^n}=\left\{\mathop{\Xi}_{k_1=0}^{1}\cdots\left(\mathop{\Xi}_{k_{n-1}=0}^{1}\left(\bigotimes_{s=0}^{n-1}\begin{bmatrix}1 & w_n^{2^s\sum_{m=1}^{n-1}k_m 2^m}\end{bmatrix}\right)\right)\Xi\right.$$

$$\left.\mathop{\Xi}_{k_1=0}^{1}\cdots\left(\mathop{\Xi}_{k_{n-1}=0}^{1}\left(\bigotimes_{s=0}^{n-1}\begin{bmatrix}1 & w_n^{2^s\left(1+\sum_{m=1}^{n-1}k_m 2^m\right)}\end{bmatrix}\right)\right)\right\}=$$

$$\left\{\mathop{\Xi}_{k_1=0}^{1}\cdots\left(\mathop{\Xi}_{k_{n-1}=0}^{1}\left(\begin{bmatrix}1 & w_n^{2^n\sum_{m=1}^{n-1}k_m 2^{m-1}}\end{bmatrix}\otimes\left(\bigotimes_{s=0}^{n-2}\begin{bmatrix}1 & w_n^{2^s\sum_{m=1}^{n-1}k_m 2^m}\end{bmatrix}\right)\right)\right)\Xi\right.$$

$$\left.\mathop{\Xi}_{k_1=0}^{1}\cdots\left(\mathop{\Xi}_{k_{n-1}=0}^{1}\left(\begin{bmatrix}1 & w_n^{2^n\sum_{m=1}^{n-1}k_m 2^{m-1}+2^{n-1}}\end{bmatrix}\otimes\left(\bigotimes_{s=0}^{n-2}\begin{bmatrix}1 & w_n^{2^s\sum_{m=1}^{n-1}k_m 2^m}\end{bmatrix}\begin{bmatrix}1 & 0\\ 0 & w_n^{2^s}\end{bmatrix}\right)\right)\right)\right\}. \tag{6.3.31}$$

Note that $w_n^{2^n\sum_{m=1}^{n-1}k_m 2^{m-1}}=\exp\left(i2\pi\sum_{m=1}^{n-1}k_m 2^{m-1}\right)=1$ and $w_n^{2^{n-1}}=\exp(i\pi)=-1$. Therefore

$$\begin{bmatrix}1 & w_n^{2^n\sum_{m=1}^{n-1}k_m 2^{m-1}}\end{bmatrix}=[1\quad 1]=[\mathbf{11}];\begin{bmatrix}1 & w_n^{2^n\sum_{m=1}^{n-1}k_m 2^{m-1}+2^{n-1}}\end{bmatrix}=[1\quad -1]=[\mathbf{1\bar{1}}] \tag{6.3.32}$$

Now, applying matrix identity Eq. 6.2.6, one can rewrite Eq. 6.3.31 as

$$\overline{\mathbf{FOUR}}_{2^n}=\left\{[\mathbf{11}]\otimes\mathop{\Xi}_{k_1=0}^{1}\cdots\left(\mathop{\Xi}_{k_{n-1}=0}^{1}\left(\bigotimes_{s=0}^{n-2}\begin{bmatrix}1 & w_n^{2^s\sum_{m=1}^{n-1}k_m 2^m}\end{bmatrix}\right)\right)\Xi\right.$$

$$\left.[\mathbf{1\bar{1}}]\otimes\mathop{\Xi}_{k_1=0}^{1}\cdots\left(\mathop{\Xi}_{k_{n-1}=0}^{1}\left(\bigotimes_{s=0}^{n-2}\begin{bmatrix}1 & w_n^{2^s\sum_{m=1}^{n-1}k_m 2^m}\end{bmatrix}\begin{bmatrix}1 & 0\\ 0 & w_n^{2^s}\end{bmatrix}\right)\right)\right\}=$$

$$\left\{[1 1]\otimes\left[\mathop{\Xi}_{k_1=0}^{1}\cdots\left(\mathop{\Xi}_{k_{n-1}=0}^{1}\left(\bigotimes_{s=0}^{n-2}\begin{bmatrix}1 & w_{n-1}^{2^s\sum_{m=0}^{n-2}k_{m+1}2^m}\end{bmatrix}\right)\right)\right]\Xi\right.$$

$$\left.[1\bar{1}]\otimes\left[\mathop{\Xi}_{k_1=0}^{1}\cdots\left(\mathop{\Xi}_{k_{n-1}=0}^{1}\left(\bigotimes_{s=0}^{n-2}\begin{bmatrix}1 & w_{n}^{2^s\sum_{m=0}^{n-2}k_{m+1}2^m}\end{bmatrix}\begin{bmatrix}1 & 0\\ 0 & w_n^{2^s}\end{bmatrix}\right)\right)\right]\right\} \tag{6.3.33}$$

and, using identities (6.2.4) and (6.2.7), as

$$\overline{\mathbf{FOUR}}_{2^n}=\left\{[1 1]\otimes\left[\mathop{\Xi}_{k_1=0}^{1}\cdots\left(\mathop{\Xi}_{k_{n-1}=0}^{1}\left(\bigotimes_{s=0}^{n-2}\begin{bmatrix}1 & w_{n-1}^{2^s\sum_{m=0}^{n-2}k_{m+1}2^m}\end{bmatrix}\right)\right)\right]\Xi\right.$$

$$\left.[1\bar{1}]\otimes\left[\mathop{\Xi}_{k_1=0}^{1}\cdots\left(\mathop{\Xi}_{k_{n-1}=0}^{1}\left[\left(\bigotimes_{s=0}^{n-2}\begin{bmatrix}1 & w_{n-1}^{2^s\sum_{m=0}^{n-2}k_{m+1}2^m}\end{bmatrix}\right)\cdot\left(\bigotimes_{s=0}^{n-2}\begin{bmatrix}1 & 0\\ 0 & w_n^{2^s}\end{bmatrix}\right)\right]\right)\right]\right\}=$$

$$\left\{[1 1]\otimes\left[\mathop{\Xi}_{k_1=0}^{1}\cdots\left(\mathop{\Xi}_{k_{n-1}=0}^{1}\left(\bigotimes_{s=0}^{n-2}\begin{bmatrix}1 & w_{n-1}^{2^s\sum_{m=0}^{n-2}k_{m+1}2^m}\end{bmatrix}\right)\right)\right]\Xi\right.$$

$$\left.[1\bar{1}]\otimes\left[\mathop{\Xi}_{k_1=0}^{1}\cdots\left(\mathop{\Xi}_{k_{n-1}=0}^{1}\left(\bigotimes_{s=0}^{n-2}\begin{bmatrix}1 & w_{n-1}^{2^s\sum_{m=0}^{n-2}k_{m+1}2^m}\end{bmatrix}\right)\right)\right]\cdot\left(\bigotimes_{s=0}^{n-2}\begin{bmatrix}1 & 0\\ 0 & w_n^{2^s}\end{bmatrix}\right)\right\}. \tag{6.3.34}$$

Finally, using Eq. 6.3.30, obtain:

$$\overline{\mathbf{FOUR}}_{2^n}=\left([1 1]\otimes\overline{\mathbf{FOUR}}_{2^{n-1}}\right)\Xi\left\{[1\bar{1}]\otimes\left[\overline{\mathbf{FOUR}}_{2^{n-1}}\cdot\left(\bigotimes_{s=0}^{n-2}\mathbf{d}_{2^{s-n}}\right)\right]\right\}. \tag{6.3.35}$$

where

$$\mathbf{d}_u=\begin{bmatrix}1 & 0\\ 0 & \exp(i2\pi u)\end{bmatrix}. \tag{6.3.36}$$

This formula can be recursively converted into a layered Kronecker matrix representation with n layers as follows:

$$\overline{\mathbf{FOUR}}_{2^n} = [\mathbf{1 1}]^{\otimes n} \Xi \left\{ \mathop{\Xi}_{r=1}^{n} \left[[\mathbf{1 1}]^{\otimes(n-r)} \otimes \left([\mathbf{1} \bar{\mathbf{1}}] \otimes \left[\overline{\mathbf{FOUR}}_{2^{r-1}} \cdot \left(\bigotimes_{s=0}^{r-2} \mathbf{d}_{2^{s-r}} \right) \right] \right) \right] \right\} \tag{6.3.37}$$

Derive now a matrix formula for Fast Fourier Transform. Applying matrix identities (6.2.1, 6.2.7 and 6.2.18) to Eq. 6.3.34, obtain:

$$\begin{aligned}
&\overline{\mathbf{FOUR}}_{2^n} = \\
&\left(\overline{\mathbf{FOUR}}_{2^{n-1}} \oplus \overline{\mathbf{FOUR}}_{2^{r-1}} \cdot \left(\bigotimes_{s=0}^{r-2} \mathbf{d}_{2^{s-r}} \right) \right) \cdot \left[\left([\mathbf{1 1}] \otimes \mathbf{I}_{2^{n-1}} \right) \Xi \left([\mathbf{1} \bar{\mathbf{1}}] \otimes \mathbf{I}_{2^{n-1}} \right) \right] = \\
&\left(\overline{\mathbf{FOUR}}_{2^{n-1}} \oplus \overline{\mathbf{FOUR}}_{2^{r-1}} \right) \cdot \left(\mathbf{I}_{2^{n-1}} \oplus \left(\bigotimes_{s=0}^{r-2} \mathbf{d}_{2^{s-r}} \right) \right) \cdot \left[\left([\mathbf{1 1}] \Xi [\mathbf{1} \bar{\mathbf{1}}] \right) \otimes \mathbf{I}_{2^{n-1}} \right] = \\
&\left(\mathbf{I}_2 \otimes \overline{\mathbf{FOUR}}_{2^{r-1}} \right) \cdot \left(\mathbf{I}_{2^{n-1}} \oplus \left(\bigotimes_{s=0}^{r-2} \mathbf{d}_{2^{s-r}} \right) \right) \cdot \left(\mathbf{h}_2 \otimes \mathbf{I}_{2^{n-1}} \right)
\end{aligned} \tag{6.3.38}$$

Applying this relationship recursively, obtain

$$\begin{aligned}
&\overline{\mathbf{FOUR}}_{2^n} = \\
&\prod_{r=0}^{n-1} \left(\mathbf{I}_{2^{(n-1-r)}} \otimes \left[\mathbf{I}_{2^r} \oplus \left(\bigotimes_{s=0}^{r-1} \mathbf{d}_{2^{s+2-n}} \right) \right] \right) \cdot \left(\mathbf{I}_{2^{(n-r-1)}} \otimes \mathbf{h}_2 \otimes \mathbf{I}_{2^r} \right),
\end{aligned} \tag{6.3.39}$$

or, finally,

$$\begin{aligned}
&\mathbf{FOUR}_{2^n} = \\
&\mathbf{BRP}_{2^n} \cdot \prod_{r=0}^{n-1} \left(\mathbf{I}_{2^{(n-1-r)}} \otimes \left[\mathbf{I}_{2^r} \oplus \left(\bigotimes_{s=0}^{r-1} \mathbf{d}_{2^{s+2-n}} \right) \right] \right) \cdot \left(\mathbf{I}_{2^{(n-r-1)}} \otimes \mathbf{h}_2 \otimes \mathbf{I}_{2^r} \right),
\end{aligned} \tag{6.3.40}$$

a matrix formula for the "decimation-in-time" Fast Fourier Transform algorithm described in Sect. 6.1 and illustrated in Fig. 6-2 (a).

Using similar matrix manipulations based on the above matrix identities one can design many other modifications of FFT ([5,6]). For instance, transposing matrices of Eq. 6.3.39 and using matrix symmetry yields an FFT algorithm with bit reversal in signal domain ("decimation in frequency" algorithm):

$$\mathbf{FOUR}_{2^n} = \left\{\prod_{r=0}^{n-1}\left(\mathbf{I}_{2^r}\otimes\mathbf{h}_2\otimes\mathbf{I}_{2^{(n-r-1)}}\right)\cdot\left(\mathbf{I}_{2^r}\otimes\left[\mathbf{I}_{2^{(n-r-1)}}\oplus\left(\bigotimes_{s=0}^{n-2-r}\mathbf{d}_{2^{s+2-n}}\right)\right]\right)\right\}\cdot\mathbf{BRP}_{2^n} \quad (6.3.41)$$

It is very instructive to compare formulas (6.3.39) and (6.3.40) for Fast Fourier Transform algorithms with formula (6.3.10, b) for Fast Walsh-Paley transform algorithm. One can see that Fast Fourier transform algorithm can be obtained from Fast Paley-Hadamard transform algorithm by appending each r-th of n factor matrices of Walsh-Paley factorized matrix with a corresponding diagonal "twiddle" matrix $\left(\mathbf{I}_{2^{(n-r-1)}}\otimes\left[\mathbf{I}_{2^r}\oplus\left(\bigotimes_{s=0}^{r-1}\mathbf{d}_{2^{-(s+2)}}\right)\right]\right)$. This adds, in each r-th computation stage, $\left(2^r-1\right)$ operations of multiplication of complex numbers. Therefore total number of operations required for computing the DFT with FFT algorithm can be evaluated as $\sum_{r=0}^{n-1}\left(2^r-1\right)=N-1-n$ of multiplications of complex number plus $N\log N$ of additions of complex number, or totally $4(N-1-n)$ multiplications and $2(N\log N+N-1-n)$ additions of real numbers.

6.3.6 DCT and DcST

Matrices of the DCT and DcST defined in Sect. 4.3.3 are, for $N=2^n$, real and imaginary parts of the matrix

$$\mathbf{DCST}_{2^n}=\left\{\exp\left(i\pi\frac{(k+1/2)r}{2^n}\right)\right\} \quad (6.3.42)$$

Applying this transform to a signal vector with real components gives a spectral vector whose real part is DCT and imaginary part is DcST spectrum of the signal.

Similarly to matrix **FOUR** of the DFT that exhibits recursive symmetry when its rows are rearranged in bit reversal order, matrix **DCST** reveals its symmetry when its columns are subjected to Hadamard permutation. For $\mathbf{DCST}_{2^n}$, introduce the corresponding auxiliary matrix $\overline{\mathbf{DCST}}_{2^n}$:

$$\overline{\mathbf{DCST}}_{2^n} = \mathbf{DCST}_{2^n}\mathbf{HADP}_{2^n}. \tag{6.3.43}$$

It was shown in [7] that Hadamard permuted matrix $\overline{\mathbf{DCST}}_{2^n}$ can be represented in a recursive form via an auxiliary matrix $\overline{\mathbf{DCST}}^{(m)}_{2^{n-s}}$, as:

$$\begin{aligned} &\overline{\mathbf{DCST}}^{(0)}_{2^n} = \overline{\mathbf{DCST}}_{2^n}; \\ &\overline{\mathbf{DCST}}^{(m)}_{2^{n-s}} = \mathbf{R}^{(m)}_{2^{n-s}} \cdot \mathbf{B}_{2^{n-s}} \cdot \left(\overline{\mathbf{DCST}}^{(2m)}_{2^{n-s-1}} \oplus \overline{\mathbf{DCST}}^{(2m+1)}_{2^{n-s-1}}\right); \end{aligned} \tag{6.3.44}$$

$$\mathbf{B}_{2^{n-s}} = \left[\left([1 1]\otimes \mathbf{I}_{2^{n-s-1}}\right) \boxminus \left([1 \bar{1}]\otimes\left[\mathbf{I}^{*}_{1} \oplus \mathbf{I}_{2^{n-s-1}-1}\right]\right)\right]^{tr} \tag{6.3.45}$$

$$\begin{aligned} \mathbf{R}^{(m)}_{2^{n-s}} = &\left([1 0]\otimes \mathbf{I}_{2^{n-s-1}}\right) \boxminus \\ &\left(\left[\mathbf{0} \oplus \left(-\bar{\mathbf{I}}_{2^{n-s-1}-1}\right)\cdot \mathbf{I}^{*}_{2^{n-s-1}}\right] ||| \left[\left(\mathbf{de}_{s,m}\, \bullet\right) \oplus 2d_{s,m} \cdot \mathbf{I}_{2^{n-s-1}-1}\right]\right) \end{aligned} \tag{6.3.46}$$

where $s = 0,1,\ldots,n$, $m = 0,1,\ldots,2^s - 1$, $\overline{\mathbf{DCST}}^{(m)}_{1} = 1$, $\mathbf{I}^{*}_{S}$ is the conjugation operator,

$$\mathbf{de}_{s,m} = d_{s,m} + ie_{sm} = \exp\left[i\frac{\pi}{2^{s+2}}\left(2\tilde{m}_{2^s} + 1\right)\right] \tag{6.3.47}$$

$(\bullet)$ is an operator of Hadamard multiplication of complex numbers:

$$a \bullet b = \left(a^{re} + ia^{im}\right)\bullet\left(b^{re} + ib^{im}\right) = a^{re}b^{re} + ia^{im}b^{im}\ , \tag{6.3.48}$$

and $\tilde{m}_{2^{s+1}}$ is the position of index m in the vector of 2^{s+1} indices after Hadamard permutation of this vector.

In order to represent the fast algorithm as a product of sparse matrices, the two inner $\overline{\mathbf{DCST}}$ matrices in Eq. 6.3.35 should be recursively substituted. Carrying out this substitution until the last stage obtain a direct formula:

$$\overline{\mathbf{DCST}}_{2^n} = \left[\prod_{s=0}^{n-1}\left(\bigoplus_{m=0}^{2^s-1} \mathbf{R}^{(m)}_{2^{n-s}} \cdot \mathbf{B}_{2^{n-s}}\right)\right]. \tag{6.3.49}$$

Fig. 6-4 illustrates the matrices set for DCST of 8 points.

For input signal vector $\mathbf{a}$ with real elements, algorithm in this formulation assumes that the signal has to be represented with complex numbers as:

$$\tilde{\mathbf{a}} = \mathbf{a} + i\mathbf{a} \tag{6.3.50}$$

Then DCT and DcST transforms are obtained as:

$$\overline{\mathbf{DCST}_{2^n}} \cdot \tilde{a} = \mathbf{DCST}_{2^n}\mathbf{HADP}_{2^n} \cdot \tilde{a} = \mathbf{DCT}_{2^n} \cdot a + i\mathbf{DcST}_{2^n} \cdot a \tag{6.3.51}$$

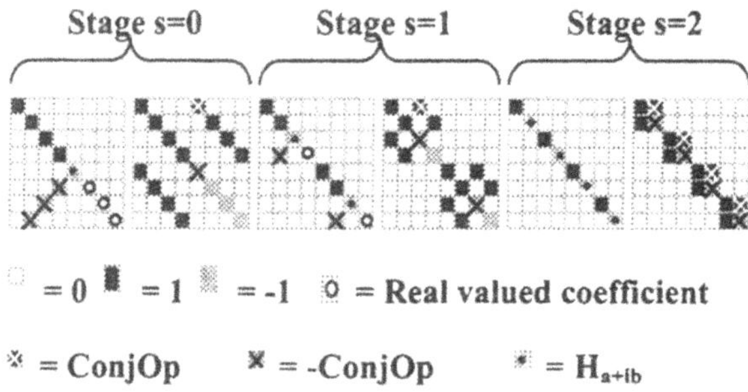

Figure 6-4. Sparse matrix structure for fast DCST

Eq. 6.3.49 can be split up into formulas for separate computation of the DCT and DcST as:

$$\overline{\mathbf{DCT}_{2^n}} = \mathbf{DCT}_{2^n}\mathbf{HADP}_{2^n} = \prod_{r=0}^{n-1}\left(\bigoplus_{r=0}^{2^r-1} \mathbf{Rc}_{2^{n-r}}^{(2^r-s-1)}\right)\left(\mathbf{I}_{2^r} \otimes \mathbf{h}_2 \otimes \mathbf{I}_{2^{n-r-1}}\right) \tag{6.3.52}$$

and

$$\overline{\mathbf{DST}_{2^n}} = \mathbf{DST}_{2^n}\mathbf{HADP}_{2^n} = \prod_{r=0}^{n-1}\left(\bigoplus_{r=0}^{2^r-1} \mathbf{Rs}_{2^{n-r}}^{(2^r-s-1)}\right)\left(\mathbf{I}_{2^r} \otimes \mathbf{h}_2 \otimes \mathbf{I}_{2^{n-r-1}}\right), \tag{6.3.53}$$

where

$$\mathbf{Rc}_{2^{n-s}}^{(m)} = \left([\mathbf{10}] \otimes \mathbf{I}_{2^{n-s-1}}\right)\Xi \left(\left[\mathbf{0} \oplus \left(-\bar{\mathbf{I}}_{2^{n-s-1}-1}\right) \cdot \mathbf{I}^{*}_{2^{n-s-1}-1}\right] [||] \left[d_{s,m} \oplus 2d_{s,m} \cdot \mathbf{I}_{2^{n-s-1}-1}\right]\right), \tag{6.3.54}$$

and

$$\mathbf{Rs}^{(m)}_{2^{n-s}} =$$
$$\left[\left(\mathbf{0} \oplus \mathbf{I}_{2^{n-s-1}}\right) \Xi \left(e_{s,m} \oplus \left(-\bar{\mathbf{I}}_{2^{n-s-1}}\right)\right)\right] [\,|\,] \left[\left(1 \oplus \mathbf{0}_{2^{n-s-1}}\right) \Xi \left(\mathbf{0} \oplus 2d_{s,m}\mathbf{I}_{2^{n-s-1}}\right)\right]. \tag{6.3.55}$$

Fig. 6-5 shows flowchart of the DCT and DcST for N=8.

To evaluate the computational complexity of the algorithm, denote the number of complex multiplications of matrix $\mathbf{M}$ by $\mu(\mathbf{M})$. Then obtain:

$$\mu\left(\overline{\mathbf{DCST}}_N\right) = \mu\left(\mathbf{R}_N\right) + 2\mu\left(\overline{\mathbf{DCST}}_{\frac{N}{2}}\right). \tag{6.3.56}$$

Considering that the input vector for each stage consists of complex numbers, the number of multiplications of real number can be found as

$$\gamma\left(\overline{DCST}_N\right) = 2\gamma\left(\mathbf{R}_N\right) + 2\gamma\left(\overline{DCST}_{N/2}\right) = N + 2\gamma\left(\overline{DCST}_N\right). \tag{6.3.57}$$

For matrix transform of size $N = 2^n$, there are n stages in the algorithm, and the total number of multiplications is then $N \log_2 N$.

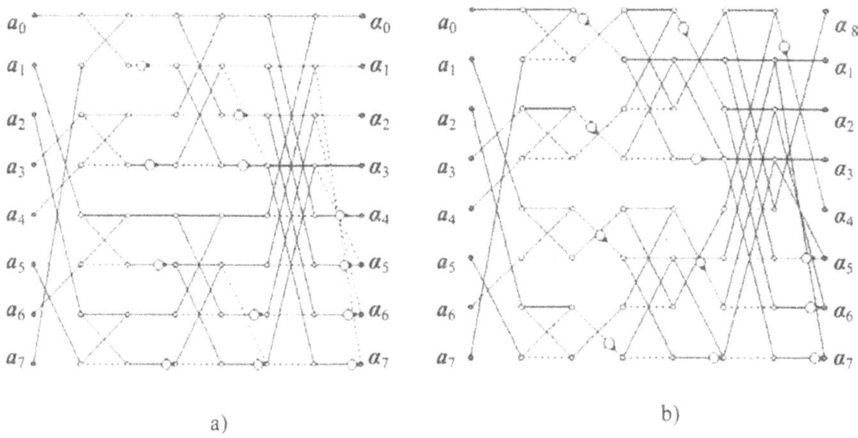

Figure 6-5. Illustrative flow diagram of the DCT (a) and DcST (b) for N=8. Solid lines denote addition at the graph nodes; dotted lines denote subtraction; circle denotes multiplication by respective coefficients $\{d_{s,m}\}$ and $\{e_{s,m}\}$ (numerical values are not shown)

6.4 PRUNED ALGORITHMS

In digital holography it often happens that either initial signal array contains many zeros or else not every sample of the signal spectrum is to be calculated or both. The latter case is typical, for instance, for digital reconstruction of holograms when one need to reconstruct only a fragment of the output image plane with the object of interest. In such cases pruned FFT algorithms can be constructed with lesser number of operations as compared with full FFT algorithms ([5]). For this purpose it suffices to multiply the transformation matrix in its factorized form on the left and on the right by diagonal "cutting" matrices containing zeros at those points of the diagonal whose positions correspond to the positions of zero-valued or unneeded elements.

If the number of nonzero signal samples and of the required spectrum samples is equal to a power of two, the pruning matrices are of Kronecker type built from elementary matrices:

$$\mathbf{P}_2 = \begin{bmatrix} 1 & 0 \\ 0 & 0 \end{bmatrix} \tag{6.4.1}$$

and identity matrix $\mathbf{I}_2$. The pruning matrix $\mathbf{P}_{2^n}^l$ of dimension $\mathbf{2}^n$ containing exactly $\mathbf{2}^l$ ones at the beginning of the main diagonal is expressed as follows:

$$\mathbf{P}_{2^n}^l = \mathbf{P}_2^{\otimes(n-l)} \otimes \mathbf{I}_2^{\otimes l} \tag{6.4.2}$$

Let ones need to compute first $\mathbf{2}^k$ of $\mathbf{2}^n$ transform coefficients of a signal with only $\mathbf{2}^l$ nonzero samples at the beginning of the signal vector. Then transform matrix should be truncated by multiplying it from both left and right sides as:

$$\mathbf{FOUR}_{2^n}^{(k,l)} = \mathbf{P}_{2^n}^k \cdot \mathbf{FOUR}_{2^n} \cdot \mathbf{P}_{2^n}^l . \tag{6.4.3}$$

Factorized representations of such a pruned FFT obtained by substituting of Eq. 6.4.2 into Eq. 6.4.3 and using the factorized DFT matrix of Eq. 6.3.40 ("decimation in frequency" algorithm) are as follows ([5]):

$$\mathbf{FOUR}_{2^n}^{(k,l)} = \left\{ \prod_{r=0}^{n-1-k} \left(\mathbf{I}_{2^r} \otimes \mathbf{S}_2^0 \otimes \mathbf{P}_{2^{n-1-r}}^l \right) \cdot \right.$$

$$\left[\mathbf{I}_{2^r} \otimes \left(\mathbf{P}^k_{2^{n-1-r}} \Xi \mathbf{P}^k_{2^{n-1-r}} \cdot \bigotimes_{s=0}^{r-1} \mathbf{d}_{2^{-(n-s-1)}}\right)\right]\Bigg\} \cdot \prod_{r-n-k}^{l-1} \left(\mathbf{I}_{2^r} \otimes \mathbf{h}_2 \otimes \mathbf{I}_{2^{n-1-r}}\right)$$

$$\left[\mathbf{I}_{2^r} \otimes \left(\mathbf{I}_{2^{n-1-r}} \Xi \bigotimes_{s=0}^{r-1} \mathbf{d}_{2^{-(n-s-1)}}\right)\right]\Bigg\} \cdot \left(\mathbf{I}_{2^l} \otimes \mathbf{S}^{tr}_{2^{\otimes(n-1)}}\right) \cdot \mathbf{BRP}_{2^n}\,;\; k+l \ge n. \quad (6.4.4)$$

and, for $k+l \le n$,

$$\mathbf{FOUR}^{(k,l)}_{2^n} = \left\{ \prod_{r=0}^{n-1-k} \left(\mathbf{I}_{2^r} \otimes \mathbf{S}^0_2 \otimes \mathbf{P}^l_{2^{n-1-r}}\right) \times \right.$$

$$\left[\mathbf{I}_{2^r} \otimes \left(\mathbf{P}^k_{2^{n-1-r}} \Xi \mathbf{P}^k_{2^{n-1-r}} \cdot \bigotimes_{s=0}^{r-1} \mathbf{d}_{2^{-(n-s-1)}}\right)\right]\Bigg\} \cdot \left(\mathbf{I}_{2^l} \otimes \mathbf{P}_2^{\otimes(n-k-l)} \otimes \mathbf{S}^{tr}_{2^{\otimes k}}\right) \cdot \mathbf{BRP}_{2^n}, \quad (6.4.5)$$

where

$$\mathbf{S}^0_2 = \begin{bmatrix} 1 & 1 \\ 0 & 0 \end{bmatrix}; \quad \mathbf{S}^{tr}_{2^{\otimes k}} = \left(\begin{bmatrix} 1 & 1 \\ 0 & 0 \end{bmatrix}^{\otimes k}\right)^{tr}. \quad (6.4.6)$$

and $\mathbf{M}^{tr}$ is transposed matrix $\mathbf{M}$.

When the disposition and the number of signal zero samples and of unnecessary spectral samples is arbitrary, explicit matrix formulas for pruned algorithms are described in [9].

For the fast algorithm

$$\mathbf{FOUR}_{2^n} = \mathbf{BRP}_{2^n} \cdot \prod_{r=0}^{n-1} \left(\mathbf{I}_{2^{n-1-r}} \otimes \left[\mathbf{I}_{2^r} \oplus \left(\bigotimes_{s=0}^{r-1} \mathbf{d}_{-2^{s-1-r}}\right)\right]\right) \cdot \left(\mathbf{I}_{2^{n-1-r}} \otimes \mathbf{h}_2 \otimes \mathbf{I}_{2^r}\right) =$$

$$\mathbf{B} \cdot \prod_{r=0}^{n-1} \mathbf{D}_r \cdot \mathbf{H}_r\,, \quad (6.4.7)$$

where

$$\mathbf{B} = \mathbf{BRP}_{2^n}\,; \mathbf{D}_r = \left(\mathbf{I}_{2^{n-1-r}} \otimes \left[\mathbf{I}_{2^r} \oplus \left(\bigotimes_{s=0}^{r-1} \mathbf{d}_{-2^{s-1-r}}\right)\right]\right); \; \mathbf{H}_r = \left(\mathbf{I}_{2^{n-1-r}} \otimes \mathbf{h}_2 \otimes \mathrm{I}_{2^r}\right) \quad (6.4.8)$$

and left and right pruning matrices $\mathbf{P}_{2^n}^{(L)}$ and $\mathbf{P}_{2^n}^{(R)}$ containing ones in the positions that correspond to nonzero input samples and to transform coefficients to be computed, pruned transform may be written as

$$\mathbf{PFOUR}_{2^n} = \mathbf{P}_{2^n}^{(L)} \cdot \left\{ \mathbf{B} \cdot \prod_{r=0}^{n-1} \mathbf{D}_r \cdot \mathbf{H}_r \right\} \cdot \mathbf{P}_{2^n}^{(R)} = \widetilde{\mathbf{B}} \cdot \prod_{r=0}^{n-1} \widetilde{\mathbf{D}}_r \cdot \widetilde{\mathbf{H}}_r , \tag{6.4.9}$$

where

$$\widetilde{\mathbf{B}} = \widetilde{P}_{2^n}^{(L)}(\mathbf{B}) \cdot \mathbf{B} \cdot \widetilde{P}_{2^n}^{(R)}(\mathbf{B}); \tag{6.4.10}$$

$$\widetilde{\mathbf{D}}_r = \widetilde{\mathbf{P}}_{2^n}^{(L)}(\mathbf{D}_r) \cdot \mathbf{D}_r \cdot \widetilde{\mathbf{P}}_{2^n}^{(R)}(\mathbf{D}_r); \tag{6.4.11}$$

and

$$\widetilde{\mathbf{H}}_r = \widetilde{\mathbf{P}}_{2^n}^{(L)}(\mathbf{H}_r) \cdot \mathbf{H}_r \cdot \widetilde{\mathbf{P}}_{2^n}^{(R)}(\mathbf{H}_r) \tag{6.4.12}$$

are stage pruned matrices and matrices $\widetilde{\mathbf{P}}_{2^n}^{(L)}(\mathbf{B})$, $\widetilde{\mathbf{P}}_{2^n}^{(R)}(\mathbf{B})$, $\widetilde{\mathbf{P}}_{2^n}^{(L)}(\mathbf{D}_r)$, $\widetilde{\mathbf{P}}_{2^n}^{(R)}(\mathbf{D}_r)$, $\widetilde{\mathbf{P}}_{2^n}^{(L)}(\mathbf{H}_r)$, and $\widetilde{\mathbf{P}}_{2^n}^{(R)}(\mathbf{H}_r)$ are defined recursively using the following notions of "reflected" cutting matrices $\mathbf{RP}^{(L)}(\mathbf{M},\mathbf{P})$ and $\mathbf{RP}^{(R)}(\mathbf{P},\mathbf{M})$. The "reflected" cutting matrices are defined as:

$$\mathbf{RP}^{(L)}(\mathbf{M},\mathbf{P}) = \bigoplus_{k=0}^{2^n-1} \left[\mathbf{Sh}\left(\mathbf{Sh}(\mathbf{M}) \cdot \mathbf{P} \cdot \begin{bmatrix} 1 \\ 1 \end{bmatrix}^{\otimes[n]} \right)^{tr} \cdot \left(\mathbf{\Delta}_{2^n}^{(2^n-1-k)} \right)^{tr} \right]; \tag{6.4.13}$$

$$\mathbf{RP}^{(L)}(\mathbf{M},\mathbf{P}) = \bigoplus_{k=0}^{2^n-1} \left[\mathbf{Sh}\left([1\,1]^{\otimes[n]} \cdot \mathbf{P} \cdot \mathbf{Sh}(\mathbf{M}) \right) \cdot \left(\mathbf{\Delta}_{2^n}^{(2^n-1-k)} \right) \right]; \tag{6.4.14}$$

where $\mathbf{Sh}(\mathbf{M})$ is a binary "shadow" matrix of matrix $\mathbf{M}$ elements of which are ones wherever elements of matrix $\mathbf{M}$ are nonzero.

Using the "reflected" cutting matrices of Eqs. 6.4.13-14, matrices $\widetilde{\boldsymbol{B}}$, $\widetilde{\boldsymbol{D}}$ and $\widetilde{\boldsymbol{H}}$ may be found recursively as following:

$$\widetilde{\mathbf{P}}^{(L)}(\mathbf{B}) = \mathbf{P}_{2^n}^{(L)};$$

$$\widetilde{\mathbf{P}}^{(L)}(\mathbf{D}_0) = \mathbf{RP}^{(R)}\left(\mathbf{P}_{2^n}^{(L)}, \mathbf{B}\right);$$

$$\widetilde{\mathbf{P}}^{(L)}(\mathbf{H}_0) = \mathbf{RP}^{(R)}\left(\widetilde{\mathbf{P}}^{(L)}(\mathbf{D}_0), \mathbf{D}_0\right);$$

$$\widetilde{\mathbf{P}}^{(L)}(\mathbf{D}_1) = \mathbf{RP}^{(R)}\left(\widetilde{\mathbf{P}}^{(L)}(\mathbf{H}_0), \mathbf{H}_0\right);$$

$$\widetilde{\mathbf{P}}^{(L)}(\mathbf{H}_1) = \mathbf{RP}^{(R)}\left(\widetilde{\mathbf{P}}^{(L)}(\mathbf{D}_1), \mathbf{D}_1\right);$$

....

$$\widetilde{\mathbf{P}}^{(L)}(\mathbf{D}_{n-1}) = \mathbf{RP}^{(R)}\left(\widetilde{\mathbf{P}}^{(L)}(\mathbf{H}_{n-2}), \mathbf{H}_{n-2}\right);$$

$$\widetilde{\mathbf{P}}^{(L)}(\mathbf{H}_{n-1}) = \mathbf{RP}^{(R)}\left(\widetilde{\mathbf{P}}^{(L)}(\mathbf{D}_{n-1}), \mathbf{D}_{n-1}\right);$$

....

$$\widetilde{\mathbf{P}}^{(L)}(\mathbf{D}_{n-1}) = \mathbf{RP}^{(R)}\left(\widetilde{\mathbf{P}}^{(L)}(\mathbf{H}_{n-2}), \mathbf{H}_{n-2}\right);$$

$$\widetilde{\mathbf{P}}^{(L)}(\mathbf{H}_{n-1}) = \mathbf{RP}^{(R)}\left(\widetilde{\mathbf{P}}^{(L)}(\mathbf{D}_{n-1}), \mathbf{D}_{n-1}\right);$$

and

$$\widetilde{\mathbf{P}}^{(R)}(H_{n-1}) = \mathbf{P}_{2^n}^{(R)};$$

$$\widetilde{\mathbf{P}}^{(R)}(\mathbf{D}_{n-1}) = \widetilde{\mathbf{P}}^{(L)}\left(\mathbf{H}_{n-1}, \widetilde{\mathbf{P}}^{(R)}(\mathbf{H}_{n-1})\right);$$

$$\widetilde{\mathbf{P}}^{(R)}(\mathbf{H}_{n-2}) = \widetilde{\mathbf{P}}^{(L)}\left(\mathbf{D}_{n-1}, \widetilde{\mathbf{P}}^{(R)}(\mathbf{D}_{n-1})\right);$$

$$\widetilde{\mathbf{P}}^{(R)}(\mathbf{D}_{n-2}) = \widetilde{\mathbf{P}}^{(L)}\left(\mathbf{H}_{n-2}, \widetilde{\mathbf{P}}^{(R)}(\mathbf{H}_{n-2})\right);$$

$$\widetilde{\mathbf{P}}^{(R)}(\mathbf{H}_{n-3}) = \widetilde{\mathbf{P}}^{(L)}\left(\mathbf{D}_{n-2}, \widetilde{\mathbf{P}}^{(R)}(\mathbf{D}_{n-2})\right);$$

$$\widetilde{\mathbf{P}}^{(R)}(\mathbf{D}_{n-3}) = \widetilde{\mathbf{P}}^{(L)}\left(\mathbf{H}_{n-3}, \widetilde{\mathbf{P}}^{(R)}(\mathbf{H}_{n-3})\right);$$

....

$$\widetilde{\mathbf{P}}^{(R)}(\mathbf{H}_0) = \widetilde{\mathbf{P}}^{(L)}\left(\mathbf{D}_1, \widetilde{\mathbf{P}}^{(R)}(\mathbf{D}_1)\right);$$

$$\widetilde{\mathbf{P}}^{(R)}(\mathbf{D}_0) = \widetilde{\mathbf{P}}^{(L)}\left(\mathbf{H}_0, \widetilde{\mathbf{P}}^{(R)}(\mathbf{H}_0)\right);$$

$$\tilde{\mathbf{P}}^{(R)}(B) = \tilde{\mathbf{P}}^{(L)}\left(\mathbf{D}_0, \tilde{\mathbf{P}}^{(R)}(\mathbf{D}_0)\right).$$

Fig. 6-6 shows an example of a flow diagram of pruned FFT for computing 9 spectral coefficients of signals with 7 nonzero input samples.

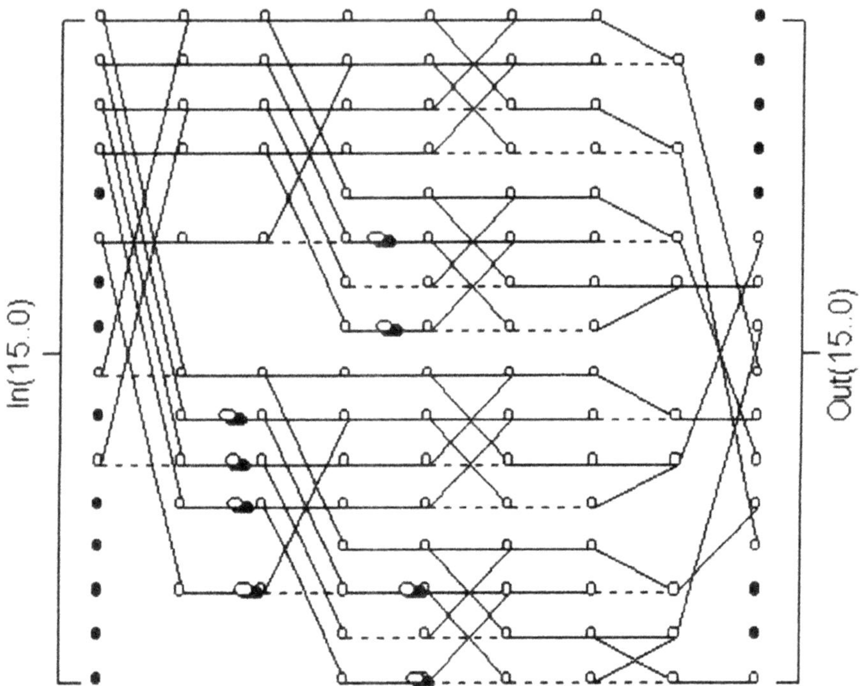

Figure 6-6. Illustrative flow diagram of a pruned DFT for N=16. Solid lines denote multiplication by 1, or j, doted lines denote multiplication by -1, or -j and solid lines with black bubbles denote multiplication by "twiddle" factors.

6.5 QUANTIZED DFT

Modifying the multiplier matrices in the matrix representation of fast algorithms one can generate many transforms with fast algorithms with some special features. The ***Quantized Discrete Fourier Transform*** (QDFT) is an example of a transform designed in this way. The QDFT approximates the Discrete Fourier Transform but requires less or no multiplication operations and lesser number of addition operations.

Consider a FFT algorithm featuring bit reversal of the result of transformation. (Eq. 6.3.39, Fig. 6-2, a).

$$\mathbf{FOUR}_{2^n} = \mathbf{BRP}_{2^n} \cdot \prod_{r=0}^{n-1} \left(\mathbf{I}_{2(n-1-r)} \otimes \left[\mathbf{I}_{2^r} \oplus \left(\bigotimes_{s=0}^{r-1} \mathbf{d}_{2^{-(s+2)}} \right) \right] \right) \cdot \left(\mathbf{I}_{2(n-r-1)} \otimes \mathbf{h}_2 \otimes \mathbf{I}_{2^r} \right). \quad (6.5.1)$$

Each r -th constituent matrix of this algorithm

$$D_r = \mathbf{I}_{2(n-1-r)} \otimes \left[\mathbf{I}_{2^r} \oplus \left(\bigotimes_{s=0}^{r-1} \mathbf{d}_{2^{-(s+2)}} \right) \right] \quad (6.5.2)$$

is a diagonal matrix that consists of ones and twiddle factors $\exp(i\theta_r)$ in its diagonal. Multiplication of complex numbers by these twiddle factors may be represented as multiplication of a vector composed of real and imaginary parts of the numbers by a matrix:

$$\mathbf{TF}_r = \begin{bmatrix} \cos\theta_r & \sin\theta_r \\ -\sin\theta_r & \cos\theta_r \end{bmatrix}. \quad (6.5.3)$$

These multiplications of complex number and computing cosine and sine functions are the most time consuming operations in the algorithm. To save time for computing cosine and sine functions, one of these functions is sometimes computed in advance for all needed arguments and stored in a look-up table. This look up table is then applied in all stages of the algorithm for finding values of both functions using the relationship $\sin\theta_r = \cos(\pi/2 - \theta_r)$.

Replace cosine and sine functions $\cos\theta_r$ and $\sin\theta_r$ in these twiddle matrices by quantized functions $\mathrm{qos}\,\theta_r$ and $\mathrm{qin}\,\theta_r = \mathrm{qos}(\pi/2 - \theta_r)$ that take limited number of quantized values as it is illustrated in Fig. 6-7. Thus, with three levels of quantization, $\mathbf{qos}(\theta_r)$ and $\mathbf{qin}(\theta_r)$ take the values (-1, 0 and 1); with five levels of quantization, the values (-1, -1/2, 0, 1/2, 1), and so on.

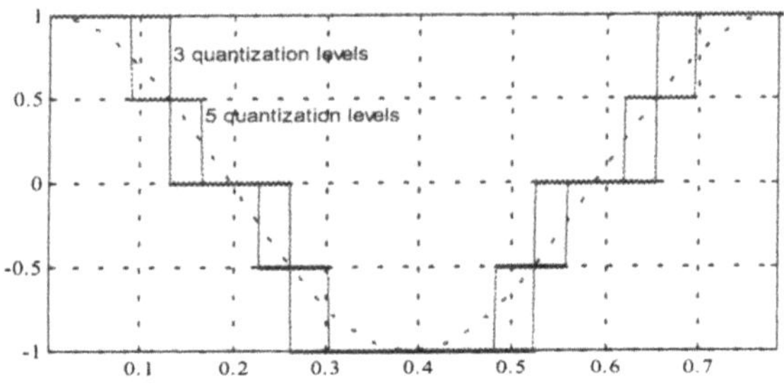

Figure 6-7. Cosine function (dashed line) and corresponding $\mathbf{qos}(.)$ functions for 3 and 5 quantization levels

This corresponds to replacing matrices $\{\mathbf{TF}_r\}$ of Eq.(6.5.3) with the matrices

$$\mathbf{QTF}_r = \begin{bmatrix} \mathbf{qos}\,\theta_r & \mathbf{qin}\,\theta_r \\ -\mathbf{qin}\,\theta_r & \mathbf{qos}\,\theta_r \end{bmatrix}. \tag{6.5.4}$$

Respectively, diagonal constituent matrices $\{\mathbf{D}_r\}$ of the algorithm of Eq. 6.5.1 that involve twiddle factors $\{\mathbf{TF}_r\}$ are replaced with diagonal matrices $\{\mathbf{QD}_r\}$ with quantized twiddle factors $\{\mathbf{QTF}_r\}$ on the corresponding places.

Since, by the above definition, $\mathbf{qin}(\theta_r) = \mathbf{qos}(\pi/2 - \theta_r)$ sum $\left(\mathbf{qin}^2(\cdot) + \mathbf{qos}^2(\cdot)\right)$ never equals to zero though it may not always be equal to one, matrices $\{\mathbf{QTF}_r\}$ are orthogonal though not necessarily orthonormal. To secure the proper normalization, introduce orthonormal quantized twiddle factor matrices:

$$\overline{\mathbf{QTF}}_r = \frac{1}{\sqrt{\mathbf{qos}^2\theta_r + \mathbf{qin}^2\theta_r}} \begin{bmatrix} \mathbf{qos}\,\theta_r & \mathbf{qin}\,\theta_r \\ -\mathbf{qin}\,\theta_r & \mathbf{qos}\,\theta_r \end{bmatrix}. \tag{6.5.5}$$

and corresponding diagonal matrices $\{\overline{\mathbf{QD}}_r\}$. In this we arrive at the orthonormal transform

$$\mathbf{FOUR}_{2^n} = \mathbf{BRP}_{2^n} \cdot \prod_{r=0}^{n-1} \overline{\mathbf{QD}_r} \cdot \left(\mathbf{I}_{2^{(n-r-1)}} \otimes \mathbf{h}_2 \otimes \mathbf{I}_{2^r} \right) \tag{6.5.6}$$

that is referred to as ***Quantized DFT***.

With three quantization levels (-1, 0, 1), the QDFT requires a lesser number of addition-subtraction operations than the DFT and does not require multiplication operation. Reduction in the number of addition-subtraction operations is roughly inverse to the fraction of the period of $\mathbf{qos}(.)$ and $\mathbf{qin}(.)$ in which they take zero values. With five levels of quantization (-1, -1/2, 0, 1/2, 1), multiplications can be replaced with the shift operations, and some addition-subtraction operations are omitted too. Therefore, in its computational complexity, the QDFT is between the Haar and the Walsh-Hadamard-Paly fast transforms; that is, it may require considerably fewer operations than FFT, while owing to its origin it can be used as an approximation of the DFT.

Graphs in Fig. 6-8 show for comparison four basis functions of the DFT and QDFT with 5 quantization levels for N=32.

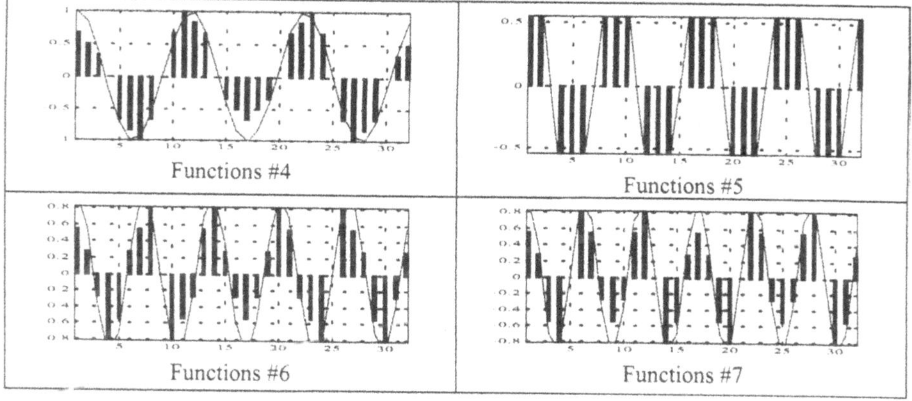

Figure 6-8. Basis functions of the DFT (linearly interpolated, solid lines) and the QDFT with 5 quantization levels (bars) for N=32

Values of basis functions of the QDFT are coarsely quantized because of quantization of cosine and sine functions in the transform factor matrices. However, the number of their quantization levels is greater than that of $\mathbf{qos}(.)$ and $\mathbf{qin}(.)$ owing to intermixing in the transform stages. Moreover, it increases with the increasing number of the transform stages, that is with the size of the transform matrix.

The accuracy with which rows of the QDFT represent those of the DFT depends on selected number of quantization levels and on position of quantization intervals. The latter can be optimized to minimize the approximation error.

The QDFT was first introduced for synthesis of computer generated holograms [6]. Its use instead of DFT was justified there by the fact that speckle noise which inevitably appears in images reconstructed from the holograms may hide the quantization effects of the QDFT. This is illustrated in Fig. 6-9 that shows images optically reconstructed from digital holograms synthesized using the QDFT instead of the DFT.

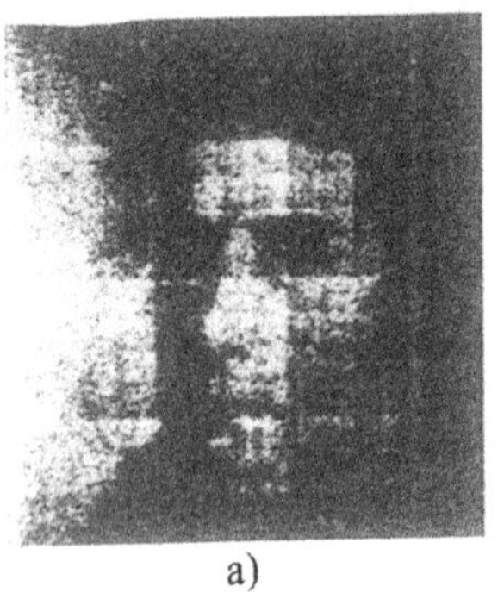

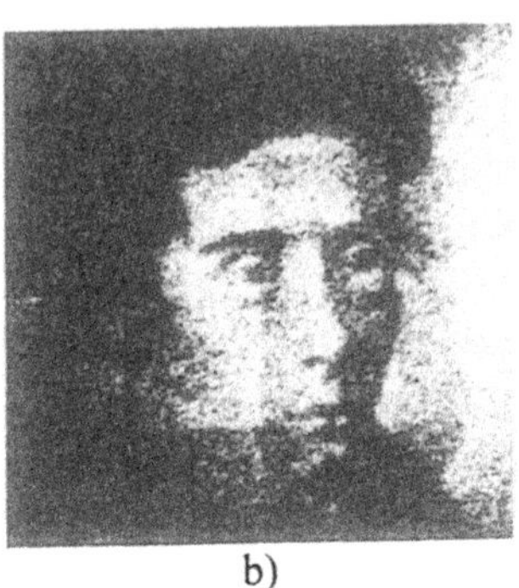

Figure 6-9. Images optically reconstructed from digital holograms synthesized with the use of the QDFT replacing the DFT: a) - the QDFT with three quantization levels; b) the QDFT with 5 quantization levels (adopted from [6])

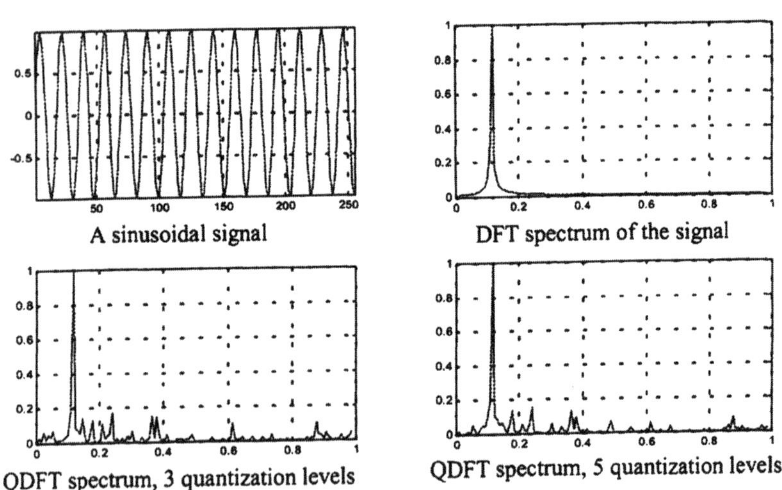

Figure 6-10. DFT and QDFT spectra of a sinusoidal signal

Among other applications of the QDFT as approximations to the DFT, one can mention spectral estimation, fringe analysis and generalized filtering in the transform domain. In the generalized filtering described in Ch. 8, samples of signal spectrum obtained by direct transform are multiplied by coefficients characterizing the filter in the selected basis, and filtered signal is reconstructed by inverse transform. Because normalizing coefficients of the QDFT basis functions can be combined with the filter coefficients, no additional normalization operations are required for the direct and inverse QDFTs in this case. An illustrative example of spectral analysis of a sinusoidal signal with the use of the DFT and QDFT is shown in Fig. 6-10.

REFERENCES

1. L. Yaroslavsky, M. Eden, Fundamentals of Digital Optics, Birkhauser, Boston, 1996
2. I. J. Good, J. Roy, Statist. Soc. Ser. B 20, 362, 1958
3. A. M. Trakhtman, V. A. Trakhtman, Fundamentals of the Theory of Discrete Signals Defined of Finite Intervals, Sov. Radio, Moscow, 1975, (In Russian)
4. N. Ahmed, K. R. Rao, Orthogonal Transforms for Digital Signal Processing, Springer Verlag, 1975
5. L. Yaroslavsky, Digital Picture Processing, An Introduction, Springer Verlag, Heidelberg, 1985
6. L. Yaroslavskii, N. Merzlyakov, Methods of Digital Holography, Plenum Press, N.Y., 1980
7. L. Yaroslavsky, A. Happonen, Y.Katiyi, Signal Discrete Signal Sinc-interpolation in DCT Domain: Fast Algorithms, in : Proceedings of International TISCP workshop on Spectral Methods and Multirate Signal Processing, SMMSP'02, Toulouse – France, 07.09.2002 - 08.09.2002, Ed. T Saramaki, K. Egiazarian &J. Astola, p. 179-185, TICSP Series #17, ISBN 952-15-08881-7, ISSN 1456-2774, TTKK Monistamo 2002
8. Y. Katiyi, L. Yaroslavsky, V/HS Structure for Transforms and their Fast Algorithms, 3rd Int. Symposium, Image and Signal Processing and Analysis, Sept. 18-20, 2003, Rome, Italy
9. Y. Katiyi, L. Yaroslavsky, Regular Matrix Methods for Synthesis of Fast Transforms: General Pruned and Integer-to-Integer Transforms, in Proc. Of International Workshop on Spectral Methods and Multirate Signal Processing, ed. By T. Saramäki, K. Egiazarian & J. Astola, SMMSP'2001, Pula, Croatia, June 16-18, 2001

Chapter 7

STATISTICAL COMPUTATION METHODS AND ALGORITHMS

7.1 MEASURING SIGNAL STATISTICAL CHARACTERISTICS

7.1.1 Measuring probability distribution and its moments and order statistics

By definition, the distribution function $P(V)$ of a random variable v is the probability that the random variable does not exceed the value V. The derivative of $P(V)$ with respect to V

$$p(v) = \left. \frac{dP(V)}{dV} \right|_{V=v}$$

is called the ***probability distribution density*** of the random variable v. Digital signals are characterized by discrete analogs of the distribution function and distribution density respectively - the relative share $R(m)$ of the samples that do not exceed the given quantized value q, and the rate $h(q)$ of the samples having the value q. The latter characteristic is referred to as signal ***distribution histogram,*** the former one as ***cumulative distribution histogram***.

A histogram describing the appearance rate of quantized values of signal samples is called a ***one-dimensional histogram***. A histogram that characterizes the rate of simultaneous appearance of groups of signal values or, for multi-component signals, groups of signal sample components is called multidimensional, or a ***multivariable distribution histogram***. For instance, distribution histogram of three (R,G,B) components of color images is three-dimensional histogram.

The following simple algorithm is available for computing the N-dimensional distribution histogram $h(\vec{q}_N)$ of N groups of M samples or components of signal $a = \{a_k^{(m)}\}$, $k = 0,1,\dots,N-1$, $m = 1,\dots,M$ with Q_m quantization levels in m-th signal component:

$$h(\vec{q}_M) = \frac{1}{N}\sum_{k=0}^{N-1}\prod_{m=1}^{M}\delta\left(q_m - a_k^{(m)}\right), \tag{7.1.1}$$

where $\delta(\cdot)$ is Kronecker delta, k is index of signal sample co-ordinates and m is index of signal components or signal samples in the groups. In applications, it is very frequently sufficient to compute and to use histograms without normalization by the number of signal samples. We will denote them $\bar{h}(\vec{q}_M)$:

$$\bar{h}(\vec{q}_M) = \sum_{k=0}^{N-1}\prod_{m=1}^{M}\delta\left(q_m - a_k^{(m)}\right). \tag{7.1.2}$$

Sometimes histograms should be computed over signal samples weighted with some weights $\{w_k\}$. We will refer to these histograms as ***weighted histograms*** $\bar{h}_w(\vec{q}_M)$:

$$\bar{h}_w(\vec{q}_M) = \sum_{k=0}^{N-1}\prod_{m=1}^{M} w_k\delta\left(q_m - a_k^{(m)}\right). \tag{7.1.3}$$

In image processing, when histogram is computed for an entire image frame or over a set of images it is referred to as ***global histogram***. Many image processing algorithms are based on histograms computed for every pixel over a set of surrounding samples called pixel spatial neighborhood (see Ch. 12). Such histograms are referred to as ***local histograms***.

If local histograms should be computed for every pixel in the process of image scanning, computations can be carried out recursively. Let local histograms be computed in a rectangular window of $(2N_1+1)\times(2N_2+1)$

pixels in the process of regular scanning image array $\{a_{k,l}^{(m)}\}$ row-wise and then column-wise on the rectangular sampling grid. For each l-th row, compute first histograms $\bar{h}_l^{(k)}(\vec{q}_m)$ of window columns:

$$\bar{h}_l^{(k)}(\vec{q}_m) = \sum_{r=k-N_1}^{r+N_1} \delta\left(q_m - a_{r,l}^{(m)}\right). \tag{7.1.5}$$

Then local histograms $\bar{h}_{k,l}(\vec{q}_m)$ can be recursively computed as

$$\bar{h}_{k,l}(\vec{q}_M) = \sum_{s=l-N_2}^{l+N_2} \bar{h}_s^{(k)}(\vec{q}_m) = \sum_{s=l-N_2-1}^{l+N_2-1} \bar{h}_s^{(k)}(\vec{q}_m) + \bar{h}_{l+N_2}^{(k)}(\vec{q}_m) - \bar{h}_{l+N_2-1}^{(k)}(\vec{q}_m) =$$
$$\bar{h}_{k,l-1}(\vec{q}_m) + \bar{h}_{l+N_2}^{(k)}(\vec{q}_m) - \bar{h}_{l+N_2-1}^{(k)}(\vec{q}_m). \tag{7.1.6}$$

The interpretation of this formula is self-evident: local histogram $\bar{h}_{k,l}(\vec{q}_M)$ in the window centered at pixel (k,l) may be obtained from the histogram $\bar{h}_{k,l-1}(\vec{q}_M)$ of the in the window in the previous position $(k,l-1)$, if it is added to the difference of column histograms $\bar{h}_{l+N_2}^{(k)}(\vec{q}_m)$ and $\bar{h}_{l+N_2-1}^{(k)}(\vec{q}_m)$ of window columns that enter window and leave it in the process of window shift from $(k,l-1)$-th to (k,l) position. Column histograms $\bar{h}_l^{(k)}(\vec{q}_m)$ may be, in their turn, also computed recursively:

$$\bar{h}_l^{(k)}(\vec{q}_m) = \sum_{r=k-N_1}^{r+N_1} \delta\left(q_m - a_{r,l}^{(m)}\right) =$$
$$\sum_{r=k-N_1-1}^{r+N_1-1} \delta\left(q_m - a_{r,l}^{(m)}\right) + \delta\left(q_m - a_{r+N_1,l}^{(m)}\right) - \delta\left(q_m - a_{r+N_1-1,l}^{(m)}\right) =$$
$$\bar{h}_l^{(k)}(\vec{q}_m) + \delta\left(q_m - a_{r+N_1,l}^{(m)}\right) - \delta\left(q_m - a_{r+N_1-1,l}^{(m)}\right) \tag{7.1.7}$$

from column histograms on the previous row. Thanks to such a recursive computation, computational complexity of computing image local histograms may not depend on the sliding window size provided memory buffers for storing column-wise window histograms are available. Otherwise it is proportional to vertical size of the sliding window when only row-wise recursive computation is implemented.

Useful statistical characteristics of signals are histogram moments. We will describe them for single component signals. n-th histogram moment $\bar{a}^{(n)}$ of signal sequence $\{a_k\}$ is defined as:

$$\overline{a}^{(n)} = \sum_{q=0}^{Q-1} q^n h(q). \qquad (7.1.8)$$

It follows from this definition and from the definition of the histogram (Eq.7.1.1) that histogram moments can be directly computed from signal samples as:

$$\overline{a}^{(n)} = \frac{1}{N}\sum_{k=0}^{N-1} (a_k)^n . \qquad (7.1.9)$$

Signal first moment is its ***mean value***. Signal second moment is called ***signal variance***. Square root of the variance of signals centered around their mean value

$$\sigma_a = \sqrt{\overline{a_0}^{(2)}} = \sqrt{\overline{a}^{(2)} - \left(\overline{a}^{(1)}\right)^2}\ ; a_0 = a - \overline{a}^{(1)}. \qquad (7.1.10)$$

is called signal ***standard deviation***.

Like the distribution histograms, the moments may be employed as global characteristics of the entire image or of an ensemble of images or as local characteristics of image fragments. Local moments of image array $\{a_{k,l}\}$ in a window of $\left(2N_1^{(W)}+1\right)\left(2N_2^{(W)}+1\right)$ pixel sliding over the image array row-wise and then column-wise:

$$\overline{a}_{k,l}^{(n)} = \frac{1}{\left(2N_1^{(W)}+1\right)\left(2N_2^{(W)}+1\right)}\sum_{r=-N_1^W}^{N_1^W}\sum_{s=-N_2^W}^{N_2^W}\left(a_{r,s}\right)^n \qquad (7.1.11)$$

are local means of signal $\left\{\left(a_{k,l}\right)^n\right\}$. Therefore recursive algorithm for computing signal local mean described in Sect. 5.1.1 can be efficiently used for computing signal local moments.

Digital signal statistical characteristic alternative to histograms and moments are ***order statistics***. Signal order statistics are defined through a notion of the ***variational row***. Variational row $VR_{\{a_k\}}$ of a signal $\{a_k\}$, $k = 0,1,..., N-1$ is obtained by ordering signal samples according to their values in a monotonically increasing sequence:

$$VR_{\{a_k\}} = \left\{a^{(1)}, a^{(2)},..., a^{(N)}\right\}: a^{(1)} \le a^{(2)} \le ... \le a^{(N)}. \qquad (7.1.12)$$

If signal samples happen to have the same quantized value they are positioned in the variational row one after another in an arbitrary order, for instance, in the order they were accessed in the process of the variational row formation. Note that coordinate indices $\{k\}$ of signal samples are ignored in the variational row.

Element $a^{(R)}$ that occupies R-th place in the variational row is called signal R-th ***order statistics***. Index R ($R = 1,2,\dots,N$) of the order statistics is called its ***rank***. The rank shows how many signal samples have value lower or equal to that of the element. First order statistics is signal minimum:

$$a^{(1)} = \mathbf{MIN}\{a_k\}. \tag{7.1.13}$$

Last order statistics in the variational row is signal maximum:

$$a^{(N)} = \mathbf{MAX}\{a_k\}. \tag{7.1.14}$$

If the number of signal samples N is odd, $(N+1)/2$-th signal order statistics that is located exactly in the middle of the variational row is called signal ***median***

$$a^{((N+1)/2)} = \mathbf{MEDN}\{a_k\}. \tag{7.1.15}$$

If N is even number, median may be defined as

$$\left(a^{(N/2-1)} + a^{(N/2-1)}\right)/2 = \mathbf{MEDN}\{a_k\}. \tag{7.1.16}$$

Rank of a sample with quantized value q_a can be found from signal histogram $\overline{h}(q)$ as

$$R(q) = \sum_{q=0}^{q_a-1} \overline{h}(q) + \overline{h}(q_a) - 1. \tag{7.1.17}$$

Signal median may be regarded as an alternative to the signal first moment, its mean value. Yet another alternative to signal mean is signal ***alpha-trimmed mean*** ([1]):

$$\overline{a}_\alpha = \frac{1}{N-\alpha} \sum_{r=\alpha}^{N-1-\alpha} a^{(r)}, \tag{7.1.18}$$

the arithmetic mean of the elements of the variational row which are minimum α elements apart from its ends.

Signal second moment, or its variance, is a parameter that characterizes spread of the signal histogram. An alternative characteristics of signal values spread based on order statistics is signal quasi-spread ([1]):

$$\mathbf{QSPR}_n(\boldsymbol{a}) = \boldsymbol{a}^{\left(\frac{N+1}{2}+n\right)} - \boldsymbol{a}^{\left(\frac{N+1}{2}-n\right)} \tag{7.1.18}$$

with $\boldsymbol{n}$ as a parameter.

In image processing, order statistics and ranks can be measured both globally over entire image or set of images and over a sliding window. In the latter case they are called local. Local histograms, ranks and order statistics are the base of nonlinear filters discussed in Ch. 12.

Local ranks are usually computed for the window central pixel. Because local ranks can be obtained from local histogram (see Eq. 7.1.17), they may be computed in a recursive way. They may also be calculated recursively not through the local histogram, but in a direct manner as well. For this purpose, it suffices to calculate the difference between the number of window elements appearing at a given scanning step with a value smaller than that of the window central pixel, and the number of such elements disappearing from the fragment, and add this difference to the rank obtained at the previous step.

7.1.2 Measuring signal correlation functions and spectra

Auto-correlation functions, cross-correlation functions and power spectra as signal characteristics appear in solving problems of image restoration and object localization with the use of linear filtering (see Chs. 8, 10, 11). Being statistical characteristics, they assume signal treatment as random processes when signals and images subjected to processing are regarded as representative or realizations of a certain signal/image ensemble or data base.

Let $\{\boldsymbol{a}_\omega(\boldsymbol{x})\}$ is an ensemble of signals specified on interval $\boldsymbol{X}$ with $\omega \in \Omega$ as an ensemble Ω parameter. The ensemble may be represented by a set of signals in a data base or it may be defined by a certain mathematical statistical model. The ***autocorrelation function*** of the signal ensemble is defined as:

$$\boldsymbol{R}_a(\xi) = \boldsymbol{AV}_\omega\left(\frac{1}{\boldsymbol{X}}\int_X \boldsymbol{a}_\omega(\boldsymbol{x})\boldsymbol{a}_\omega(\boldsymbol{x}+\xi)d\boldsymbol{x}\right) \tag{7.1.19}$$

where $\mathbf{AV}_{\omega}(\cdot)$ is an operator of averaging over the ensemble. ***Cross-correlation function*** of two signals $a_{\omega}(x)$ and $b_{\omega}(x)$ is defined similarly as:

$$R_{a,b}(\xi) = \mathbf{AV}_{\omega}\left(\frac{1}{X}\int_{X} a_{\omega}(x) b_{\omega}(x+\xi) dx\right). \tag{7.1.20}$$

Fourier transform of signal autocorrelation function

$$P_a(f) = \int_{-\infty}^{\infty} R_a(\xi)\exp(i2\pi f\xi) d\xi \ . \tag{7.1.21}$$

is called signal ensemble ***power spectrum***. Alternatively, signal power spectrum may be found directly from the signal through its Fourier Transform as

$$P_a(f) = \mathbf{AV}_{\omega}\left(\left|\int_{-\infty}^{\infty} a_{\omega}(x)\exp(i2\pi fx) d\xi\right|^2\right). \tag{7.1.22}$$

In this case signal correlation function is computed as Fourier transform of its power spectrum:

$$R_a(\xi) = \int_{-\infty}^{\infty} P_a(f)\exp(-i2\pi f\xi) d\xi \ . \tag{7.1.23}$$

Discrete representation of correlation functions and spectra through signal samples may be obtained similarly to discrete representations of signals discussed in Ch. 4. For signals $\mathbf{a}_{\omega}$ and $\mathbf{b}_{\omega}$ specified by their samples $\{a_{\omega,k}\}$ and $\{b_{\omega,k}\}$, $k = 0,1,...,N-1$, autocorrelation function, cross-correlation function and power spectrum are computed, respectively, as:

$$R_a(n) = \mathbf{AV}_{\omega}\left(\frac{1}{N}\sum_{k=0}^{N-1} a_{\omega,k} a^*_{\omega,k+n}\right), \tag{7.1.24}$$

$$R_{a,b}(n) = \mathbf{AV}_{\omega}\left(\frac{1}{N}\sum_{k=0}^{N-1} a_{\omega,k} b^*_{\omega,k+n}\right); \tag{7.1.25}$$

$$P_a(r) = \frac{1}{\sqrt{N}} \sum_{n=0}^{N-1} R_a(n) \exp\left(i2\pi \frac{nr}{N} \right), \tag{7.1.26}$$

where ω , as before, is a signal ensemble parameter, or, alternatively, as

$$P_a(r) = \mathrm{AV}_\omega \left[\left| \frac{1}{\sqrt{N}} \sum_{n=0}^{N-1} a_{\omega,k} \exp\left(i2\pi \frac{kr}{N} \right) \right|^2 \right] \tag{7.1.27}$$

and

$$R_a(n) = \frac{1}{\sqrt{N}} \sum_{n=0}^{N-1} P_a(r) \exp\left(-i2\pi \frac{nr}{N} \right) \tag{7.1.28}$$

Obviously, in these computations boundary effects associated with the finite length of signals should be treated in ways similar to those discussed in Sect. 5.1.4 for digital filtering. The definitions of signal power spectrum given by Eqs. 7.1.26 and 7.1.27 assume signal periodical replication. If signal correlation function is computed through the DFT of its power spectrum it also will assume signal periodical replication.

Boundary effects in evaluation of signal correlation functions and spectra may be rather severe. It is especially true in image processing. One of a popular solutions of this problem is evaluation signal power spectrum and correlation functions using signal and spectrum windowing as follows:

$$P_a(r) = \mathrm{AV}_\omega \left[\left| \frac{1}{\sqrt{N}} \sum_{n=0}^{N-1} W_k^{(a)} a_{\omega,k} \exp\left(i2\pi \frac{kr}{N} \right) \right|^2 \right] \tag{7.1.29}$$

and

$$R_a(n) = \frac{1}{\sqrt{N}} \sum_{n=0}^{N-1} W_r^{(p)} P_a(r) \exp\left(-i2\pi \frac{nr}{N} \right), \tag{7.1.30}$$

where $W_r^{(a)}$ and $W_r^{(p)}$ are window functions that are equal to one in the middle and the largest part of the interval $0 \div N$ and decay more or less rapidly to its ends. They are discrete analogs of optical apodization masks.

Among the different possible window functions the most known are Hamming window:

$$W_n = 0.54 + 0.46 \cos[\pi (n - N/2)/N] \tag{7.1.31}$$

and Kaiser window:

$$W_n = \frac{I_0\left(\alpha\sqrt{1-[(n-N/2)/N]^2}\right)}{I_0(\alpha)}, \tag{7.1.32}$$

where $I_0(\cdot)$ is the modified zero order Bessel function of the first kind and α is a parameter.

Hamming and Kaiser window functions do not decay to zero at ends of interval $0 \div N$. They are applicable as spectral windows. As signal window functions, they are applicable for signals with zero mean value. For images, window functions that decay to zero are more appropriate. Two possible examples are functions:

$$W_n = \begin{cases} 0.5 + 0.5\cos[\pi(K-n)/K], & n = 0,1,..,K \\ 1, & n = K+1,...,N-K-2 \\ 0.5 + 0.5\cos[\pi(n-N+1+K)/K], & n = N-K-1,...,N-1 \end{cases} \tag{7.1.33}$$

and

$$W_n = \frac{2}{\pi}\operatorname{atan}\left[\beta\exp\left(-\frac{n-N/2}{2\sigma^2}\right)\right] \tag{7.1.34}$$

with K, σ β and as spread and speed of decay parameters.

For 2-D signals and images, window functions may be built as separable to product of 1-D functions. Better results with lower boundary effect artifacts can be achieved with inseparable circularly symmetrical window functions.

Yet another option for evaluating signal power spectra with reduced boundary effects is the use of the DCT as a replacement for the DFT in the same way and on the same base as as it was recommended in Sect. 5.2.2 for signal convolution.

In image processing it is very frequently necessary to evaluate signal correlation functions and spectra with sub-pixel accuracy, that is, not only in the sampling points but in intermediate points as well. This can be achieved by methods of signal discrete interpolation discussed in Ch. 9.

7.1.3 Measuring parameters of random interferences in sensor and imaging systems

One of the tasks in processing images, holograms and interferograms is suppressing random interferences introduced by the imaging, holographic or interferometric sensoric systems. For this, one usually needs to know parameters of the interferences. Sometimes the required data are provided by manufacturers of the systems. In practice, however, such data are often lacking, and parameters of signal interferences need to be derived directly from signals to be processed.

At first sight, this problem may seem to be internally contradictory, because in order to evaluate parameters of noise, one must separate the noise from the signal, which may be done only by knowing the parameters of this noise. A way out of this "vicious circle" is not in separating the signal from noise in order to obtain statistical characteristics of the latter, but rather, in separating signal and noise characteristics based on measurements of the corresponding characteristics of the noisy signal.

Luckily, in most practical cases random interferences, statistically speaking, are very simple objects; that is, they are described by a very small number of parameters. For instance, additive signal independent white Gaussian noise is fully specified by only two parameters: by its mean and variance. Because of this, noise parameter estimation problem may often be solved by rather simple means. Basic principle is choosing those signal characteristics in which signal distortion by noise manifests itself in an abnormal behavior that can be detected in the easiest way.

All detection methods are based on an a priori assumption for certain smoothness and regularity of the characteristics of undistorted signal that are violated when signal contains noise. Two practical detection methods are the prediction method, and the voting method for detecting anomalies.

The prediction method requires that for every element of the analyzed sequence, the difference between its actual value and the value predicted by previously investigated elements be found. If this difference exceeds some given threshold, a decision is made that an abnormal burst has been encountered. The prediction depth, the method of obtaining a predicted value and the threshold value are method parameters that are to be specified a priori.

The simplest prediction technique employs the value of the previous sequence element as the predicted value of the analyzed sequence element. Prediction may also be carried out by means of summing several previous values taken with a certain weight on each. In this case the optimal values of the weight coefficients are found from the conditions of preserving the first, the second and higher differences of the analyzed sequence, or from some auto-regression models built empirically from signal data base.

The voting method implements ideas of robust estimation ([1]). The essence of the method is in considering each element of the analyzed sequence together with certain number of its neighbors. This set of elements is ordered in a variational row and checked whether or not the value of the given element belongs to a certain predefined number of extreme (maximal or minimal) value of the variational row. If the answer is yes, then a decision on the presence of an abnormally high (or, alternatively, abnormally low) value in the given element is made. The voting method is therefore based on the assumption that deviations, if any, from local monotony of a "normal" characteristic under study are insignificant. The number of neighbors and the detection threshold are to be specified a priori from the study of a "normal" behavior of the selected characteristic of the undistorted signal.

We will illustrate these approaches using examples of parameter estimation of additive broad band and narrow-band noise, impulse noise, and quantization noise.

7.1.3.1 Parameter estimation of additive signal-independent broad band noise in images.

Additive statistically signal-independent noise with normal distribution is fully specified by its standard deviation and auto correlation function. If the noise is uncorrelated or weakly correlated (broad band), as is often the case, then its variance and correlation function may be found through the use of the following simple algorithm based on measuring abnormalities in the covariance function of the observed image.

Let it is known that observed signal $\{a_{\omega_a,k}\}$ is contaminated by adding to a “pure” signal $\{\bar{a}_{\omega_a,k}\}$ of a signal ensemble Ω_a zero mean signal independent noise $\{n_{\omega_n,k}\}$ $\omega_n \in \Omega_n$ of noise ensemble with correlation function $R_n(k)$:

$$a_{\omega_a,k} = \bar{a}_{\omega_a,k} + n_{\omega_n,k} \,. \tag{7.1.35}$$

Correlation function of signal $\{a_{\omega_a,k}\}$ is:

$$R_a(r) = AV_{\omega_a}\left[AV_{\omega_n}\left(\frac{1}{N}\sum_{k=0}^{N-1} a_{\omega,k}a^{*}_{\omega,k+r}\right)\right] =$$

$$AV_{\omega_a}\left(\frac{1}{N}\sum_{k=0}^{N-1} \bar{a}_{\omega_a,k}\bar{a}^{*}_{\omega_a,k+r}\right) + AV_{\omega_n}\left(\frac{1}{N}\sum_{k=0}^{N-1} n_{\omega_n,k}n^{*}_{\omega_n,k+r}\right) +$$

$$\frac{1}{N}\sum_{k=0}^{N-1} AV_{\omega_a}\left(a_{\omega_a,k}\right) AV_{\omega_n}\left(n^*_{\omega_n,k+r}\right) + \frac{1}{N}\sum_{k=0}^{N-1} AV_{\omega_a}\left(a_{\omega_a,k}\right) AV_{\omega_n}\left(n^*_{\omega_n,k+r}\right) =$$
$$R_{\bar{a}}(r) + R_n(r) + \varepsilon_n . \qquad (7.1.36)$$

One can see from Eq. 7.1.36 that, owing to noise additivity and statistical independence of the signal, the correlation function of the observed signal $R_a(r)$ is a sum of the correlation function the noise-less signal $R_{\bar{a}}(r)$, the correlation function $R_n(r)$ of the noise and of the realization of some "random" process ε_n. The latter describes the error in measuring the noise correlation function that are in practice carried out over its finite-size realization rather then over the entire ensemble. The variance of the random process ε_n is known to be inversely proportional to the volume of the realization used for the measurement. Normally, in image processing hundreds of thousands of pixels may be involved in the measurement. Therefore, the variance of random error ε_n in (7.1.36) is usually small enough and $R_n(r)$ may be, with a good accuracy, estimated as

$$R_n(r) = R_{\bar{a}}(r) - R_{\bar{a}}(r). \qquad (7.1.37)$$

Consider, first, the uncorrelated noise case when,

$$R_n(r) = \sigma_n^2 \delta(r) = \begin{cases} \sigma_n^2, r = 0 \\ 0, \ \ r > 0 \end{cases}, \qquad (7.1.38)$$

where σ_n^2 is the noise variance. In this case, the correlation function of the observed image deviates from the correlation function of the noise-less image in the coordinate origin only, and the difference is equal to the noise variance:

$$\sigma_n^2 = R_{\bar{a}}(0) - R_{\bar{a}}(0). \qquad (7.1.39)$$

For all other values r the magnitude of $R_{\bar{a}}(r)$ may serve as an estimate of $R_a(r)$. Assuming that $R_a(r)$ is a monotonic function in the vicinity of $r = 0$, one can use $R_{\bar{a}}(r) = R_a(r)$ for $r > 0$ to extrapolate its value $\overline{R}_a(0)$ for $r = 0$ and then use the extrapolated value to compute noise variance from Eq. 7.1.39.

A similar approach may be used to estimate the variance of a weakly correlated (broad band) noise, i.e. a noise whose correlation function $R_n(r)$

is nonzero only in a small vicinity around the coordinate origin. In this vicinity, the values of the noise-less image correlation function may be quite satisfactorily obtained by extrapolating $R_{\bar{a}}(r)$ from points where $R_n(r)$ is known to be zero. Fig. 7-1 illustrates this idea on an example of measuring noise in interferograms.

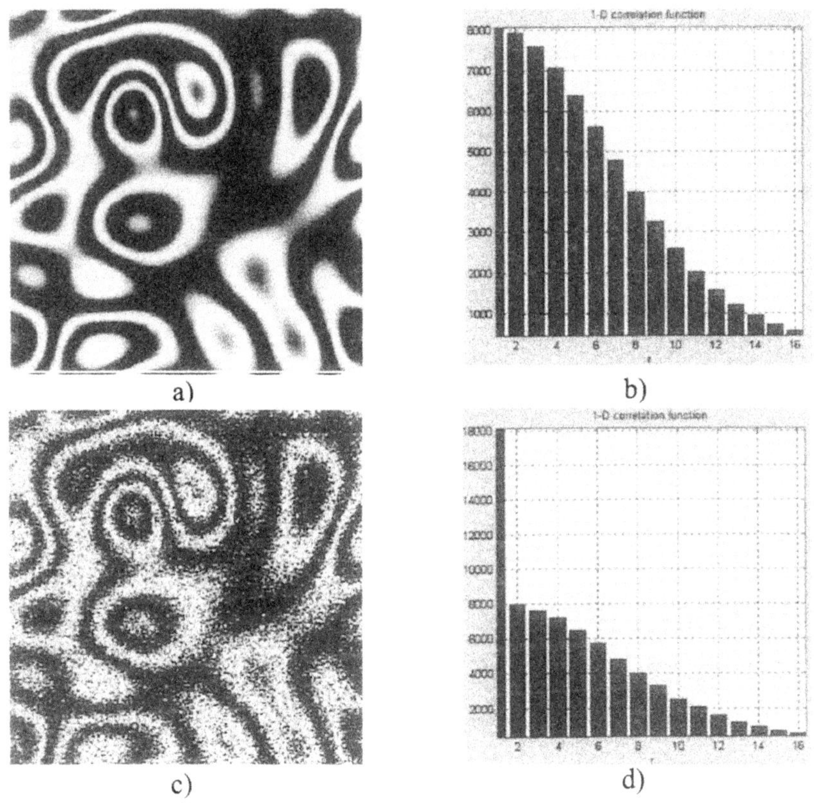

Figure 7-1. Measurement noise level in an interferogram: a) - noise-less interferogram; b) - its 1-D correlation function; c) - noisy interferogram; noise variance 10000; d) - 1-D autocorrelation function of the noisy interferogram. One can clearly see an anomalous peak in the correlation function of the noisy interferogram.

7.1.3.2 Estimating intensities and frequencies of spectral components of periodic and similar narrow-band noise.

Additive periodic (moiré) interferences are very characteristic for electronic imaging systems and image sensors. A feature typical for this type of interference is that their Fourier spectrum contains just a few components. On the other hand, the spatial spectrum of noise-free images is generally a more or less smooth function. Therefore, the presence of a narrow-band noise in images manifests itself in the form of abnormally large and localized peaks in image power spectra and it is image power spectrum that has to be analyzed for the noise diagnostics.

In contrast to the case considered above in which the broad band noise produces anomalous peaks in the correlation function only in the vicinity of the coordinate origin, the locations in spectral domain of peaks of noise spectrum are usually unknown and have to be determined. This may be achieved with the help of the prediction or voting techniques described earlier.

We will explain the diagnostics of the periodical noise in images using as an example an image shown in Fig. 7.2 a).

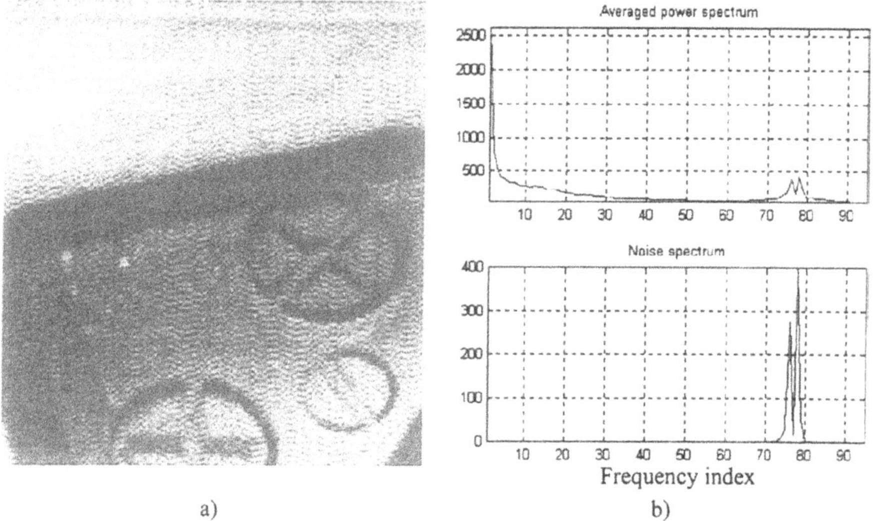

Figure 7-2. Diagnostics of periodical interferences: a) - noisy image; b) averaged column power spectrum (top) and the result of estimation of spectrum of the interference (bottom)

The image is distorted by a periodical interference. The top graph of Fig.7-2, b) shows image 1-D power spectrum obtained by computing squared absolute values of DFT spectra of image columns and summing up spectra of all columns. One can easily see a peak in image high frequencies that can be attributed to the interference. In order to detect this peak and mark spectral components affected by the interference, one can analyze this spectrum beginning from low frequencies employing the prediction or voting techniques. By virtue of noise additivity, the intensity of each distorted spectral component $P_a(r)$ is equal to the sum of the intensity $P_n(r)$ of the corresponding spectral component of noise and of that $P_{\bar{a}}(r)$ of noise-free signal and that. The latter may be estimated by interpolation between those image spectral components that are found not affected by the interference assuming smoothness of noise-free image power spectrum. Power spectrum of the interference may then be found as:

$$P_n(r) = P_n(r) - \hat{P}_{\bar{a}}(r). \tag{7.1.40}$$

This is illustrated in Fig. 7-2, b, bottom graph.

The same approach with analysis of image power spectra may be employed for diagnostics of additive broad band noise in interferograms with a spatial carrier, illustrated by an example shown in Fig. 7-3, a). In spectra of noisy interferograms, noiseless interferograms will manifest themselves in concentrated peaks in spectral domain (see Fig. 7-3, b)) that may be easily detected and marked. Then noise variance may be straightforwardly evaluated over the area in spectral domain outside the marked signal peaks.

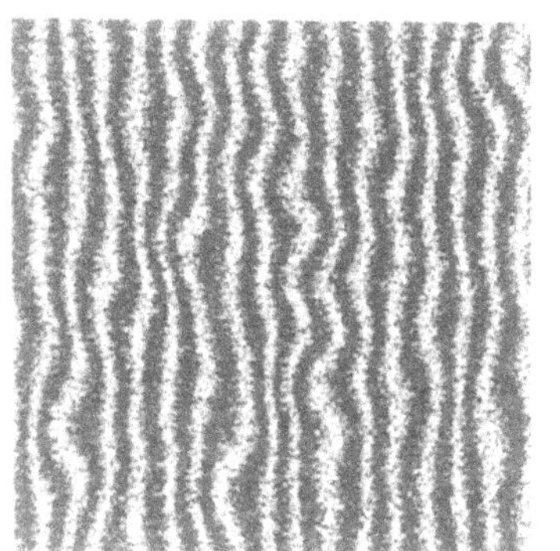

Figure 7-3. Noisy interferogram with a spatial carrier (left) and its power spectrum displayed as an image (right).Vertical and horizontal directions in this image correspond to vertical and horizontal frequencies, respectively while brightness represents spectrum intensity.

7.1.3.3 Parameter estimation of impulse noise, quantization noise.

As it was indicated in Sect. 2.8.1, the major statistical characteristic of impulse noise is the probability of signal sample distortion. It may be estimated through the relative number of the picture elements affected by the noise bursts. These noise bursts may be found with the help of the prediction and voting techniques applied to the analysis of the sequence of noisy image sample values in a small neighborhood of every pixel. Thus, the procedure for estimating the impulse noise parameters may be combined with that of filtering. Impulse noise diagnostics and filtering are treated in more details in Chapter 8.

Quantization noise is determined by the number of signal quantization levels. To find the number of these levels, it suffices to calculate the distribution histogram of the signal values and find the number of such values that corresponds to non-zero samples of the histogram. In a similar way, employing the prediction or voting techniques to find the dips in the histogram and measuring the depth of such dips, we may estimate the probability of failure of any quantization level which is characteristic of some digital imaging systems. Fig. 7-4 shows, as an illustration, an interferogram quantized to 256 gray levels and its histogram. One can easily see in gaps between bars in the histogram that indicate that in reality the interferogram has only about half of 256 levels allocated in the quantization.

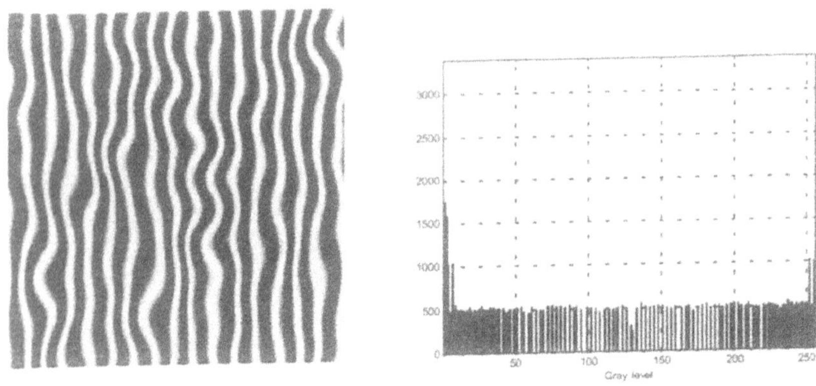

Figure 7-4. An interferogram quantized to 256 gray levels and its histogram

7.2 DIGITAL STATISTICAL MODELS AND MONTE CARLO METHODS

7.2.1 Principles of generating pseudo-random numbers. Generating independent uniformly distributed pseudo-random numbers

Digital modeling is widely used for modeling optical signals and systems. Modeling may be either deterministic or statistical. In statistical modeling, the final result is obtained through averaging of the characteristics and data acquired in a set of experiments. In deterministic modeling, the result of each model experiment is treated individually; no averaging over a set of experiments is assumed. Therefore, unlike deterministic models, statistical digital models include generators of signals that are regarded random.

In statistical modeling, "random" images and objects are generated in the form of arrays of so called "pseudo-random" numbers with pre-defined statistical properties. This means that in this type of modeling it is not the set of specific values of the array elements that are specified but, rather, only certain parameters or characteristics averaged over all realizations of this sequence. The choice of parameters depends upon the nature of the modeling problem we are dealing with. Even though in digital modeling any concrete sequence or realization may be reproduced with absolute accuracy, the effect of any specific set of values of the sequence elements on the modeling result is not considered. Statistical simulation with the use of pseudo-random numbers is referred to as ***Monte-Carlo simulation***.

As a rule, description and specification of images and wave fields as random objects are carried out with the help of such commonly known statistical characteristics as one- and multi-dimensional probability distribution functions, correlation functions and power spectra. The question we consider next is how one can generate realizations of pseudo random sequences with pre-specified statistical characteristics. A general method rests upon an assumption that pseudo-random sequences with arbitrary statistical characteristics may be generated from elementary sequence of statistically independent pseudo-random numbers by means of applying to it appropriate linear or non-linear transformations.

The most widely used algorithm for generating statistically independent pseudo-random numbers is a recursive one: each new number ξ_k in the sequence is generated from the number ξ_{k-1} obtained on the previous recursion step as following:

$$\xi_k = \left(C_1 \xi_{k-1} + C_2\right) \mathbf{mod}\, C_3, \qquad (7.2.1)$$

where C_1, C_2 and C_3 are constant parameters of the algorithm [2]. For generating 2-D pseudo-random arrays of numbers, the arrays are successively filled in row-wise / column-wise.

The initial number, or seed, generally has a little effect upon the quality of the sequence thus obtained. However the algorithm is very sensitive to the parameter especially to parameter C_3 that plays a decisive role in determining the period of repeatability of the generated numbers. Provided appropriately selected parameters, the algorithm generates pseudo-random numbers that do not exhibit noticeable correlations and have approximately uniform distribution. We will refer to this algorithm as "***primary pseudo-random number generator***". It is usually used to generate pseudo-random numbers with any special statistical properties.

7.2.2 Generating pseudo-random numbers with special statistical properties

In applications, pseudo-random numbers with different distribution functions or/and different correlation functions are needed. They are usually generated by means of appropriate transformations of the numbers obtained with primary random number generator.

7.2.2.1 Generating pseudo-random numbers with special distribution function

Pseudo-random numbers $\{\xi\}$ with known distribution density $p(\xi)$ can be converted into numbers $\{\eta\}$ with distribution density $p(\eta)$ by element-wise nonlinear transformation $\eta = F(\xi)$ that is defined by the differential equation:

$$p(\eta)dF = p(\xi)d\xi\ . \tag{7.2.2}$$

Flow diagram of the algorithm is shown in Fig. 7-5.

If primary random number generator produces numbers with uniform distribution in the range $\xi_{min} \div \xi_{max}$ output pseudo-random numbers will have distribution

$$p(\eta) = \frac{1}{(\xi_{max} - \xi_{min})} \frac{d\xi}{dF}\ . \tag{7.2.3}$$

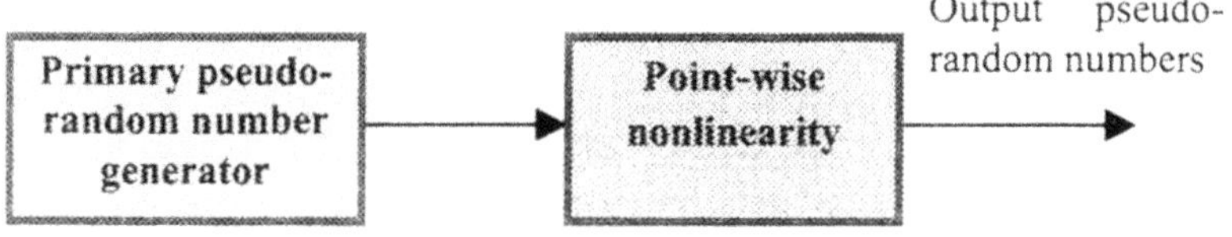

Figure 7-5. Flow diagram of an algorithm for generating pseudo-random numbers with a given distribution function

For example, exponential transformation

$$F(\xi) = \exp(\xi) \tag{7.2.4}$$

enables generating, from uniformly distributed pseudo-random number, pseudo-random number distributed in accord with the hyperbolic law:

$$p(\eta) = \frac{1}{\ln(\eta_{max}/\eta_{min})}\frac{1}{\eta}. \tag{7.2.5}$$

with the range $(\eta_{min} = \exp(\xi_{min}), \eta_{max} = \exp(\xi_{max}))$.

An important sub-class of element-wise transformations, one that accomplishes binarization of values by thresholding

$$F(\xi) = \begin{cases} 1, & \xi \le Thr; \\ 0, & otherwise \end{cases} \tag{7.2.6}$$

is useful for generation binary pseudo-random numbers. For standard primary pseudo-random generator with uniform distribution in the range $(0 \div 1)$, parameter Thr of the algorithm is determined by the required probability $P(1)$ of ones: $Thr = P(1)$.

If transformation function $F(\xi)$ remains the same with all samples of the input sequence (array) of pseudo-random numbers, than the output sequence (array) is stationary, or homogeneous. Inhomogeneous arrays of pseudo-random numbers may be obtained by arranging to alter the transformation function in a certain controlled way.

7.2.2.2 Generating pseudo-random numbers with normal distribution and specified correlation functions and power spectra

Pseudo-random numbers with normal distribution function and special correlation function and spectra are of a special importance in statistical simulation of noise phenomena and in statistical modeling images and wave fields. The major method for generating such numbers is linear filtering of arrays of numbers generated by the primary random number generator according to flow diagram shown in Fig. 7-6.

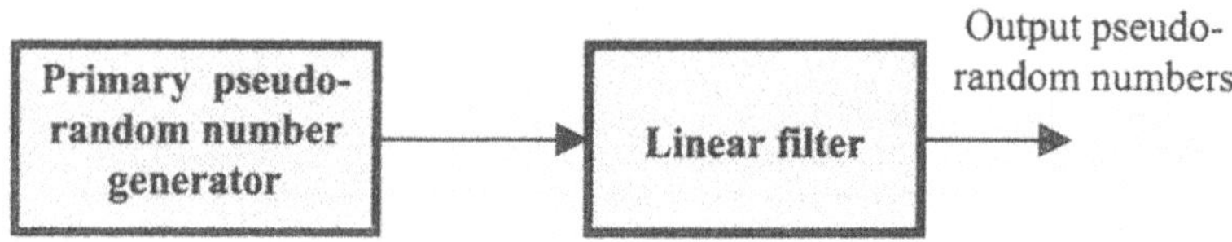

Figure 7-6. Flow diagram of an algorithm for generating pseudo-random numbers with a given correlation function

Correlation function of pseudo-random numbers $\{\eta_k\}$

$$\eta_k = \sum_n h_n \xi_{k-n} \tag{7.2.7}$$

obtained by filtering non-correlated pseudo-random numbers $\{\xi_k\}$ with correlation function $R_\xi = \sigma_\xi^2 \delta(n)$ using digital filter with PSF $\{h_n\}$ is may be found as:

$$R_\eta(n) = AV_\xi(\eta_k \eta_{k+n}) = AV_\xi\left(\sum_m \sum_l h_m h_l \xi_{k-m} \xi_{k+n-l}\right) =$$

$$\sum \sum h_m h_l AV_\xi(\xi_{k-m} \xi_{k-l}) = \sum_m \sum_l h_m h_l \sigma_\xi^2 \delta(m+n-l) = \sigma_\xi^2 \sum_m h_m h_{m+n}, \tag{7.2.8}$$

where $\delta(\cdot)$ is the Kronecker delta-function. Linear filtering results also in "normalization" of the distribution function of numbers $\{\eta_k\}$. By virtue of the central limit theorem of the probability theory, the higher is the number of terms involved in weighted summation of initial pseudo-random numbers for computing each of the numbers $\{\eta_k\}$ according to Eq. 7.2.7, the closer is their distribution function to the Gaussian (normal) one.

As it was described in Sect. 5.2.1, convolution is frequently more efficiently implemented as cyclic convolution in the DFT domain. This suggests an alternative algorithm for generating correlated pseudo-random numbers described by flow diagram in Fig. 7-7. This algorithm enables generating groups of correlated pseudo-random numbers with specified power spectrum and distribution function that, for large group size perfectly approximate the Gaussian (normal) distribution function. The computational complexity of the algorithm is $O(\log_2 N)$ operations per one pseudo-random number of the group of N numbers,.

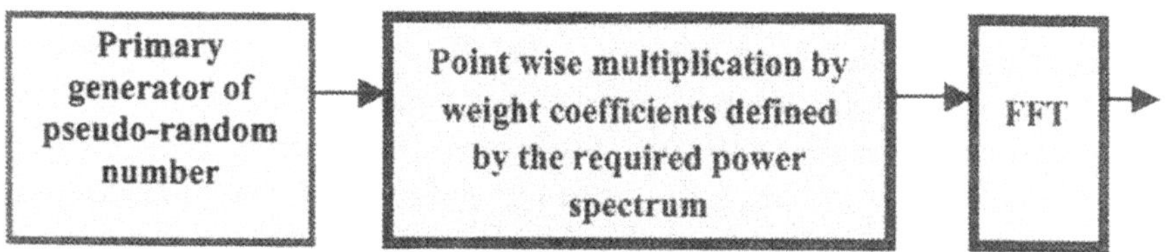

Figure 7-7. Flow diagram of the algorithm for generation pseudo-random numbers with normal distribution and given correlation function

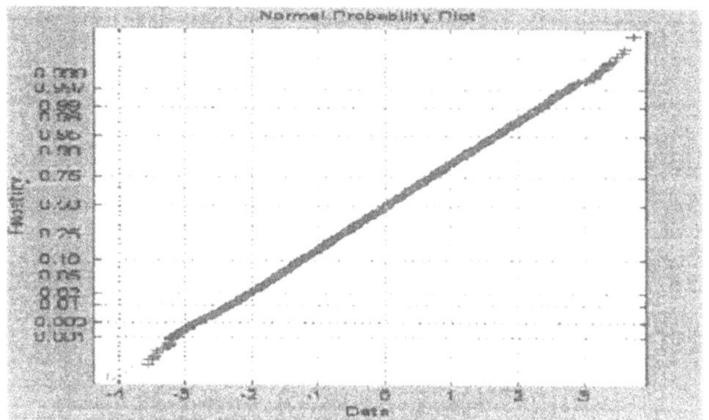

Figure 7-8. Empirical distribution function of 65000 pseudo-random numbers with Gaussian distribution generated by the FFT method

In order to explain the algorithm, let $\{\xi_k^{(re)}\}$ and $\{\xi_k^{(im)}\}$ be two arrays of N non-correlated pseudo-random numbers. Form out of them an array of complex numbers $\{\xi_k^{(re)} + i\xi_k^{(im)}\}$, multiply these numbers by some weight coefficients $\{w_k\}$ and compute DFT of the result:

$$\eta_r = \eta_r^{(re)} + i\eta_r^{(im)} = \frac{1}{\sqrt{N}} \sum_{k=0}^{N-1} w_k \left(\xi_k^{(re)} + i\xi_k^{(im)}\right) \exp\left(i2\pi \frac{kr}{N}\right) \qquad (7.2.9)$$

By virtue of the central limit theorem of the probability theory, distribution function of pseudo-random numbers $\{\eta_r^{(re)}\}$ and $\{\eta_r^{(im)}\}$ tends, with the increase of N, to the Gaussian (normal) one for practically whatever distribution function of the initial pseudo-random numbers $\{\xi_r^{(re)}\}$ and $\{\xi_r^{(im)}\}$. In practice of Monte Carlo simulation in image processing and holography, the dimension N of arrays amounts to tens and hundreds of thousands. Therefore the quality of approximation of normal distribution function is usually extremely high. Fig. 7-8 illustrates this feature of the method. Graph on the figure shows distribution function of 65000 pseudo-random numbers generated by the described method. The graph is plotted coordinates in which true Gaussian distribution corresponds to the diagonal straight line. One can see from the graph that the empirical distribution slightly deviates from the diagonal only for probabilities of the order of $\mathbf{10^{-4}}$.

For zero mean and mutually non-correlated numbers $\{\xi_k^{(re)}\}$ and $\{\xi_k^{(im)}\}$ with the same variance $\sigma_\xi^2 = \mathbf{AV}_\xi\left(\xi_k^{(re)}\right) = \mathbf{AV}_\xi\left(\xi_k^{(im)}\right)$, correlation functions of numbers $\{\eta_r^{(re)}\}$ and $\{\eta_r^{(im)}\}$ and their mutual correlation function are determined by weight coefficients $\{w_k\}$:

$$\boldsymbol{R}_{\eta^{(re)}}(r_1,r_2) = \mathbf{AV}_\xi\left(\eta_{r_1}^{(re)}\eta_{r_2}^{(re)}\right) = \boldsymbol{R}_{\eta^{(im)}}(r_1,r_2) = \mathbf{AV}_\xi\left(\eta_{r_1}^{(im)}\eta_{r_2}^{(im)}\right) =$$

$$\frac{1}{N}\mathbf{AV}_\xi\left\{\sum_{k=0}^{N-1} w_k\left[\xi_k^{(re)}\cos\left(2\pi\frac{kr_1}{N}\right) + \xi_k^{(im)}\sin\left(2\pi\frac{kr_1}{N}\right)\right]\times\right.$$

$$\left.\sum_{k=0}^{N-1} w_k\left[\xi_k^{(re)}\cos\left(2\pi\frac{kr_2}{N}\right) - \xi_k^{(im)}\sin\left(2\pi\frac{kr_2}{N}\right)\right]\right\} =$$

$$\frac{\sigma_\xi^2}{N}\sum_{k=0}^{N-1} w_k^2\cos\left[2\pi\frac{k(r_1-r_2)}{N}\right]; \tag{7.2.10}$$

$$\boldsymbol{R}_{\eta^{(re)},\eta^{(im)}}(r_1,r_2) = \mathbf{AV}_\xi\left(\eta_{r_1}^{(re)}\eta_{r_2}^{(im)}\right) = \frac{\sigma_\xi^2}{N}\sum_{k=0}^{N-1} w_k^2\sin\left[2\pi\frac{k(r_1-r_2)}{N}\right], \tag{7.2.11}$$

where $\sigma_\xi^2 = \mathbf{AV}_\xi\left(\xi_k^{(re)}\right) = \mathbf{AV}_\xi\left(\xi_k^{(img)}\right)$.

Select weight coefficients $\{w_k\}$ such that $\{w_k = w_{N-k}\}$. Then obtain:

$$\boldsymbol{R}_{\eta^{(re)}}(r_1,r_2) = \frac{\sigma_\xi^2}{N}\sum_{k=0}^{N-1} w_k^2\exp\left[2\pi i\frac{k(r_1-r_2)}{N}\right] \tag{7.2.12}$$

and

$$R_{\eta^{(re)},\eta^{(im)}}(r_1, r_2) = 0 . \qquad (7.2.13)$$

Therefore, squared weight coefficients $\{w_k\}$ define power spectrum of numbers $\{\eta_r^{(re)}\}$ and $\{\eta_r^{(im)}\}$ and these sets of numbers are mutually non-correlated.

As it was already mentioned, the computational complexity of the FFT method is $O(\log_2 N)$ operations per one of N numbers assuming using a FFT for computing the DFT. We will refer to this method as ***FFT method for generating correlated numbers with Gaussian distribution***. An important advantage of the FFT method, in addition to its capability to very accurately imitate Gaussian distribution with low computational complexity, is the fact that it is not very sensitive to the distribution function of the primary pseudo-random numbers and to their possible correlations. These correlations, if they exist, will not propagate to output and may cause only certain inhomogeneity of the resulted arrays of correlated Gaussian numbers.

7.2.3 Generating pseudo-random images

One of the applications of the above-described methods for generating pseudo-random numbers with specified statistical characteristics is generating pseudo-random images that imitate natural texture images. a constructive approach to solving this problem is an algorithmic one illustrated in Fig. 7-9 ([3]). According to this approach, any texture image may be created by means of an appropriately designed transformation of a primary generating texture produced by the primary pseudo-random number generator. We will illustrate this approach using several examples.

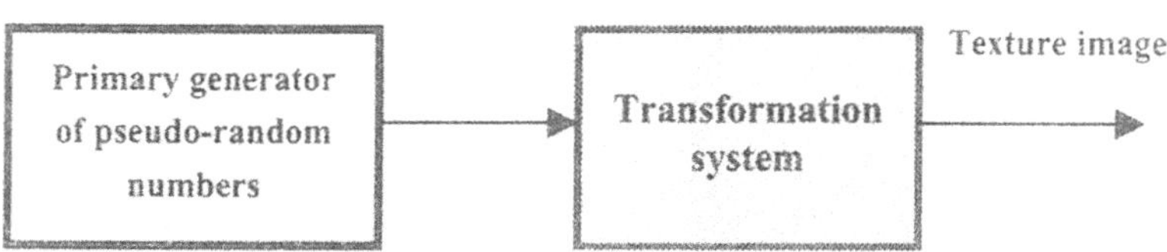

Figure 7-9. An algorithmic approach to synthesis of texture images

Texture images shown in Fig. 7-10 were generated by linear filtering of a primary array of uniformly distributed uncorrelated pseudo-random numbers by the above-described FFT method. Image (a) was produced by linear filtering with circular frequency response shown in the black box at the side

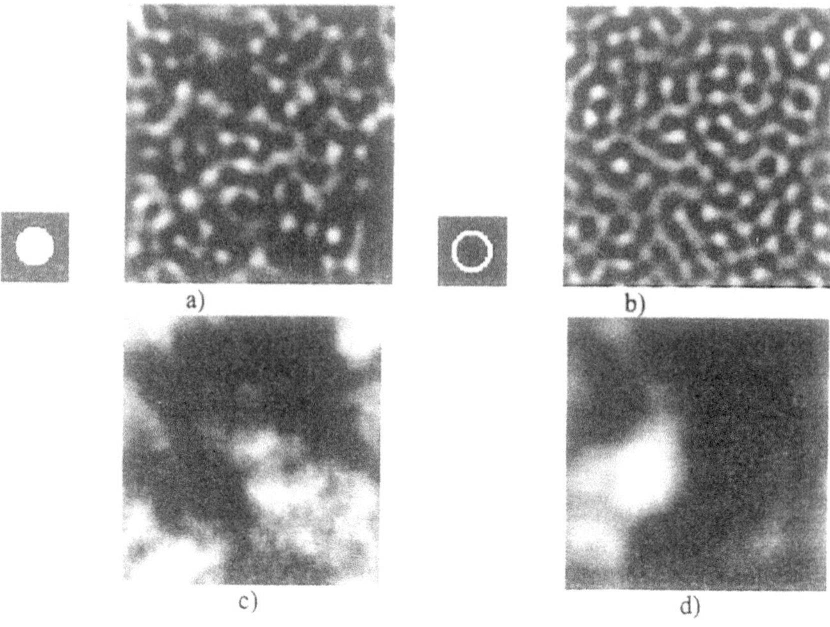

Figure 7-10. Correlated texture images generated by the FFT method.

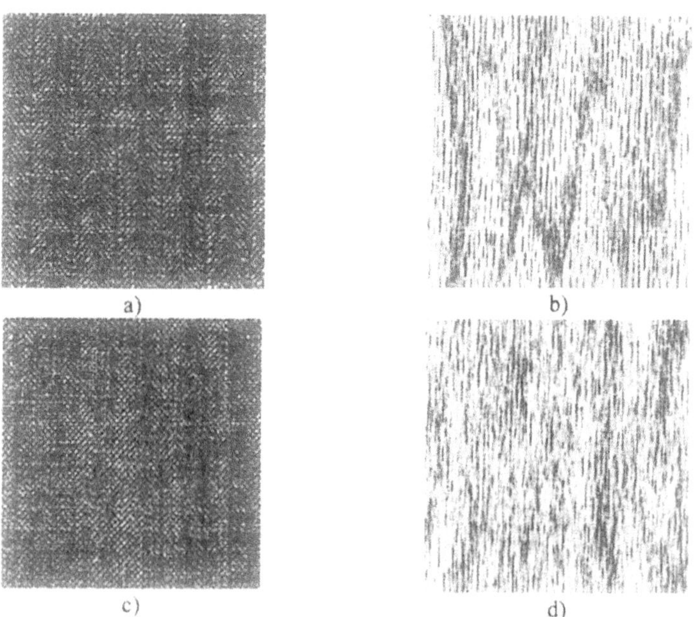

Figure 7-11. Imitating natural textures: a), b) - images of natural textures; c), d) - generated images

of the image. The radius of the circle is equal to 0.1 of the highest spatial frequency. Image (b) was produced using linear filter with frequency response in a form of a ring shown in the black box. External radius of the ring is 0.1 and internal radius is 0.075 of the highest spatial frequency of the image.

Images (c) and (d) exemplify fractal images with power spectrum that decays with spatial frequency index r as r^{-1} and $r^{-1.5}$ respectively. Such images can be used to imitate clouds of different density.

Images shown in Fig. 7-11, (c) and (d) show textures that imitate natural texture image of textile (a) and wood (b), correspondingly. They were obtained by assigning power spectra of natural texture images to images generated from primary independent random numbers.

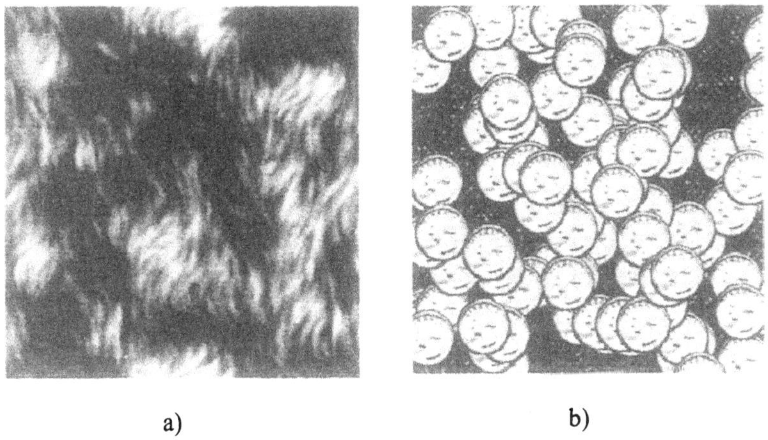

a) b)

Figure 7-12. Inhomogeneous and composite synthetic texture images

Image shown in Fig. 7-12, (a) represents an example of a spatially inhomogeneous texture generated by the method of recursive filtering in sliding window described in Sect. 5.4. In each position of the window, central pixel of the window is generated by inverse FFT after assigning to primary pseudo-random numbers in the window an anisotropic power spectrum. Orientation of the anisotropy was controlled by another texture image of the type shown in Fig. 7-10, (a).

Image in Fig. 7-12, (b) is a complex texture image composed of two textures. The background texture is a synthetic texture similar to that shown in Fig. 7-9, (c) that imitates textile. The foreground image was obtained by convolution, with a template in a form of a coin, of a binary texture image with probability of ones $4 \cdot 10^{-4}$.

7.2.4 Generating correlated phase masks. Kinoform and programmed diffuser holograms

Correlated phase masks are frequently needed in digital holography. Two typical examples are ***kinoform*** and ***programmed diffuser holograms***.

Kinoform is a phase only computer generated hologram ([4]). In synthesis of kinoform, a pseudo-random array of numbers in the range $(-\pi, \pi)$ is generated and recorded on a phase media as a hologram phase profile. The array should be selected in such a way as to reconstruct from the kinoform, in Fourier plane of an optical reconstruction setup, a required image, that is, a certain given distribution of light energy.

"Programmed diffuser" method for synthesis of computer generated display holograms simulates diffuse light scattering from an object surface by assigning to the object wave front a correlated pseudo-random phase component with power spectrum that is determined by a given scattering directivity pattern of the object diffuse surface ([5]). Diffuse surface directivity pattern is the distribution, in the far (Fourier) zone of diffraction, of intensity of radiation scattered by the surface uniformly illuminated by a plane wave.

Let $\{\vartheta_{k,l}\}$ be an array of numbers in the range $(-\pi, \pi)$ representing the required phase mask and $\{P(r,s)\}$ is a discrete representation of the required distribution of light intensity in the Fourier domain. In both above cases, as well as in general in the synthesis of correlated phase masks, it is required that

$$P(r,s) = C \cdot \left|\mathbf{DFT}\left(\left\{\exp\left(i2\pi\vartheta_{k,l}\right)\right\}\right)\right|^2, \tag{7.2.14}$$

where C is an appropriate normalizing constant. Given $\{P(r,s)\}$, $\{\vartheta_{k,l}\}$ should be found as a solution of Eq. 7.2.14. In general, the exact solution of the equation may not exist. However, one can always replace it by a solution $\{\hat{\vartheta}_{k,l}\}$, taken from a certain class of phase distributions, that minimizes an appropriate measure $D(\cdot,\cdot)$ of deviation of $C \cdot \left|\mathbf{DFT}\left\{\exp\left(i2\pi\hat{\vartheta}_{k,l}\right)\right\}\right|^2$ from $\{P(r,s)\}$:

$$\left\{\hat{\vartheta}_{k,l}\right\} = \underset{\{\vartheta_{k,l}\}}{\arg\min}\left[D\left(\{P(r,s)\}, C \cdot \left|\mathbf{DFT}\left[\left\{\exp\left(i2\pi\hat{\vartheta}_{k,l}\right)\right\}\right]\right|^2\right)\right]. \tag{7.2.15}$$

Pseudo-random number generator may serve as a source for realizations of the phase distributions and a solution of Eq. 7.2.15 may be searched through an iterative procedure ([6]) according to a flow diagram shown in Fig. 7-13. One can also include in this process, when it is necessary,

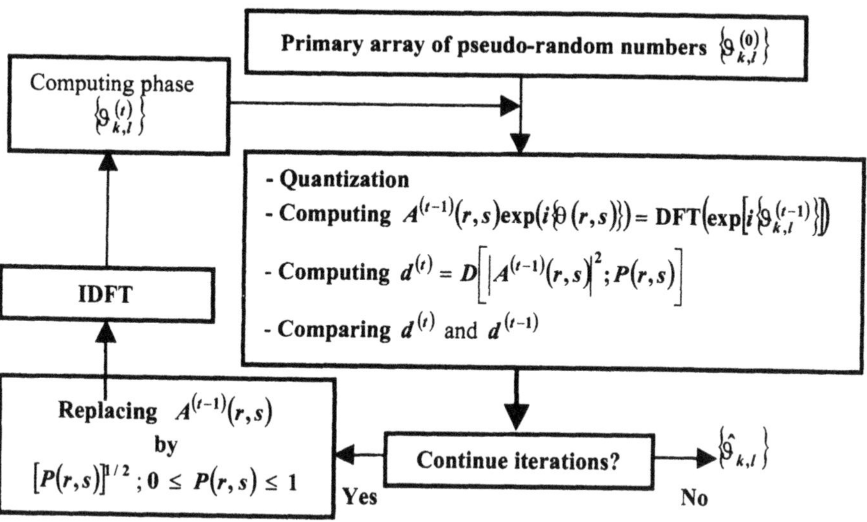

Figure 7-13. Flow diagram of an iterative algorithm for generating pseudo-random phase mask with a specified power spectrum.

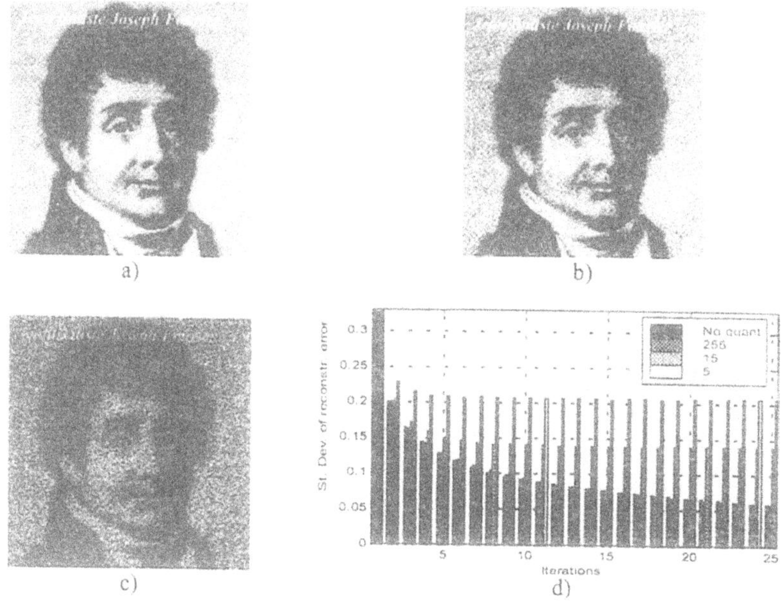

Figure 7-14. Examples of images reconstructed from kinoforms: a) original image used for the synthesis of kinoforms; b) and c) - reconstructed images for kinoforms with 255 and 5 phase quantization levels, respectively; d) plots of standard deviation of the reconstruction error as a function of the number of iterations for no quantization, 255, 15 and 5 phase quantization levels

quantization of the phase distribution to a certain specified number of quantization levels.

Flow diagram of Fig. 7-13 assumes that iterations are carried out over one realization of the array of primary pseudo-random numbers. In principle, one can repeat the iterations for different realizations and then select of all obtained phase masks the one that provides the least deviation from the required distribution $\{P(r,s)\}$.

Illustrative examples of images reconstructed, in computer simulation, from kinoforms synthesized by the described method with 255 and 5 phase quantization levels are shown in Fig. 7-14, b) and c) along with original image (Fig. 7-14, a) used as an amplitude mask $[P(r,s)]^{1/2}$. In this simulation, mean squared error criterion was used as a measure of deviation of the reconstructed image from the original one. On the graph of Fig. 7-13, d), one can also evaluate quantitatively the reconstruction accuracy and the speed of convergence of iterations. The graphs show that iterations converge quite rapidly, especially for coarser quantization and that the reconstruction accuracy for 255 phase quantization levels can be practically the same as that for kinoform without phase quantization. Coarse quantization of the kinoform phase worsens the reconstruction quality substantially.

In the programmed diffuser method, correlated random numbers are needed to simulate, in the synthesis of holograms, a property of object surfaces to non-uniformly scatter irradiation and in this way to provide a visual clue about the object shape in images reconstructed from holograms. The method assumes the observation scheme The method assumes the object observation scheme shown in Fig. 7-15 and defines the object by two array of numbers. The first array $A(x,y)$describes object's reflectivity as a function of coordinates (x,y) in a plane tangent to the object surface and

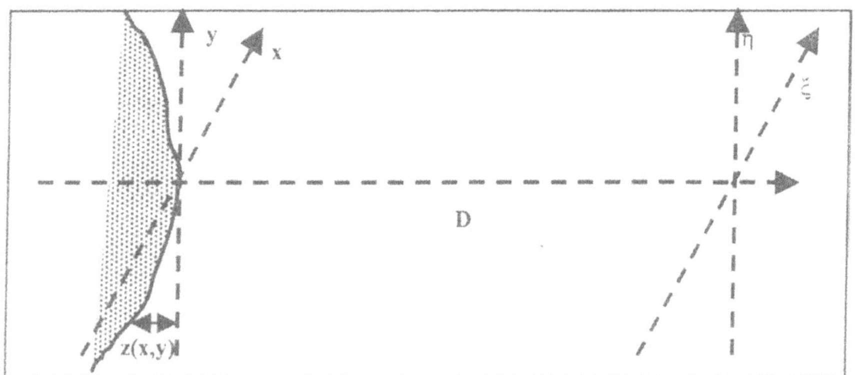

Figure 7-15. Scheme of observation of diffusely scattering objects

perpendicular to the direction to the observation (hologram) plane (ξ,η).. The second array describes object's surface depth $z(x,y)$ with respect to this plane. In addition, an array of correlated pseudo-random numbers $PrDiff(x,y)$ is generated, as it was described above, with power spectrum defined by the required directivity pattern of the object surface. Then the Fourier hologram is computed for the object wave front

$$\tilde{A}(x,y) = A(x,y)\exp\{i\,[z(x,y) + PrDiff(x,y)]\} \qquad (7.2.16)$$

Flow diagram of the computation is shown in Fig. 7.16. Fig. 7-17 illustrates results of reconstruction of a hologram synthesized in this way for a uniformly pained conic surface. As one can see, the hologram is not uniform. Since the object scatters most of radiation perpendicularly to its surface, most of the hologram energy is concentrated along a circle. When fragments of the hologram along the circle are reconstructed, that is when hologram is viewed through windows placed along the circle, as in Fig. 7-17, c, corresponding different aspects of the cone are seen. When viewing window is situated out of the circle, practically nothing can be seen. Fig. 7-18 provides yet another example of reconstruction of a programmed diffuser hologram synthesized for an object in a form of a hemisphere on which numbers 0 and 2 are painted on its four different sides.

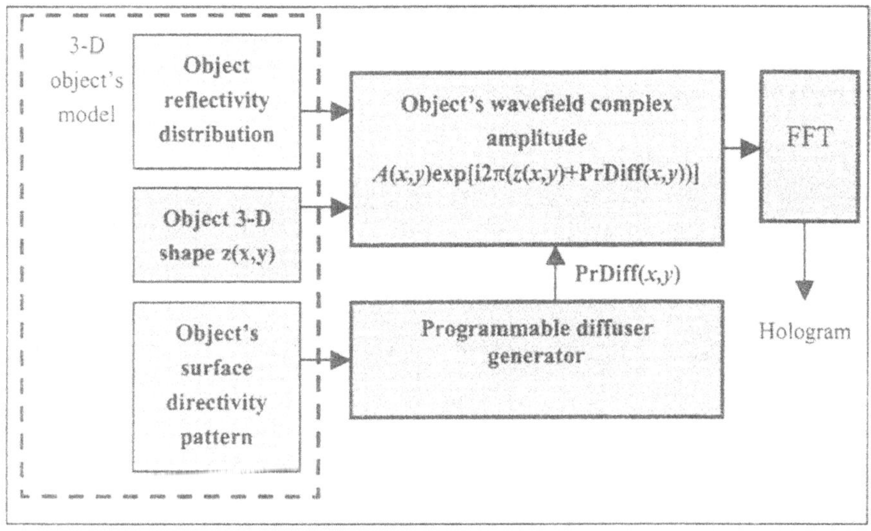

Figure 7-16. Flow diagram of synthesis of programmed diffuser holograms.

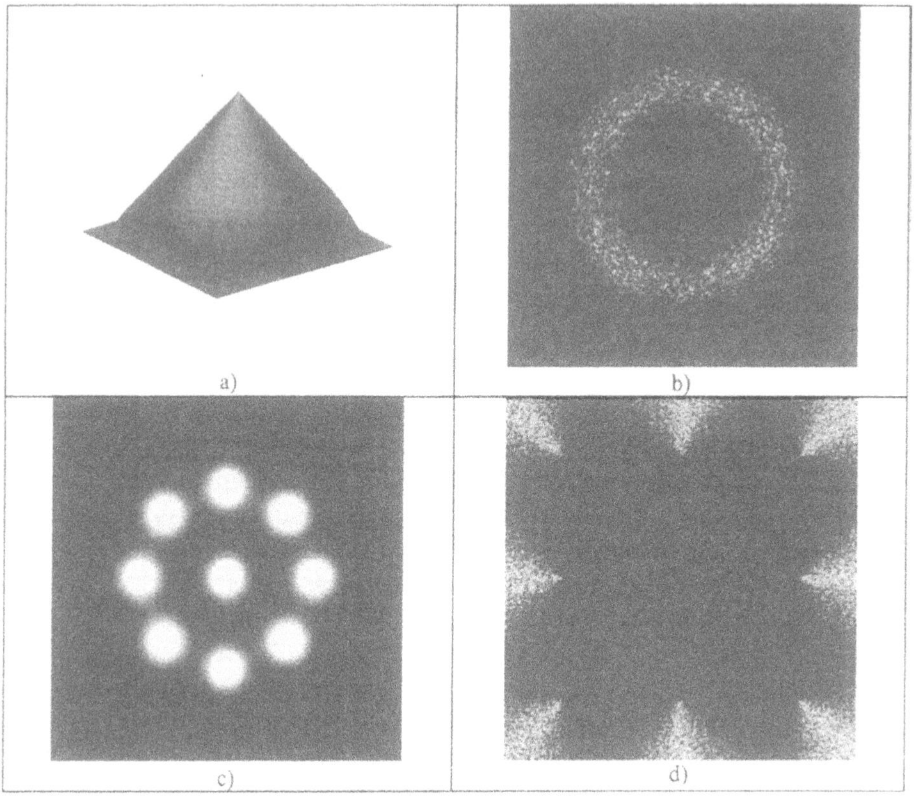

a) b)

c) d)

Figure 7-17. Illustration of reconstruction of holograms with programmed diffuser: Cone phase profile of the object (a), its programmable diffuser Fourier hologram (b), nine hologram reconstruction windows (c) and corresponding reconstructed images (d) that show views of the object from corresponding 9 hologram windows

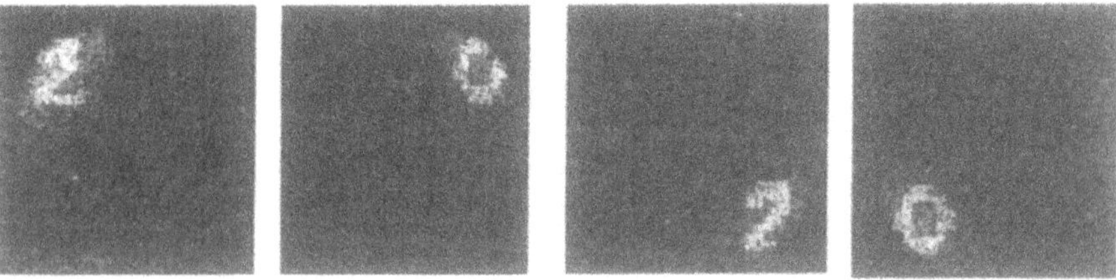

Figure 7-18. Another example of images reconstructed from a programmed diffuser hologram: an object in form of a hemisphere with numbers 0 and 2 on its four sides as it is reconstructed from corresponding four corners (top left, top right, bottom left and bottom right) of its hologram.

7.3 STATISTICAL (MONTE CARLO) SIMULATION. CASE STUDY: SPECKLE NOISE PHENOMENA IN COHERENT IMAGING AND DIGITAL HOLOGRAPHY

Iterative method for generating correlated phase masks described in Sect. 7.2.4 belongs to a family of methods known as ***Monte Carlo methods***. Monte Carlo methods represent a power tool for optimization, simulation and testing systems and processing algorithms. They assume working with multiple realizations of pseudo-random numbers or arrays of pseudo-random numbers and evaluating the results on average over those realizations. In this section we will provide a more extended illustration of Monte Carlo methods using as an example studying statistical properties of speckle noise in coherent imaging such as ultrasound imaging, synthetic aperture radars and digital holographic imaging methods.

Speckle noise imposes fundamental limitation on image quality in coherent radiation based imaging and metrology systems. Appearance of the speckle noise is associated with properties of physical objects to diffusely scatter irradiation and with the fact that holograms are, in reality (contrary to the name "holography" which assumes recording whole wave field), incapable of distortion less recording the object whole wave field. In recording the wave field, a number of signal distortions inevitably occur owing to technical limitations inherent in recording materials and hologram sensors. These are limitation of the hologram area, limitation of the recording media or sensor's dynamic range and hologram signal quantization needed for digital reconstruction of holograms, to name a few.

In most applications, diffuse properties of objects can be treated statistically. Statistical theory of speckle noise was developed with regard to only limited resolution power of coherent imaging devices, or, equivalently, to limitation of hologram area ([7], see also Sect. 2.8.2). It is based on an assumption that the spread of the point spread function of the imaging system is much broader then the correlation interval of microscopic variations of object surface that cause diffuse scattering of irradiation. This theory is, therefore, valid only asymptotically as much as central limit theorem of the probability theory can be applied.

In many cases, especially in acoustic and microwave holography this asymptotic assumption is not always applicable. Moreover, in treating speckle noise problem one should also consider sources of the recorded wave front (hologram) deterioration other then limitation of the hologram area, for instance, such as those mentioned above. Analytical treatment of the relationship between such hologram distortions and statistical characteristics of speckle noise caused by these distortions is most frequently very problematic. A solution may be found by means of Monte Carlo simulation.

7.3.1 Computer model

A flow diagram of a Monte Carlo computer model for studying speckle noise phenomena is shown in Fig.7-15. In what follows we present, as an illustration of the model application, some simulation results of hologram reconstruction of an object with an arbitrary wave front intensity distribution (image brightness) and a rough surface that is modeled by arrays of pseudo-random uncorrelated samples of the phase uniformly distributed in the range $(-\pi,+\pi)$.

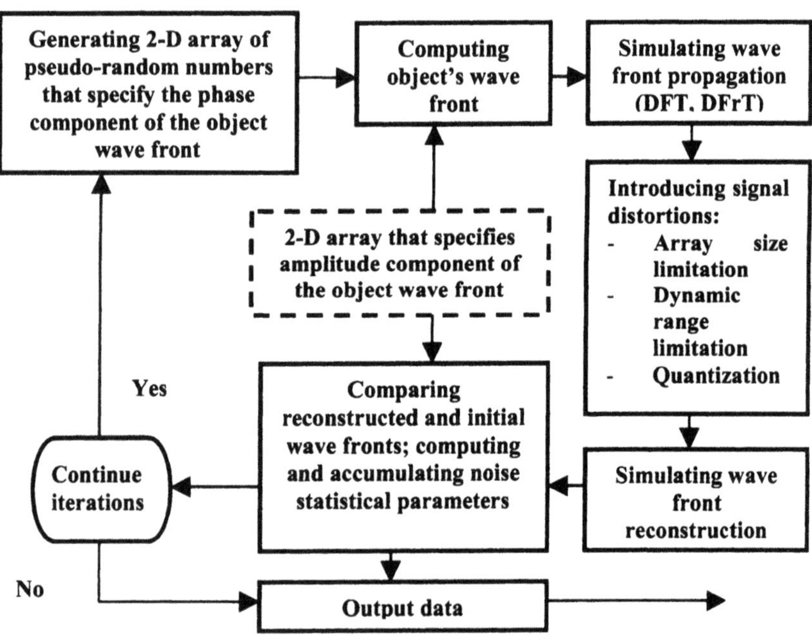

Figure 7-19. Flow diagram of a computer Monte Carlo model for studying speckle noise phenomena

Influence of three types of the wave front distortions on the speckle noise variance will be illustrated on the base of simulation results ([8]):

- limitation of the wave front area;

- limitation of the dynamic range of wave front orthogonal components, i.e., real and imaginary parts of the corresponding complex numbers;
- quantization of the orthogonal components.

Limitation of the wave front area is implemented in the computer model via zeroing array of numbers $\{\gamma_{r,s}\}, r,s = 0,1,...,N-1$, that represent the wave front. It is specified in terms of the fraction of the array area used for reconstruction to the entire array area N^2.

Limitation of the hologram dynamic range is specified in the simulation in units *Thr* of standard deviation σ of real and imaginary parts of wave front orthogonal components $\{\gamma_{r,s}^{(re)}\}$ and $\{\gamma_{r,s}^{(im)}\}$ as

$$\hat{\gamma}_{r,s}^{(re,im)} = \begin{cases} \gamma_{r,s}^{(re,im)}, & \textit{if } \left|\gamma_{r,s}^{(re,im)}\right| \le Thr \cdot \sigma \\ sign\left(\gamma_{r,s}^{(re,im)}\right) \cdot Thr \cdot \sigma, & \textit{otherwise} \end{cases} \tag{7.3.1}$$

where $\mathbf{sign}(\cdot)$ is the signum-function:

$$\mathbf{sign}(\gamma) = \gamma / |\gamma|, \tag{7.3.2}$$

Quantization of the wave front orthogonal components is carried out in the simulation in the range $[-4\sigma, 4\sigma]$ on a nonlinear scale defined as (***P***-th law quantization)

$$\tilde{\gamma}_{r,s}^{(re,im)} = 4\sigma \cdot \mathbf{sign}\left(\bar{\gamma}_{r,s}^{(re,im)}\right)\left\{\mathbf{round}\left(Q\left(\left|\bar{\gamma}_{r,s}^{(re,im)}\right|/4\sigma\right)^P\right)/Q\right\}^{1/P}, \tag{7.3.3}$$

where $\{\bar{\gamma}_{r,s}^{(re,im)}\}$ – are wave front orthogonal components after their limitation within the dynamic range $[-4\sigma, 4\sigma]$; $\mathbf{round}(\cdot)$ is an operation of rounding to the nearest integer, Q –is the number of quantization levels assigned for quantization within positive/negative halves of the dynamic range $[0, 4\sigma]$. Fig. 7-16 shows examples of images with speckle noise generated using the described model.

7.3.2 Simulation results

Fig. 7-17 shows simulation results obtained for limitation of area of the wave field area. The upper band in the image on Fig. 7-17 a) is the original distribution of the object wave front magnitude generated in a form of a step wedge function. Lower bands show magnitudes of images reconstructed from the wave front after limiting its area by portions of 1/8-th of the entire area. They clearly demonstrate appearance of the speckle noise and the fact that the higher is area limitation degree the more correlated is speckle noise.

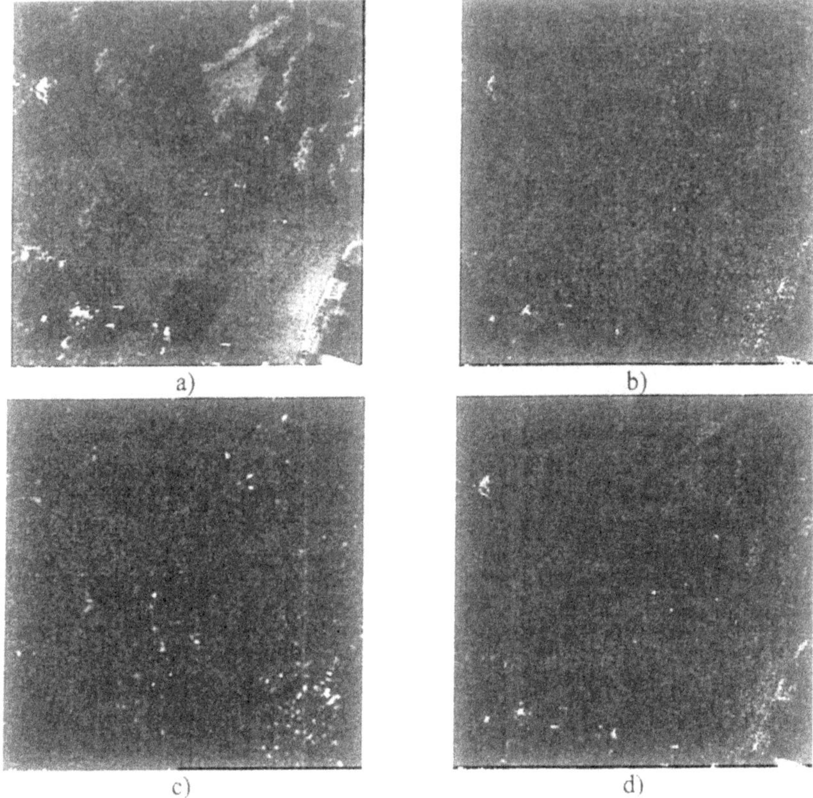

Figure 7-20. Illustrative examples of simulated images: a) - original image; b) - image reconstructed in far diffraction zone from 0.9 of area of the wave front; c) - image reconstructed in far diffraction zone from 0.5 of area of the wave front; d) - image reconstructed in far diffraction zone after limitation of the wave front orthogonal components in the range $\pm\sigma$.

Graphs in Fig. 7.17 b) provide quantitative information regarding speckle contrast as a function of the area limitation degree (for the definition of speckle contrast see Sect. 2.8.2, Eq. 2.8.33). The graphs are plotted for different magnitudes of the object wave front. They show first of all that speckle contrast does not, statistically, depend on the magnitude of the object wave front: curves for different magnitudes practically overlap. This confirms that speckle noise that is caused by the limitation of the wave front area is multiplicative with respect to magnitude of the reconstructed wave front. The graphs show also that speckle contrast asymptotically, for severe limitation, is equal to one as it is predicted by the theory (see Sect. 2.8.2).

However, for moderate limitation, speckle contrast is lower then one and it decreases with the increase of the fraction of the wave front used for object reconstruction. Quite naturally, speckle disappears if image is reconstructed from the whole wave field, which is the case of infinitely high resolving power of the imaging system.

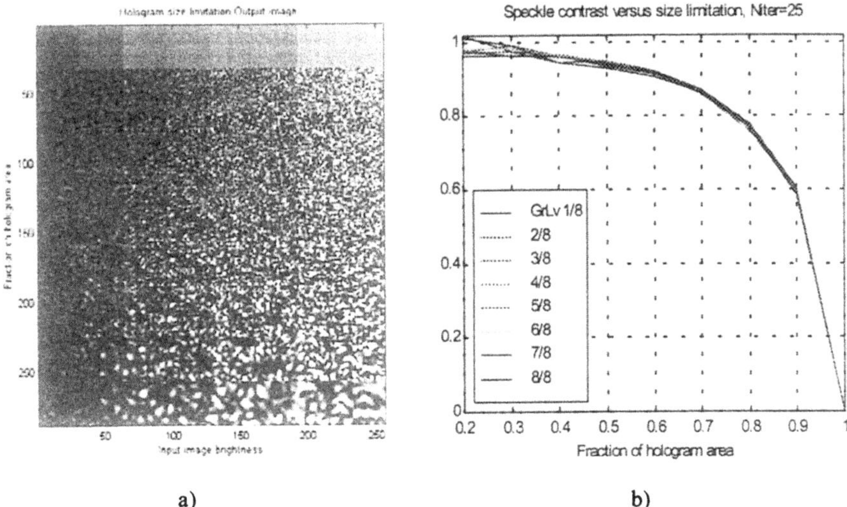

a) b)

Figure 7-21. Limitation of wave front area: a) reconstructed images, b) - speckle contrast as a function of the area limitation degree

Measurements of speckle noise distribution density confirm these observations. They reveal that while for severe limitation of the wave front area the distribution density $\boldsymbol{P(n)}$ of speckle noise values $\boldsymbol{n}$ is the theoretical exponential distribution:

$$\boldsymbol{P(n)=\frac{1}{2\sigma_n^2}\exp\left(-\frac{n}{2\sigma_n^2}\right)}, \qquad (7.3.4)$$

For smaller limitation degree, when the above-mentioned theoretical limiting assumptions are not applicable, it degenerates to the Rayleigh distribution:

$$\boldsymbol{P(n)=\frac{n}{2\sigma_n^2}\exp\left(-\frac{n^2}{2\sigma_n^2}\right)} \qquad (7.3.5)$$

and then, for very small limitation, even to the normal distribution.

$$P(n) = \frac{1}{\sqrt{2\pi\sigma_n^2}} \exp\left(-\frac{(n-\bar{n})^2}{2\sigma_n^2}\right) \tag{7.3.6}$$

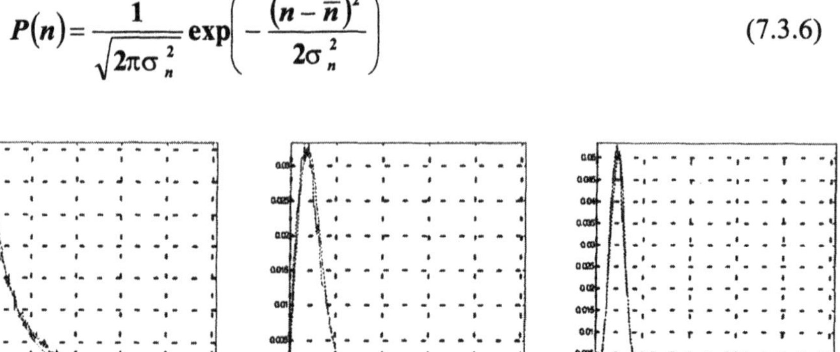

a) b) c)

Figure 7-22. Distribution density functions of speckle noise caused by the limitation of the wave front area: a) - exponential distribution, high limitation (30%); b) - Rayleigh distribution; "Medium" limitation (90%); c) - Normal distribution, very low limitation (97%). Oscillating curves show empirical distribution functions, plane curves - theoretical models of Eqs. 4-6, respectively.

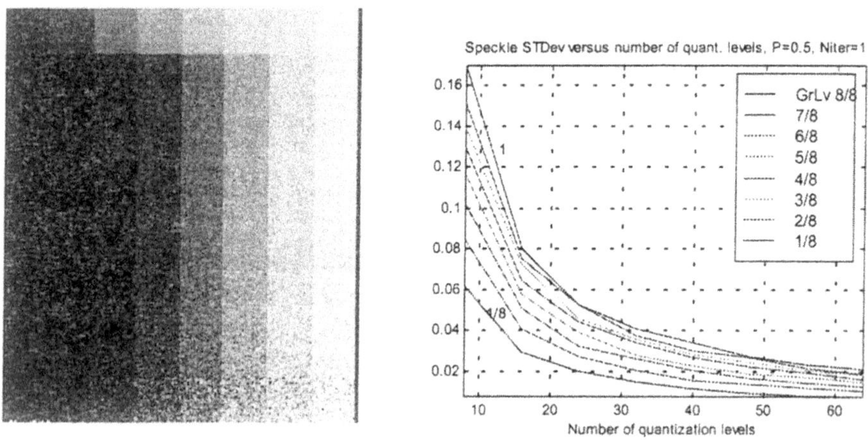

Figure 7-23. Speckle noise caused by quantization of orthogonal components

Simulation of the limitation of wave front orthogonal components shows that, contrary to the limitation of wave front area, limitation of the hologram dynamic range and quantization result in speckle noise whose statistical relationship with the hologram signal is of more involved nature. It can not be uniquely characterized as being additive or multiplicative. In particular, simulation of quantization of the wave front orthogonal components reveals that the model of additive Gaussian noise can be used with respect to the magnitude of the reconstructed wave front (square root of the intensity):

$$\boldsymbol{I}^{1/2} = \bar{\boldsymbol{I}}^{1/2} + \sigma_n \boldsymbol{n}, \tag{7.3.7}$$

where $\boldsymbol{I}$ is intensity of the reconstructed wave front, $\bar{\boldsymbol{I}}$ its mean value, $\boldsymbol{n}$ is uncorrelated random process with normal distribution function that models speckle in the reconstructed wave front and σ_n is standard deviation of the speckle that is inversely proportional to the number o quantization levels $\boldsymbol{Q}$:

$$\sigma_n \propto \frac{1}{\boldsymbol{Q}}. \tag{7.3.8}$$

The simulation allows also to determine the optimal value of the parameter $\boldsymbol{P}$ of non-uniform $\boldsymbol{P}$-th low quantization of wave front orthogonal components (Eq. 7.3.2) that minimizes standard deviation of the speckle noise. Bar diagrams in Fig. 7-20 show that this optimum lies in interval $\boldsymbol{P} = \mathbf{0.3} \div \mathbf{0.5}$.

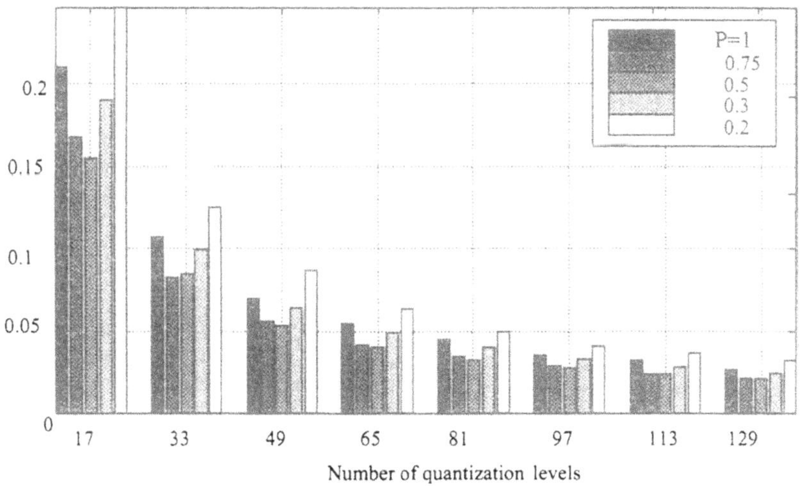

Figure 7-24. Optimization of nonlinearity index $\boldsymbol{P}$ in $\boldsymbol{P}$-th low quantization of wave front orthogonal components.

References

1. P. J. Huber, Robust Statistics, John Wiley and Sons, N.Y., 1981
2. W. H. Press, S. A. Teukolsky, W. T. Vetterling, B. P. Flannery, Numerical Receipes in C. The Art of Scientific Computing, Second Edition, Cambridge University Press, 1995
3. L. Yaroslavsky, M. Eden, Fundamentals of Digital Optics, Birkhauser, Boston, 1995
4. L. B. Lesem, P. M. Hirsch, J. A. Jordan, Kinoform, IBM Journ. Res. Dev., 13, 150, 1969
5. L. P. Yaroslavskii, N. S. Merzlyakov, Methods of Digital Holography, Consultant Bureau, N.Y., 1980
6. N. C. Gallagher, B. Liu, Method for Computing Kinoforms that Reduces Image Reconstruction Error, Applied Optics, v. 12, No. 10, Oct. 1973, p. 2328
7. Goodman, J.W.,. Statistical Properties of Laser Speckle Patterns In Laser Speckle and Related Phenomena, J.C. Dainty, ed. Springer Verlag, Berlin, 1975
8. L. Yaroslavsky, A. Shefler, Statistical characterization of speckle noise in coherent imaging systems, in: Optical Measurement Systems for Industrial Inspection III, SPIE's Int. Symposium on Optical Metrology, 23-25 June 2003, Munich, Germany, W. Osten, K. Creath, M. Kujawinska, Eds., SPIE v. 5144, pp. 175-182

Chapter 8

SENSOR SIGNAL PERFECTING, IMAGE RESTORATION, RECONSTRUCTION AND ENHANCEMENT

8.1 MATHEMATICAL MODELS OF IMAGING SYSTEMS

Imaging systems always have certain technical limitations in their design and implementations and generate images that are not as perfect as they would be if there were no implementation limitations. Deviations of real images from perfect ones may be treated as distortions introduced by imaging systems to hypothetical perfect, or " ideal" signals. Correction of these distortions is the primary goal of image processing.

Methods for distortion correction are based on the canonical model of imaging systems shown in Fig. 8.1.

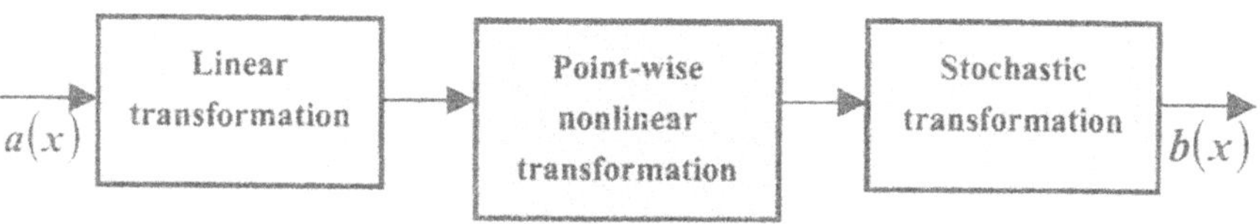

Figure 8-1. A canonical model of imaging systems

The model represents image formation as a combination of signal linear transformations, point-wise nonlinear transformations and stochastic transformations that are applied to a hypothetical perfect signal $a(x)$ and jointly determine the system's output signal $b(x)$.

Linear transformation are specified in terms of the system point spread function or frequency response. Frequency response of the ideal imaging system is assumed to be uniform for all frequencies in a certain base band. Frequency responses of real imaging usually more or less rapidly decay on high frequencies. This results in image distortions such as image blur.

Point-wise nonlinear are specified in terms of the system transfer functions. It is assumed that the ideal system has a linear transfer function. Deviations of system transfer functions form linear one cause distortions of gray scale nonlinear distortions.

Stochastic transformations model signal random distortions and cause random interferences, or noise in output images. Stochastic transformations are specified by statistical noise models described in Sect. 2.8.

The processing goal is estimating the perfect signal $a(x)$ given distorted signal $b(x)$ produced by the system. This problem is frequently referred to as the ***inverse problem***. Signal transformations that are applied to system's output signals $\{b(x)\}$ to produce an estimate $\hat{a}(x)$ of the "ideal" signal are called ***signal recovery*** or, in application to image processing, ***image restoration***. Similar problem is ***image reconstruction***. This term is usually refers to image formation in transform domain imaging such as tomography and holography.

If the system does not introduce any random distortions and system's parameters such as point spread function and transfer function are known, the inverse problem has a trivial solution: signal $b(x)$ has to be subjected to transformations inverse to those introduced by the system.

However, one can not, in general, neglect random or similar uncontrolled distortions such as round-off errors in digital processing. In reality, applying inverse transformation may result in artifacts that may even be useless. To solving the inverse problem in such cases, the statistical approach that explicitly accounts for signal random distortions appears to be the most appropriate.

Generally, signal recovery, image restoration and reconstruction is indivisible procedure that should account for and correct all distortions. However, in practice this process is divided into separate steps carried out in the order inverse to that distortion factors have in the model.

8.2 LINEAR FILTERS FOR IMAGE RESTORATION.

8.2.1 Transform domain MSE optimal scalar Wiener Filters

In this section we consider image restoration for a reduced imaging system model that disregards point-wise nonlinear transformation in imaging systems and treats image distortions as a combination of distortions caused by linear filtering and of those caused by action of random interferences.

The design of the optimal restoration procedure requires specifying a criterion for evaluating the restoration quality. Let $\{b_k\}$ be a set of N signal samples ($k = 0,1,...,N-1$) at the output of the imaging system, $\{a_k\}$ be a set of the system's input signal samples that model perfect, or "ideal" signal and $\{\hat{a}_k\}$ be a set of restored signal samples. For the sake of generality, we will consider the set $\{a_k\}$ as a realization taken from a statistical ensemble Ω_A or data base of input signals and the set $\{b_k\}$ as a realization of a signal ensemble generated by ensembles Ω_A and Ω_N of input signals and of random interferences.

Define the restoration procedure performance measure as a squared difference between restored and perfect signals averaged over the available set of signal samples and over statistical ensembles Ω_A and Ω_N. We will call this measure ***mean squared restoration error*** (MSE). The restoration procedure $\mathbf{R}\{b_k\} = \{\hat{a}_k\}$ that minimizes this difference:

$$\{\hat{a}_k\} = \underset{\mathbf{R}\{b_k\}=\{\hat{a}_k\}}{\arg\min}\left\{\mathbf{AV}_{\Omega_A}\mathbf{AV}_{\Omega_N}\left(\sum_{k=0}^{N-1}|a_k - \hat{a}_k|^2\right)\right\}. \tag{8.2.1}$$

will be referred to as to ***MSE-optimal filtering***.

For the implementation of the MSE-optimal filtering, we will restrict ourselves to linear filtering. In general, linear filtering of a discrete signal may be described as multiplication of a vector of input signal samples $\mathbf{B} = \{b_k\}$ by a filter matrix $\mathbf{H}$ (vector filter, see Ch. 4, Eq. 4.1.7):

$$\hat{\mathbf{A}} = \mathbf{H} \cdot \mathbf{B}, \tag{8.2.2}$$

where $\hat{\mathbf{A}} = \{\hat{a}_k\}$ is a vector of filter output signal samples

For signals of N samples, a general vector filter matrix $\mathbf{H}$ has dimensions $N \times N$. Specification of such a filter requires determining N^2

filter coefficients, and the filtering itself requires performing N^2 operations per N signal samples. In image processing, the computational complexity of both determination of filter coefficients and of the filtering may be become too high because of high dimensionality of image arrays. Fast transforms that may be computed for $O(N\log N)$ operations (see Ch. 6) allow to radically decrease the filter design and the implementation complexity. Therefore in what follows we will consider only scalar filtering in a domain of orthogonal transforms that can be computed with fast algorithms. This class of filters may be described by the equation:

$$\hat{\mathbf{A}} = \mathbf{T}^{-1} \cdot \mathbf{H}_d \cdot \mathbf{T} \cdot \mathbf{B}, \tag{8.2.3}$$

where, $\mathbf{T}$ and $\mathbf{T}^{-1}$ are, correspondingly, direct and inverse orthogonal transforms, $\mathbf{H} = \mathbf{diag}\{\eta_r\}$, $(r = 0,1,..., N-1)$ is a diagonal filter matrix. Such a scalar filtering implies the following relationship between filter output and input signal samples $\{\hat{\alpha}_r\} = \mathbf{T} \cdot \hat{\mathbf{A}}$ and $\{\beta_r\} = \mathbf{T} \cdot \mathbf{B}$:

$$\hat{\alpha}_r = \eta_r \beta_r. \tag{8.2.4}$$

In the assumption of orthogonality of the transform $\mathbf{T}$, one can, by virtue of the Parceval relationship (Eq. 2.1.28), modify the filter optimality condition defined by Eq. 8.2.1 in the following way:

$$\{\hat{\alpha}_r\} = \underset{\{\eta_r\}}{\arg\min}\left\{\mathbf{AV}_{\Omega_A}\mathbf{AV}_{\Omega_N}\left(\sum_{r=0}^{N-1}|\alpha_r - \hat{\alpha}_r|^2\right)\right\} =$$
$$\underset{\{\eta_r\}}{\arg\min}\left\{\mathbf{AV}_{\Omega_A}\mathbf{AV}_{\Omega_N}\left(\sum_{r=0}^{N-1}|\alpha_r - \eta_r\beta_r|^2\right)\right\}. \tag{8.2.5}$$

By computing derivatives over sought variables and equaling them to zero, one can obtain from Eq. 8.2.5 that optimal scalar filter coefficients may be found as cross-correlation coefficients between spectral coefficients β_r and α_r of the input and perfect signals:

$$\eta_r = \frac{\mathbf{AV}_{\Omega_A}\mathbf{AV}_{\Omega_N}\left(\alpha_r\beta_r^*\right)}{\mathbf{AV}_{\Omega_A}\mathbf{AV}_{\Omega_N}\left(|\beta_r|^2\right)}. \tag{8.2.6}$$

We will refer to MSE optimal linear filters and, in particular, to MSE optimal scalar filters defined by Eq. 8.2.6 as to ***Wiener filters***. This name gives a credit to Norbert Wiener for his pioneer works in the theory of statistical methods of signal restoration ([1])[1].

In order to implement optimal scalar Wiener filter, one should therefore know cross-correlation $\left\{\mathbf{AV}_{\Omega_A}\mathbf{AV}_{\Omega_N}\left(\alpha_r \beta_r^*\right)\right\}$ between filter input signal and perfect signal spectral coefficients and power spectrum $\left\{\mathbf{AV}_{\Omega_A}\mathbf{AV}_{\Omega_N}\left(\left|\beta_r\right|^2\right)\right\}$ of the input signal in the selected basis. The statistical approach we adopted that assumes averaging of the restoration error over statistical ensembles of perfect signals and of filter input signals implies that these statistical parameters should be measured in advance for these ensembles or over the data bases.

8.2.2 Empirical Wiener filters for image denoising

Consider a ASIN-model (Sect. 2.8.1) in which filter input signal samples $\{\boldsymbol{b}_k\}$ are obtained as a sum of perfect signal samples $\{\boldsymbol{a}_k\}$ and samples $\{\boldsymbol{n}_k\}$ of signal independent zero mean random noise:

$$\boldsymbol{b}_k = \boldsymbol{a}_k + \boldsymbol{n}_k \,. \tag{8.2.7}$$

In spectral domain, the same relationship holds for signal and noise spectral coefficients:

$$\beta_r = \alpha_r + \nu_r , \tag{8.2.8}$$

where $\{\nu_r\} = \mathbf{T}\{\boldsymbol{n}_k\}$. For this model one can obtain that

$$\mathbf{AV}_{\Omega_A}\mathbf{AV}_{\Omega_N}\left(\alpha_r \beta_r^*\right) = \mathbf{AV}_{\Omega_A}\mathbf{AV}_{\Omega_N}\left[\alpha_r\left(\alpha_r^* + \nu_r^*\right)\right] = \mathbf{AV}_{\Omega_A}\left(\left|\alpha_r\right|^2\right) \tag{8.2.9}$$

and

$$\mathbf{AV}_{\Omega_A}\mathbf{AV}_{\Omega_N}\left(\left|\beta_r\right|^2\right) = \mathbf{AV}_{\Omega_A}\mathbf{AV}_{\Omega_N}\left[\left(\alpha_r + \nu_r\right)\left(\alpha_r^* + \nu_r^*\right)\right] = \mathbf{AV}_{\Omega_A}\left(\left|\alpha_r\right|^2\right) + \mathbf{AV}_{\Omega_A}\left(\left|\nu_r\right|^2\right) \tag{8.2.10}$$

[1] It will be just to give also a credit to Andrey N. Kolmogorov who developed the similar theory for discrete signals ([2]).

because for zero mean noise $\mathbf{AV}_{\Omega_N}(\nu_r^*)=\mathbf{AV}_{\Omega_N}(\nu_r)=\mathbf{0}$. Therefore scalar Wiener filter for suppressing additive signal independent noise is defined through its coefficients $\{\eta_r\}$ as

$$\eta_r=\frac{\mathbf{AV}_{\Omega_A}\left(|\alpha_r|^2\right)}{\mathbf{AV}_{\Omega_A}\left(|\alpha_r|^2\right)+\mathbf{AV}_{\Omega_N}\left(|\nu_r|^2\right)}. \tag{8.2.11}$$

One can give to this formula a clear physical interpretation. Define ***signal-to-noise ratio*** as:

$$\boldsymbol{SNR}_r=\frac{\boldsymbol{AV}_{\Omega_A}\left(|\alpha_r|^2\right)}{\boldsymbol{AV}_{\Omega_N}\left(|\nu_r|^2\right)}. \tag{8.2.12}$$

Then obtain:

$$\eta_r=\frac{\boldsymbol{SNR}_r}{1+\boldsymbol{SNR}_r} \tag{8.2.13}$$

which means that scalar Wiener filter weight coefficients are defined, for each signal spectral coefficient, by the signal-to-noise ratio for this coefficient. The lower is signal-to-noise ratio for a particular signal spectral component, the lower will be the contribution of this component to the filter output signal.

In order to implement scalar Wiener filter one have to measure in advance power spectra $\mathbf{AV}_{\Omega_A}\left(|\alpha_r|^2\right)$ and $\mathbf{AV}_{\Omega_N}\left(|\nu_r|^2\right)$ of perfect signals and of noise in the selected basis. Noise power spectrum may be known from the specification certificate of the imaging device. Otherwise it may be measured in noisy input signals using methods described in Sect 7.1.2. As for the prefect signal power spectrum, it is most frequently not known. However, one can, using Eq. 8.2.10, attempt to estimate it from the power spectrum $\mathbf{AV}_{\Omega_A}\mathbf{AV}_{\Omega_N}\left(|\beta_r|^2\right)$ of input noisy signals as

$$\mathbf{AV}_{\Omega_A}\left(|\alpha_r|^2\right)=\mathbf{AV}_{\Omega_A}\mathbf{AV}_{\Omega_N}\left(|\beta_r|^2\right)-\mathbf{AV}_{\Omega_A}\left(|\nu_r|^2\right). \tag{8.2.14}$$

The latter has to be estimated from the observed signal spectrum $|\beta_r|^2$. Denote this estimate as $\overline{|\beta_r|^2}$. This empirical estimate made by averaging over available realization of input images may, when used as a replacement for $\mathbf{AV}_{\Omega_A}\mathbf{AV}_{\Omega_N}\left(|\beta_r|^2\right)$, give negative values for some spectral coefficients because of the limited depth of the averaging. Since power spectra can not assume negative values, the following modified spectrum estimation may be adopted:

$$\mathbf{AV}_{\Omega_A}\left(|\alpha_r|^2\right)=\mathbf{max}\left[\overline{|\beta_r|^2}-\mathbf{AV}_{\Omega_A}\left(|\nu_r|^2\right);\ \ \mathbf{0}\right]. \tag{8.2.15}$$

In this way we arrive at the filter:

$$\eta_r=\mathbf{max}\left[\frac{\overline{|\beta_r|^2}-\mathbf{AV}_{\Omega_A}\left(|\nu_r|^2\right)}{\overline{|\beta_r|^2}};\ \ \mathbf{0}\right]. \tag{8.2.16}$$

We will refer to this filter as to the ***empirical Wiener filter***.

If the imaging system noise is known to be white noise with variance σ_n^2, the empirical Wiener filter takes the form:

$$\eta_r=\mathbf{max}\left[\frac{\overline{|\beta_r|^2}-\sigma_n^2}{\overline{|\beta_r|^2}};\ \ \mathbf{0}\right]. \tag{8.2.17}$$

As a zero order approximation to the input images power spectrum, power spectrum $|\beta_r|^2$ of a single input image subjected to filtering may be used. In this case empirical Wiener filter weight coefficients are found as:

$$\eta_r=\mathbf{max}\left[\frac{|\beta_r|^2-\sigma_n^2}{|\beta_r|^2};\ \ \mathbf{0}\right]. \tag{8.2.18}$$

Note that such an empirical Wiener filter is adaptive because its weight coefficients depend on the spectrum of the image to which it will be applied.

Weight coefficients of scalar Wiener filters assume values in the range between zero and one. A version of the empirical Wiener filter of Eq. 8.2.18 with binary weight coefficients:

$$\eta_r=\begin{cases}\mathbf{1},\ \boldsymbol{if}\ |\beta_r|^2\geq \boldsymbol{Thr}\\ \mathbf{0},\quad \boldsymbol{otherwise}\end{cases} \tag{8.2.19}$$

where ***Thr*** is a rejecting threshold is called the ***rejecting filter***. As it follows from Eq. 8.2.18, the rejecting threshold has a value of the order of magnitude of the noise variance σ_n^2. Rejecting filters eliminate from the input images spectra all components for which signal-to-noise ratio is lower then a certain threshold.

A version of the empirical Wiener filter and of the rejecting filters that are implemented with wavelet transform image decomposition is known as ***wavelet shrinkage*** filtering ([3]). Empirical Wiener filtering according to Eq. 8.2.18 an a wavelet jargon is called "***soft thresholding***". Rejecting filtering according to Eq. 8.2.19 is called "***hard thresholding***". Fig. 8-2 shows flow diagram of signal denoising by the wavelet shrinkage

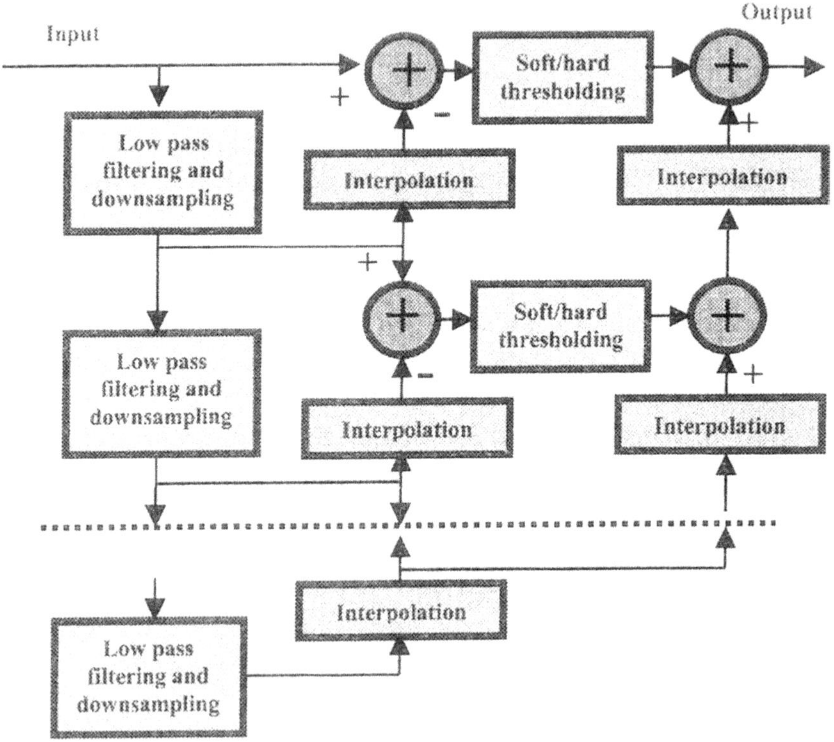

Figure 8-2. Wavelet shrinkage: signal denoising in wavelet transform domain

Described empirical Wiener filters for signal denoising are particularly very efficient if signal and/or noise spectra are well concentrated and are separated in the transform domain. A typical example of such a situation is filtering of narrow band noise, whose spectrum has only a few components in the transform domain. Figs. 8-3 through 8-5 illustrate examples of such a narrow band noise filtering. Fig. 8-3 demonstrates filtering periodical noise pattern in an image. Such interferences frequently appear in images digitized by frame grabbers from analog video. Left column in Fig. 8.3 shows input and filtered images. Right column shows averaged spectra of input and output image rows. One can clearly see anomalous peaks of noise spectrum in input image spectrum that are eliminated in the output image spectrum after applying empirical Wiener filtering designed using methods of noise spectrum estimation described in Sect. 7.1.3.2. Note that the filtering is carried out in this example as 1-D row-wise filtering in DFT domain. This type of interferences can also be successfully filtered in Walsh transform domain ([4]).

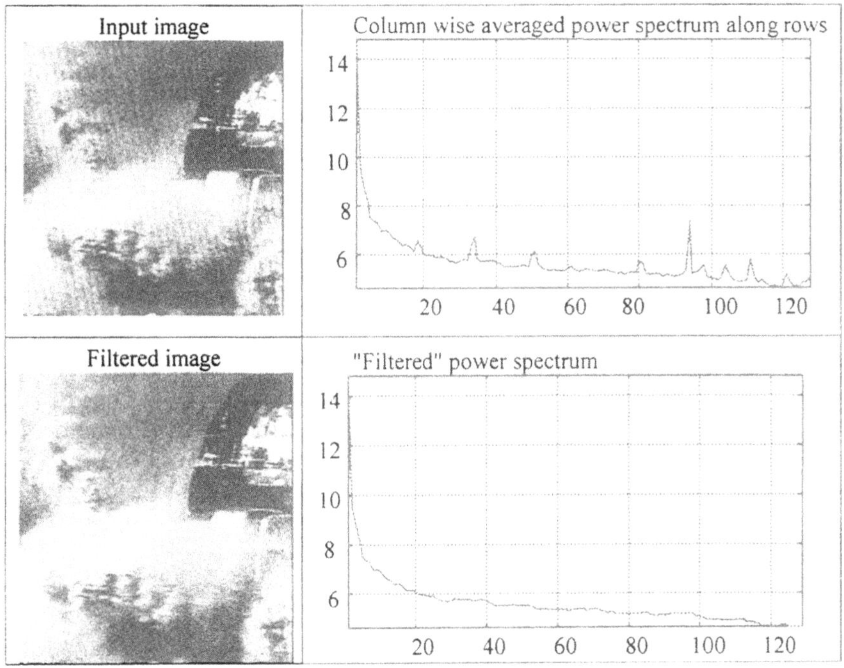

Figure 8-3. Filtering periodic interferences

Fig. 8-4 illustrates similar 2-D rejecting filtering for eliminating periodical noise components in a digitized Fresnel hologram. Fig. 8-5 shows yet another example of empirical Wiener filtering narrow-band interferences. Banding noise in the initial image shown in Fig. 8-5 is characteristic for imaging systems with mechanical scanning. This particular image was produced by an atomic force microscope. Banding noise randomly changes image dc component (row-wise mean value) in the direction of scanning.

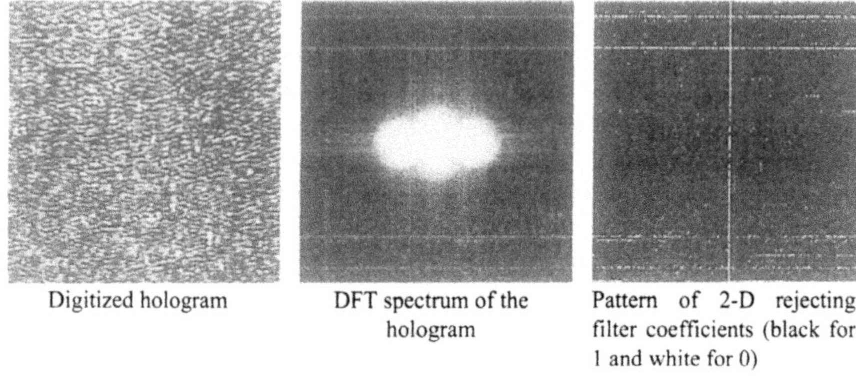

Figure 8-4. Periodical noise in a Fresnel hologram and a pattern of rejecting filter coefficients for removing the noise. The noise manifests itself in horizontal and vertocal bands in DFT spectrum of the hologram. (The hologram was kindly granted by Dr. T. Kraus, BIAS, Germany)

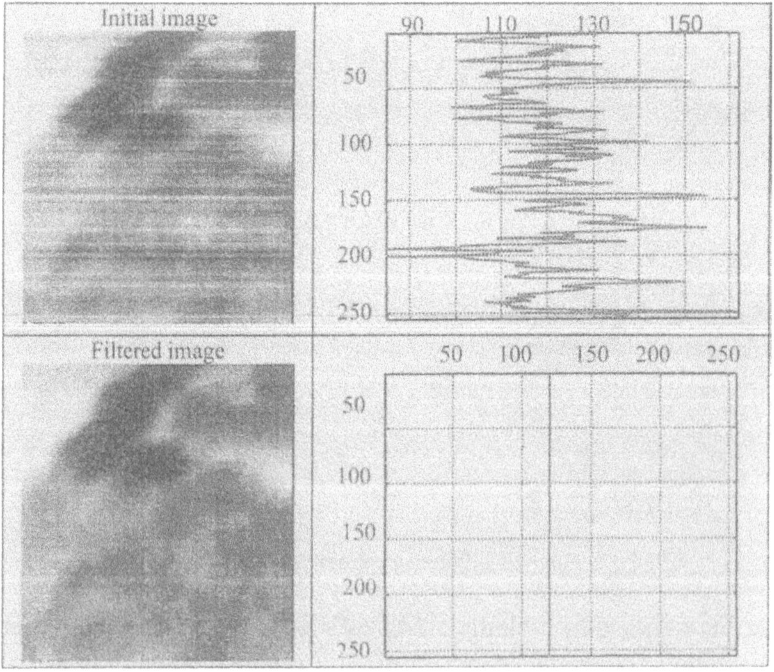

Figure 8-5. Filtering banding noise

Upper right plot in Fig. 8-5 shows row-wise mean values, or row-wise image Radon transform, (horizontal coordinate) as a function of the row number (vertical coordinate). Taking as an estimate of perfect mean row-

wise values a mean value over all rows (shown with a straight line in the bottom right plot in Fig. 8-5), one can estimate banding interference on every row by subtracting this estimate from the observed mean values. Subtracting the found values from all pixels in the corresponding row eliminates the noise.
Wiener filtering of wide band noise and especially white noise is less efficient. When filtering of white noise, Wiener filter tends to weaken low energy signal spectral components. However usually these components are exactly the components that are the most important because they carry information about signal changes such as at edges in images. Moreover, Wiener filtering converts input white noise into output correlated noise though with a reduced variance. As one can see from Eq. 8.2.11, in case of the intensive input white noise, power spectrum of the residual noise is roughly proportional to the signal power spectrum which means that the residual noise becomes, statistically, signal alike. This may hamper subsequent image analysis. In particular, it is well known that human vision is more sensitive to correlated noise that to white noise of the same intensity. Therefore Wiener filtering for image denoising may even worsen images.

Fig. 8-6 gives an example of computer simulation of Wiener filtering of a test image with additive white noise. In the simulation, one can implement the ideal Wiener filter because both signal and noise spectra are known. The empirical Wiener filter was implemented in this example according to Eq. 8.2.18. As one can see from the figure, ideal Wiener filtering does improve image quality. For the empirical Wiener filter, improvement is much less appreciable. The difference image between initial noisy image and the result of the empirical Wiener filtering (restoration error) shows that the filtering, along with noise suppression, destroys image edges.

8.2.3 Image deblurring, inverse filters and aperture correction.

Consider now an imaging system model that accounts also for signal linear transformations in imaging systems. Suppose that the linear transformation unit can be modeled as a scalar filter with filter coefficients $\{\lambda_r\}$ in a selected basis. For DFT basis, this assumption is just, to the accuracy of boundary effects, for shift invariant linear filtering (see Ch. 5). In this case coefficients $\{\lambda_r\}$ are samples of imaging system frequency response. In ideal imaging systems, they should all be equal to unity. In reality, they decay with the frequency index $\boldsymbol{r}$, which results, in particular, in image blur. Processing aimed at correcting this type of distortions is frequently referred to as ***image deblurring***.

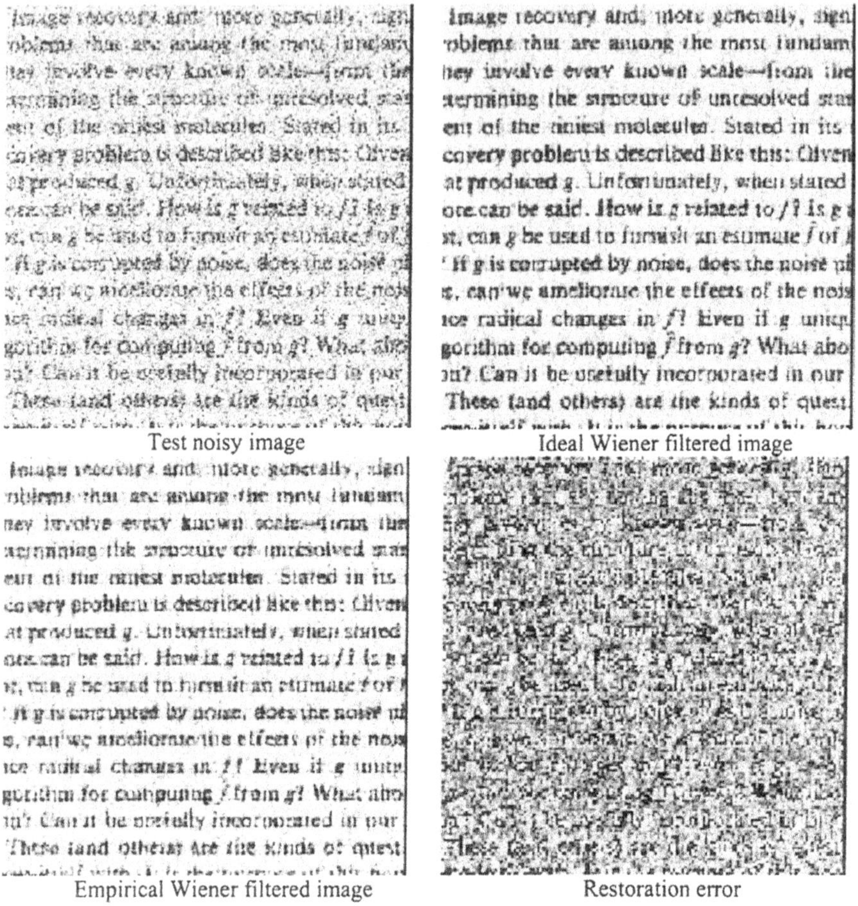

Test noisy image

Ideal Wiener filtered image

Empirical Wiener filtered image

Restoration error

Figure 8-6. Ideal Wiener and Empirical Wiener filtering for image denoising

For such systems, we have in the transform domain:

$$\beta_r = \lambda_r \alpha_r + \nu_r , \tag{8.2.20}$$

Using Eqs. 8.2.6, one can obtain that the scalar Wiener image restoration filter is defined in this case by the equation:

$$\eta_r = \frac{1}{\lambda_r} \frac{\boldsymbol{SNR}_r}{1 + \boldsymbol{SNR}_r} , \tag{8.2.21}$$

where $\boldsymbol{SNR}$ is signal-to-noise ratio at the output of the linear filter unit of the imaging system model:

$$\boldsymbol{SNR_r} = \frac{|\lambda_r|^2 \mathbf{AV}_{\Omega_A}\left(|\alpha_r|^2\right)}{\mathbf{AV}_{\Omega_N}\left(|\nu_r|^2\right)}. \tag{8.2.22}$$

Correspondingly, the general empirical Wiener filter, empirical Wiener filter with zero order approximation to perfect signal spectrum and the rejecting filter for aperture correcting and image deblurring will be in this case as follows:

$$\eta_r = \mathbf{max}\left[\frac{\mathbf{1}}{\lambda_r}\frac{\overline{|\beta_r|^2} - \mathbf{AV}_{\Omega_A}\left(|\nu_r|^2\right)}{\overline{|\beta_r|^2}};\ \mathbf{0}\right]; \tag{8.2.23}$$

$$\eta_r = \mathbf{max}\left[\frac{\mathbf{1}}{\lambda_r}\frac{\overline{|\beta_r|^2} - \sigma_n^2}{\overline{|\beta_r|^2}};\ \mathbf{0}\right] \tag{8.2.24}$$

and

$$\eta_r = \begin{cases} \dfrac{\mathbf{1}}{\lambda_r}, \boldsymbol{if}\ |\beta_r|^2 \geq \boldsymbol{Thr} \\ \mathbf{0}, \quad \boldsymbol{otherwise} \end{cases}. \tag{8.2.25}$$

All these filters may be treated as two filters in cascade: the filter with coefficients

$$\eta_r^{inv} = \frac{\mathbf{1}}{\lambda_r} \tag{8.2.26}$$

usually called the ***inverse filter*** and signal denoising filters described by Eqs. 8.2.17 - 19. Inverse filters compensate weakening signal frequency components in the imaging system while denoising filters prevent from excessive amplification of noise and perform what is called "***regularization***" of inverse filters. As one can see from Eq. 8.2.22, weight coefficients of denoising filters for small $\{\lambda_r\}$ fall faster than weight coefficients of the inverse filter grow.

One of the most immediate applications of inverse filters is correcting distortions caused by finite size of apertures of image sensors, image discretization devices and image displays. We will refer to this processing as to ***aperture correction***.

Let an image sensor and discretization device is an array of light sensitive elements with a square aperture of size $d^{(d)} \times d^{(d)}$ (Fig. 8-7). Then frequency response of the individual sensor elements is

$$H(f_x, f_y) = \int_{-d^{(d)}/2}^{d^{(d)}/2} \exp(i2\pi f_x x) dx \int_{-d^{(d)}/2}^{d^{(d)}/2} \exp(i2\pi f_y y) dx =$$

$$\frac{\sin(\pi f_x d^{(d)}/2)}{\pi f_x d^{(d)}/2} \frac{\sin(\pi f_y d^{(d)})}{\pi f_y d^{(d)}} = \mathrm{sinc}(\pi f_x d^{(d)}) \mathrm{sinc}(\pi f_x d^{(d)}), \qquad (8.2.27)$$

or, in dimensionless coordinates $\{\bar{f}_x = f_x \Delta x, \bar{f}_y = f_y \Delta x\}$, where Δx is the discretization interval in both coordinates,

$$H(\bar{f}_x, \bar{f}_y) = \mathrm{sinc}(\pi \bar{f}_x d^{(d)} / \Delta x) \mathrm{sinc}(\pi \bar{f}_x d^{(d)} / \Delta x). \qquad (8.2.28)$$

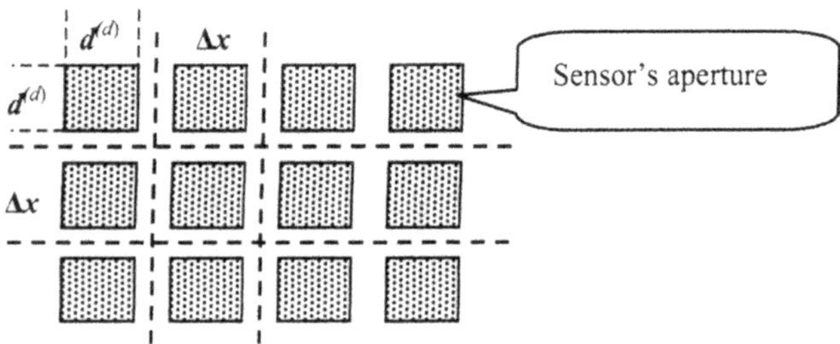

Figure 8-7. Arrangement of light sensitive elements in image sensor arrays

The discrete frequency response is then:

$$\lambda_{r,s} = \lambda_r \lambda_s = \mathrm{sinc}(\pi \bar{d}^{(d)} \lambda_r / N_x) \cdot \mathrm{sinc}(\pi \bar{d}^{(d)} \lambda_s / N_y), \qquad (8.2.29)$$

where $\bar{d}^{(d)} = d^{(d)} / \Delta x$.

If the image display device has also a square aperture of size $d^{(r)} \times d^{(r)}$, the overall imaging discrete system frequency response is:

$$\lambda_{r,s} = \lambda_r \lambda_s = \mathrm{sinc}(\pi \bar{d}^{(d)} \lambda_r / N_x) \cdot \mathrm{sinc}(\pi \bar{d}^{(r)} \lambda_r / N_x) \times$$
$$\mathrm{sinc}(\pi \bar{d}^{(d)} \lambda_s / N_y) \mathrm{sinc}(\pi \bar{d}^{(r)} \lambda_s / N_y). \qquad (8.2.30)$$

Parameters $d^{(d)}$, $d^{(r)}$ and Δx are imaging system design parameters that may be know system's certificate. They can be used for correcting image distortions by processing images in computer.

Fig. 8-8 illustrates an example of the aperture correction of an air photograph. Right image in this figure is obtained by applying to the left image an inverse filter for the system's frequency response defined by Eq. 8.2.20 with $\bar{\boldsymbol{d}}^{(d)} = \bar{\boldsymbol{d}}^{(r)} = \mathbf{1}$.

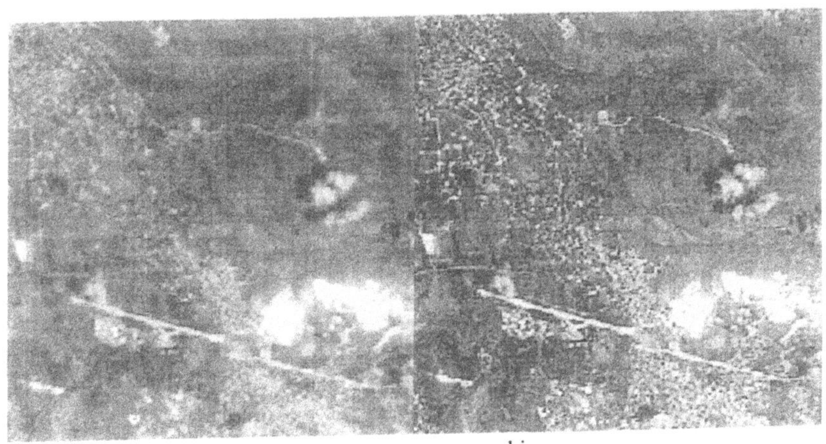

Figure 8-8. Aperture correction: initial (left) and aperture corrected (right) images

In synthesis of computer generated holograms, aperture correction is required to compensate masking images reconstructed from holograms by the frequency response of the hologram recording device (see Sect. 13.3). This masking may result in substantial reducing image contrast on a periphery of image plane. The aperture correction can be implemented by multiplying object image, before the hologram computation, by the function inverse to the hologram recording device frequency response. This method is illustrated in Fig. 8-9.

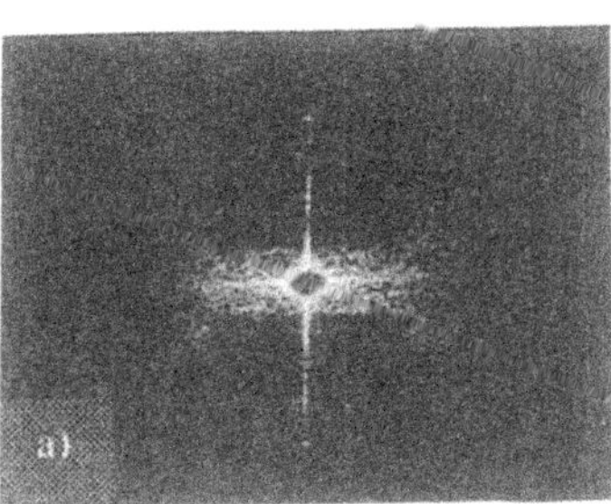

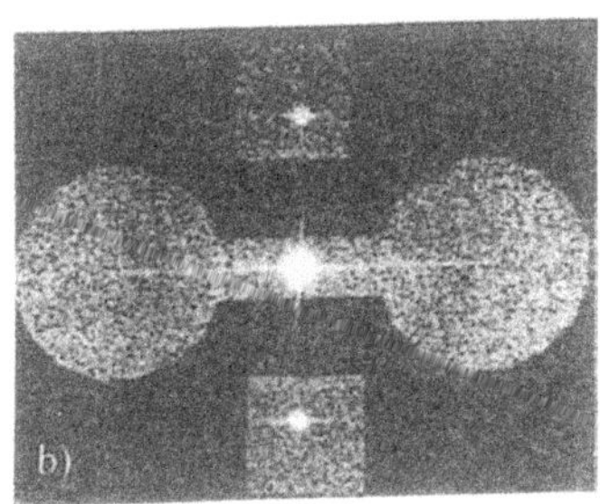

Figure 8-9. Effect of spatial masking of reconstructed images due to the finite size of the hologram recording aperture (a) and the result of its compensation (b) (adopted from [5])

8.3 SLIDING WINDOW TRANSFORM DOMAIN ADAPTIVE SIGNAL RESTORATION

8.3.1 Local adaptive filtering

In the derivation of the MSE optimal scalar linear filters, the size of input signal vectors was a free parameter. In the filter implementation, one should select the fashion in which signals are to be processed. Given an image to be processed, filtering can be designed and carried out either over the entire set of available image samples or fragment-wise. We will refer to the former as to ***global filtering*** and to the latter as to local filtering.

There is a number of arguments in favor of "local" filtering versus "global" one:

- "Global" filtering assumes signal "stationarity". For scalar linear filtering, this means signal homogeneity in terms of signal power spectra: spectra measured over different signal fragments should not substantially deviate one from another. One hardly can regard such signals as images as being spatially homogeneous. Fig. 8-10 illustrates image spatial inhomogeneity on an example of a noisy "Lena" image. One can see that while noise in the image is spatially homogeneous, individual fragments of the image substantially differ one from another visually and, obviously, in their power spectra as well.

- Adaptive filter design assumes empirical evaluation of signal spectra. Global evaluation of image spectrum disregards spectra variations due to image inhomogeneity. This obviously results in neglecting local image information in favor of global one which usually contradicts processing goals.

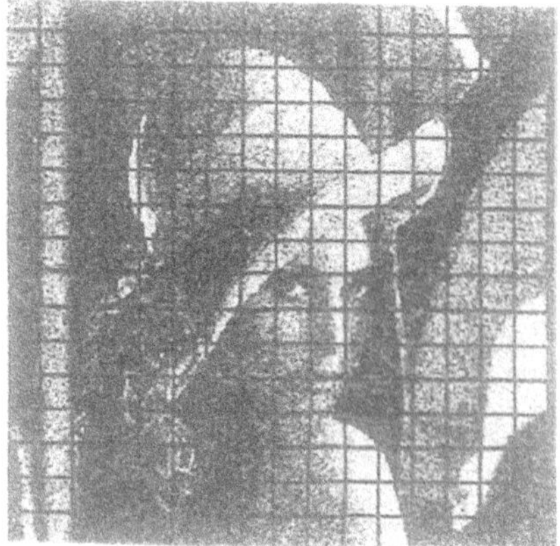

Figure 8-10. Noisy image: global versus local analysis

- Objects to be recognizable visually have to contain sufficiently large number of resolution cells (pixels). As an immediate illustration of this fact one can recall that, for the reproduction of printed characters, one needs a array of 8x8 or more pixels. The same and even to a greater degree holds for "texture" images. Texture identification is also possible only if texture area contains sufficiently large numbers of pixels. This means that image can be regarded as a composition of object domains with the linear size from several to several tens of resolution cells.

- It is well known that when viewing image, eye optical axis permanently jumps "randomly" over the field of view. Human visual acuity is very uneven over the field of view. The field of view of the human vision is about 30°. Resolving power of the vision is about one angular minute. However such a relatively high resolution power is concentrated only within a small fraction of the field of view that has size of about 2° ([6]). Therefore, the area of acute vision is about 1/15-th of the field of view. For images of 512x512pixels this means window of roughly 33x33 pixels.

The theoretical framework for local filtering is provided by local criteria of processing quality introduced in Sect. 2.1.2. For optimal local MSE scalar filters, filter coefficients $\left\{\eta_r^{(k)}\right\}$ are, according to the local criteria, defined by the equation:

$$\hat{\alpha}_r^{(k)} = \underset{\{\eta_r\}}{\mathbf{arg\,min}}\left\{\mathbf{AV}_{\Omega_\Lambda}\mathbf{AV}_{\Omega_N}\left(\sum_{r=0}^{W-1}\left|\alpha_r^{(k)} - \eta_r^{(k)}\beta_r^{(k)}\right|^2\right)\right\}, \tag{8.3.1}$$

where W is it the filter window size and the size of the area over which filtering performance is evaluated, k is an index of the window position, $\left\{\beta_r^{(k)}\right\}$ are spectral coefficients of the filter window samples, $\left\{\alpha_r^{(k)}\right\}$ are spectral coefficients of filter window samples of a hypothetical perfect signal. Solutions obtained in Sect. 8.2.2 and 8.2.3 for empirical Wiener filters for image restoration may then be extended to local filtering by replacing in Eqs. 8.2.16 - 19 and 8.2.23-26 signal spectra with corresponding spectra of the signal window samples.

The most straightforward way to implement local filtering is to perform it in a hopping window. This is exactly the way of processing implemented in most popular audio and image coding methods such as JPEG and MPEG. "Hopping window" processing being very attractive from the computational complexity point of view suffers however from "blocking effects" - artifacts in form of discontinuities at the edges of the hopping window. Obviously, an ultimate solution of the "blocking effects" problem would be sliding window processing. Sliding window filtering is local adaptive because, for empirical

Wiener filters, filter coefficients depend on local spectra of image fragments,.

In the sliding window filtering, one should, for each position of the filter window, compute transform coefficients of the signal vector within window, design on this base the filter, modify accordingly transform coefficients and then compute inverse transform. With sliding window, inverse transform need not, in principle, be computed for all signal vector within window since only the central sample of the window has to be determined in order to form the output signal in the process of image scanning with the filter window. Fig. 8-11 illustrates this process.

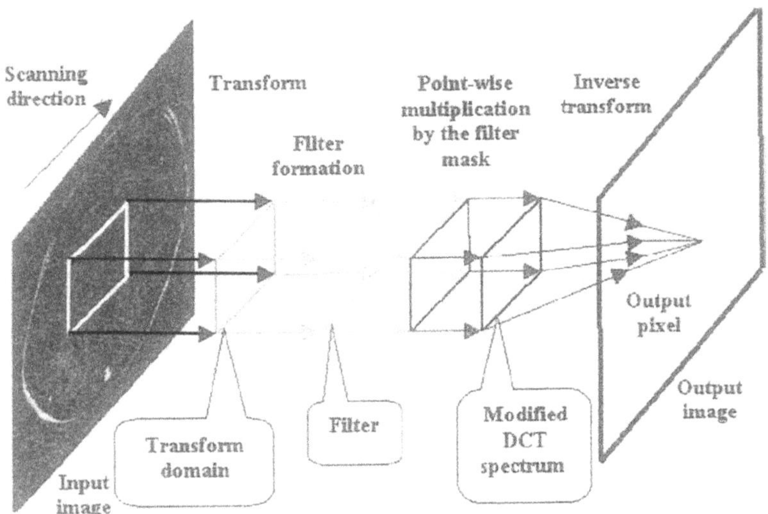

Figure 8-11. The principle of local adaptive filtering in sliding window

8.3.2 Sliding window transform domain DCT filtering

The selection of orthogonal transforms for the implementation of the filters is governed by

- the required accuracy of approximation of general linear filtering with scalar filtering;
- the convenience of formulating a priori knowledge regarding image spectra in the chosen base;
- by the accuracy of the empirical spectrum estimation from the observed data that is required for the adaptive filter design;
- by the computational complexity of the filter implementation.

Among all known transforms, Discrete Cosine Transform proved to be one of the most appropriate ones for sliding window transform domain filtering ([7-9]). DCT exhibits good energy compaction capability which is a key feature for the efficiency of filters. Being advantageous to DFT in terms of energy compaction capability, DCT can also be regarded as a good substitute for DFT in signal/image restoration tasks with imaging system specification in terms of their frequency responses. DCT is suitable for multi component signal/image processing. The use of DCT in sliding window has low computational complexity owing to the recursive algorithms for computing DCT in sliding windows described in Sect. 5.1.3. In addition note that, if the window size N_w is an odd number, the inverse DCT transform of local spectrum $\beta_r^{(k)}$ for computing window central pixel $a_{k+(N_w-1)/2}$

$$a_{k+(N_w-1)/2} = \beta_0^{(k)} + 2\sum_{r=1}^{W-1} \beta_r^{(k)} \cos(\pi r/2) = \beta_0^{(k)} + 2\sum_{s=1}^{(W-1)/2} \beta_{2s}^{(k)}(-1)^s \quad (8.3.2)$$

involves only signal spectrum coefficients with even indices. Therefore only those spectral coefficients have to be computed and the computational complexity of sliding window filtering in DCT domain is $O[(W+1)/2]$ operations for 1-D filtering and $O[(W+1)^2/4]$ operations for 2-D filtering in a square window of $W \times W$ pixels.

Fig. 8-12 shows an example of local adaptive empirical Wiener rejecting filtering in DCT domain according to Eq. (8.2.19) of a noisy electrocardiogram in the window of 25 samples. Upper plot shows initial electrocardiogram. Image in the second row of the figure shows local spectra of the electrocardiogram for each window position. Image gray levels represent here amplitudes of spectral coefficients, vertical coordinate represents spectral component indices, horizontal coordinate represents window position indices. Such a signal representation is its ***time-frequency representation***. Image in the third row displays the same spectra after rejecting signal spectral components that do not exceed the selected threshold. The bottom plot shows the filtering.

Fig. 8-13 illustrates local adaptive rejecting filtering of a piece-wise constant test image. Upper left image is th initial test image. Upper right image is the input image with noise added. Left bottom image is the result of the filtering. Right bottom image shows a map of the rejecting filter "transparency" for each window position; i.e., the ratio of the number of ones in the array of the filter coefficients to the window size. One can see from this image that, when filter window seats on an image patch where image gray values are constant, the filter is almost completely opaque and preserves only local dc image component. On the contrary, on the boundaries of the patches, the filter is almost completely transparent thus preserving edges of the patches. This feature of the local adaptive filters can be even more vividly seen from plots shown in Fig. 8-14 of one of the image rows. The edge preserving capability is an important advantage of the local adaptive filtering comparing to the global filtering.

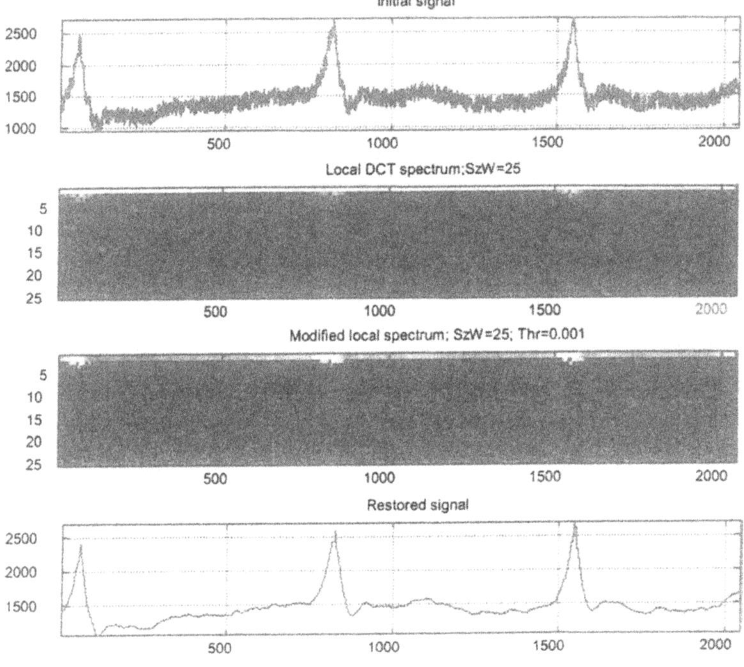

Figure 8-12. An example of denoising 1-D signal (electrocardiogram)

An example of such a denoising of a real MRI image is shown in Fig. 8-15. Left image in this figure is the initial image; central image is the result of the filtering and right image is a difference image between initial and filtered ones. From the difference image one can see that it looks almost chaotic and does not contain any visible details of the initial image.

Local adaptive filtering allows also denoising images with signal dependent noise such as speckle noise. The characteristic feature of the speckle noise is that its standard deviation is proportional to the signal (see Sect. 2.8.2). For filtering speckle noise with local adaptive filters, one should set parameters σ_n and ***Thr*** of the filters of Eqs. 8.2.17 through 19 to be proportional to the image local dc component $\beta_0^{(k)}$. Fig. 8-16 shows an example of such a denoising of an ultrasound image.

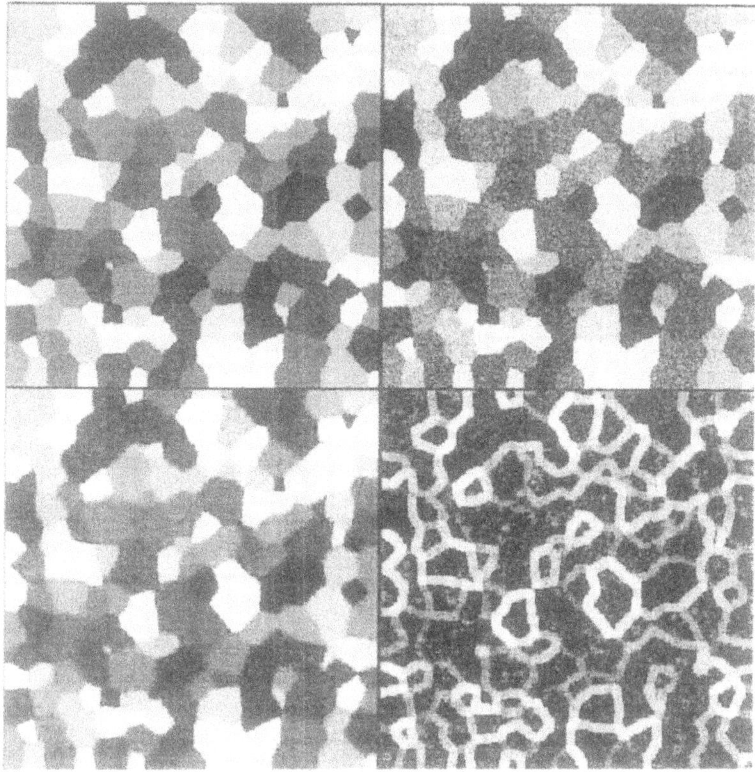

Figure 8-13. Sliding window DCT domain denoising

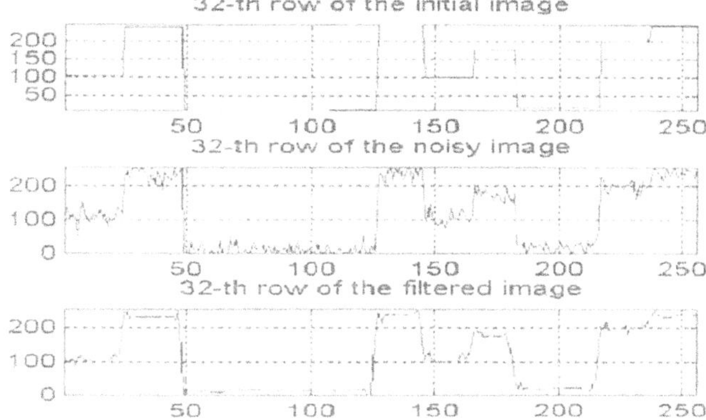

Figure 8-14. A typical row of the initial, noisy and filtered images

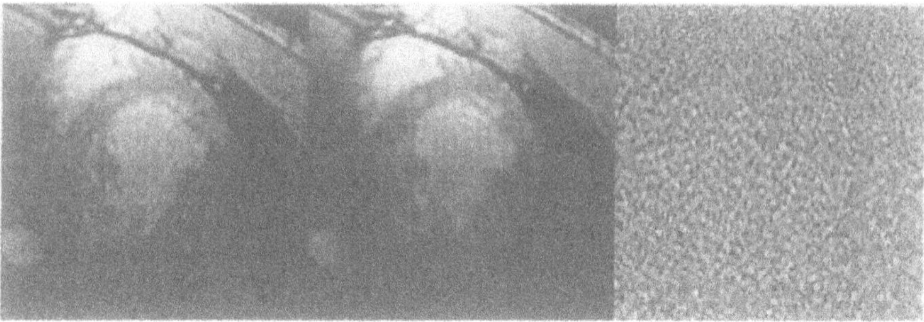

Figure 8-15. Denoising MRI image. Left to right: initial noisy image, filtered image, and difference between initial and filtered images

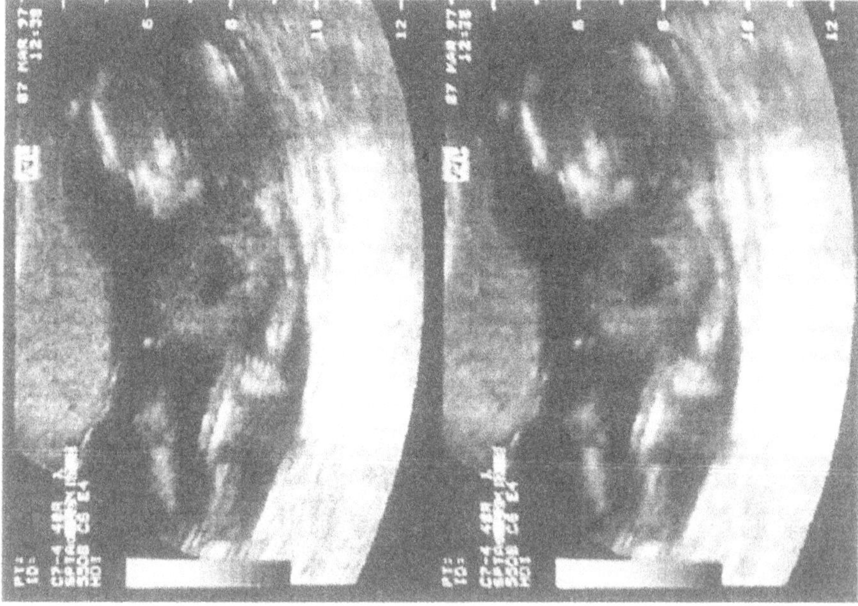

Figure 8-16. An example of local adaptive filtering speckle noise in ultrasound images: initial (left) and filtered (right) images

8.3.3 Hybrid DCT/wavelet filtering.

An important practical issue in using local adaptive filters is selecting the filter window size. This selection is governed by the typical size of image details that should be preserved in the filtering.

In general, optimal window size depends on the image and may vary even within the image. This requires filtering in multiple windows with an appropriate combination of the filtering results.

One can avoid filtering in multiple windows by combining multi resolution property of the wavelet shrinkage and local adaptivity of DCT filtering in the hybrid filtering method ([10]). In the hybrid filtering hard or soft thresholding in each scale level of the wave let filtering shown in a flow diagram of Fig. 8.2 is replaced by local adaptive filtering in DCT domain with a fixed window size. Owing to multi scale image representation in wave let transform, this replacement imitates parallel filtering of the input image with a set of windows accordingly to the scales selected. Flow diagram of the hybrid wave let/DCT filtering is shown in Fig. 8.17.

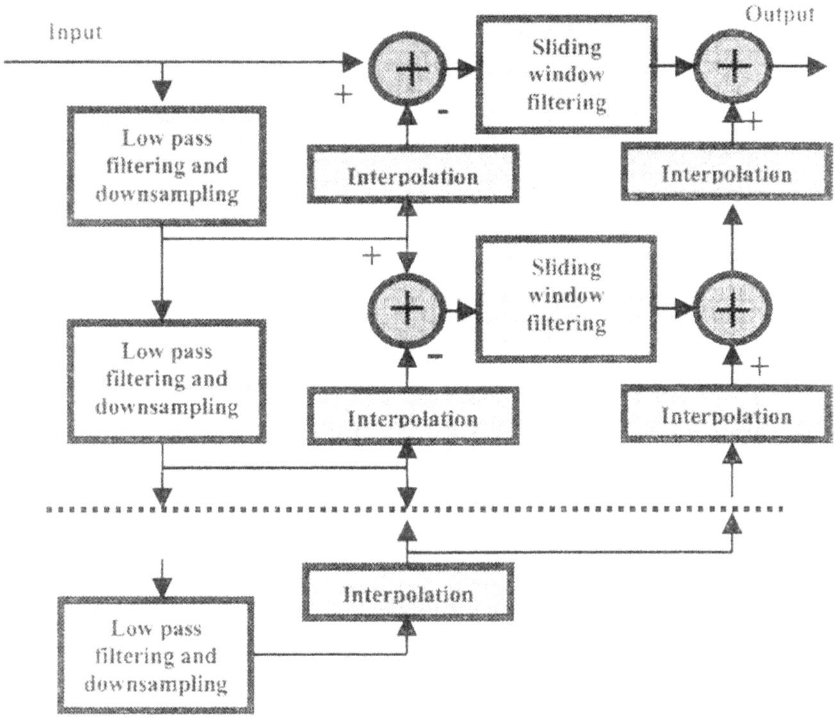

Figure 8-17. Flow diagram of the hybrid sliding window DCT domain and wavelet denoising

For high resolution images, DCT filtering in moving window requires window size 3x3, otherwise the filtering may result in loss of tiny image details. This requirement limits the noise suppression capability of the local adaptive filtering that increases with the window size. In the hybrid filtering, when window 3x3 in the first scale is used, in the scale 2 effective window size is 6x6, in the scale 3 it is 9x9 and so on. This promises an increase to the filtering noise suppression capability.

In DCT filtering in the window of $\mathbf{3}\times\mathbf{3}$ pixels, only the following four basis function are involved:

$$\varphi_{0,0}=\begin{bmatrix}1&1&1\\1&1&1\\1&1&1\end{bmatrix};\varphi_{0,2}=\frac{\sqrt{2}}{2}\begin{bmatrix}-1&-1&-1\\+2&+2&+2\\-1&-1&-1\end{bmatrix};$$

$$\varphi_{2,0}=\frac{\sqrt{2}}{2}\begin{bmatrix}-1&+2&-1\\-1&+2&-1\\-1&+2&-1\end{bmatrix};\quad \varphi_{2,2}=\frac{1}{2}\begin{bmatrix}-1&-2&-1\\-2&+4&-2\\-1&-2&-1\end{bmatrix}. \qquad (8.3.3)$$

The second and the third functions represent what is called vertical and horizontal ***Laplacian operators***. The last function can be decomposed into a sum of diagonal Laplacians:

$$\varphi_{2,2}=\frac{1}{2}\begin{bmatrix}-1&-2&-1\\-2&+4&-2\\-1&-2&-1\end{bmatrix}=\frac{1}{2}\left\{\begin{bmatrix}+2&-1&-1\\-1&+2&-1\\-1&-1&+2\end{bmatrix}+\begin{bmatrix}-1&-1&+2\\-1&+2&-1\\+2&-1&-1\end{bmatrix}\right\}. \qquad (8.3.4)$$

Therefore, DCT filtering in the $\mathbf{3}\times\mathbf{3}$ window may be replaced by filtering in the domain of four directional Laplacians. Experimental experience proves that, for high resolution images, such an implementation is advantageous to simple DCT since it produces less filtering artifacts. Its use in hybrid filtering promises additional advantages because it is equivalent to the corresponding increase of number of effective basis functions. Examples of the effective set, for the input image plane, of these directional Laplacians in four scales are shown in Fig.8-18.

Results of the experimental comparison noise filtering capabilities of the sliding window DCT filtering, wavelet shrinkage filtering and of the hybrid filtering reported in [10] are summarized in Table 8-1. For the wavelet shrinkage filter, a code of image pyramid from University of Pennsylvania package [11] was used in the experiments. For all filters, optimal parameters were experimentally found that minimize MSE between initial noise free image and filtered ones. For DCT moving window filtering, window size and parameter ***Thr*** were optimized. For wavelet shrinkage and hybrid filter, parameters ***Thr*** were optimized and the best of the following types of wavelets was chosen: Haar wavelet ("haar"), Binomial coefficient filter ("binom3", "binom9" and "binom13), Daubechies wavelet ("daub2" and "daub4") and Symmetric Quadrature Mirror Filters ("qmf5", "qmf9" and "qmf13"). The table summarizes results obtained for four of 12 test images (a test piece-wise constant image, "Lena" image, an MRI image and an air photograph) of 256x256 pixels with 256 quantization level and standard deviation of additive Gaussian noise of 13. For DCT moving window filtering, optimal window size is shown in brackets. For the wavelet shrinkage, the best wavelet filter kernel is indicated in brackets.

Figure 8-18. Effective basis functions in four scales of the hybrid DCT/wavelet filtering

Table 8-1. Standard deviation of the difference between initial noise free and filtered images for different filtering methods (Standard deviation of additive noise is 13)

Filter		**Piece-wise constant image**	**Lenna image**	**MRI**	**Air photo**
DCT	Hard	*8.1* *(3 × 3)*	*9.4* *(3 × 3)*	*6.7* *(5 × 5)*	*8.2* *(3 × 3)*
	Soft	*7.5* *(3 × 3)*	*8.6* *(5 × 5)*	*6.3* *(7 × 7)*	*7.7* *(3 × 3)*
Wave let shrinkage.	Hard	*8.6* *(binom5)*	*10.1* *(binom5)*	*8.5* *(binom5)*	*9.3* *(binom5)*
	Soft	*8.4* *(binom5)*	*9.0* *(qmf13)*	*7.8* *(binom5)*	*8.1* *(binom5)*
Hybrid	Hard	*8.7* *(binom5)*	*9.4* *(binom5)*	*6.6* *(binom5)*	*8.2* *(binom5)*
	Soft	*7.9* *(binom5)*	*8.6* *(binom5)*	*6.2* *(binom5)*	*7.5* *(binom5)*

Fig. 8-19 compares filtering results for two of the test images. To ease visual comparison, 64x64 pixel image fragments are shown.

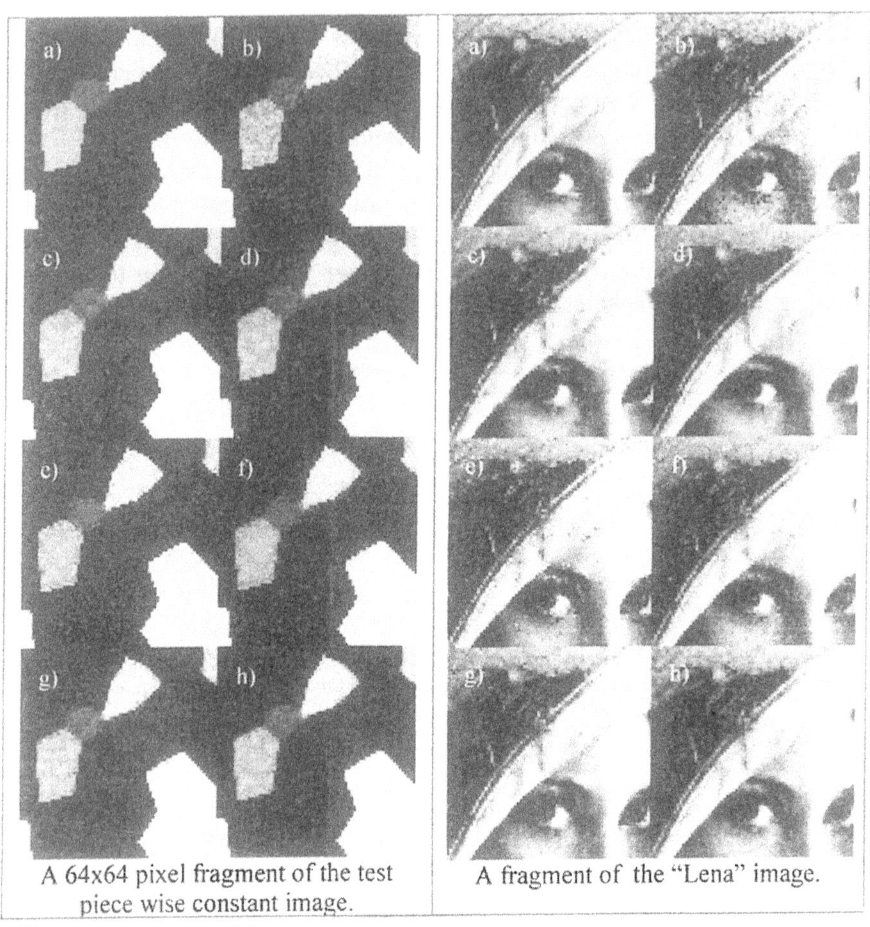

Figure 8-19. Sliding window DCT domain, wave let and hybrid filtering for image denoising: a) initial images; (b) fragments of the noisy images, (c, d) DCT filtering with hard and soft threshold respectively. (e, f) wavelet shrinkage filtering with hard and soft threshold respectively. (g, h) hybrid filtering with hard and soft threshold respectively.

In conclusion note that above described sliding window DCT, wavelet and hybrid filtering methods may be treated as a sub-band decomposition. Sliding window filtering in DCT domain is filtering in sub-bands uniformly distributed in the base band with the number of the sub-bands equal to $(W+1)/2$ in 1-D case. Wavelet shrinkage is filtering in sub-bands arranged in the base band in a logarithmic scale with the number of sub-bands equal to the number of the resolution scales. Hybrid filtering combines logarithmic arrangement of sub-bands of wavelets with uniform arrangement of sub-sub-bands within wavelet sub-bands. Fig. 8-20 illustrates this interpretation on an example of wavelet filtering in 4 scales and sliding window filtering in DCT domain in the window of 3 samples.

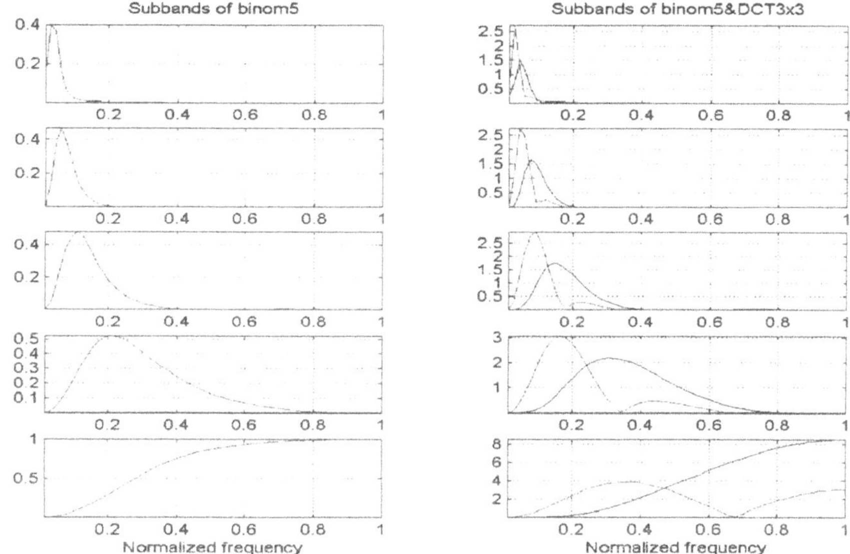

Figure 8-20. Wavelet domain (Binom5) and hybrid sliding window DCT domain processing as sub-band decompositions

8.4 MULTI-COMPONENT IMAGE RESTORATION.

The above theory of MSE optimal scalar filtering can be extended to restoration of multi component images such as color and multi spectral images, video sequences, images of same objects generated by different imaging devices. Such a processing is sometimes called ***data fusion***.

Consider an M component imaging system and assume that each image of M components can be described by the model of Eq. (8.2.20):

$$\left\{\beta_r^{(m)} = \lambda_r^{(m)}\alpha_r^{(m)} + \nu_r^{(m)}\right\}, \tag{8.4.1}$$

where m is component index, $m = 1,2,...,M$, $\left\{\beta_r^{(m)}\right\}$ and $\left\{\alpha_r^{(m)}\right\}$ are spectral coefficients of the observed and perfect image components, $\left\{\lambda_r^{(m)}\right\}$ are component linear transformation spectral coefficients and $\left\{\nu_r^{(m)}\right\}$ are additive noise spectral coefficients in m-th component. Suppose that image components are restored by a linear combination of scalar filtered input image components:

$$\left\{\hat{\alpha}_r^{(m)} = \sum_{l=1}^{M}\eta_r^{(m,l)}\beta_r^{(l)}\right\}, \tag{8.4.2}$$

where $\left\{\eta_r^{(m,l)}\right\}$ are restoration scalar linear filter coefficients.

From the MSE criterion

$$\hat{\alpha}_r^{(m)} = \underset{\left\{\eta_r^{(m,l)}\right\}}{\mathbf{arg\,min}}\left\{\mathbf{AV}_{\Omega_A}\mathbf{AV}_{\Omega_N}\left(\sum_{r=0}^{N-1}\left|\alpha_r^{(m)} - \sum_{l=1}^{M}\eta_r^{(m,l)}\beta_r^{(l)}\right|^2\right)\right\} \tag{8.4.3}$$

one can obtain the following optimal restoration filter coefficients ([12]):

$$\eta_r^{(m,l)} = \frac{1}{\lambda_r^{(l)}}\frac{AV_{\Omega_A}\left(\alpha_r^{(m)}\alpha_r^{(l)}\right)}{AV_{\Omega_A}\left(\left|\alpha_r^{(l)}\right|^2\right)}\frac{SNR_r^{(l)}}{1+\sum_{m=1}^{M}SNR_r^{(m)}}. \tag{8.4.4}$$

where

$$\left\{SNR_r^{(m)} = \frac{\left|\lambda_r^{(m)}\right|^2\mathbf{AV}_{\Omega_A}\left(\left|\alpha_r^{(m)}\right|^2\right)}{\mathbf{AV}_{\Omega_n}\left(\left|\nu_r^{(m)}\right|^2\right)}\right\} \tag{8.4.5}$$

are signal-to-noise ratios in image component spectral coefficients.

An important special case of multi component image restoration is averaging of multiple images of the same object obtained with the same sensor. The averaging

$$\hat{a}_k = \frac{\sum_{m=1}^{M} b_k^{(m)} / \sigma_n^{(m)}}{\sum_{m=1}^{M} 1/\sigma_n^{(m)}}, \tag{8.4.6}$$

where $\{b_k^{(m)}\}$ are samples of image components and $\{\sigma_n^{(m)}\}$ are standard deviation of uncorrelated additive noise in image components, follows from Eq. 8.4.2 and 8.4.44 in the assumption that $\{\alpha_r^{(m)} = \alpha_r\}$, $\{\lambda_r^{(m)} = \mathbf{1}\}$ and $\mathbf{AV}_{\Omega_n}\left(|\nu_r^{(m)}|^2\right) \propto \left(\sigma_n^{(m)}\right)^2$.

It may very frequently happen that different images in the available set of images contain the object in different position. In this case images should be aligned before averaging. This conclusion follows in particular, from Eq. 8.4.4 in which when DFT basis is used

$$\frac{AV_{\Omega_A}\left(\alpha_r^{(m)}\alpha_r^{(l)}\right)}{AV_{\Omega_A}\left(|\alpha_r^{(l)}|^2\right)} = \mathbf{exp}\left(i2\pi\frac{k_0^{(m,l)}r}{N}\right) \tag{8.4.7}$$

where, according to the shift theorem for DFT, $\{k_0^{(m,l)}\}$ are mutual displacement between m-th anf l-th image components.

When mutual displacements of images are known, images can be aligned through their resampling with an appropriate interpolation. Methods of image resampling and other geometrical transformation are discussed in Ch. 9. If displacements $\{k_0^{(m,l)}\}$ are known with a sub-pixel accuracy, in the averaging samples of images will be interlaced according to the mutual shifts of the images. Therefore the resulting image will have accordingly higher sampling rate than that of the original image set and, hence, will have higher resolving power. This resolution increase is frequently referred to as ***super-resolution*** from multiple images.

Mutual image displacements are a special case of general image geometric transformations. Obviously, image alignment before averaging should allow for all transformations.

When mutual image displacements or corresponding parameters are not known, they have to determined. Determination of image displacements or other geometrical parameters is an example of a general problem of image parameter estimation. It is discussed in Ch. 10. In particular, it is shown in Ch. 10, that, for the model of additive uncorrelated Gaussian noise in image components, mutual image displacements can be determined from the shifts of the position of image cross-correlation peaks from their positions in autocorrelation. Image averaging with such an image correlational alignment is called ***correlational averaging*** or ***correlational accumulation*** ([13]). The correlation accumulation has found many practical applications, especially in high resolution electrocardiography and similar biomedical signal processing ([14]) and for processing electron micrographs ([15]).

8.5 FILTERING IMPULSE NOISE

Image distortions in form of impulse noise usually take place in image transmission via digital communication channels. Signal transmission errors in such channels result in losses, with certain probability, of individual image samples and in replacing their values by random values. Impulse noise may also appear as a residual noise after applying to noisy images the additive or speckle noise filtering. In this case, impulse noise is formed from large outbursts of additive or speckle noise that the filtering failed to remove.

Impulse noise is described by an ImpN-model (see Sect. 2.8.1):

$$\boldsymbol{b}_{k,l} = \left(\mathbf{1} - \boldsymbol{e}_{k,l}\right) \cdot \boldsymbol{a}_{k,l} + \boldsymbol{e}_{k,l} \boldsymbol{n}_{k,l}, \tag{8.5.1}$$

where $\{\boldsymbol{b}_{k,l}\}$ are samples of a perfect image contaminated by impulse noise, $\{\boldsymbol{a}_{k,l}\}$ are the perfect image samples, $\{\boldsymbol{e}_{k,l}\}$ are a samples of a binary random sequence that symbolizes the presence ($\boldsymbol{e}_{k,l} = \mathbf{1}$) or the absence ($\boldsymbol{e}_{k,l} = \mathbf{0}$) of impulse noise in the $(\boldsymbol{k},\boldsymbol{l})$-th sample, and $\{\boldsymbol{n}_{k,l}\}$ are samples of yet another random sequence that replace signal samples in cases when impulse interference take place.

From this model it follows that filtering impulse noise assumes two stages:

- detecting signal samples that have been lost due to the noise and
- estimating values of those lost samples.

Impulse noise appears in images as isolated noise samples (see, for instance, Fig. 2.21, a). Hence, the task of detecting impulse noise samples may be treated as a special case of the task of object localization in clutter images treated in Ch. 11. In this chapter, the design of optimal linear filters is discussed that maximize their response to the target object with respect to the standard deviation of the signal for non--target image component. For detecting impulse noise samples, individual pixels replaced by noise are the localization target. As follows from results obtained in Ch. 11, optimal linear filters in this case can be approximated by convolution of two ***Laplacians*** (see Sect. 11.6). Detection of impulse noise samples can be implemented by comparing of the output of such a filter in every pixel with a certain threshold. The reliability of the detection can be increased if it is implemented in an iterative fashion with decreasing the detection threshold in course of iterations.

As for estimating image samples that have been found lost, they can be replaced by values predicted from those samples in their vicinity that are not marked as replaced by noise. The prediction can be implemented in different ways. The simplest is weighted averaging of these samples.

This reasoning lies in the base of the following simple algorithm for filtering impulse noise:

$$\hat{a}_{k,l}^{(t)} = \begin{cases} \hat{a}_{k,l}^{(t-1)}, & if \left|\hat{a}_{k,l}^{(t-1)} - \overline{\hat{a}_{k,l}^{(t-1)}}\right| > DetThr^{(t)} \\ \overline{\hat{a}_{k,l}^{(t-1)}}, & otherwise \end{cases}, \tag{8.5.2}$$

where $\left\{\hat{a}_{k,l}^{(t)}\right\}$ are filtered image samples on t-th iteration, $t = 0,...,N_{it}$, $\left\{\hat{a}_{k,l}^{(0)} = b_{k,l}\right\}$, $\overline{\hat{a}_{k,l}^{(t-1)}}$ are “predicted” sample values found by averaging of neighbor pixels:

$$\overline{\hat{a}_{k,l}^{(t-1)}} = \frac{1}{8}\left[\sum_{n=-1}^{1}\sum_{m=-1}^{1}\hat{a}_{k-n,l-m}^{(t-1)} - \hat{a}_{k,l}^{(t-1)}\right], \tag{8.5.3}$$

and $DetThr^{(t)}$ is a detection threshold that is changing in course of iterations as

$$DetThr^{(t)} = DetThr_{max} - \left(\frac{DetThr_{max} - DetThr_{min}}{N_{it} - 1}\right)(t-1). \tag{8.5.4}$$

As a practical recommendation, the number of filtering iterations can be set from 3 to 5, the upper threshold $DetThr_{max}$ is takenof the order of $1/3 \div 1/4$ of the image dynamic range and the lower threshold should be selected commensurable with the local contrast of image details that have to be preserved from false detecting as noise. For good quality digital images with gray values in the range $[0 \div 255]$, this threshold can be set as $15 \div 20$.

In order to avoid iterations, this algorithm can be replaced by a recursive filtering in a sliding window in which detection and estimation for each pixel is carried out from the analysis of image samples that have been processed on the previous filter window position. For filtering along image rows (index l), this filter implementation can be described as:

$$\hat{a}_{k,l} = \begin{cases} b_{k,l}, & if \left|a_{k,l} - \overline{\hat{a}_{k,l}}\right| > DetThr \\ \overline{\hat{a}_{k,l}}, & otherwise \end{cases}, \tag{8.5.5}$$

where predicted values $\overline{\hat{a}_{k,l}}$ are computed as in DPCM coding (see Sect. 3.7.2, Eqs. 3.7.12, 3.7.13).

Fig. 8-21 illustrates filtering impulse noise in an image with probability of impulse noise 0.3 by the iterative and recursive filters. One can see that the filtering makes unreadable image to be readable.

a)

b)

c)

Figure 8 21. Filtering a printed text image corrupted by impulse noise with iterative and recursive filters: (a) - noisy image with probability of impulse nooise 0.3; (b) - result of iterative filtering according to Eq. 8.5.2 after 4 iterations; (c) - result of recursive filtering according to Eq. 8.5.5

The described filters are not the only options for filtering impulse noise. In general, for filtering impulse noise, nonlinear filters are required. Some nonlinear filters for filtering impulse noise are described in Sect. 12.3.

In conclusion of this section, consider ways of numerical evaluation of the impulse noise filtering. MSE criterion used in the design of filters for suppressing additive noise is not applicable for impulse noise. One can easily see it from comparison of images distorted with additive and impulse noise with the same MS-error with respect to the noiseless image (Fig. 8-22).

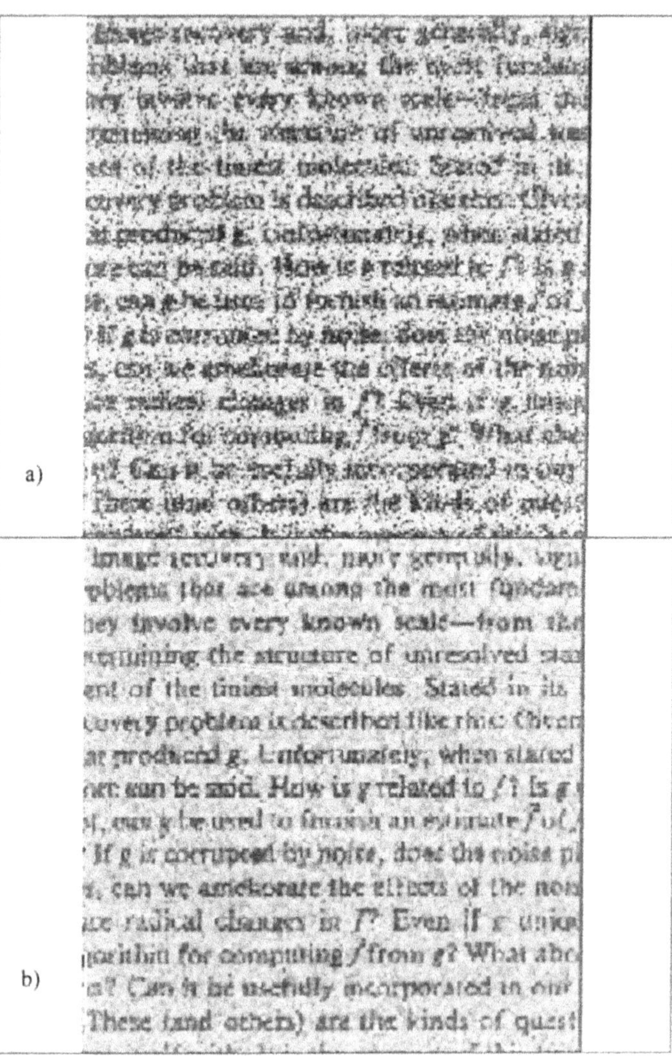

Figure 8-22. Images distorted by additive noise (a) and impulse noise (b) with the same MS-error with respect to the noiseless image

The evaluation criterion for impulse noise filtering must take into account the following three factors.

- Distortions due to missing noisy samples and the probability of missing that determines contribution of these distortions to quality corruption of the filtered image.
- Distortions due to false noise detection and the probability of false alarms that determines contribution of these distortions to quality corruption of the filtered image Probability of false noise detection.
- Estimation error for those pixels that were correctly detected as replaced by noise.

Each of these errors can be evaluated in terms of their mean square values. However, their contribution to the total evaluation is of more involved nature and depends on application.

8.6 METHODS FOR CORRECTING GRAY SCALE NONLINEAR DISTORTIONS

In the imaging system model of Fig. 8-1, gray scale distortions are attributed to the point-wise nonlinear transformation unit. Usually in correcting gray scale distortions, noise described by the stochastic transformations is ignored. In such an assumption, for correcting image nonlinear transformations in imaging systems it is sufficient to apply to images the corresponding inverse transformation. However, in digital processing one should allow for quantization effects and for the fact that any nonlinear transformation of quantized signals reduces the number of signal quantization levels. The phenomenon of sticking of quantization levels together as the result of their nonlinear transformation is illustrated in Fig. 8-23.

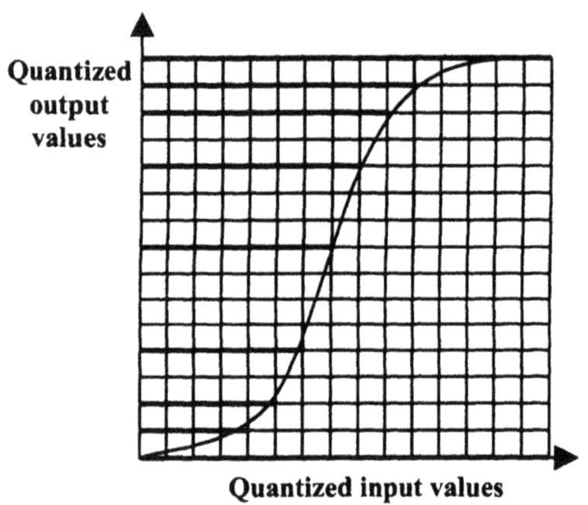

Figure 8-23. Nonlinear signal transformation and sticking quantization levels together. Thin lines on the figure outline 16 uniform quantization intervals. The curve shows transfer function of a nonlinear transformation. Bold lines indicate 9 quantization levels left after the nonlinear transformation of 16 quantization levels of input values

In principle, for generating correcting nonlinear transformations that can be applied to quantized distorted input signal, methods of signal optimal scalar quantization described in Sect. 3.6 should be applied. Consider a diagram shown in Fig. 8-24. Let be known the nonlinear imaging system transfer function to be corrected as well as nonlinear compressing and expanding functions used for signal non-uniform compressor-expander quantization. Project boundaries of quantization intervals of uniform quantizer at the output of the compressing transformation to its input and then further to the input of the nonlinear transformation to be corrected , as it is indicated by arrows in upper left and upper right quarters of the diagram in Fig. 8.24. This provides boundaries of the quantization intervals of the perfect signal that has to be restored from the distorted and quantized input signal. Within these boundaries, find now optimal quantized representative values of the perfect signal that minimize the quantization error. Finally, project these values back to the input scale of the expanding transformation as indicated by arrows in bottom right quarter of the diagram. Obtained values determine the needed output values of the correcting look up table (shown by bold circles in the diagram), that have to assigned to quantized values of the distorted signal.

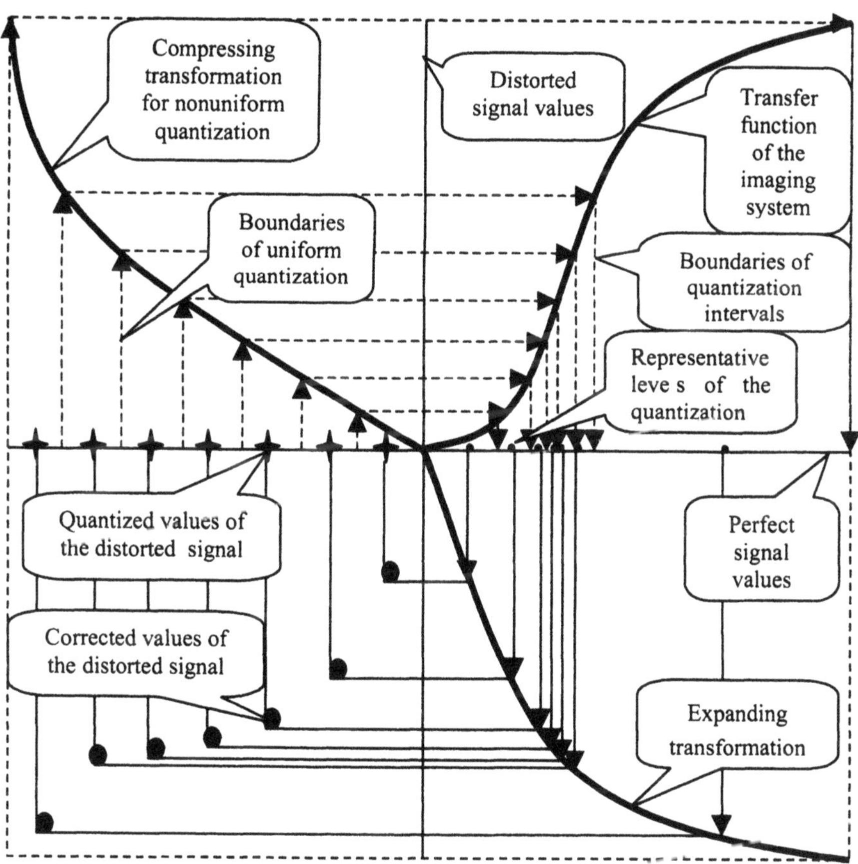

Figure 8-24. Principle of correcting nonlinear distortions with an account for effects of quantization

In practical implementations, more simple methods are frequently used. On of the simple methods to minimize quantization artifacts is randomization of quantization errors by adding to the quantized signal a pseudo-random noise uniformly distributed within the corresponding quantization intervals. Fig. 8-25 illustrates this method on an example of correcting low image contrast by stretching its dynamic range. False contours that are visible in the corrected image (Fig. 8-22, b) may be made invisible (Fig. 8-25, c) by adding to the corrected image pseudo-random noise uniformly distributed in the stretched quantization intervals.

Low contrast image

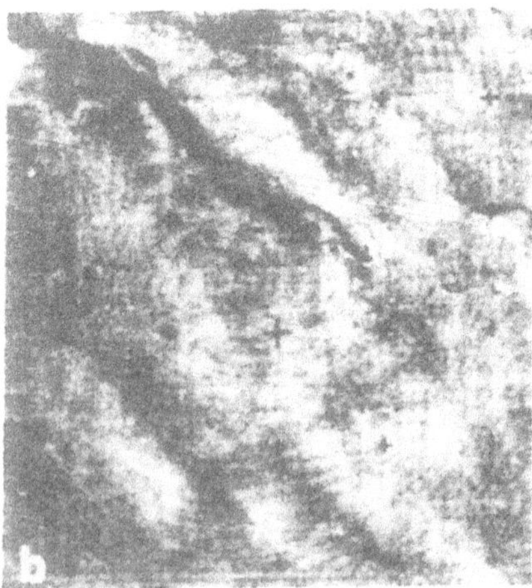

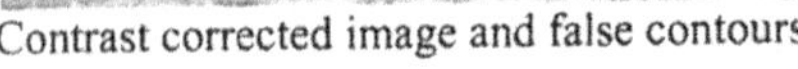

Contrast corrected image and false contours

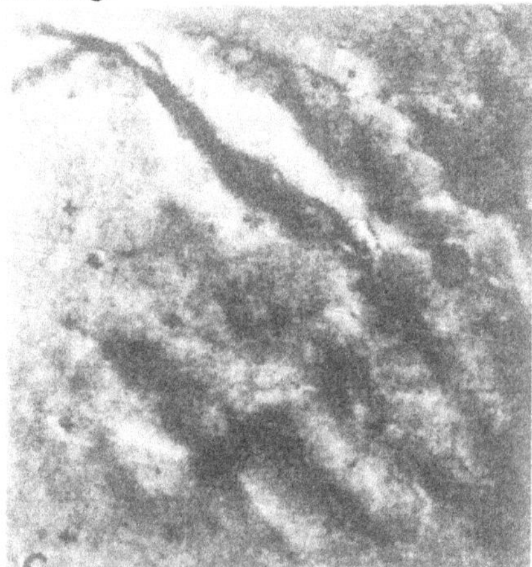

Destroying false contours by adding noise to the corrected image

Figure 8-25. Image contrast correction with adding pseudo-random noise for randomization of false contours

In applications, it very frequently happens that the nonlinear signal distortion transfer function that has to be corrected is not known. In some cases it can be determined directly from an analysis of the distorted signal. A typical example to illustrate such an opportunity is correcting nonlinear distortions of photographic recording of interferograms.

Perfect interferogram signal is a phase modulated sinusoidal signal. If the interferogram to be corrected contains many periods of the sinusoidal signal, distribution function of its phase can be regarded to be uniform. One can analytically compute the distribution function of a sinusoidal signal with uniformly distributed phase. Therefore, the correcting nonlinear transformation can be found as a transformation that converts the histogram of the distorted interferogram into the histogram defined by the distribution function of the sinusoidal signal with uniformly distributed phase. We will refer to such a transformation as to ***histogram matching***. Fig. 8-26 illustrates the histogram matching for correcting nonlinear distortions of an interferogram.

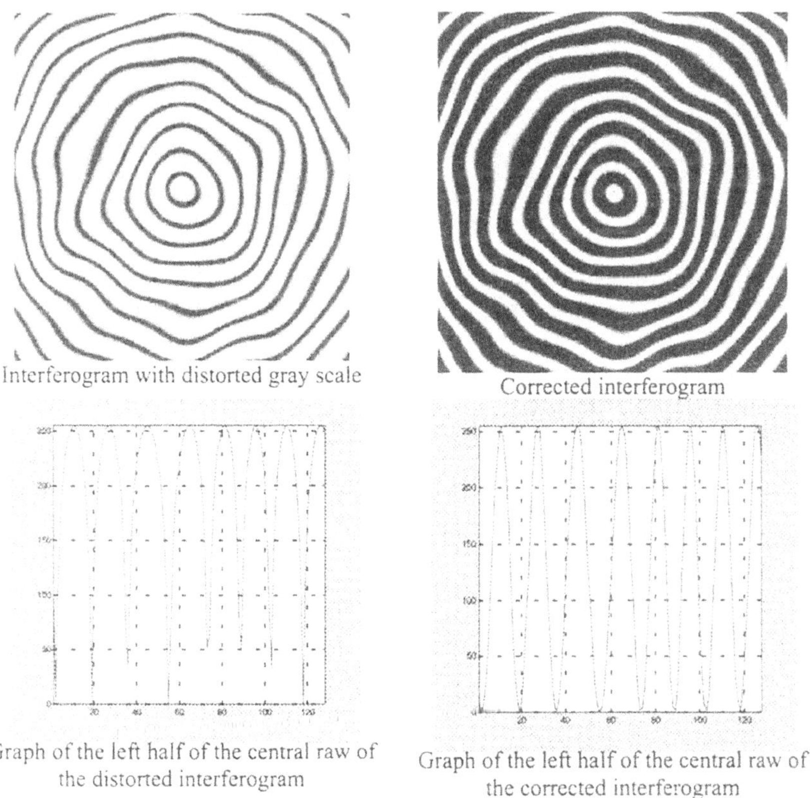

Figure 8-26. An example of correcting unknown nonlinear distortion of an interferogram

Yet another example of an application in which the histogram matching can be used for correcting unknown nonlinear image distortion is creating image mosaics from sets of images that are mutually overlapping.

Algorithmically, the histogram matching can be implemented through the ***histogram equalization***, a nonlinear transformation that converts images with whatever histograms into images with a uniform histogram. For the histogram matching, it is sufficient to find histogram equalization transformation for image to be transformed and that for the image with the target histogram. The required histogram matching transformation can then be obtained by combining histogram equalization transformation for the image to be transformed and the transformation inverse to the histogram equalization transformation for the target histogram.

Histogram equalization is also a very useful method for image enhancement (see Sect. 8.7.2).

8.7 IMAGE RECONSTRUCTION

As it was already mentioned, the term "image reconstruction" is conventionally used to designate processing aimed at image formation in transform imaging methods such as tomography and holography. In principle, image reconstruction requires applying to image sensor data transform inverse to that used to collect these data. However, in reality one cannot implement this ideal scheme. Two major reasons for that are: (i) artifacts due to discretization involved in data collecting for digital reconstruction in computer and (ii) sensor signal distortions due to sensor's noise, limited resolution, nonlinear distortions and quantization. Therefore, similarly to the inverse filtering for image restoration (Sect. 8.2.3), data inverse transformation for image reconstruction has to be supplemented with a processing that eliminates these artifacts and correcting distortions to a maximal possible degree.

Correcting processing is, generally, nonlinear and space variant. Therefore, it should be applied both before and after the inverse transform (see Fig. 8-27). Note that the inverse transform can also be explicitly incorporated with correcting in the form of transform ***"regularization"*** as it done in some tomographic reconstruction algorithms.

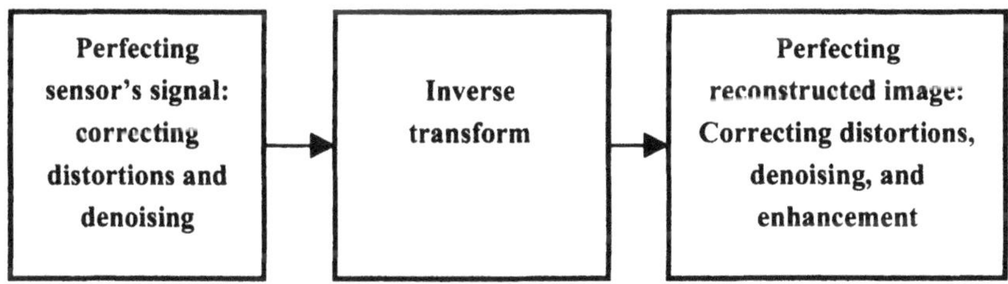

Figure 8-27. Principle of image reconstruction by inverse filtering via inverse transform

We will illustrate image reconstruction with pre-inverse transform and post- inverse transform processing using as examples image reconstruction from projections and from optical holograms.

Image reconstruction by back projecting its projections is a straightforward inversion of image projecting in the tomographic process. Back projection is a very simple operation that sometimes can be implemented in pure analog, for instance optical, set-ups, with a very high speed. This is however, not the inverse transform required for the exact reconstruction. Images reconstructed in this way are blurred and need restoration. Left image in the second row of Fig. 8-28 illustrates such an image restoration from parallel projections (right image in the first row) of a test image shown in the left image in the first row. Inverse filtering applied to images reconstructed by back projection can substantially improve image quality and approximate exact tomographic reconstruction especially if it is supplemented with a post-processing for removing inverse filtering artifacts.

To enable the inverse filtering, one should measure point spread function of the tomographic machine with back projection reconstruction. This can be done using specially phantom objects such as for instance, a point object. The system PSF uniquely specify tomographic reconstruction systems as they can be treated as linear ones.

Let $\mathbf{V}$ be a vector of observed image samples, $\tilde{\mathbf{V}}$ be a result of its reconstruction in a given imperfect tomographic system, $\hat{\mathbf{V}}$ be the corrected result, $\mathbf{V}_c$ be a calibrating test image for measuring the system point spread function, and $\tilde{\mathbf{V}}_c$ is its copy reconstructed by the system. The inverse filtering processing can be described, in a matrix form, as following:

$$\hat{\mathbf{V}} = \mathbf{IF} \cdot \tilde{\mathbf{V}}, \tag{8.7.1}$$

where $\mathbf{IF}$ is a restoration linear operator obtained as:

$$\mathbf{IF} = \left[\mathbf{V}_c \left(\tilde{\mathbf{V}}_c\right)^{-1}\right] \tag{8.7.2}$$

In order to reduce the computational complexity of the restoration, all transformations can be carried out in the domain of Discrete Fourier Transform using Fast Fourier Transform algorithm:

$$\hat{\mathbf{V}} = \mathbf{IFFT}\left\{\mathbf{\Phi}_{IF} \bullet \mathbf{FFT}\left(\tilde{\mathbf{V}}\right)\right\}, \tag{8.7.3}$$

where $\mathbf{FFT}(\cdot)$ and $\mathbf{IFFT}(\cdot)$ are direct and inverse Fast Fourier Transforms and $\mathbf{\Phi}_{IF}$ is discrete frequency response of the inverse filter. It can be found from $\mathbf{FFT}(\mathbf{V}_c)$ and $\mathbf{FFT}\left(\tilde{\mathbf{V}}_c\right)$ as

$$\mathbf{IF} = \mathbf{FFT}(\mathbf{V}_c) \bullet \left[\mathbf{FFT}\left(\tilde{\mathbf{V}}_c\right)\right]^{-1}. \tag{8.7.4}$$

As the calibrating test image, it is most natural to choose a desired overall point-spread function of the system.

If the tomographic system can be modeled as a space invariant one (such as, for instance, systems with parallel beams) the above transforms should be applied "globally" to the entire volume within applicability of the space invariance assumption. Otherwise they should be applied to selected regions of interest small enough to fulfill the requirement for space invariance with an acceptable approximation.

Right image in the second row in Fig. 8-28 illustrates the result of such a restoration by the inverse filtering applied to the left image. The filtering was implemented in the domain of Discrete Fourier Transform. As a "desired" system point spread function, a Gaussian-shaped impulse in coordinates $\{k,l\}$ of pixel indices was chosen:

$$\mathbf{V}_c = \left\{\exp\left[-\left(k^2 + l^2\right)/2\sigma_c^2\right]\right\}, \tag{8.7.5}$$

with a spread parameter σ_c was set to have magnitude of about pixel size.

Frequency response of the "inverse filter" was generated, similarly to Wiener deblurring filter of Eq. 8.2.21, as

$$\overline{\mathbf{IF}} = \frac{\mathbf{IF}^*}{|\mathbf{IF}|^2 + \varepsilon^2 \cdot \mathbf{var}\left(|\mathbf{IF}|^2\right)}, \tag{8.7.6}$$

where $\mathbf{IF}$ is defined by Eq. 8.7.4, $\tilde{\mathbf{V}}_c$ is the result of back projection reconstruction of the desired system point spread function $\mathbf{V}_c$, $\mathbf{FFT}(\cdot)$ is an operator of Fast Fourier Transform, ε^2 is a regularizing parameter playing the role of the parameter $\boldsymbol{SNR}$ in Eq. 8.2.21 and $\mathbf{var}(\cdot)$ is an operator of computing mean square value.

Right image in the second row in Fig. 8-28 demonstrates the results of such an inverse filtering of the left image reconstructed by back projection. The image is intentionally over-contrasted to make reconstruction artifacts visible. These artifacts can be substantially reduced by non-linear post-filtering. Left and right images in the bottom row in Fig. 8-28 illustrate this option. Left image was obtained using iterative sigma-filter (See Sect. 12.3.1). Right image shows difference non-filtered and filtered images.

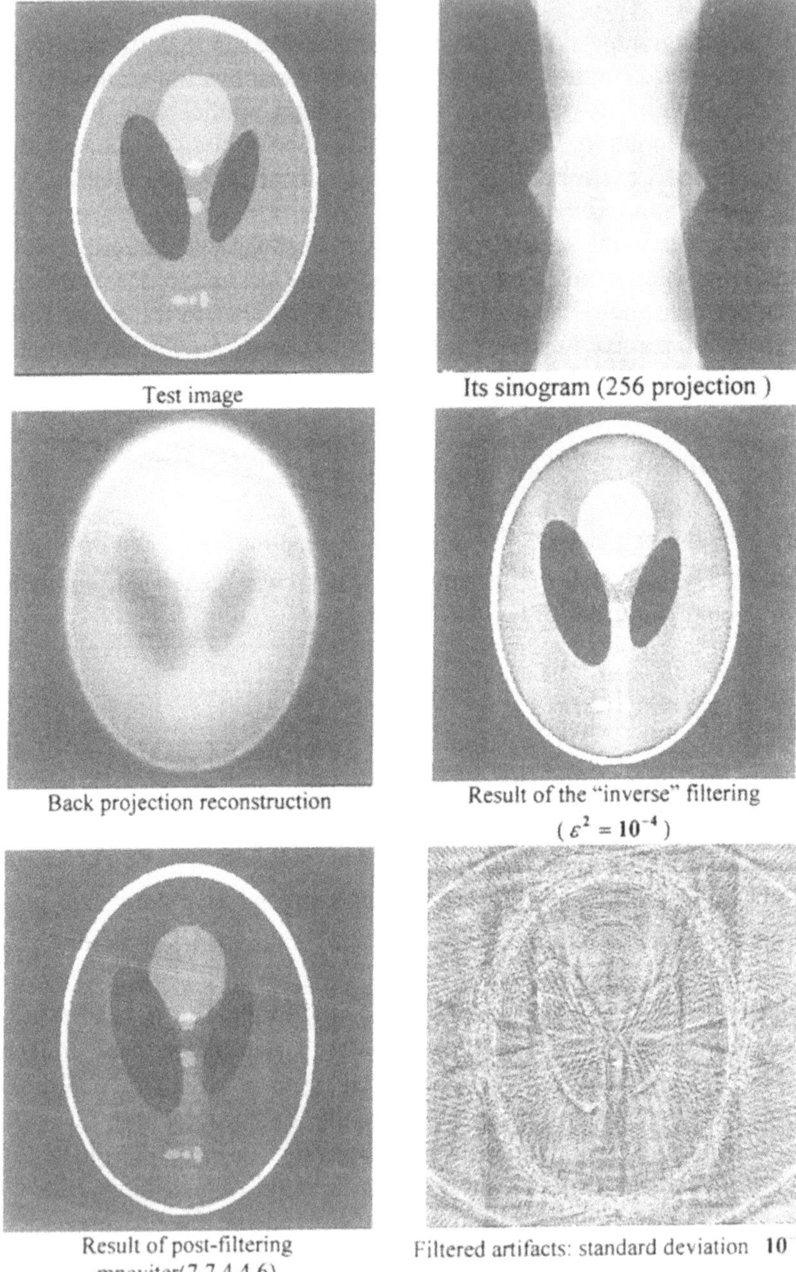

Test image

Its sinogram (256 projection)

Back projection reconstruction

Result of the "inverse" filtering ($\varepsilon^2 = 10^{-4}$)

Result of post-filtering mneviter(7,7,4,4,6)

Filtered artifacts: standard deviation 10^{-2}

Figure 8-28. Illustration of the inverse filtering for image restoration from back projection reconstruction

Similar "pre-processing - inverse transform - post-processing" technology is adopted also for digital reconstruction of holograms (Fig. 8.27).

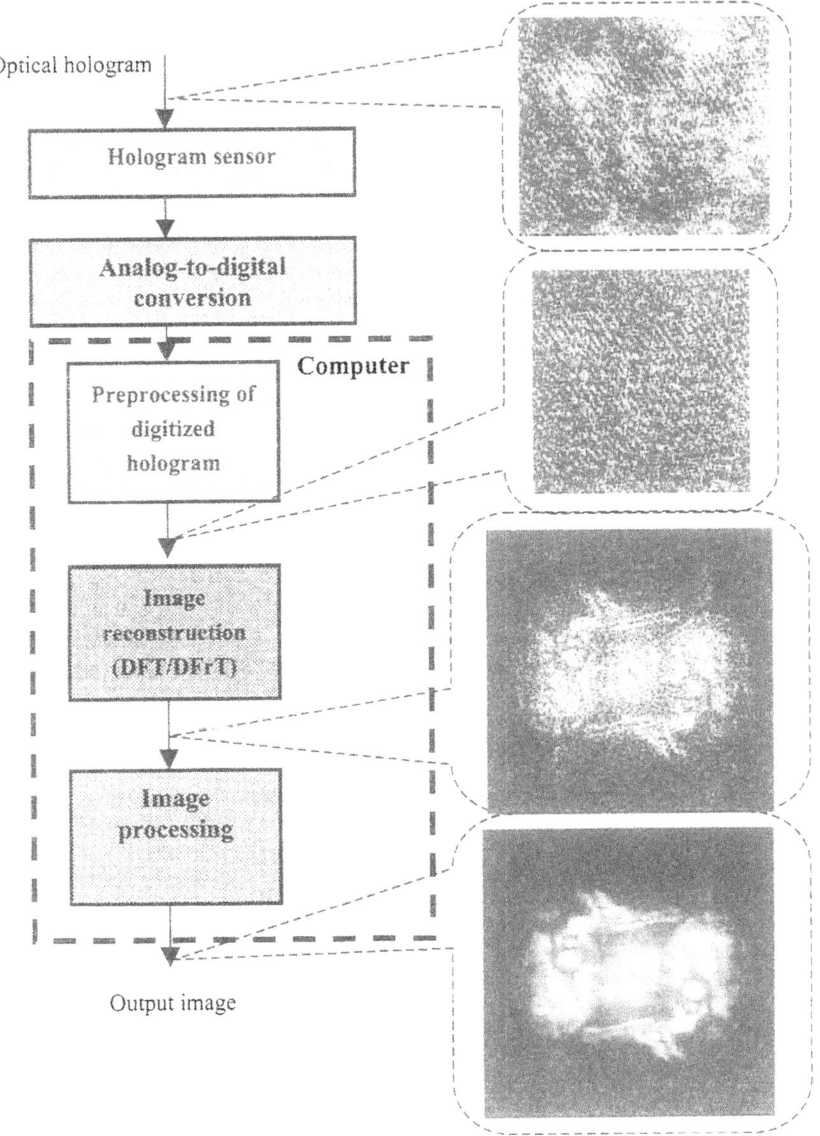

Figure 8-29. Basic stages of the digital reconstruction of hologram

Hologram preprocessing is aimed at correcting signal distortions in hologram sensor and removing hologram components that may deteriorate image reconstruction quality. In Sect. 8.2.2, Fig. 8.4, we illustrated filtering periodic interferences that sometimes take place in the process of digitization of holograms in frame grabbers when analog video cameras are used as hologram sensors.

A frequent purpose of hologram preprocessing is also eliminating zero order diffraction spot in the reconstructed image. Zero order diffraction spot appears in images reconstructed from Fourier and Fresnel holograms due to the presence in the recorded hologram a constant or slowly changing component (see Sects. 4.4.2 and 13.3). A simple method to achieve this is subtraction from the hologram signal its local mean value computed over a sliding window ([5]).

$$\tilde{\Gamma}_{r,s} = \Gamma_{r,s} - \frac{1}{(2N_1+1)(2N_2+1)} \sum_{k=-N_1}^{N_1} \sum_{l=-N_2}^{N_2} \Gamma_{r-k,s-l}, \tag{8.7.8}$$

where $\{\Gamma_{r,s}\}$ is an array of hologram samples with (r,s) being the sample indices. As it is described in Sect. 5.1.1, local mean in a sliding window can be very efficiently computed recursively. In this application, window size is usually 3×3 to 7×7 pixels. Fig. 8-30, center, shows result of reconstruction of the hologram shown in Fig. 8-4 after removing zero-order diffraction spot. Right image in this figure shows result of the hologram reconstruction after both removing zero-order diffraction spot and periodic interferences.

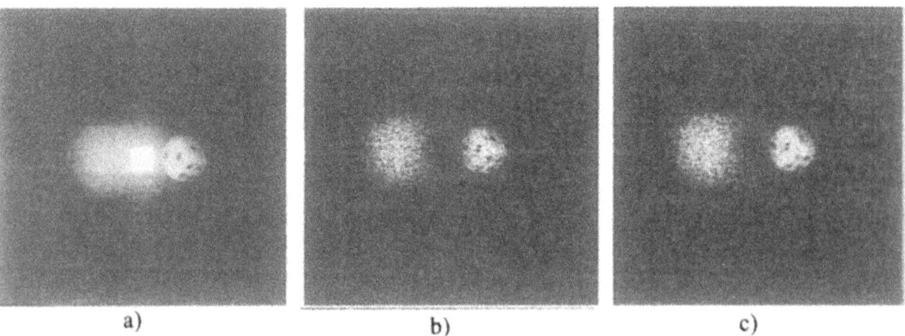

Figure 8-30. Reconstruction of hologram without pre-processing (a) and with preprocessing for removing zero order diffraction spot (b) and for removing, in addition, periodic interferences (c)

Fig. 8.31 illustrates elimination of zero order diffraction spot and image post-processing in digital reconstruction of a hologram recorded in two exposures by the method of phase-shifted digital holography.

Normally, phase-shifting method for digital recording optical holograms assumes taking three or four exposures of the interference pattern of object and reference beams with phase shifts of the reference beam $\mathbf{0}, \pi / \mathbf{2}, \pi$ for three exposures and $\mathbf{0}, \pi / \mathbf{2}, \pi, 3\pi / \mathbf{2}$ for four exposures ([16]). This gives the following three or four recorded holograms, correspondingly:

$$\begin{aligned}
\Gamma_0 &= |\mathbf{Ref}|^2 + |\mathbf{Obj}|^2 + 2|\mathbf{Ref}|\mathbf{Obj}_{re};\\
\Gamma_{\pi/2} &= |\mathbf{Ref}|^2 + |\mathbf{Obj}|^2 + 2|\mathbf{Ref}|\mathbf{Obj}_{im};\\
\Gamma_{\pi} &= |\mathbf{Ref}|^2 + |\mathbf{Obj}|^2 - 2|\mathbf{Ref}|\mathbf{Obj}_{re}.
\end{aligned} \tag{8.7.9}$$

and

$$\begin{aligned}
\Gamma_0 &= |\mathbf{Ref}|^2 + |\mathbf{Obj}|^2 + 2|\mathbf{Ref}|\mathbf{Obj}_{re};\\
\Gamma_{\pi/2} &= |\mathbf{Ref}|^2 + |\mathbf{Obj}|^2 + 2|\mathbf{Ref}|\mathbf{Obj}_{im};\\
\Gamma_{\pi} &= |\mathbf{Ref}|^2 + |\mathbf{Obj}|^2 - 2|\mathbf{Ref}|\mathbf{Obj}_{re};\\
\Gamma_{3\pi/2} &= |\mathbf{Ref}|^2 + |\mathbf{Obj}|^2 - 2|\mathbf{Ref}|\mathbf{Obj}_{im},
\end{aligned} \tag{8.7.10}$$

where $|\mathbf{Ref}|^2$ is intensity distribution in the hologram plane of the reference beam, $|\mathbf{Obj}|^2$ is the intensity distribution of the object wave front and $\mathbf{Obj}_{re}$ and $\mathbf{Obj}_{im}$ are orthogonal (real and imaginary) components of the object wave front. By means of an appropriate combination of the recorded holograms, one can separate orthogonal components of the hologram for reconstruction. For instance, for four exposures:

$$|\mathbf{Ref}|\mathbf{Obj}_{re} = (\Gamma_0 - \Gamma_\pi)/\mathbf{4}; \quad |\mathbf{Ref}|\mathbf{Obj}_{im} = (\Gamma_{\pi/2} - \Gamma_{3\pi/2})/\mathbf{4}. \tag{8.7.11}$$

With two exposures that produce two holograms, Γ_0 and $\Gamma_{\pi/2}$, the reconstruction that uses these two holograms as real and imaginary parts of the object hologram, gives a reconstructed image hidden behind a heavy zero order diffraction spot due to hologram components $|\mathbf{Ref}|^2$ and $|\mathbf{Obj}|^2$ as it is seen in Fig. 8.31, a. Hologram preprocessing described by Eq. 8.7.8 allows to substantially eliminate influence of these components (Fig. 8.31, b). Filtering speckle noise and remaining after this filtering impulse noise provides further improvement of quality of the reconstructed image (Figs. 8. 31, c and d).

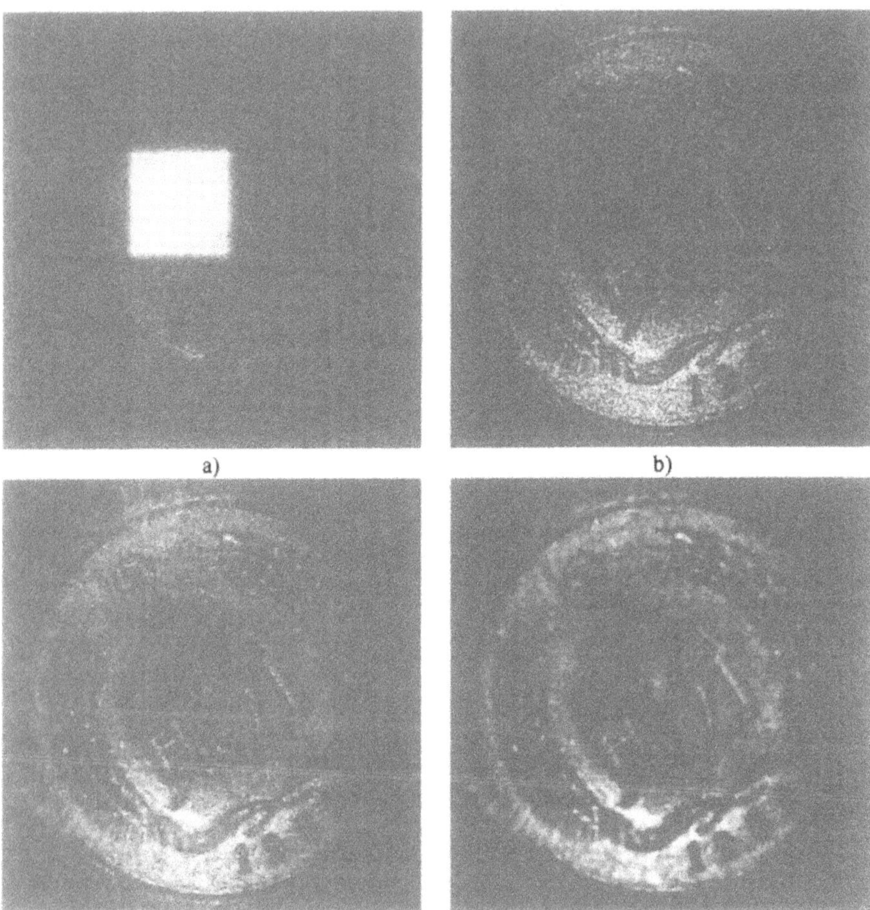

Figure 8-31. Removing zero-order Reconstructed image denoising; (a) - image reconstructed directly from two exposure holograms; (b) - image reconstructed from two exposure holograms after preprocessing according to Eq. 8.7.8; (c) - reconstructed image after filtering speckle noise using local adaptive filtering algorithm (Sect. 8.3.2); (d) - result of subsequent filtering remaining impulse noise (Sect. 8.5)

8.8 IMAGE ENHANCEMENT

8.8.1 Image enhancement as an image processing task. Classification of image enhancement methods

Image enhancement is a processing aimed at assisting image visual analysis. In visual image analysis, the user that observes images plays a role of a decision making machine while images may be regarded as sets of features that are needed for the decision making and that are represented in a special way perceivable to the machine.

Human visual system has certain limitations in its capability to perceive information carried by images. Visual system can not detect brightness contrasts that are lower then a certain contrast sensitivity threshold. It has also a limited resolving power. It fails to analyze images corrupted by noise, especially by impulse noise or highly correlated noise patterns. From the other side, visual system perceives colors, it has 3-D capability through stereo vision, it is capable of efficient detecting changing images in time. One can say that visual system has five channels per pixel to perceive information: three for RGB colors and additional two for stereo and for dynamical vision. Image enhancement is intended to convert images into a form that makes use of capabilities of human visual system to perceive information to their highest degree.

Theoretically, image enhancement methods may be regarded as an extension of image restoration methods. However, in contrast to image restoration, image enhancement frequently requires intentional distorting image signal such as exaggerating brightness and color contrasts, deliberate removing certain details that may hide important objects, converting gray scale images into color, stereo and movie ones to display three-four-five component features, etc. In this sense image enhancement is image preparation or enrichment in the same meaning these words have in mining.

An important peculiarity of image enhancement as image processing is its interactive nature. The best results in visual image analysis can be achieved if it is supported by a feedback from the user to the image processing system.

Image enhancement methods date back to the beginning of the last century photography and microscopy methods such as unsharp masking, solarization, phase contrast ([17]). Electronic television enabled using more sophisticated methods for manipulating image contrasts, brightness, color and geometry. However only digital computers equipped with appropriate display devices enabled implementation of truly interactive image processing. Flexibility of programmable digital computers make them the ideal vehicle for image enhancement.

From the point of view of implementations, image enhancement methods can be classified into four categories: gray scale, or histogram manipulation methods, image spectra manipulation methods, geometrical transformation methods and display methods.

Gray scale, or histogram manipulation methods implement different modifications of image display transfer function. We will review them in Sect. 8.8.2. Image spectra manipulation methods implement modification of image spectra in different bases, most frequently DFT and DCT ones. They are reviewed in Sect. 8.8.3. Image geometrical transformation are used in image enhancement to facilitate representation of 3-D objects in 2-D images. Methods for image geometrical transformations are discussed in Ch. 9. Special information display methods for image enhancement are briefly outlined in Sect. 8.8.4.

8.8.2 Gray level histogram modification methods

One of the features that determine the capability of vision to detect and analyze objects in images is contrast of objects in their brightness with respect to their background. While globally image may utilize the entire dynamic range of the display device, locally, within small windows image signal has very frequently much lower dynamic range. Small local contrasts may make it difficult to detect and analyze objects.

The simplest and the most straightforward way for amplification of local contrasts is stretching local dynamic range of image signal in a sliding window to the entire dynamic range of the display device. The two computationally inexpensive methods for stretching local contrasts are max-min stretching and local dynamic range normalization by local mean and variance. They are described correspondingly as

$$\hat{a}_{k,l} = \frac{a_{k,l} - \min_{k,l}}{\max_{k,l} - \min_{k,l}}, \tag{8.8.1}$$

and

$$\hat{a}_{k,l} = \begin{cases} a_{\max}, & if \quad \hat{\tilde{a}} \geq a_{\max} \\ \hat{\tilde{a}}, & if \quad a_{\min} \leq \hat{\tilde{a}} \leq a_{\max} \\ a_{\min}, & if \quad \hat{\tilde{a}} \leq a_{\min} \end{cases}, \tag{8.8.2}$$

where

$$\hat{\hat{a}} = g\frac{a_{k,l} - \mathbf{mean}_{k,l}^{(W_1,W_2)}}{\mathbf{stdev}_{k,l}} + \bar{a}\,, \qquad (8.8.3)$$

$a_{k,l}$ and $\hat{a}_{k,l}$ are input and output image samples in the window position (k,l), $\mathbf{max}_{k,l}$, $\mathbf{min}_{k,l}$, are maximal and minimal signal values within the window, $a_{\max}$ and $a_{\min}$ are boundaries of the display device dynamic range,

$$\mathbf{mean}_{k,l}^{(W_1,W_2)} = \frac{1}{(2W_1+1)(2W_2+1)}\sum_{m=-W_1}^{W_2}\sum_{n=W_1}^{W_2} a_{k-m,l-n}\,, \qquad (8.8.4)$$

is local mean value and

$$\mathbf{stdev}_{k,l} = \sqrt{\frac{1}{(2W_1+1)(2W_1+1)}\sum_{m=-W_1}^{W_1}\sum_{n=-W_2}^{W_2}\left(a_{k-m,l-n} - \mathbf{mean}_{k,l}^{(W_1,W_2)}\right)^2}\,, \qquad (8.8.5)$$

is local standard deviation within the window.

Local dynamic range normalization by local mean and variance is illustrated in Fig. 8.32, upper row. In Sect. 11.4 it is shown that this operation is also a useful image preprocessing operation that improves the discrimination capability of the optimal adaptive correlator for target detection in a complex background.

Alternative methods of amplification of local contrasts are methods of the direct modification of image histograms: global and ***local histogram equalization*** and ***P***-histogram equalization ([4, 18]). Histogram equalization converts an image $\{a_{k,l}\}$ with histogram $h(q)$ into an image $\{\hat{a}_{k,l}\}$ with a uniform histogram by means of the following gray level transformation:

$$\hat{a}_{k,l} = (q_{\max} - q_{\min})\frac{\sum_{q=q_{\min}}^{a_{k,l}} h(q)}{\sum_{q=q_{\min}}^{a_{\max}} h(q)} + q_{\min}\,, \qquad (8.8.6)$$

where $q_{\min}$ and $q_{\max}$ are minimal and maximal quantized values of image signal (for 256 quantization levels these are 0 and 255).

It follows from Eq. 8.8.5 that the slope of the histogram equalization transfer function $\hat{a}_{k,l}(a_{k,l})$ is proportional to the histogram value:

$$\frac{\Delta \hat{a}_{k,l}}{\Delta a_{k,l}} \propto h(q)\Big|_{q=a_{k,l}}. \tag{8.8.7}$$

This means that histogram equalization results in amplification of contrasts that, for each gray level, is proportional to its histogram value, or to the gray level ***cardinality***.(see Sect. 12.1).

P-histogram equalization defined by the equation

$$\hat{a}_{k,l} = (q_{\max} - q_{\min}) \frac{\sum_{q=q_{\min}}^{a_{k,l}} [h(q)]^P}{\sum_{q=q_{\min}}^{q_{\max}} [h(q)]^P} + q_{\min}, \tag{8.8.8}$$

is a natural generalization of the histogram equalization (Kim/Yaro) that makes it more flexible through the use of a user-defined parameter P. If $P = 0$, P-histogram equalization is equivalent to the automatic stretching image dynamic range from $(a_{\min}, a_{\max})$ to $(q_{\min}, q_{\max})$. $P = 1$ corresponds to histogram equalization. Intermediate values $0 \le P \le 1$ result in more soft contrast enhancement that is frequently better perceived visually.

The described histogram modification techniques can be applied both globally and locally. In the latter case, the modification is local adaptive.

Histogram equalization and P-histogram equalization are illustrated in Fig. 8-32 (middle and bottom rows). Fig 8-33 compares global histogram equalization and P-histogram equalization and their corresponding transfer functions (compare them with the image histogram).

Note that the P-histogram equalization may be regarded as an implementation of the nonlinear pre-distortion for optimal compander quantization (Eqs. 3.6.22-23).

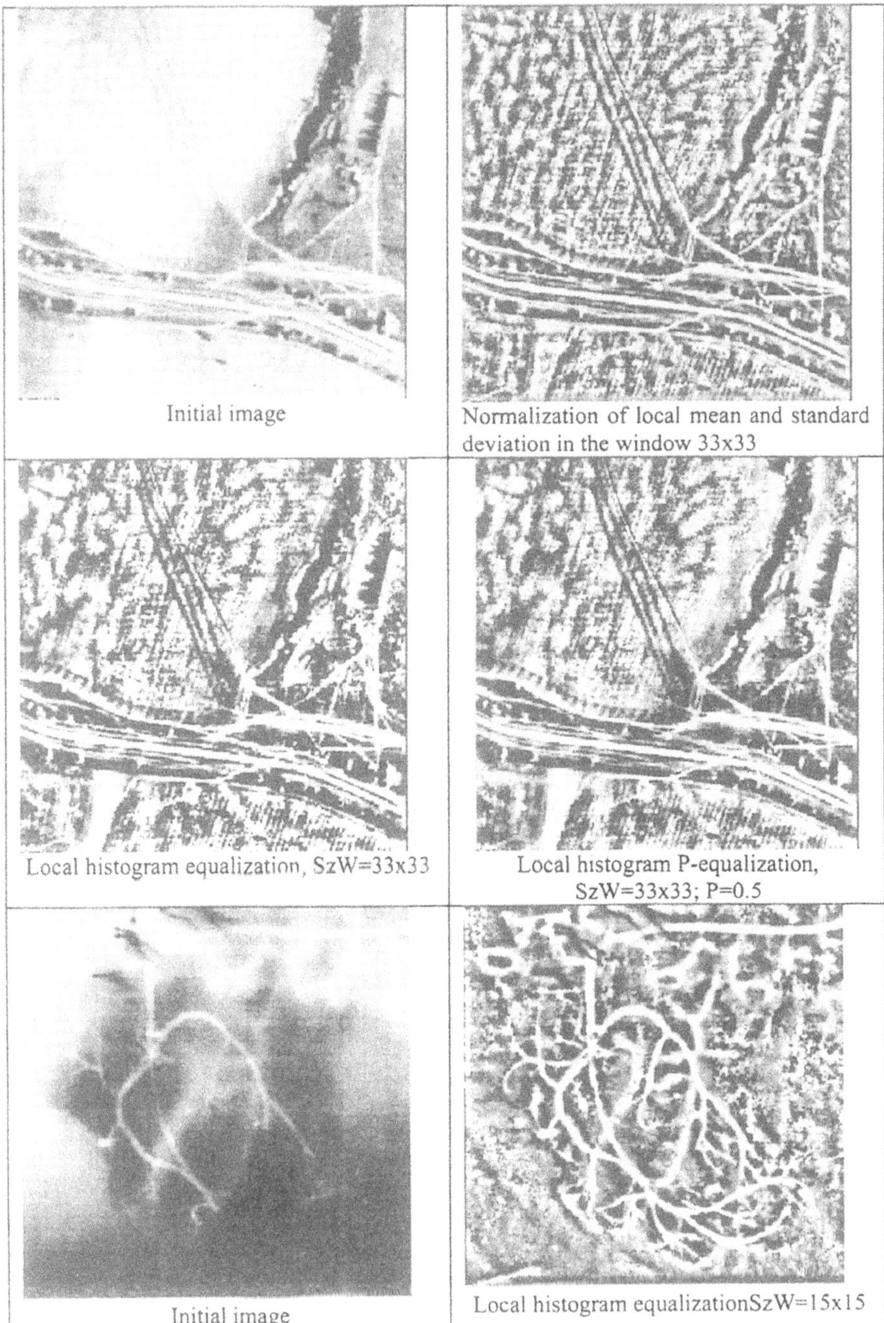

Figure 8-32. Local histogram modification by means of standardization of local mean and variance (uper row) and by *P*-histogram equalization (middle and bottom rows)

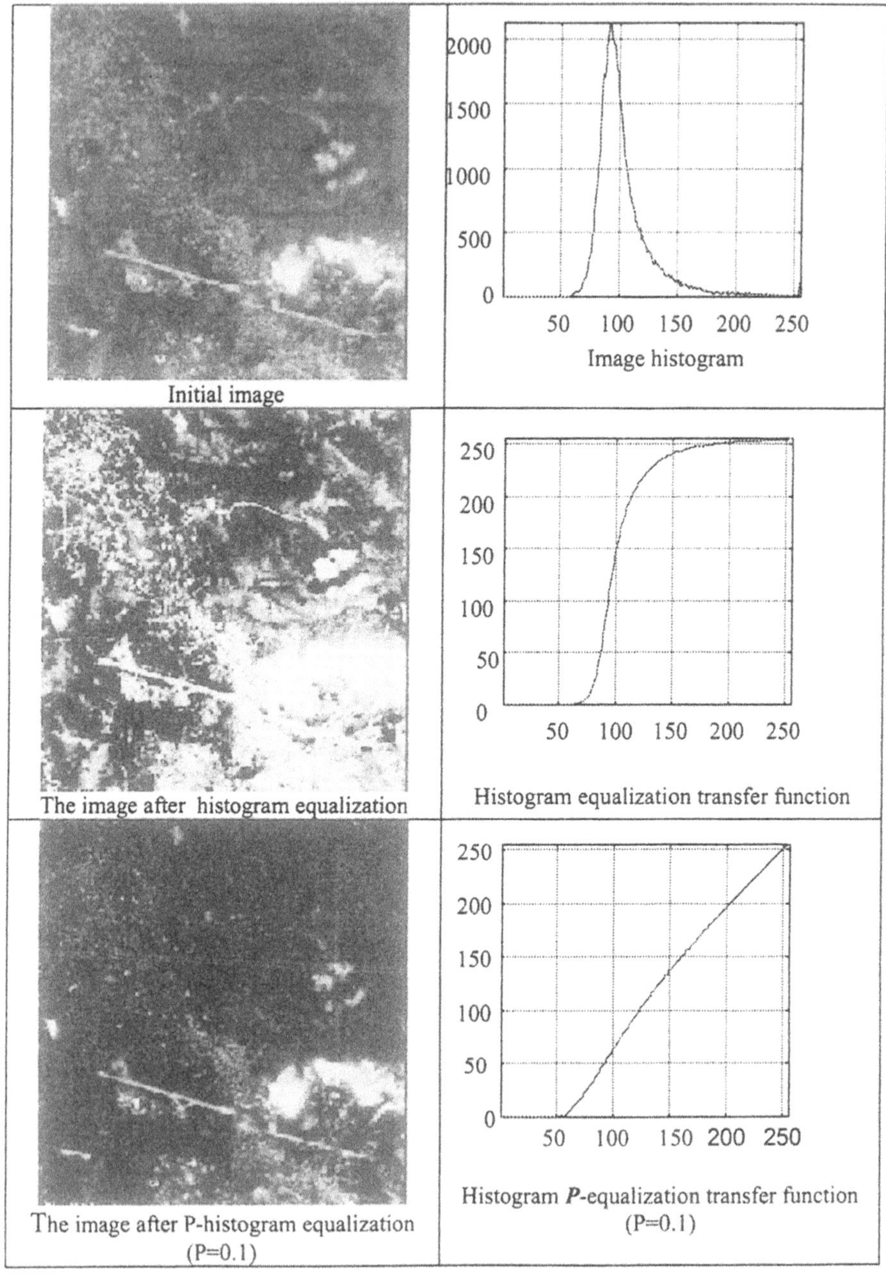

Figure 8-33. Global histogram modification by means of histogram equalization and P-histogram equalization. Left column (from top to bottom): initial image, result of histogram eqalization and P-histogram equalization (P=0.1). Right column, from top to bottom: histogram of the initial image and corresponding gray level transformation transfer functions

8.8.3 Image spectra modification methods

Two image spectra manipulation methods for local contrast enhancement are the most simple in the implementation: unsharp masking and nonlinear spectra coefficients transformations. They are implemented, respectively, in signal domain and transform domain filtering.

Unsharp masking is defined by the equation:

$$\hat{a}_{k,l} = a_{k,l} + g\left(a_{k,l} - \mathbf{mean}_{k,l}^{(W_1,W_2)}\right), \tag{8.8.9}$$

where $\mathbf{mean}_{k,l}^{(W_1,W_2)}$ is the image local mean in the window of $(2W_1+1)(2W_2+1)$ pixels defined by Eq. 8.8.3 and g is a user defined local contrast amplification parameter. Window size parameters W_1 and W_2 are also user defined parameters that are commensurate with the size of the objects to be enhanced.

Modification of image spectra that results from applying unsharp masking is determined by the unsharp masking frequency response. As it follows from Eq. 8.8.9, it is, for image of $N_1 \times N_2$ samples, equal to:

$$\eta_{r,s} = 1 + g\left[1 - \mathbf{sincd}(2W_1 - 1; N_1; r)\mathbf{sincd}(2W_2 - 1; N_2; r)\right] \tag{8.8.10}$$

where $\mathbf{sincd}(\cdot;\cdot;\cdot)$ is the discrete sinc-function defined by Eq. 4.3.18. Fig. 8.-34 illustrates 1-D sections of unsharp masking frequency responses for windows of 3 and 15 pixels.

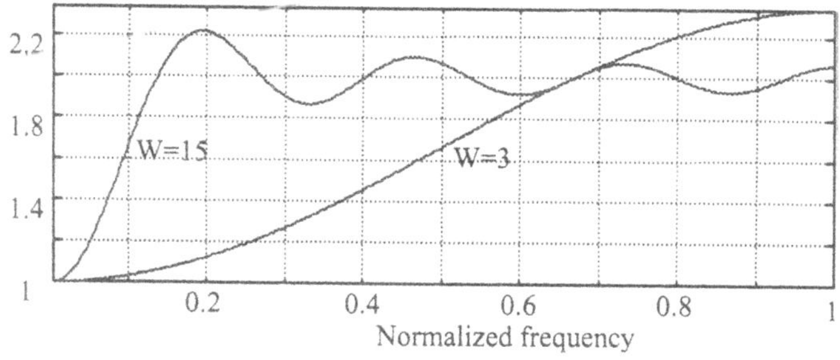

Figure 8-34. 1-D sections of unsharp masking frequency responses for window sizes 3 and 15 pixels with amplification coefficient g=1

Unsharp masking is a non-adaptive local contrast enhancement procedure. With unsharp masking, the degree of amplification of image high frequencies is the same for any image. An alternative and adaptive method for modification of image spectra that results in local contrast enhancement is ***P***-th law nonlinear modification of absolute values of image spectra coefficients. It is described by the equation

$$\hat{\alpha}_{r,s} = \frac{\alpha_{r,s}}{\left|\alpha_{r,s}\right|}\left|\alpha_{r,s}\right|^{P}, \tag{8.8.9}$$

where $\{\alpha_{r,s}\}$ and $\{\ddot{\alpha}_{r,s}\}$ are initial and transformed image spectral coefficients in a selected basis, respectively, and $\boldsymbol{P}$ is a user defined parameter. As a transform, usually DFT or DCT are used. When $\mathbf{0} \le \boldsymbol{P} < \mathbf{1}$, this modification redistributes energy of spectral coefficients in favor of low energy coefficients. The degree to which individual coefficients are amplified depends now on the image spectrum.

Such a spectrum modification may be applied globally to the spectrum of the entire image and locally in a user specified sliding window to spectra of the window samples in each position of the window. In the latter case, the modification is local adaptive. In Fig. 8-35 one can compare image local contrast enhancement by means of unsharp masking and global and local nonlinear modification of image DCT spectrum.

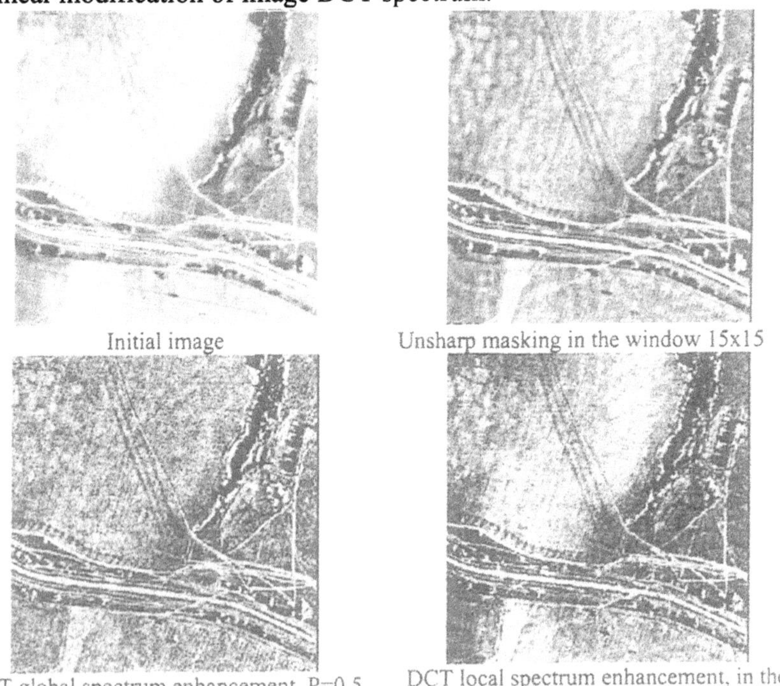

Figure 8-35. Unsharp masking and nonlinear spectrum modification for image enhancement

8.8.4 Using color, stereo and dynamical vision for image enhancement

The basic principle of using color, stereo and dynamical capabilities of vision is straightforward. Image is processed to produce several output images that represent certain several features of the input image. For instance, this processing may be image sub-band decomposition, image local histogram ***P***-equalization using several different window sizes, edge enhancement and extraction with different algorithm parameters (see some of the algorithms described in Ch. 12), object detection, to name a few. The obtained set of images may then be used to generate animated artificial

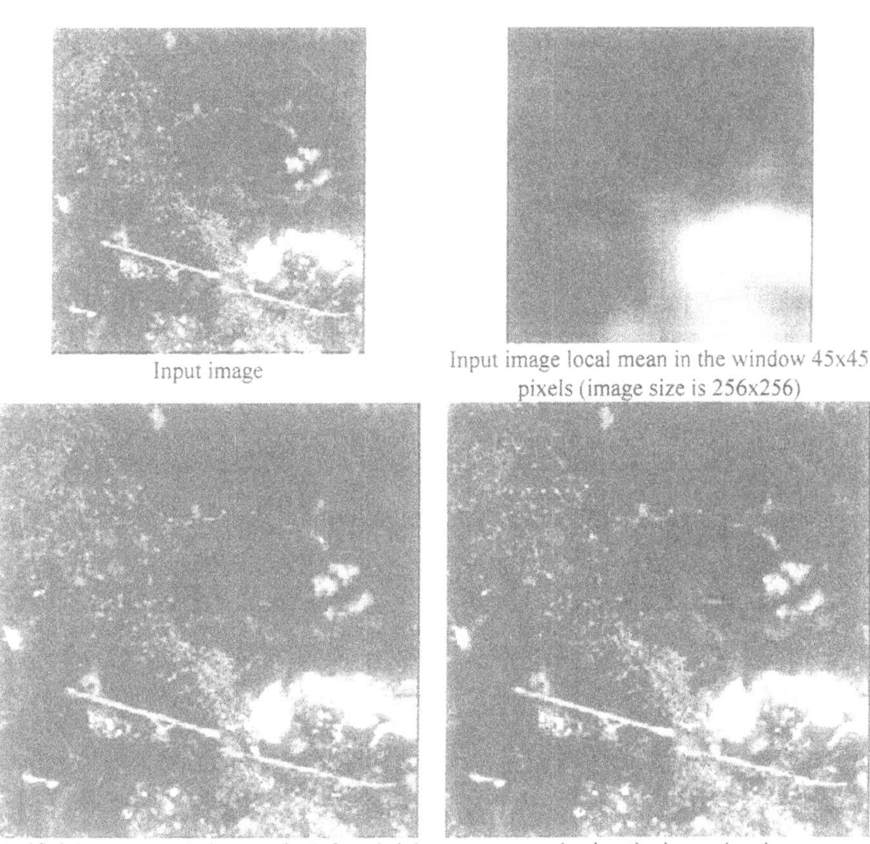

Figure 8-36. Stereo visualization of an image and its local mean. One can observe 3_d image using a stereoscope or by looking at the images with squinted eyes. In the latter case, one should try to squint eyes until two pairs of images seen with each eye overlap into 3 images. Then the central of those three images with be seen as a 3-D image.

movie by using generated images as movie frames, or to generate artificial color images by representing combination of three generated images as red, green and blue components of the color image that is displayed for visual analysis, or to generate artificial stereoscopic images for left and right eyes from couples of the obtained images. In the latter case, one of the processed images or the initial image is treated as a "reflectance map" of an artificial 3-D surface and another as its "depth map". Using the reflectance map and the depth map, one can generate a pair of images for right and left eyes by introducing to every pixel of the "reflectance map" a horizontal shift proportional the value of the "depth map" for this pixel. This can be done using image resampling methods described in Ch. 9. Fig. 8-36 illustrates an example of such an artificial stereo image generated from air photograph of 256x256 pixels treated as a "reflectance map" and an image of its local means in the window of 45x45 pixels treated as a "depth map".

References

1. N. Wiener, The interpolation, extrapolation and smoothing of stationary times series, Wiley, New York, 1949
2. A. N. Kolmogorov, Sur l'interpolation de suits stationaires, C. R. Acad. Sci., 208, 2043-2045
3. D.L. Donoho and I.M. Johnstone, Ideal spatial adaptation by wavelet shrinkage, Biometrica, 81(3): 425-455, 1994
4. L. Yaroslavsky and M. Eden, Fundamentals of Digital Optics, Birkhauser, Boston, 1996
5. L. Yaroslavsky, N. Merzlyakov, Methods of Digital Holography, Consultant Bureau, N.Y., 1980
6. M.D. Levine, Vision in Man and Mashine, , McGraw-Hill, 1985, pp. 110-130
7. L.P. Yaroslavsky, Local Adaptive Filters for Image Restoration and Enhancement, In: *Int. Conf. on Analysis and Optimization of Systems. Images, Wavelets and PDE's*, Paris, June 26-28, 1996, Eds.: M.-O. Berger, R. Deriche, I. Herlin, J. Jaffre, J.-M. Morel, Springer Verlag, Lecture Notes in Control and Information Sciences, 219., 1996, pp. 31-39.
8. L. Yaroslavsky, "Local Adaptive Image Restoration and Enhancement with the Use of DFT and DCT in a Running Window", *Proceedings, Wavelet Applications in Signal and Image Processing IV*, SPIE Proc. Series, v. 2825, pp. 1-13, 6-9 August 1996, Denver, Colorado.
9. L. Yaroslavsky, Image Restoration, enhancement and target location with local adaptive filters, in: International Trends in Optics and Photonics, ICOIV, ed. by T. Asakura, Springer Verlag, 1999, pp. 111-127
10. B.Z.Shaick, L. Ridel, L. Yaroslavsky, A hybrid transform method for image denoising, EUSIPCO2000, Tampere, Finland, Sept. 5-8, 2000
11. ftp://ftp.cis.upenn.edu/pub/eero/matlabPyrTools.tar.gz
12. L. P. Yaroslavsky, H. J. Caulfield, Deconvolution of multiple images of the same object, Applied Optics, v. 33, No. 11, pp. 2157-2162, 1994
13. L. Yaroslavsky, M. Eden, Correlational Accumulation as a Method for Signal Restoration, Signal Processing, v. 39, pp. 89-106, 1994
14. N. El Sherif, G. Turitto, eds., High Resolution Electrocardiography, Futura Mount Cisco, N.Y., 1992
15. B. L. Trus, M. Unser, T. Pan and A.C. Steven, Digital Image Processing of Electron Micrographs: the PIC system II", Scanning Microscopy, v. 6, 1992, pp. 441-451
16. I. Yamaguchi, J. Kato, S. Ohta, J. Mizuno, Image formation in phase shifting digital holography and application to microscopy, Applied Optics, v. 40, No. 34, pp. 6177-618
17. Photography for the Scientist, Ch. E. Engel, ed., Academic Press, London, 1968, pp. 632
18. V. Kim, L.P. Yaroslavsky, Rank Algorithms for Picture Processing, Computer Vision, Graphics and Image Processing, v. 35, 1986, p. 234-258.

Chapter 9

IMAGE RESAMPLING AND GEOMETRICAL TRANSFORMATIONS

9.1 PRINCIPLES OF IMAGE RESAMPLING

Image resampling is required in many image processing applications. It is a key issue in signal and image differentiating and integrating, image geometrical transformations and re-scaling, target location and tracking with sub-pixel accuracy, Radon Transform and tomographic reconstruction, 3-D image volume rendering and volumetric imaging.

When image geometrical transformations are required one have to either map, according to the required transformation, samples of the output image onto input image co-ordinate system or map input image samples onto the output image coordinate system. Because the latter method does not allow to keep arrangement of output image samples on a fixed rectangular grid, the first method, "backward co-ordinate mapping" one, is adopted. This principle is illustrated in Fig. 9-1. As one can see from the figure, computing transformed image samples requires resampling of the input image in a new sampling grid defined by the mapping.

Image resampling assumes, that a continuous model of the input image should be obtained from its available samples as it is described by Eq. 3.3.1, (b) and then this model should be re-sampled in the new sampling raster defined by the backward mapping. In digital processing one can approximate this model by generating, from available signal samples, a new set of samples that correspond to sub-sampling the initial signal in a more dense grid, the grid density being defined by the required approximation accuracy. This can be done in the following way.

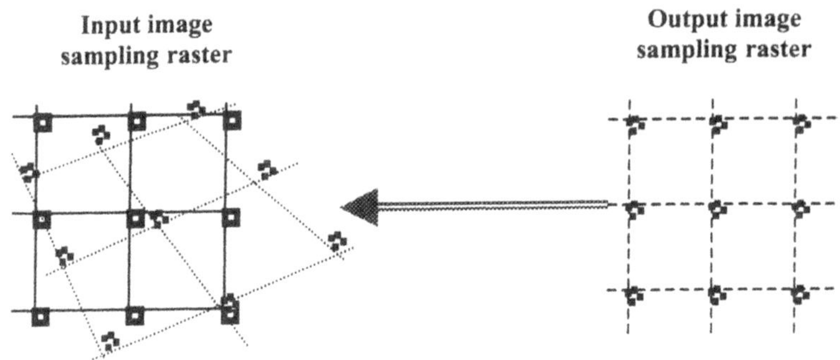

Figure 9-1. The principle of image resampling for geometrical transformations: backward coordinate mapping

Consider first a 1-D model. Let $\{a_k\}$ be a set of available signal samples. Generate an L-times extended set $\{\tilde{a}_k\}$ of samples by complementing each available signal sample with L of zero samples:

$$\{\tilde{a}_k\} = a_{k_N}\delta(k_L); \; k = Lk_N + k_L; \; k_N = 0,1,...,N-1; \; k_L = 0,1,...,L-1; \tag{9.1.1}$$

Then reconstruction of the continuous signal from its samples described by Eq. 3.3.1, b can be, in digital processing, implemented by interpolating zero valued samples in the set $\{\tilde{a}_k\}$ by means of a digital convolution:

$$\tilde{\tilde{a}}_k = \sum_n \tilde{a}_{k-n} h_n^{\text{int}}, \tag{9.1.2}$$

with an interpolation kernel $\{h_n^{\text{int}}\}$. In this way a signal is generated with L times larger number of samples then that of the original signal. We will refer to this procedure as to ***signal zooming***, or L-zooming.

Resampling the interpolated zoomed signal to a new sampling grid can, in its turn, also be implemented by a digital convolution

$$\hat{a}_{k_N} = \sum_n \tilde{\tilde{a}}_{Lk_N - n} h_n^{\text{smpl}}. \tag{9.1.3}$$

with a sampling kernel $\{h_n^{\text{smpl}}\}$. Eq. 9.1.3 is a digital counterpart of Eq. 3.3.1, a) that describes sampling continuous signals.

Digital convolution assumes one or another method for signal extension outside the interval defined by its available samples (See 5.1.4). In view of the advantages of computing digital convolution in DFT domain (see Sect. 5.2) using fast transforms, let's assume cyclic extension of signal $\{\tilde{a}_k\}$ that implies that the convolutions defined in Eqs. 9.1.2 and 3 are cyclic convolutions:

$$\tilde{\tilde{a}}_k = \sum_{n=0}^{LN-1} \tilde{a}_{(k-n) \bmod LN} h_n^{\text{int}} ; \tag{9.1.4}$$

$$\tilde{\tilde{a}}_k = \sum_{n=0}^{LN-1} \tilde{a}_{(k-n) \bmod LN} h_n^{\text{smpl}} \tag{9.1.5}$$

As far as resampling is cyclic convolution based, resampling errors can be analyzed by the analysis of DFT frequency responses of the interpolation and sampling kernels. From discrete sampling theorem (Sect. 4.3.2) it follows that, among cyclic convolution interpolation kernels, kernel

$$h_n^{\text{int}} = \text{sincd}(N, N, n) = \frac{\sin(\pi n)}{N \sin(\pi n / N)} \tag{9.1.6}$$

is the only one that assures preservation of initial signal spectral components and the absence of aliasing terms in spectra of the interpolated signals. We will call this interpolation method and its versions ***the discrete sinc-interpolation***.

2-D signal interpolation can be implemented as either separable or inseparable digital convolution. When carried out in signal domain, separable digital convolution is computationally much more efficient then inseparable one (Sect. 5.1.2). If the convolution is implemented in transform domain, the use of inseparable convolution kernels is possible without any loss of the computational efficiency. Inseparable interpolation kernels are more natural for image processing. In Sect. 9.3 DFT and DCT based discrete sinc-interpolation algorithms will be described that, in principle, enable the implementation of inseparable convolution kernels.

9.2 NEAREST NEIGHBOR, LINEAR AND SPLINE INTERPOLATION METHODS

The most known discrete signal interpolation methods are nearest neighbor interpolation, linear (bilinear, for 2-D signals) interpolation and spline interpolation. For L-zooming signal of N samples, interpolation kernel $\{h_n^{int(0)}\}$ for the nearest neighbor interpolation is defined by the equation:

$$h_n^{int(0)} = \mathrm{rect}(n/L) = \begin{cases} 1, & 0 \le n \le L-1 \\ 0, & L \le n \le NL-1 \end{cases}. \tag{9.2.1}$$

Module of its frequency response $\left\{\left|\eta_r^{int(0)}\right|\right\} = \left|\mathbf{DFT}_{LN}\left\{h_n^{int(0)}\right\}\right|$ that characterizes distortions of the original signal frequency components at the result of its zooming, is described by the equation

$$\left|\eta_r^{int(0)}\right| \propto \left|\mathrm{sincd}(L; LN; r)\right| = \left|\frac{\sin(\pi r/N)}{\sin(\pi r/LN)}\right|. \tag{9.2.2}$$

Graph of $\left\{\left|\eta_r^{int(0)}\right|\right\}$ for $L=16$, $N=128$ is shown in Fig. 9-2 a centered around the symmetry point r=1024. Central lobe of the graph describes distortions of the original signal frequency components while its side lobes describe aliasing artifacts. Fig. 9-2, b) illustrates image 8-zooming and pixelation aliasing artifacts characteristic for the nearest neighbor interpolation.

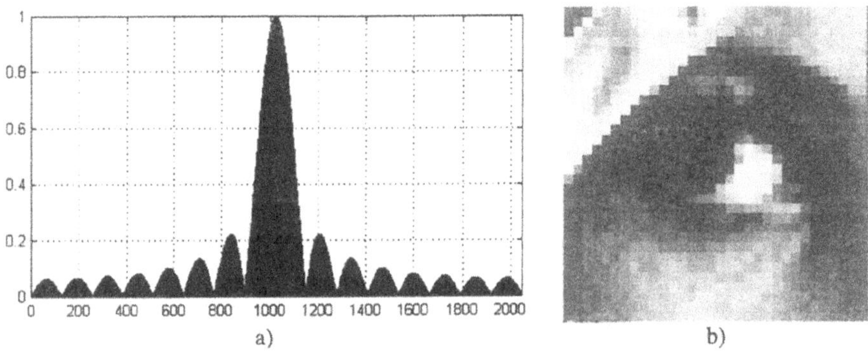

Figure 9-2. Frequency response of the nearest neighbor signal interpolation for 16-zooming signals of 128 samples (a) and an example of image 8-zooming

The principle of linear and separable linear interpolation along two coordinates (bilinear interpolation) is illustrated in Fig. 9-3 for interpolation of signal samples shifted with respect to available samples by u-th fraction of the sampling interval (u-th and v th fractions of the 2-D discretization interval, for bilinear interpolation).

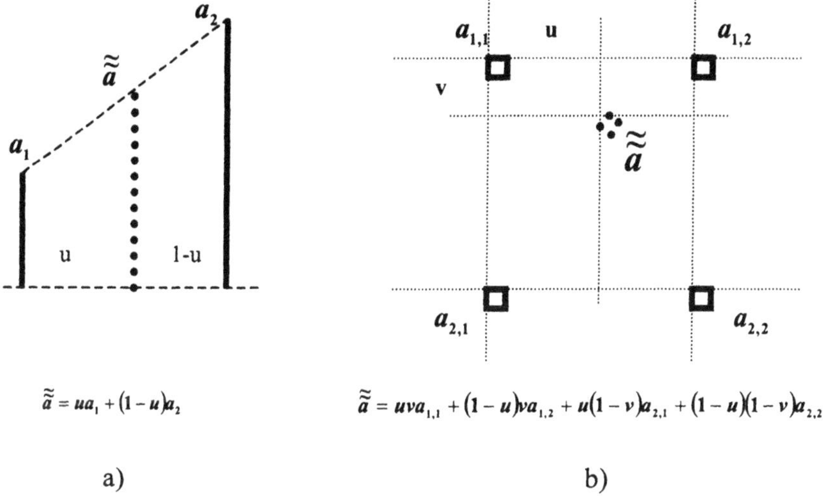

Figure 9-3. The principle of linear (a) and bilinear (b) interpolation. Available signal samples are shown in bold lines. Interpolated samples are shown in doted lines

For L-zooming signal of N samples, interpolation kernel $\{h_n^{int(1)}\}$ for linear interpolation is defined by the equation:

$$h_n^{int(1)} = \mathrm{tri}(n/L) = \begin{cases} n/L, & 0 \le n \le L-1 \\ 2-n/L, & L \le n \le 2L-1 \\ 0, & 2L \le n \le NL-1 \end{cases}. \quad (9.2.3)$$

This kernel is a convolution of the nearest neighbor interpolation kernel of Eq. 9.2.1 with itself. Therefore by virtue of the convolution theorem for DFT, frequency response of linear interpolation is a squared frequency response of the nearest neighbor interpolation:

$$\left|\eta_r^{int}\right| \propto \left|\mathrm{sincd}(L;LN;r)\right|^2 = \left|\frac{\sin(\pi r/N)}{\sin(\pi r/LN)}\right|^2. \quad (9.2.4)$$

It is shown for 16-zooming signal of 128 samples in Fig. 9-4 (a) centered around the symmetry point r=1024. An example of image 8-zooming with bilinear interpolation is shown in Fig. 9-4, b). As one can see on the figures, aliasing effects are much less pronounced with linear (bilinear) interpolation while distortions of original signal frequency components within the base band is more severe.

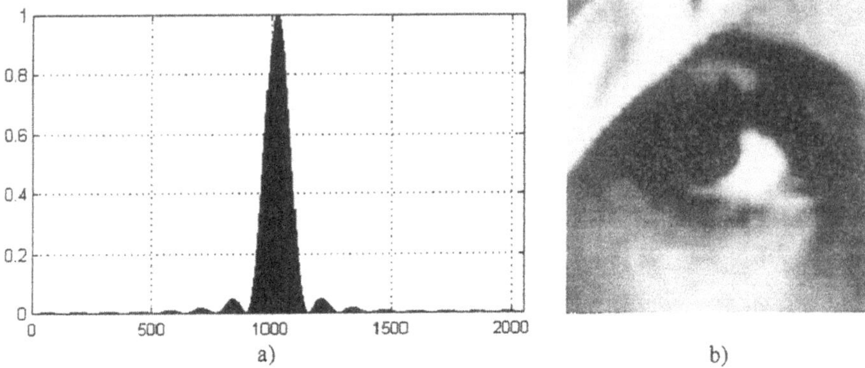

Figure 9-4. Frequency response of linear interpolation for 16-zooming signals of 128 samples (a) and an example of image 8-zooming with bilinear interpolation

Nearest neighbor and linear interpolations are zero order and first order special cases of spline (B-spline) interpolation ([1]). In the nearest neighbor interpolation, one sample of the original signal is involved in calculation of each sample of the zoomed signal. In the linear interpolation, two samples of the original signal are involved. In R-th order spline interpolation, $R+1$ samples are involved. Interpolation kernel in R-th order B-spline is defined as ([1])

$$h_{spl}^{(R)}(x)=\sum_{t=0}^{R+1}\frac{(-1)^t(R+1)}{(R+1-t)!t!}\max\left[0,\left(\frac{R+1}{2}+x-t\right)^R\right] \tag{9.2.5}$$

where x is a continuous variable measured in the units of the discretization interval. For signal L-zooming, $x=n/L$. A cubic B-spline is often used in practice:

$$h_{spl}^{(3)}(x)=\begin{cases}2/3-|x|^2/2(2-|x|), & 0\le|x|<1\\(2-|x|)^3/6, & 1\le|x|<2\\0, & 2\le|x|\end{cases} \tag{9.2.6}$$

A natural requirement to the interpolation kernels $h^{int}(x)$ as functions of a continuous variable x is that they should be equal to one for $x = 0$ and to zero for other integer x. Splines of the order $R > 1$ do not benefit from this property. To be true interpolation kernels, they should be appropriately modified to meet this requirement. The modified splines are called cardinal splines ([1]).

Fig. 9-5 shows graph of the cardinal cubic spline interpolation kernel and its frequency response for 16-zooming signals of 128 samples. One can notice much lower then for the above methods level of side lobes and closer approximation of the ideal low pass filter that ensures lower level of aliasing artifacts and distortions of signal frequency components.

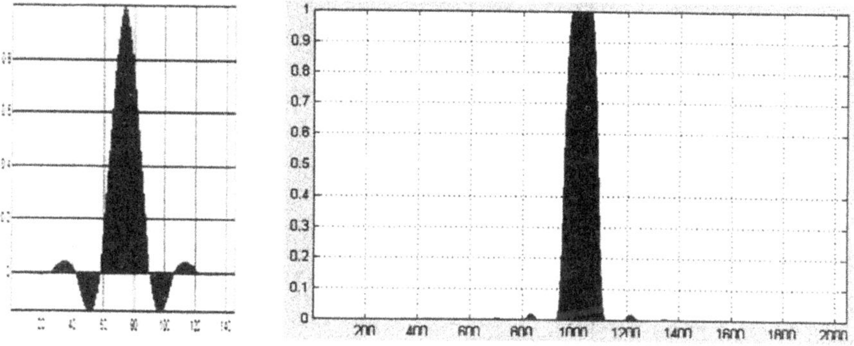

Figure 9-5. Cardinal cubic spline interpolation kernel (a) and frequency response (b) for 16-zooming signals of 128 samples

9.3 ALGORITHMS OF DISCRETE SINC-INTERPOLATION

9.3.1 Discrete sinc-interpolation by zero padding signal DFT spectrum

Zero padding discrete sinc-interpolation method follows straightforwardly from the discrete sampling theorem discussed in Sect. 4.3.2. For a signal of N samples, it enables generating L equidistant intermediate signal samples per each initial one by padding signal DFT spectrum with, depending on whether N is odd or even number, $(L-1)N$ or $(L-1)N \pm 1$ zeros.

For odd N, the algorithm is described as (see Eq. 4.3.49):

$$\tilde{a}_k = \mathbf{IDFT}_{LN}\left\{\left[1 - rect\frac{r-(N+1)/2}{LN-N-1}\right] \bullet \mathbf{DFT}_N\{a_k\}\right\} = \frac{1}{\sqrt{L}}\sum_{n=0}^{N-1} a_n \,\mathrm{sincd}(N;N;(k/L-n)), \qquad (9.3.1)$$

where $r = 0,1,\ldots, LN-1$ is index of DFT signal spectrum, $\mathbf{IDFT}_{LN}\{\ \}$ is operator of inverse Discrete Fourier Transform of the size LN and $\mathbf{DFT}_N\{\ \}$ is operator of DFT of size N and dot $\bullet$ signifies element-wise multiplication of the arrays.

For even N, zero padding should be carried out in either of two ways (see Eqs. 4.3.54 and 55):

$$\tilde{a}_k = \mathbf{IDFT}_{LN}\left\{\left[1 - rect\frac{r-N/2}{LN-N}\right] \bullet \mathbf{DFT}_N\{a_k\}\right\} = \frac{1}{\sqrt{L}}\sum_{n=0}^{N-1} a_n \,\mathrm{sincd}(N-1;N;(k/L-n)), \qquad (9.3.2)$$

or

$$\tilde{a}_k = \mathbf{IDFT}_{LN}\left\{\left[1 - rect\frac{r-N/2-1}{LN-N-2}\right] \cdot \mathbf{DFT}_N\{a_k\}\right\} = \frac{1}{\sqrt{L}}\sum_{n=0}^{N-1} a_n \,\mathrm{sincd}(N-1;N;(k/L-n)). \qquad (9.3.3)$$

Fig. 9-6 illustrates zero padding for an even N .Upper plot on the figure shows DFT spectrum of a signal of 8 samples to be 4-zoomed. The left plot in the second row is a plot of the zero padded spectrum with discarding central sample $\alpha_{N/2}$ of the initial spectrum. To the right from it, the corresponding interpolation kernel $\mathbf{sincd}(N-1;N;x)$ is shown. The third row illustrates zero padding with replication of the initial spectrum central sample $\alpha_{N/2}$ and the corresponding interpolation kernel $\mathbf{sincd}(N+1;N;x)$.

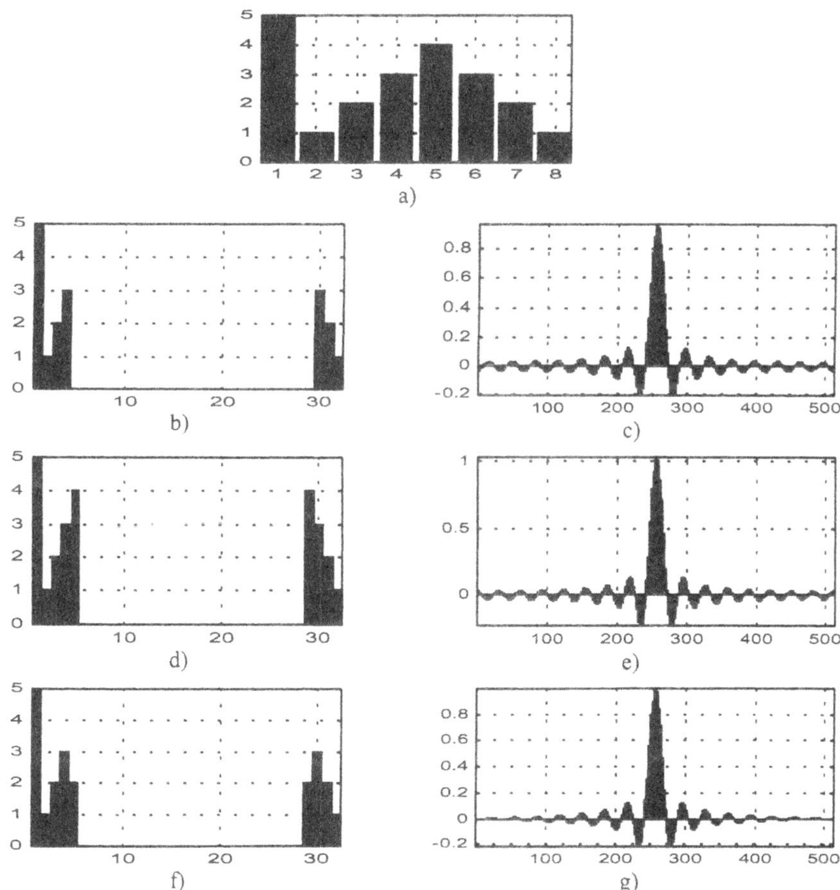

Figure 9-6. Zero padding signal spectrum and corresponding interpolation kernels

Distorting $N/2$-th signal spectrum sample by discarding or replicating it in the above methods of zero padding can be reduced if this spectrum component is halved and repeated twice in the spectrum zero padding as it is illustrated in in Fig. 9-6 e). Obviously, such method corresponds to interpolation

$$\tilde{\tilde{a}}_k = \frac{1}{\sqrt{L}} \sum_{n_1=0}^{N-1} a_n \,\mathrm{sincd}\left[\pm 1; N; (k/L-n)\right], \qquad (9.3.4)$$

with interpolation kernel

$$\mathrm{sincd}(\pm 1;N;x) = \left[\mathrm{sincd}(N+1;N;x) + \mathrm{sincd}(N-1;N;x)\right]/2. \qquad (9.3.5)$$

It is illustrated in Fig.9-6 on bottom right plot.

As one can see, this interpolation kernel converges to zero substantially faster then kernels $\mathrm{sincd}(N-1;N;x)$ and $\mathrm{sincd}(N+1;N;x)$. Therefore, it is less liable to boundary effects. This why such type of the discrete sinc-interpolation is recommended as a practical solution for signals with even number of samples.

An important issue in practical applications of discrete signal resampling is its computational complexity. Signal zooming by the zero padding method can be implemented with the use of FFT algorithms for computing direct and inverse Discrete Fourier Transforms. In this case, signal L-zooming requires $N \log N$ complex operations for direct DFT and $NL \log NL$ operations for inverse DFT, or totally $\log NL + (\log N)/L$ operations per output signal sample. In principle, zero padding signal zooming requires, also an additional intermediate buffer for $(L-1)N$ zero samples. Note, however, that computational complexity of the zero padding zooming can be greatly reduced and the need in the intermediate buffer can be avoided with the use of pruned FFT algorithms for inverse DFT.

9.3.2 DFT based discrete sinc-interpolation algorithm for signal arbitrary translation

Zero padding method of the discrete sinc-interpolation has several drawbacks. It is inflexible because it allows only integer zooming factors. Practically, zooming factor often should be chosen to be integer power of two to allow using radix 2 FFT algorithms that are most wide spread. It is also very inefficient when only signal translations are required as, for instance, in image rotation algorithm (see Sect. 3.4.1). Unless pruned FFT algorithms are used that are very rarely available in standard software packages, it involves excessive computations owing to the need to apply an IFFT to the extended zero padded signal spectrum.

Much more efficient way to implement the discrete sinc-interpolation follows from the general invertibility property of shifted DFT described by Eq. 4.3.17 ([2]). It follows from this equation that one can generate an arbitrarily shifted discrete sinc-interpolated copy of a signal if signal is reconstructed from its DFT spectrum by $\boldsymbol{ISDFT}(p,0)$. According to this equation, for signal $\{a_n\}$ of N samples, its discrete sinc-interpolated copy $\{a_n^{(p)}\}$ translated with respect to the original signal by p units of the discretization interval can be obtained as

$$\left\{a_n^{(p)}\right\} = \left\{\sum_{k=0}^{N-1} a_k \,\mathrm{sincd}[K;N;k-n-p]\right\} =$$
$$\left\{\mathbf{ISDFT}_K^{(p,0)}\left\{\mathbf{DFT}_N\left\{a_k \exp\left(-i\pi\frac{K-1}{N}k\right)\right\}\right\}\right\}\times$$
$$\left\{\exp\left(i\pi\frac{K-1}{K}(n+p)\right)\right\} =$$
$$\left\{\mathbf{IDFT}_K^{(p,0)}\left\{\mathbf{DFT}_N\left\{a_k \exp\left(-i\pi\frac{K-1}{N}k\right)\right\}\cdot\left\{\exp\left(-i2\pi\frac{pr}{N}\right)\right\}\right\}\right\}\times$$
$$\left\{\exp\left(i\pi\frac{K-1}{K}(n+p)\right)\right\} \qquad (9.3.6)$$

We will refer to this signal transformation as to ***signal p-shift***. Flow diagram of this algorithm for signal p-shift is shown in Fig. 9-7.

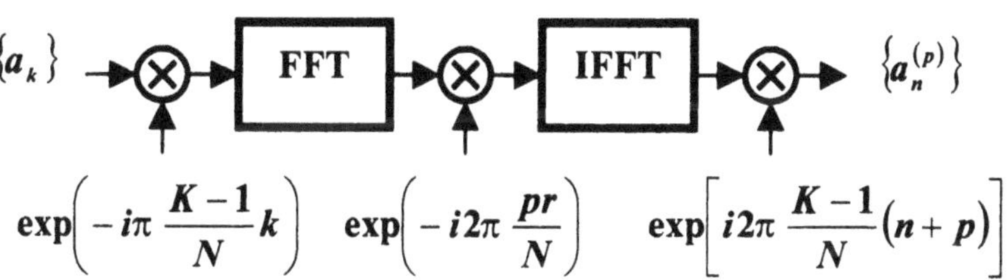

Figure 9-7. Flow diagram of an algorithm for signal p-shift

When $K = N$, the algorithm implements exact reversible discrete sinc-interpolation. For the implementation of the interpolation described by Eq.

9.3.4 with $\mathbf{sincd}(\pm 1;N;x)$ as the interpolation kernel, signal spectrum sample with index $N/2$ should be halved before computing an IFFT. As it was indicated in Sect. 9.3.1, this interpolation is less liable to boundary effects.

The described algorithm involves input signal modulation and output signal demodulation by exponential functions $\exp[-i\pi(K-1)k/N]$ and $\exp[-i\pi(K-1)(n+p)/N]$. One can eliminate additional computations needed for the modulation and demodulation, which are especially excessive when signal samples are real numbers. In this case the scheme converts into the convolution in the DFT domain (See Eq. 5.2.2) with exponential shift factor $\eta_r = \exp[-i2\pi pr/N]$ playing the role of the convolution kernel spectrum. To make the convolution kernel real, one should modify this factor to secure its symmetry $\eta_r^{(p)} = \left(\eta_{N-r}^{(p)}\right)^*$ dictated by Eq. 4.3.41 for DFT spectra of signals with real samples. In this way we arrive at the following algorithm:

$$\left\{a_k^{(p)}\right\} = \left\{\sum_{n=0}^{N-1} a_n \,\mathbf{sincd}[K;N;k-n-p]\right\} = \mathbf{IDFT}\left\{\mathbf{DFT}\{a_k\}\cdot\left\{\eta_r^{(p)}\right\}\right\}, \quad (9.3.7)$$

where, for odd N, $K = N$,

$$\eta_r^{(p)} = \begin{cases} \exp(-i2\pi pr/N), & r = 0,1,\ldots,(N-1)/2-1 \\ \left(\eta_{N-r}^{(p)}\right)^*, & r = (N+1)/2+1,\ldots,N-1 \end{cases}; \quad (9.3.8)$$

and, for even N, there are the same three options as for the zero padding discrete sinc-interpolation method:

option $K = N-1$,

$$\eta_r^{(p)} = \begin{cases} \exp(-i2\pi pr/N), & r = 0,1,\ldots,N/2-1 \\ 0, & r = N/2 \\ \left(\eta_{N-r}^{(p)}\right)^*, & r = N/2+1,\ldots,N-1 \end{cases} \propto \mathbf{DFT}\{\mathbf{sincd}[N\text{-}1;N;(k-p)]\}; \quad (9.3.9)$$

option $K = N+1$,

$$\eta_r^{(p)} = \begin{cases} \exp(-i2\pi pr/N), & r = 0,1,...,N/2-1 \\ 2\cos(2\pi pr/N), & r = N/2 \\ \left(\eta_{N-r}^{(p)}\right)^*, & r = N/2+1,...,N-1 \end{cases} \propto$$
$$\mathbf{DFT}\{\mathrm{sincd}[N+1;N;(k-p)]\}; \quad (9.3.10)$$

and option $K = \pm 1$,

$$\eta_r^{(p)} = \begin{cases} \exp(-i2\pi pr/N), & r = 0,1,...,N/2-1 \\ \cos(2\pi pr/N), & r = N/2 \\ \left(\eta_{N-r}^{(p)}\right)^*, & r = N/2+1,...,N-1 \end{cases} \propto$$
$$\mathbf{DFT}\{\mathrm{sincd}[\pm 1;N;(k-p)]\}. \quad (9.3.11)$$

Flow diagram of the algorithm for even N and the most recommended option $K = \pm 1$ is shown in Fig. 9-8.

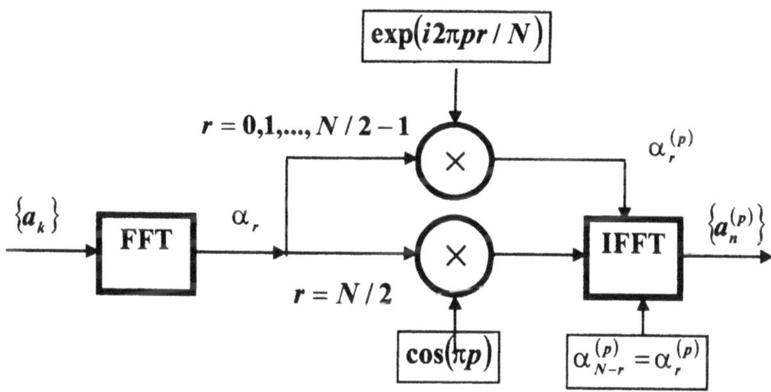

Figure 9-8. DFT based discrete sinc-interpolation algorithm for p-shifting signals of even number N of samples (option $K = \pm 1$).

We will refer to this algorithm as to ***DFT based discrete sinc-interpolation algorithm***. Computational complexity of the algorithm, with the use a FFT for direct and inverse DFT, is $O(2N\log N)$ complex operations or $O(2\log N)$ per signal sample which is substantially less that the complexity of the zero-padding algorithm. The algorithm is also completely free of other above-mentioned drawbacks of the zero-padding algorithm.

9.3.3 DCT based discrete sinc-interpolation algorithm for signal arbitrary translation

DFT based discrete sinc-interpolation described is liable to boundary effects caused by its implementation as a cyclic convolution. As it was described in Sect. 5.2.2, boundary effects can be substantially reduced by the implementation of digital convolution in the DCT domain. Interpolation equation for signal $\{a_k\}$ of N samples can be in this case obtained from Eqs. 5.2.10 and 5.2.12 as:

$$\{a_k^{(p)}\}=\left\{\sum_{n=0}^{N-1} a_n \,\mathrm{sincd}[K;N;k-n-p]\right\}=\mathbf{ISDFT}_{1/2,0}\{\mathbf{DCT}\{a_k\}\cdot\{\eta_r(p)\}\}=$$

$$\frac{1}{\sqrt{2N}}\left\{\alpha_0^{DCT}\eta_0+2\sum_{r=1}^{N-1}\alpha_r^{DCT}\eta_r^{re}\cos\left[\pi\frac{(k+1/2)}{N}r\right]-\right.$$

$$\left.2\sum_{r=1}^{N-1}\alpha_r^{DCT}\eta_r^{im}\sin\left[\pi\frac{(k+1/2)}{N}r\right]\right\}, \tag{9.3.12}$$

where

$$\{\alpha_r^{DCT}=\mathbf{DCT}\{a_k\}\}, \tag{9.3.13}$$

$$\{\eta_r(p)=\eta_r^{re}(p)+i\eta_r^{im}(p)\}=\mathbf{DFT}\{h_{\mathrm{int}}^{(p)}(k)\}=\left\{\frac{1}{\sqrt{N}}\sum_{k=0}^{N-1}h_{\mathrm{int}}^{(p)}\exp\left(i2\pi\frac{kr}{N}\right)\right\}, \tag{9.3.14}$$

and

$$h_{\mathrm{int}}^{(p)}=\begin{cases}0\ , & n=0,...,[N/2]-1\\ \mathrm{sincd}[K;N;k-[N/2]-n-p], & n=[N/2],...,[N/2]+N-1\\ 0\ , & n=[N/2]+N,...,2N-1\end{cases} \tag{9.3.15}$$

with $K=N$ for odd N and $K=\pm 1$ for even N.

By virtue of the relationships (9.3.9-11), coefficients $\eta_{2r}(p)$ with even indices can be found directly as

$$\eta_{2r}(p)=\frac{1}{\sqrt{N}}\exp\left(i2\pi\frac{pr}{N}\right). \tag{9.3.16}$$

Therefore one needs to additionally compute from Eq. 9.3.14 only terms $\{\eta_{2r+1}(p)\}$ with odd indices. Flow diagram of this algorithm for generating p-shifted sinc-interpolated copy of signal is shown in Fig. 9.9

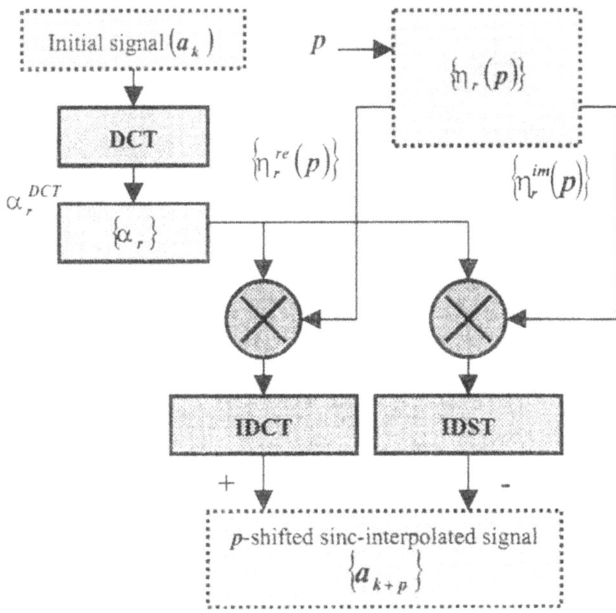

Figure 9-9. Flow diagram of a DCT based algorithm for p-shift signal of *N* samples

We will refer to this algorithm as to ***DCT based discrete sinc-interpolation*** ([3]).

Figs. 9-10 and 11 illustrate signal and image zooming with DFT based and DCT based discrete sinc-interpolation. Zooming was carried out by repetitive application of signal shifts as it is described in Sect. 9.4.2. One can notice heavy boundary effects in form of oscillations that appear when DFT based interpolation is used and that practically completely disappear when the DCT based interpolation is used instead.

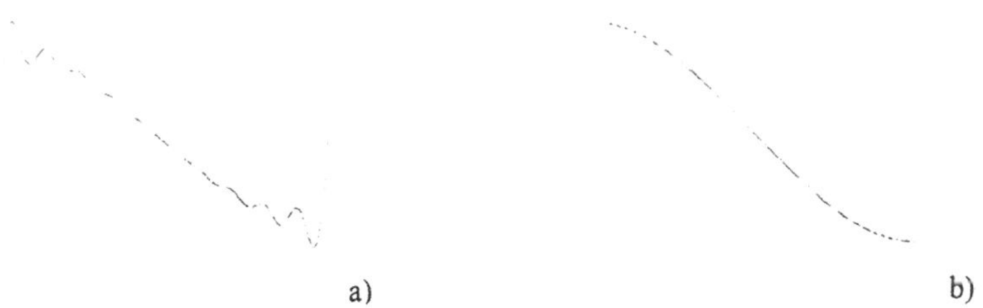

Figure 9-10. Signal zooming with DFT based (a) and DCT based (b) discrete sinc-interpolations

Figure 9-11. Zooming image fragment with DFT based (upper right) and DCT based (bottom right) discrete sinc-interpolations

The DCT based discrete sinc-interpolation is computationally efficient if regular (equiduistant) signal resamplimg is required. In this case it enables computing p-shifted sinc-interpolated copy of the signal of N samples with the complexity of $O(N \log N)$ operations or $O(\log N)$ operations per signal sample. Above-described DFT based discrete sinc-interpolation algorithm has the same, by the order of magnitude, complexity. However fast algorithms for DCT, IDCT and IDST require less computations than FFT and IFFT involved in the DFT domain algorithm (see Sect. 6.4). This, along with less liability to boundary effects makes the DCT based algorithm preferable in applications. In conclusion note that both DFT and DCT based algorithms can implement, with the same computational complexity, not only dicsrete sinc-interpolation, but interpolation with arbitrary interpolation kernel as well. One only needs to specify the kernel by its DFT spectrum.

9.3.4 Sliding window adaptive discrete sinc-interpolation algorithms

When, as it frequently happens in signal and image resampling tasks, required signal sample shifts are different for different samples, above described discrete sinc-interpolation algorithms have no efficient computational implementation. However, in such applications they can be implemented in sliding window ([4]). In processing signal in sliding window, only those shifted and interpolated signal samples that correspond to the window central sample have to be computed in each window position from signal samples within the window. Interpolation function in this case is a windowed discrete sinc-function whose extent is equal to the window size rather that to the whole signal size required for the perfect discrete sinc-interpolation. Fig. 9.12 illustrates frequency responses and PSF of the corresponding interpolating low pass filters of two different window sizes for signal 3-zooming. As one can see from the figure, interpolation filter frequency responses deviate from a rectangle function, a frequency response of the ideal low pass filter.

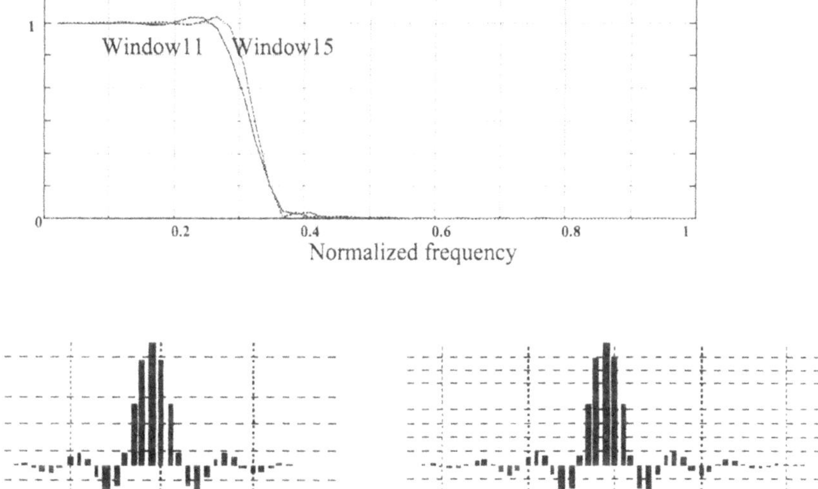

Figure 9-12. Windowed discrete sinc-functions for windows of 11 and 15 samples and the corresponding interpolator frequency responses for signal 3-zooming

Such an implementation of the discrete sinc-interpolation can be regarded as a variety of direct convolution interpolation methods. In terms of the interpolation accuracy it has no special advantages over other direct convolution interpolation methods such as spline oriented ones. However it offers features that are not available with other methods. These are: (i) signal resampling with arbitrary shifts and simultaneous signal restoration and enhancement and (ii) local adaptive interpolation with "super resolution".

For signal resampling with simultaneous restoration/enhancement, the sliding window discrete sinc-interpolation should be combined with local adaptive filtering. As it was described in Sect. 8.6, local adaptive filters, in each position $\boldsymbol{k}$ of the window of $\boldsymbol{W}$ samples (usually an odd number), compute transform coefficients $\{\beta_r = \boldsymbol{T}\{\boldsymbol{a}_n\}\}$ of the signal $\{\boldsymbol{a}_n\}$ in the window ($n, r = 1,2,...,W$) and nonlinearly modify them to obtain coefficients $\{\hat{\alpha}_r(\alpha_r)\}$. These coefficients are then used to generate an estimate $\hat{a}_k$ of the window central pixel by inverse transform $T_k^{-1}\{\ \}$ computed for the window central pixel as

$$\hat{a}_k = T_k^{-1}\{\hat{\alpha}_r(\beta_r)\}, \tag{9.3.17}$$

Such a filtering can be implemented in the domain of any transform though the DCT has proved to be one of the most efficient. Therefore one can, in a straightforward way, combine the sliding window DCT domain discrete sinc-interpolation signal resampling (Eq. 9.3.12) and filtering for signal restoration and enhancement described in Sect. 8.6:

$$\{a_k^p\} = ISDFT_{1/2,0}\{\hat{\alpha}_r^{DCT}(\alpha_r^{DCT})\cdot\eta_r(p)\}. \tag{9.3.18}$$

Fig. 9-13 shows flow diagram of such a combined algorithm for signal restoration/enhancement and fractional p-shift. It is assumed in the diagram that signal p-shift is implemented according to the flow diagram of Fig. 9-9.

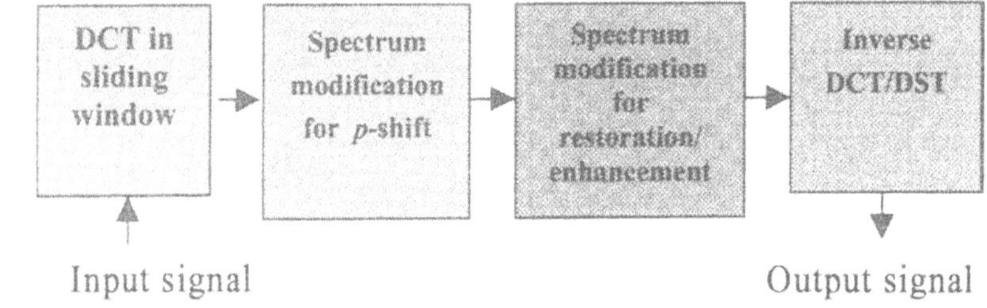

Figure 9-13. Flow diagram of signal resampling combined with denoising/enhancement

Fig. 9-14 illustrates application of the combined filtering/interpolation for image irregular-to regular resampling combined with denoising. In this example, left image is distorted by known displacements of pixels with respect to regular equidistant positions and by additive noise. In the right image, these displacements are compensated and noise is substantially reduced with the above-described sliding window algorithm.

One can further extend the applicability of this method to make interpolation kernel transform coefficients $\{\eta_r(p)\}$ in Eq. (9.3.18) to be adaptive to signal local features that exhibit themselves in signal local DCT spectra ([4]):

$$\left\{a_k^p\right\} = \boldsymbol{ISDFT}_{1/2,0}\left\{\hat{\alpha}_r^{DCT}\left(\beta_r^{DCT}\right)\cdot\eta_r\left(p,\left\{\beta_r^{DCT}\right\}\right)\right\}. \tag{9.3.19}$$

The adaptivity may be desired in such applications as, for instance, resampling images that contain gray tone images in a mixture with graphical data. While the discrete sinc-interpolation is completely perfect for gray tone images, it may produce undesirable oscillating artifacts in graphics.

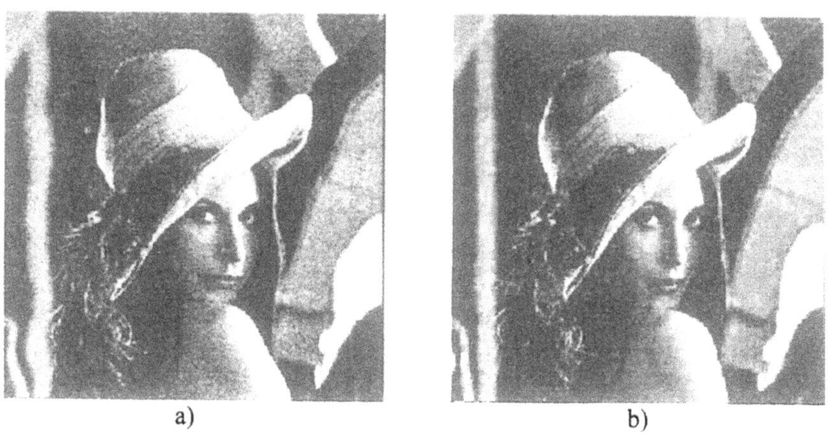

Figure 9-14. An example of image denoising and resampling from irregular to regular sampling grid: a)- noisy image sampled in a irregular sampling grid; b)- Denoised and resampled (rectified) image

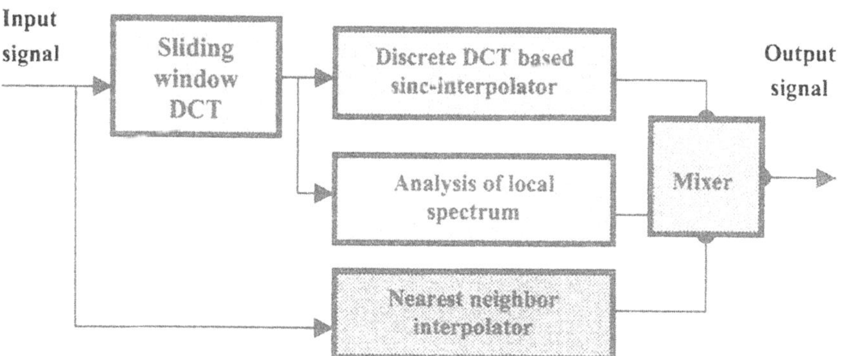

Figure 9-15. Flow diagram of local adaptive resampling

The principle of local adaptive interpolation is schematically presented on Fig. 9-15. It assumes that modification of signal local DCT spectra for signal resampling and restoration in the above-described algorithm is supplemented with the spectrum analysis for generating a control signal. This signal is used to select, in each sliding window position, the discrete sinc-interpolation or another interpolation method such as, for instance, nearest neighbor one.

Fig. 9-16 illustrates non adaptive and adaptive sliding window sinc-interpolation on an example of a shift, by an interval equal 16.54 of discretization intervals, of a test signal composed of a sinusoidal wave and rectangle impulses. As one can see from the figure, non adaptive sinc-interpolated resampling of such a signal results in oscillations at the edges of rectangle impulses. Adaptive resampling implemented in this example switches between the sinc-interpolation and the nearest neighbor interpolation whenever energy of high frequency components of signal local spectrum is higher than a certain threshold level. As a result, rectangle impulses are re-sampled with "super-resolution". Fig. 9-17 illustrates, for comparison, zooming a test signal by means of the nearest neighbor, linear and bi-cubic spline interpolations and the above-described adaptive sliding window DCT sinc-interpolation. One can see from these figures that interpolation artifacts seen in other interpolation methods are absent when the adaptive sliding widow interpolation was used

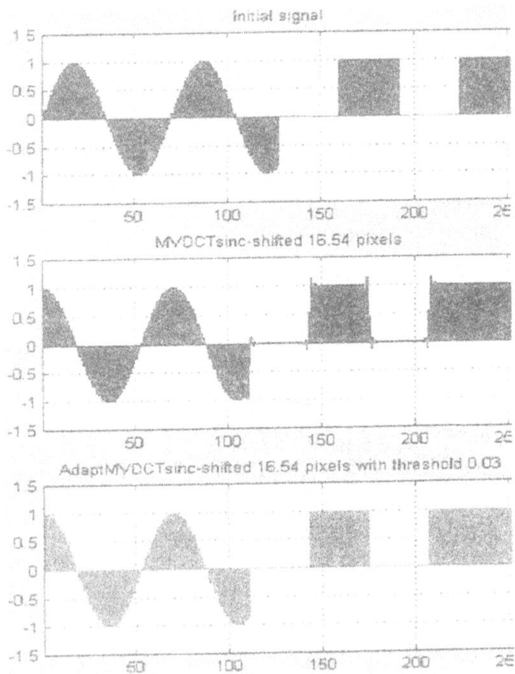

Figure 9-16. Fractional shift of a signal (upper plot) by non-adaptive (middle plot) and adaptive (bottom plot) sliding window DCT sinc-interpolations. One can notice disappearance of oscillations at the edges of rectangle impulses when interpolation is adaptive

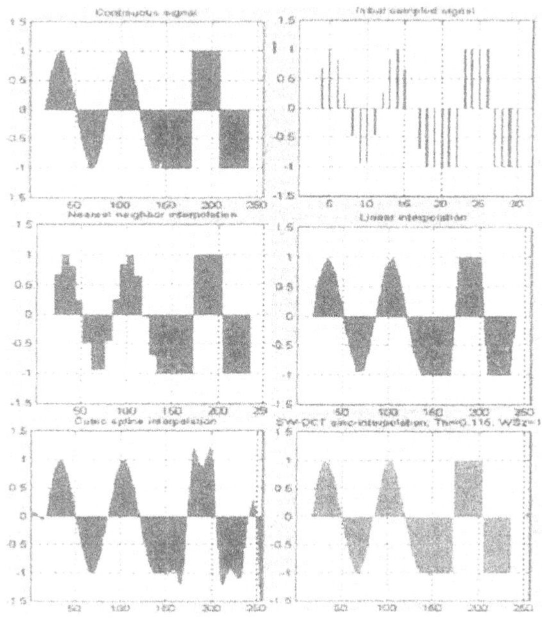

Figure 9-17. Signal zooming with non-adaptive and adaptive discrete sinc-interpolation. First row: a model of a continuous signal (left plot) and its sampled version (right plot). Second row: initial signal reconstructed by nearest neighbor (left plot) and linear (right plot) interpolation. Third row: signal reconstruction using non-adaptive (left plot) and adaptive (right plot) discrete sinc-interpolation. One can see that smooth sinusoidal signal and rectangular impulses are reconstructed with much less artifacts when adaptive interpolation is used.

Non adaptive and adaptive sliding window sinc-interpolation are also illustrated and compared in Fig. 9-18 for rotation of an image that contains gray tone and graphic components. One can clearly see oscillations at sharp boundaries of black and white squares in the result of rotation without adaptation. Adaptive sinc-interpolation does not show these artifacts because of switching between the sinc-interpolation and the nearest neighbor interpolation at the sharp boundaries.

Computational complexity of the sliding window DCT sinc-interpolation evaluated in terms of the number of multiplication and summation operations per signal sample is proportional to the window size because of the availability of recursive algorithms for computing the DCT in sliding window described in Sect. 5.4.

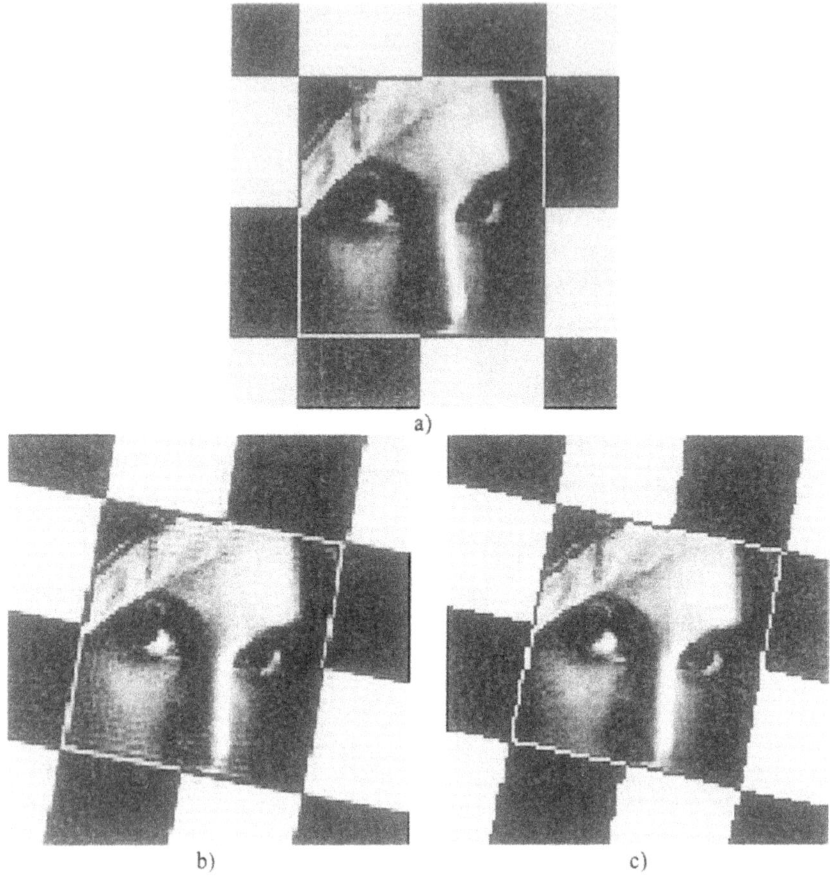

Figure 9-18. Image (upper) rotation with sliding window non-adaptive (bottom left) and adaptive DCT sinc-interpolations (bottom right).

9.4 Application examples

9.4.1 Signal and image resizing and localization with sub-pixel accuracy

One of the immediate applications of above-described discrete sinc-interpolation methods for generating p-shifted signals is signal zooming and resizing. With the use of these algorithms, signal L-zooming can be achieved as a combination of (L-1) successive signal shifts with steps $p = t / L;\ \ t = 1,...,L-1$ as it is shown in schematic diagram of Fig. 9-19a.

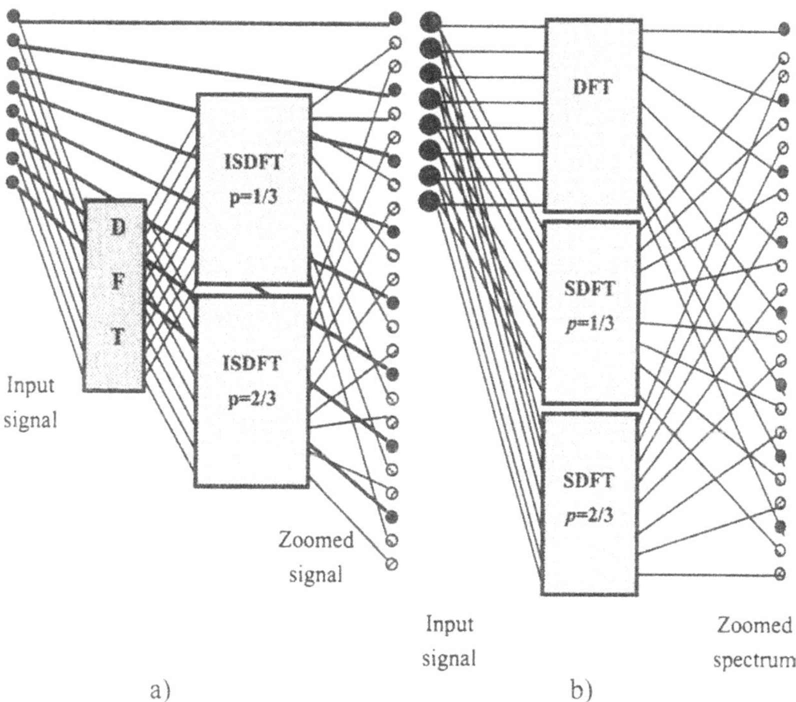

Figure 9-19. Schematic diagrams of signal zooming (a) and spectral analysis with sub-sample resolution (b) using SDFTs for zooming and subsampling factor 3

Image zooming using this algorithm can be implemented as a separable zooming along both image coordinates.

A natural application of image zooming is image resizing and arbitrary geometrical transformations. A zoomed image can be regarded as an approximation to its continuous prototype, the approximation accuracy being defined by the zooming factor. Then one can re-sample the zoomed image to the required sampling grid. If the number of pixels of output image is not less then the number of pixels of the initial image, the simplest nearest neighbor interpolation can be used for the resampling. One can also use the combination of the discrete sinc-interpolated zooming and bilinear interpolation for the resampling in order to reduce the zoom factor needed for the given resampling accuracy. Principle of the "zooming-resampling" method of image geometrical transformation is illustrated in Fig. 9-20.

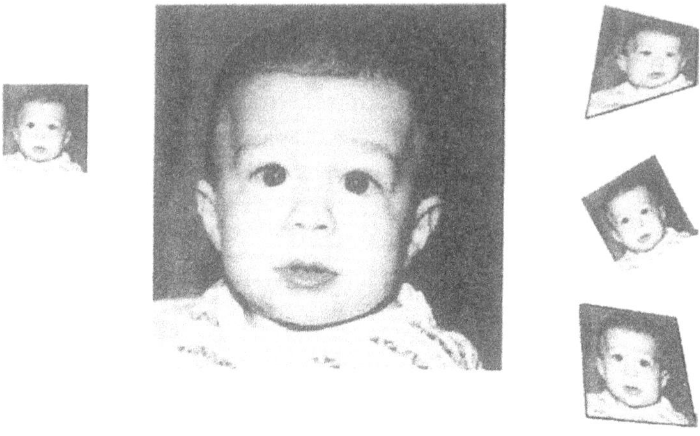

Figure 9-20. Principle of the "zooming-resampling" method of image geometrical transformations

Image geometrical transformation by resampling zoomed image can be very computationally efficient when many different geometrical transformation are required for the same image as, for instance, it is frequently the case in localization of target object of unknown size or/and orientation.

When the number of output image pixels is less then that of the initial image, resampling should be carried out with an appropriate low pass filtering to avoid aliasing artifacts. The importance of proper low-pass filtering for image resizing is illustrated in Fig. 9-21. In this example, image is resized by reducing with reduction factor 77/256 and then it was magnified with magnification factor 256/77.

The same idea of using SDFT for signal p-shift with the discrete sinc interpolation can be employed for object localization with sub-pixel accuracy for signal spectral analysis with sub-sample resolution.

For object localization with sub-pixel accuracy, correlation peak (See Ch. 10) should be sub-sampled. In this case, only signal sub-samples in the vicinity of the highest correlation peak are required. Hence, inverse SDFTs in the flow diagram of Fig. 9-19, a) can be carried out with pruned algorithms described in Sect. 6.4.

Schematic diagram of the signal spectral analysis with sub-sample resolution is show in Fig. 9-18, (b). Fig. 9-21 shows an example of such an analysis for a sinusoidal signal $\mathbf{sin}(2\pi kF/N)(k=0,...,255;F=36.7)$.

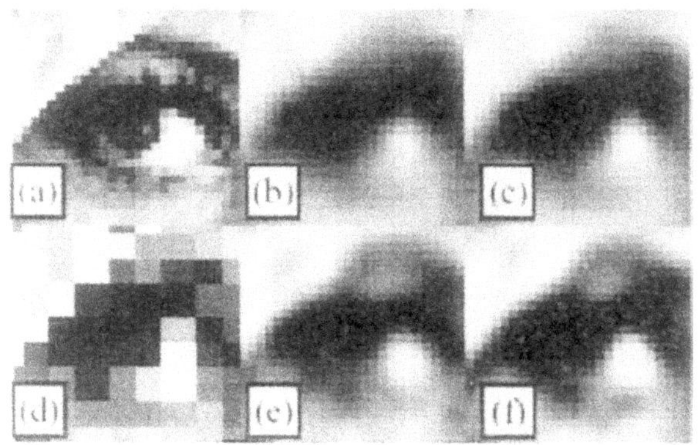

Figure 9-21. Image resizing with different resampling filters: (a)- initial image; b) – bilinear interpolation; c) – bicubic interpolation; d) – nearest neighbor interpolation; e) – sincd-interpolation in window 11x11; f) – sincd-interpolation in window 31x31

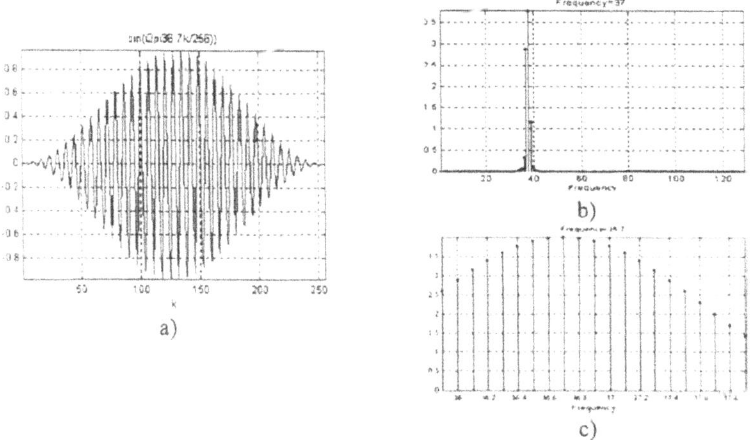

Figure 9-22. Spectral analysis with sub-sample accuracy. a) - a sinusoidal signal b) - its DFT spectrum; c) - interpolated spectrum peak

9.4.2 Fast algorithm for image rotation with discrete-sinc-interpolation

Image rotation by an angle θ as a geometrical transformation of signal co-ordinates can be described as a multiplication of signal coordinates (x, y) vector by a rotation matrix $\mathbf{ROT}_\theta$:

$$\mathbf{ROT}_\theta \begin{bmatrix} x \\ y \end{bmatrix} = \begin{bmatrix} \cos\theta & -\sin\theta \\ \sin\theta & \cos\theta \end{bmatrix} \begin{bmatrix} x \\ y \end{bmatrix} \qquad (9.4.1)$$

In digital processing, physical coordinates (x, y) are represented, given discretization intervals $(\Delta x, \Delta y)$, by integer indices of pixels $\{k, l\}$:

$$\begin{bmatrix} x \\ y \end{bmatrix} = \begin{bmatrix} k\Delta x \\ l\Delta y \end{bmatrix}, \qquad (9.4.2)$$

and the rotation matrix is applied to the vector of indices:

$$\mathbf{ROT}_\theta \begin{bmatrix} k \\ l \end{bmatrix} = \begin{bmatrix} \cos\theta & -\sin\theta \\ \sin\theta & \cos\theta \end{bmatrix} \begin{bmatrix} k \\ l \end{bmatrix}. \qquad (9.4.3)$$

Eq. 9.4.3 describes the resampling rule that should be applied to input image to generate its rotated copy. In order to simplify computations, one can factorize rotation matrix into a product of three matrices each of which modifies only one co-ordinate ([5]):

$$\mathbf{ROT}_\theta = \begin{bmatrix} \cos\theta & -\sin\theta \\ \sin\theta & \cos\theta \end{bmatrix} = \begin{bmatrix} 1 & -\tan(\theta/2) \\ 0 & 1 \end{bmatrix} \begin{bmatrix} 1 & 0 \\ \sin\theta & 1 \end{bmatrix} \begin{bmatrix} 1 & -\tan(\theta/2) \\ 0 & 1 \end{bmatrix} \qquad (9.4.4)$$

This implementation of image rotation is known as a ***three pass rotation algorithm***. It assumes, on each pass, only signal shifts along one of the co-ordinates: along x on the first pass, along y on the second pass and again along x on the third pass. Above described DFT or DCT based signal translation algorithms are ideally suited for the implementation of the shifts.

The work of the algorithm is illustrated in Fig. 9-23 a,b. One can see that with the use of interpolation algorithms that are implemented as cyclic convolution, rotation entails characteristic aliasing artifacts at image boundaries. In order to avoid them, one should inscribe the image into an array of correspondingly larger size.

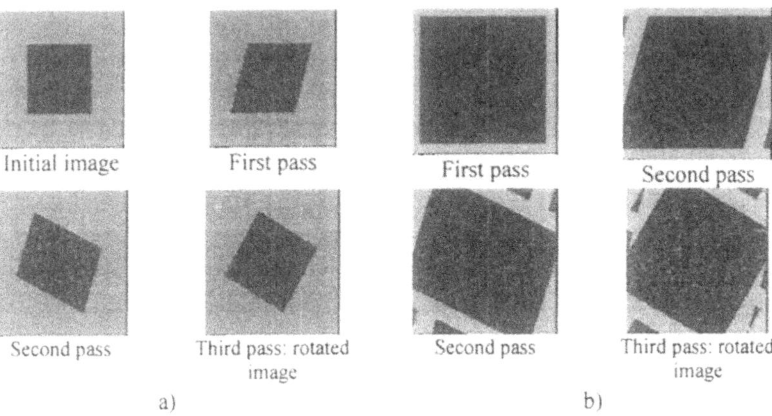

Figure 9-23. The principle of the three pass image rotation algorithm: a) - rotation without aliasing; b) - aliasing artifacts owing to the cyclicity of the convolution

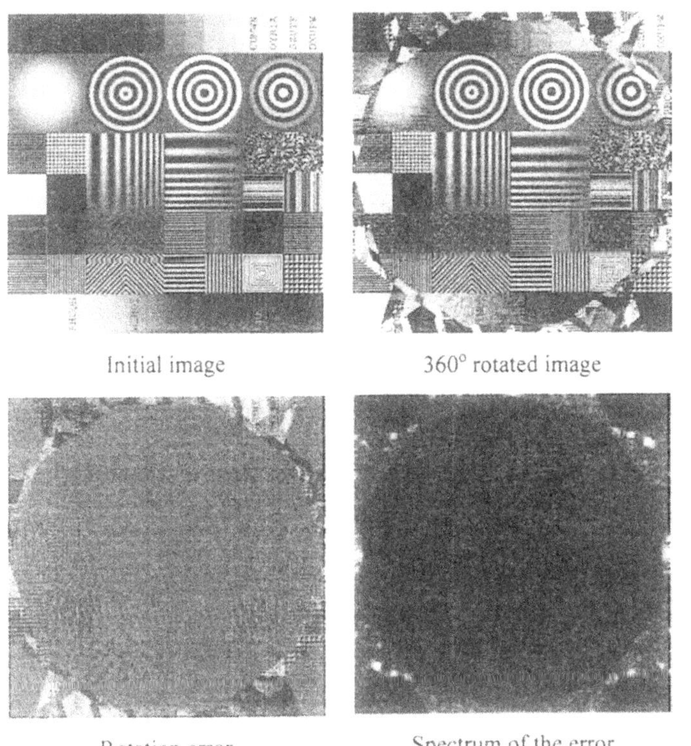

Figure 9-24. Boundary effects and rotation error in image rotation with discrete sinc-interpolation

Fig. 9-24 illustrates boundary effects and rotation error on an example of rotation of a test image (left, top row) by 360° in 10×36° steps. Right image in the bottom row displays spectrum of the central, not subjected to boundary effects, part of the difference image between the initial and rotated images (left, bottom row image). Spectrum amplitude is represented in the image by image gray scale levels and the spectrum is centered so that image borders correspond to its highest frequencies. One can see that the discrete sinc-interpolation secures preservation of image spectral components within the base band. All distortions occurred at the image highest spectral component that is halved in the interpolation algorithm and outside the image base-band. The latter errors can be attributed to boundary effects in each of the rotation steps

9.4.3 Signal differentiating and filtered back projection method for tomographic reconstruction

Signal differentiating is an operation that requires measuring infinitesimal increments of signals and their arguments. Therefore, in digital signal processing computing signal derivative assumes one of another method of building continuous models of signals specified by their samples, that is one or another method of discrete signal interpolation. Because differentiating is a linear operation, methods of computing signal derivatives from their samples can be conveniently compared in the Fourier transform domain.

Let Fourier transform spectrum of continuous signal $a(x)$ is $\alpha(f)$:

$$a(x)=\int_{-\infty}^{\infty}\alpha(f)\exp(-i2\pi fx)df \tag{9.4.5}$$

Then Fourier spectrum of its derivative will be $(-i2\pi f)\alpha(f)$:

$$\dot{a}(x)=\frac{d}{dx}a(x)=\int_{-\infty}^{\infty}[(-i2\pi f)\alpha(f)]\exp(-i2\pi fx)df \tag{9.4.6}$$

From Fourier transform convolution theorem it follows that signal differentiating can be regarded as signal linear filtering with filter frequency response

$$H_{diff}=-i2\pi f \tag{9.4.7}$$

Let now signal $a(x)$ be represented by its samples $\{a_k\}$, $k = 0,1,...,N-1$ and let $\{\alpha_r\}$ be a set of DFT coefficients of discrete signal $\{a_k\}$:

$$a_k = \frac{1}{\sqrt{N}}\sum_{r=0}^{N-1}\alpha_r \exp\left(-i2\pi\frac{kr}{N}\right) \tag{9.4.8}$$

Then, from the discrete representation of integral Fourier transform discussed in Sect. 4.3.1 one can conclude that discrete representation of differentiating equation Eq. 9.4.6 is

$$\dot{a}_k = \frac{1}{\sqrt{N}}\sum_{r=0}^{N-1}\left[\eta_r^{diff}\alpha_r\right]\exp\left(-i2\pi\frac{kr}{N}\right), \tag{9.4.9}$$

where $\{\dot{a}_k\}$ are samples of derivative of signal $a(x)$ and $\{\eta_r^{diff}\}$ are defined as:

$$\eta_r^{diff} = \begin{cases} -i2\pi r/N, & r = 0,1,...,N/2-1 \\ -\pi/2, & r = N/2 \\ i2\pi(N-r)/N, & r = N/2+1,...,N-1 \end{cases} \tag{9.4.10,a}$$

for even N and

$$\eta_r^{diff} = \begin{cases} -i2\pi r/N, & r = 0,1,...,(N-1)/2-1 \\ i2\pi(N-r)/N, & r = (N+1)/2,...,N-1 \end{cases} \tag{9.4.10,b}$$

for odd N. Note that the coefficient $\eta_{N/2}^{(diff)}$ is chosen in Eq. 9.4.10, a) according to Eqs. 9.3.4 and 9.3.11.

Eq. (9.4.9) describes digital filtering in the DFT domain with discrete frequency response $\{\eta_r^{diff}\}$ and suggests the following algorithm of computing derivative of signals specified by their samples:

$$\{\dot{a}_k\} = \mathbf{IDFT}\left(\{H_r^{diff}\}\cdot \mathbf{DFT}(\{a_k\})\right) \tag{9.4.11}$$

Digital filter described by Eq. (9.4.11) is called ***discrete ramp-filter***. Obviously, such an implementation implicitly assumes the discrete sinc-interpolation of the signal. Its computational complexity, with the use of a FFT, is $O(\log N)$.

Alternative methods of digital computing signal derivatives are those implemented by signal direct discrete convolution in the signal domain:

$$\dot{a}_k = \sum_{n=0}^{N_h-1} h_n^{diff} a_{k-n} \,. \tag{9.4.12}$$

The following simplest differentiating kernels of two and four samples are recommended in manuals on numerical methods in mathematics ([6])

$$h_n^{diff(1)} = [-1, 1] \tag{9.4.13 a}$$

and

$$h_n^{diff(2)} = [-1/12, 8/12, -8/12, 1/12] \,. \tag{9.4.14 b}$$

Both are based on an assumption that, in the vicinity of signal samples signals can be expanded into Taylor series. The first one, obviously, assumes linear interpolation. One can compare these and above-described ramp-filtering differentiating method by their frequency responses. Applying N-point Discrete Fourier transform to Eqs. 9.4.13 and 14, obtain:

$$\eta_r^{diff(1)} = \frac{\sin(\pi r / N)}{2} \exp[i\pi (r + 1/2)] \tag{9.4.13 b}$$

and

$$\eta_r^{diff(2)} = \frac{\sin(3\pi r / N) + 8\sin(\pi r / N)}{2} \exp[i\pi (r + 1/2)] \,. \tag{9.4.14 b}$$

Absolute values of these frequency responses are shown in Fig. 9-25 along with the frequency response of DFT based differentiating filter (Eq. 9.4.10). One can see from the figure that they entail certain distortions of signal spectra on signal high frequencies.

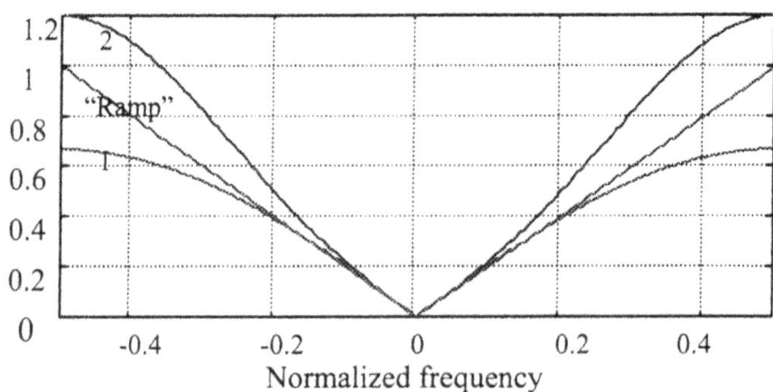

Figure 9-25. Frequency responses of differentiating filters described by Eqs. 9.4.13 (curve 1), Eq. 9.4.14 (curve 2) and Eq. 9.4.10 ("Ramp")

Signal differentiating by ramp-filtering is used, in particular, in digital image reconstruction from its projections by the filtered back projection method (Sect. 2.5, Eqs. (2.5-6, 7)). The ***discrete filtered back projection image algorithm*** of image reconstruction that uses the ramp-filter and image rotation algorithms described in Sect. 9.4.1 is as following

$$\left\{\hat{a}_{k,l}\right\} = \sum_{s=0}^{N_\theta - 1} \mathbf{ROT}_\theta \left\{\left\{\dot{p}_k^{(\theta_s)}\right\} \otimes \left\{\vec{1}_l\right\}\right\}; \quad k,l = 0,1,...,N-1 \tag{9.4.15}$$

where

$$\left\{\dot{p}_k^{(\theta_s)}\right\} = \mathbf{IDFT}\left(H_r^{diff} \cdot \mathbf{DFT}\left(\left\{p_k^{(\theta_s)}\right\}\right)\right) \tag{9.4.16}$$

is a vector-row of samples of ramp-filtered image projection $\left\{p_k^{(\theta_s)}\right\}$ obtained for angle θ_s, $s = 0,1,...,N_\theta - 1$, $\left\{\vec{1}_l\right\}$ is a vector-column of N ones and $\otimes$ symbolizes matrix Kronecker product.

9.4.4 Signal integrating and reconstructing surfaces from their slope

Digital signal integrating is yet another operation that requires interpolation of discrete signals. It has numerous applications in optical metrology. One of the most typical applications is surface reconstruction from their slope measured by optical sensors (see, for instance, [7]).

Similarly to signal differentiating, signal integrating assumes operating with infinitesimal signal increments and can also be treated as signal convolution. In the Fourier transform domain, it is described as

$$\bar{a}(x) = \int a(x)dx = \int_{-\infty}^{\infty} \frac{i}{2\pi f} \alpha(f) \exp(-i2\pi fx)dx =$$

$$\int_{-\infty}^{\infty} H_{\text{int}}(f)\alpha(f)\exp(-i2\pi fx)df. \tag{9.4.17}$$

where $a(x)$ and $\alpha(f)$ are signal and its Fourier transform spectrum, respectively, and

$$H_{\text{int}}(f) = \frac{i}{2\pi f} \tag{9.4.18}$$

is the integrating filter frequency response.

In digital processing, integrating filtering described by Eq. 9.4.17 can be implemented in the DFT domain as:

$$\{\overline{a}_k\} = \mathbf{IDFT}\{\eta_r^{(\mathrm{int})}\,\mathbf{DFT}\{a_k\}\}, \tag{9.4.19}$$

where $\{a_k\}$, $k = 0,1,...,N-1$ is input signal to be integrated and

$$\eta_r^{(\mathrm{int})} = \begin{cases} 0, & r = 0 \\ -\dfrac{N}{i2\pi r}, & r = 1,2...,N/2-1, \\ -\dfrac{1}{2\pi}, & r = N/2 \end{cases}$$

$$\eta^*_{N-r},\ r = N/2+1,...,N-1 \tag{9.4.20a}$$

for even N and

$$\eta_r^{(\mathrm{int})} = \begin{cases} 0, & r = 0 \\ -\dfrac{N}{i2\pi r}, & r = 1,2,...,(N-1)/2 \end{cases};$$

$$\eta^*_{N-r},\ r = N/2+1,...,N-1 \tag{9.4.20b}$$

for odd N. Similarly to digital differentiating, digital signal integrating according to Eq. 9.4.19 automatically implies signal discrete sinc-interpolation. Note the coefficient $\eta_{N/2}^{(\mathrm{int})}$ is chosen in Eq. 9.4.20, a) according to Eqs. 9.3.4 and 9.3.11.

A number of other numerical integration methods have been described in the literature. Most known numerical integration methods are the Newton-Cotes quadrature rules ([6,8]). The three first rules are the trapezoidal, the Simpson and the 3/8 Simpson ones. In all the methods, the value of the integral in the first point is not defined because it affects to the result constant bias and should be arbitrarily chosen. When it is chosen to be equal to zero, the trapezoidal rule is:

$$\overline{a}_1^{(T)} = 0,\quad \overline{a}_k^{(T)} = \overline{a}_{k-1}^{(T)} + \frac{1}{2}\left(a_{k-1} + a_k\right). \tag{9.4.21}$$

In the Simpson rule, the second point needs to be evaluated by the trapezoidal rule, then

$$\bar{a}_1^{(S)} = 0,\ \bar{a}_k^{(S)} = \bar{a}_{k-2}^{(S)} + \frac{1}{3}\left(a_{k-2} + 4a_{k-1} + a_k\right). \tag{9.4.22}$$

In the 3/8-Simpson rule, the second and the third points need to be evaluated by the trapezoidal rule and the Simpson rule respectively. Then

$$\bar{a}_0^{(3/8S)} = 0,\ \bar{a}_k^{(3/8S)} = \bar{a}_{k-3}^{(3/8S)} + \frac{3}{8}\left(a_{k-3} + 3a_{k-2} + 3a_{k-1} + a_k\right). \tag{9.4.23}$$

In these integration methods, a linear, a quadratic, and a cubic interpolation, respectively, are assumed between the sampled slope data. In the cubic spline interpolation, a cubic polynomial is evaluated between every couple of point ([9]), and then, an analytical integration of these polynomials is performed.

As in the case of digital differentiating, we will compare the integration methods in terms of their frequency responses. Frequency responses for trapezoidal, Simpson and 3/8 Simpson's integrators can be found from Eqs. 9.4.21 - 23 as follows:

$$\bar{a}_k^{(Tr)} - \bar{a}_{k-1}^{(Tr)} = \frac{1}{2}\left(a_{k-1} + a_k\right) \tag{9.4.24}$$

$$\bar{a}_k^{(S)} - \bar{a}_{k-2}^{(S)} = \frac{1}{3}\left(a_{k-2} + 4a_{k-1} + a_k\right) \tag{9.4.25}$$

$$a_k^{(3/8S)} - a_{k-3}^{(3/8S)} = \frac{3}{8}\left(a_{k-3} + 3a_{k-2} + 3a_{k-1} + a_k\right) \tag{9.4.26}$$

By N point Discrete Fourier Transform of these expressions we obtain

$$\bar{\alpha}_r^{(T)}\left(1 - \exp(i2\pi r/N)\right) = \frac{1}{2}\alpha_r\left[1 + \exp(i2\pi r/N)\right]; \tag{9.4.27}$$

$$\bar{\alpha}_r^{(S)}\left(1 - \exp(i4\pi r/N)\right) = \frac{1}{3}\alpha_r\left[1 + 4\exp(i2\pi r/N) + \exp(i4\pi r/N)\right]; \tag{9.4.28}$$

$$\overline{\alpha}_r^{(3/8S)}\left[1-\exp(i6\pi r/N)\right]=$$
$$\frac{3}{8}\alpha_r\left[1+3\exp(i2\pi r/N)+3\exp(i4\pi r/N)+\exp(i6\pi r/N)\right], \qquad (9.4.29)$$

Therefore the frequency responses of these integrators are, respectively:

$$\eta_r^{(T)}=\frac{\overline{\alpha}_r^{(Tr)}}{\alpha_r}=\begin{cases}0, & r=0,\\ -\dfrac{\cos(\pi r/N)}{2i\sin(\pi r/N)}, & r=1,\dots,N-1\end{cases}, \qquad (9.4.30)$$

$$\eta_r^{(S)}=\frac{\overline{\alpha}_r^{(S)}}{\alpha_r}=\begin{cases}0, & r=0\\ -\dfrac{\cos(2\pi r/N)+2}{3i\sin(2\pi r/N)}, & r=1,\dots,N-1\end{cases}, \qquad (9.4.31)$$

$$\eta_r^{(3S)}=\frac{\overline{\alpha}_r^{(3S)}}{\alpha_r}=\begin{cases}0, & r=0\\ -\dfrac{\cos(3\pi r/N)+3\cos(\pi r/N)}{i\sin(3\pi r/N)}, & r=1,\dots,N-1\end{cases}. \qquad (9.4.32)$$

In Fig. 9-27, absolute values of the frequency responses of the DFT based method of integration (Eq. 9.4.20) and of the Newton-Cotes rules are represented with a frequency coordinate normalized to its maximal value. Because the absolute values of frequency responses are symmetric, only half of the curves are shown.

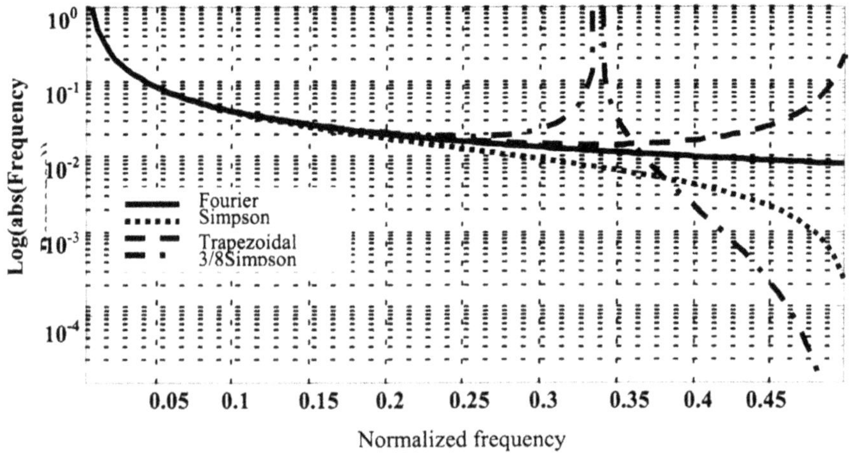

Figure 9-26. Frequency responses of four types of integrators

From the figure one can see that the integration methods are low pass filters, and that they are almost identical in low frequencies while in higher frequencies 3/8-Simpson, Simpson and trapezoidal integration methods tend to introduce integration errors. The frequency response of the 3/8-Simpson rule tends to infinity for the 2/3 of the maximum frequency, and the frequency response of the Simpson rule has almost the same tendency for the maximum frequency. This means, in particular, that round off computation errors and noise that might be present in input data will be amplified by Simpson and 3.8-Simpson in these frequencies.

9.4.5 Polar-to-Cartesian coordinate conversion, interpolation of DFT spectra of projections and the direct Fourier method of image reconstruction from projections

Polar-to-Cartesian and Cartesian-to-polar coordinate transformation is often needed in optical metrology, digital holography and image processing because optical sensors and recording devices are frequently work in polar coordinate systems while computer algorithms are almost always assume Cartesian coordinate system. It is sometimes useful also as a pre-processing step to match image natural geometry to square sampling grid oriented algorithms. A typical example is processing image slices in tomography illustrated in Fig. 9-27.

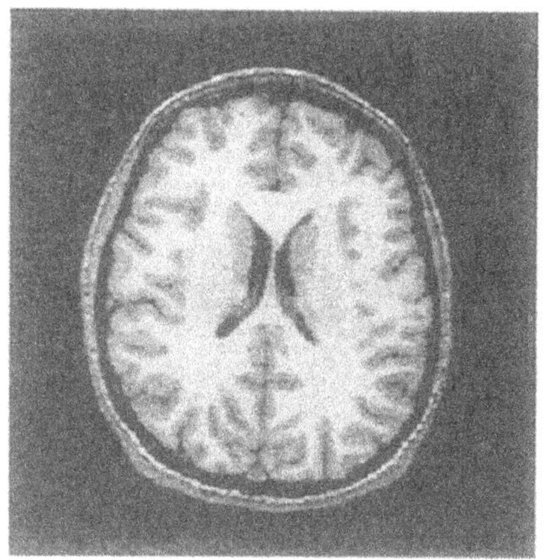

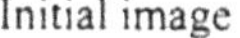

Initial image

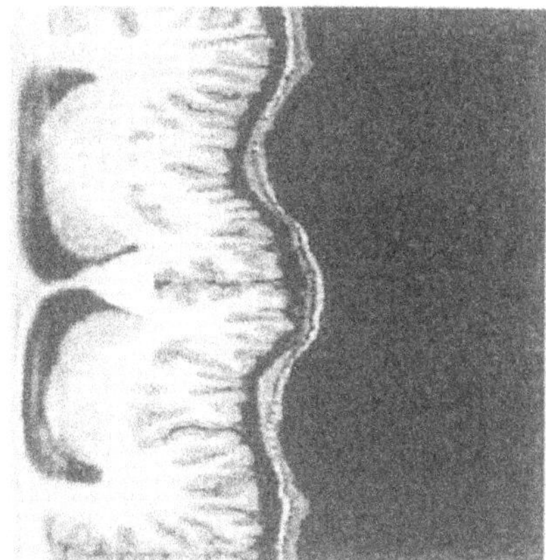

Same image in a polar coordinate system

Figure 9-27. An example of Cartesian-to-polar coordinate conversion of an MRI tomographic slice

Polar-to-Cartesian co-ordinate conversion is a key step in the direct Fourier algorithm of image reconstruction from projections described in Sect. 2.5. The zooming-resampling method of image geometrical transformations is a natural possible candidate for the polar-to-Cartesian coordinate conversion needed for the algorithm because 1-D DFT of projections can be combined with DFT based zooming along the radial coordinate.

Flow diagram of the image reconstruction algorithm with polar-to-Cartesian coordinate conversion by zooming-resampling method is shown in Fig. 9-28 (a). Fig. 9-29, (b) illustrates resampling from polar to Cartesian coordinate by the nearest neighbor interpolation.

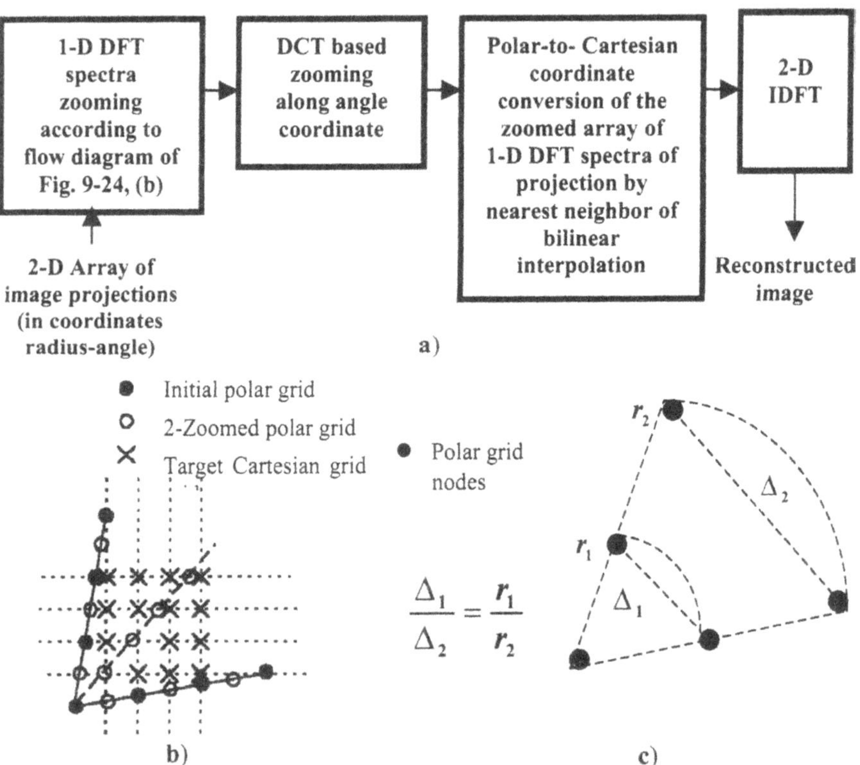

Figure 9-28. Polar to Cartesian grid conversion

Zooming factor is an important parameter of Polar-to-Cartesian co-ordinate conversion using the zooming-resampling method. The higher is zooming factor the better is resampling accuracy and the higher is computational complexity. One can reduce the computational complexity of polar-to-Cartesian coordinate conversion according to the flow diagram of

Fig. 9-28, a) by choosing variable zooming factor for zooming the array along the angle coordinate. As it follows from the geometry of polar coordinate system (Fig. 9-28, c)), zooming along the angular coordinate for smaller radial coordinates can be made with smaller zooming factor without compromising interpolation accuracy.

This principle of variable zooming is explained in Fig. 9-29 ([10]). Fig. 9-30 illustrates an example of image reconstruction with this method.

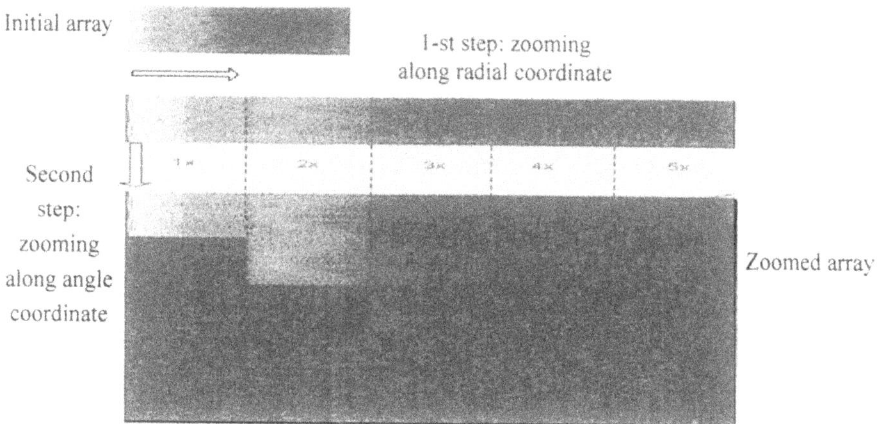

Figure 9-29. Zooming spectra of projections with variable zooming factor

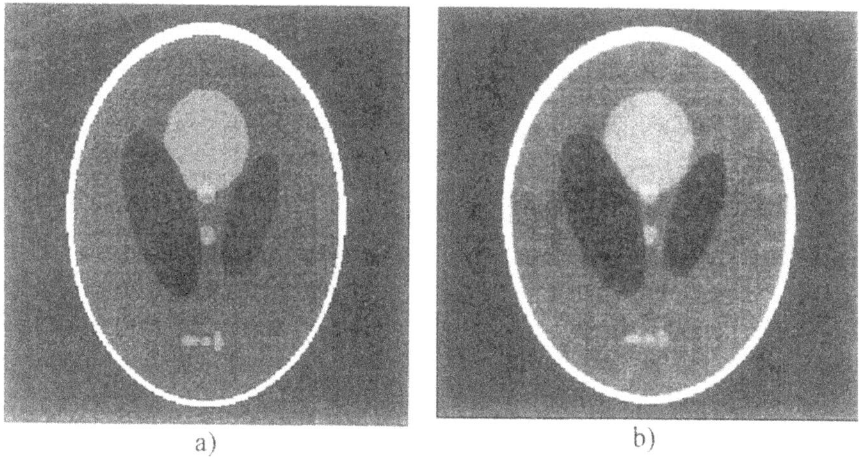

Figure 9-30. Shepp-Logan phantom (a) and a result of its reconstruction from 180 projections with the DCT based interpolation and variable zooming factor 1 to 5 (b)

References

1. P. Thévenaz, T. Blu, M. Unser, Image interpolation and resampling, Handbook of Medical Imaging, Processing and Analysis, I. N. Bankman, Ed., Academic Press, San Diego CA, USA, pp. 393-420, 2000
2. L.P. Yaroslavsky, Efficient algorithm for discrete sinc-interpolation, Applied Optics, Vol. 36, No.2, 10 January, 1997, p. 460-463
3. L. Yaroslavsky, Fast signal sinc-interpolation and its applications in signal and image processing, IS&T/SPIE's 14th Annual Symposium Electronic Imaging 2002, Science and Technology, Conference 4667 "Image Processing: Algorithms and Systems", San Jose, CA, 21-23 January 2002. Proceedings of SPIE vol. 4667
4. L. Yaroslavsky, Boundary Effect Free and Adaptive Discrete Signal Sinc-interpolation Algorithms for Signal and Image Resampling, Appl. Opt., v. 42, No. 20, 10 July 2003, p. 4166-4175
5. M. Unser, P. Thevenaz, L. Yaroslavsky, Convolution-based Interpolation for Fast, High-Quality Rotation of Images, IEEE Trans. on Image Processing, Oct. 1995, v. 4, No. 10, p. 1371-1382
6. J.H. Mathews, K.D. Fink, Numerical methods using MATLAB, Prentice Hall, Inc, 1999
7. S. C. Irick, R. Krishna Kaza, and W. R. McKinney, "Obtaining three-dimensional height profiles from a two-dimensional slope measuring instrument", Rev. Sci. Instrum. 66 (1995), 2108-2111
8. W.H. Press, B.P. Flannery, S.A. Teukolsky, W.T. Vetterling. Numerical recipes. The art of scientific computing. Cambridge University Press, Cambridge, 1987
9. C. Elster, I. Weingärtner, "High-accuracy reconstruction of a function f(x) when only df(x)/dx or $d^2f(x)/dx^2$ is known at discrete measurements points", Proc. SPIE, v. 4782
10. L. Yaroslavsky, Y. Chernobrodov, DFT and DCT based Discrete Sinc-interpolation Methods for Direct Fourier Tomographic Reconstruction, 3-d Int. Symposium, Image and Signal Processing and Analysis, Sept. 18-20, 2003, Rome, Italy

Chapter 10

SIGNAL PARAMETER ESTIMATION AND MEASUREMENT. OBJECT LOCALIZATION

10.1 PROBLEM FORMULATION. OPTIMAL STATISTICAL ESTIMATES

Measurement of physical parameters of objects is one of the most fundamental tasks in image processing. It is required in many applications. Typical examples are measuring the number of objects, their orientations, dimensions and coordinates. A special case of this problem is also object recognition when it is required to determine object index in the list of possible objects. Usually measurement devices and algorithms are designed for the use for arbitrary images from multitude of images generated by image or hologram sensors. Therefore the most appropriate approach to solving this problem is statistical one.

Let $a(x,\rho)$ be an object signal that depends on a certain parameter ρ, scalar or vectorial, that uniquely specifies the signal, $P_{\Re}(\rho)$ is a probability distribution over an ensemble $\rho \in \Re$ of parameter values (probability of values if they are discrete of probability density if the parameter is a continuous variable). Let also $b_{\omega}(x)$ be a realization of signal that is generated by the sensor in response to an object signal and is available for measuring the parameter. Realizations $\{b_{\omega}(x)\}$ belong to a statistical ensemble $\omega \in \Omega$ generated by the ensemble of parameter values and by random factors such as sensor's noise that specify a relationship between signals $\{a(x,\rho)\}$ and $\{b_{\omega}(x)\}$. Define parameter estimation as a transformation that maps signals $\{b_{\omega}(x)\}$ onto a a set of estimates of parameter values:

$$\hat{\rho} = \mathbf{ESTIM}(b_{\omega}(x)). \qquad (10.1.1)$$

Statistically best estimation transformation is the transformation that minimizes deviations, measured in certain statistical deviation measure, of the parameter estimates from their true values.

To begin with, consider the case of object recognition, when parameter ρ takes discrete values from a set of values $\{\rho_q\}$, numbered by an integer index q. A natural measure of the estimation transformation performance is in this case probability P_{err} of misrecognition. For any given realization $b_\omega(x)$ of the sensor's signal, parameter value ρ_q may be expected with probability $P[\rho_q/b_\omega(x)]$. This probability is called ***a posteriori probability*** of parameter value ρ_q. If, for signal $b_\omega(x)$, the estimation device generates estimation $\hat{\rho} = \rho_q$, error occurs with the probability of all other parameter values which is equal to $1 - P[\rho_q/b_\omega(x)]$. On average over the ensemble Ω of realizations of $b_\omega(x)$, the probability of misrecognitions will then be

$$\mathbf{AV}_\Omega(P_{err}) = \mathbf{AV}_\Omega\{1 - P[\rho_q/b_\omega(x)]\} = 1 - \mathbf{AV}_\Omega\{P[\rho_q/b_\omega(x)]\}, \quad (10.1.2)$$

where $\mathbf{AV}_\Omega(\cdot)$ symbolizes the averaging over the sensor's signal ensemble. Because all values involved in the averaging are non-negative, the probability of misrecognitions is minimal when, for all $b_\omega(x)$, parameter estimate $\hat{\rho}$ is selected that has maximal a posteriori probability:

$$\hat{\rho}_{opt} = \arg\max_{\{\rho_q\}} P\{\rho_q/b(x)\}. \quad (10.1.3)$$

Such an estimate is called ***MAP-estimate*** and the described approach to the optimization of parameter estimation is called ***Bayes approach***. According to Bayes rule:

$$P[\rho_q/b(x)]P[b(x)] = P[b(x)/\rho_q]P(\rho_q), \quad (10.1.4)$$

where $P[b(x)]$ is probability density of the sensor signal ensemble $\{b(x)\}$, $P\{b(x)/\rho_q\}$ is the probability that object signal $a(x,\rho_q)$ generates sensor signal $b(x)$ and $P(\rho_q)$ is ***a priori probability*** of the parameter. Therefore one can find optimal estimate of the parameter as

$$\hat{\rho}_{MAP} = \arg\max_{\{\rho_q\}} \frac{P\{b(x)/\rho_q\}P(\rho_q)}{P[b(x)]}, \quad (10.1.5)$$

or, as $P[b(x)]$ does not depend on $\{\rho_q\}$, as

$$\hat{\rho}_{MAP} = \arg\max_{\{\rho_q\}} P\{b(x)/\ \rho_q\}P(\rho_q). \tag{10.1.6}$$

This equation is the basic formula for the design of optimal Bayes parameter estimating devices. It requires knowledge of the conditional probability $P\{b(x)/\ \rho_q\}$ that depends on the relationship that exists between sensor signals $\{b_\omega(x)\}$ and object signals $\{a(x,\rho_q)\}$ and of a priori probabilities $\{P(\rho_q)\}$ of parameter values. The latter may be unknown. Estimates

$$\hat{\rho}_{ML} = \arg\max_{\{\rho_q\}} P\{b(x)/\ \rho_q\} \tag{10.1.7}$$

that ignore a priori probabilities (or, which is equivalent, assume that a priori probability distribution is uniform) are called Maximum Likelihood, or ***ML-estimates***.

Eqs. 10.1.6 and 7 were obtained in the assumption that parameter to be estimated is scalar and takes values from a discrete set of values. It is quite obvious that they may be in a straightforward way extended to vectorial parameters in which case probabilities involved in the equations are corresponding multi-variate probabilities and to parameters with continuum of values in which case probabilities should be replaced by corresponding probability densities.

In what follows we apply the described Bayes approach to solving a particular problem parameter estimation, that of target location, or measurement of coordinates of a target object in images. However the obtained results may also contribute to more deep understanding of the methods of solving other problems of image parameter estimation such as problems of image segmentation, object recognition, fitting 2-D curves ("snakes"), 3-D geometrical models and many others. All these problems may be formulated as model fitting that may be reduced to the estimation of the model parameters.

10.2 LOCALIZATION OF AN OBJECT IN THE PRESENCE OF ADDITIVE WHITE GAUSSIAN NOISE

10.2.1 Optimal localization device. Correlator and matched filter.

Consider a discrete model in which samples $\{b_k\}$ of the observed image signal can be regarded as a sum of samples $\{a_k(x_0, y_0)\}$ of a target object signal with unknown coordinates $\{x_0, y_0\}$ and samples $\{n_k\}$ of the noise:

$$b_k = a_k(x_0, y_0) + n_k. \tag{10.2.1}$$

Assume that noise samples $\{n_k\}$ are statistically independent on the signal $\{a_k(x_0, y_0)\}$, are uncorrelated and have a Gaussian probability distribution with zero mean and variance σ_n^2. This model describes the simplest situation in which the only disturbance interfering with object localization is the noise of the signal sensor which can frequently be regarded as additive, Gaussian, signal-independent and uncorrelated. Typical practical tasks to which such a model corresponds are, for instance, those of localization of constellations in stellar navigation and tracing target objects observed on a unform background.

For the model of Eq.(10.2.1) conditional probability of sensor signal samples $\{b_k\}$ for target signal located in coordinates (x_0, y_0) is equal

$$P(\{\{b_k\}/a_k(x_0, y_0)\}) = P\{n_k = b_k - a_k(x_0, y_0)\}. \tag{10.2.2}$$

Therefore the optimal object coordinate MAP estimate is as follows:

$$\{(\hat{x}_0, \hat{y}_0)\} = \underset{(x_0, y_o)}{\arg\max}\{P(\{a_k(x_0, y_o)\})P(\{n_k = b_k - a_k(x_0, y_0)\})\} \tag{10.2.3}$$

As samples $\{n_k\}$ are assumed to be uncorrelated Gaussian values with variance σ_n^2,

$$P(\{n_k = b_k - a_k(x_0, y_0)\}) \propto \prod_{k=0}^{N-1} \exp\left\{-\frac{1}{2\sigma_n^2}[b_k - a_k(x_0, y_0)]^2\right\} \tag{10.2.4}$$

where N is the number of signal and noise samples. Substitute Eq.(10.2.4) into Eq.(10.2.3) and, taking into account that $\exp(\cdot)$ is a monotone function, obtain for MAP-estimate:

$$\{(\hat{x}_0, \hat{y}_0)\} = \underset{(x_0, y_0)}{\arg\min} \left\{ \sum_{k=0}^{K-1} \left| b_k - a_k(x_0, y_0) \right|^2 - 2\sigma_n^2 \ln P(x_0, y_0) \right\}. \tag{10.2.5}$$

Finally, because $\sum_{k=0}^{K-1} |b_k|^2$ and $\sum_{k=0}^{K-1} |a_k(x_0, y_0)|^2$ do not depend on the coordinates (x_0, y_0) we have :

$$\{(\hat{x}_0, \hat{y}_0)\} = \underset{(x_0, y_0)}{\arg\max} \left\{ \sum_{k=0}^{K-1} b_k a_k(x_0, y_0) + \sigma_n^2 \ln P(x_0, y_0) \right\}. \tag{10.2.6}$$

Correspondingly, ML-estimate of target coordinates may be obtained as

$$\{(\hat{x}_0, \hat{y}_0)\} = \underset{(x_0, y_0)}{\arg\max} \left\{ \sum_{k=0}^{K-1} b_k a_k(x_0, y_0) \right\}. \tag{10.2.7}$$

According to the theory of discrete representation of signal transforms (Ch. 4):

$$\sum_{k=0}^{K-1} b_k a_k(x_0, y_0) = \frac{1}{\Delta x \Delta y} \int_{-\infty}^{\infty} \int_{-\infty}^{\infty} b(x, y) a(x - x_0, y - y_0) dx dy, \tag{10.2.8}$$

where $b(x, y)$ and $a(x - x_0, y - y_0)$ are the continuous signals corresponding to the sample sets $\{b_k\}$ and $\{a_k(x_0, y_0)\}$, respectively, and $\Delta x, \Delta y$ are discretization intervals corresponding to the chosen sampling rate for the discrete model of Eq.(10.2.1). Then obtain MAP- and ML-estimations for continuous signals, respectively, as:

$$\{(\hat{x}_0, \hat{y}_0)\} = \underset{(x_0, y_0)}{\arg\max} \left\{ \int_{-\infty}^{\infty} \int_{-\infty}^{\infty} b(x, y) a(x - x_0, y - y_0) dx dy + \mathrm{N}_0 \ln P(x_0, y_0) \right\}, \tag{10.2.9}$$

$$\{(\hat{x}_0, \hat{y}_0)\} = \underset{(x_0, y_0)}{\arg\max}\left\{\int_{-\infty}^{\infty}\int_{-\infty}^{\infty} b(x,y)a(x - x_0, y - y_0)dxdy\right\}. \qquad (10.2.10)$$

where

$$N_0 = \sigma_n^2 \Delta x \Delta y \qquad (10.2.11)$$

is spectral density of the noise.

Thus, the optimal ML-estimator should compute the cross-correlation function between the object signal $a(x - x_0, y - y_0)$ and the observed signal $b(x, y)$ and take the coordinates of the maximum in the correlation pattern as the estimate. Schematic diagram of such an estimator is shown in Fig. 10-1.

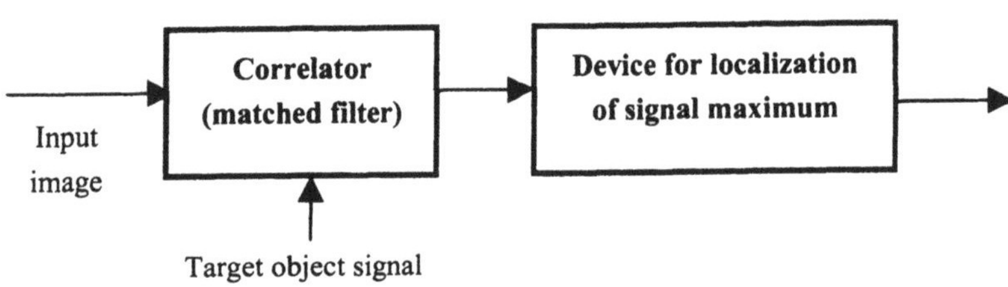

Figure 10-1. Schematic diagram of the optimal device for localization of objects in images

The optimal MAP-estimator also consists of a correlator and a decision-making device locating the maximum in the correlation pattern. The only difference between ML- and MAP-estimators is that, in the MAP-estimator, the correlation pattern is biased by the appropriately normalized logarithm of the object coordinates' a priori probability distribution.

The correlation operation for a mixture of signal and noise with a copy of the signal is often called ***matched filtering*** Correspondingly, the filter that implements this operation is called ***matched filter***. The correlation may be implemented in the frequency domain by multiplication of the input signal spectrum by frequency response of the matched filter which,, according to the properties of Fourier transform, is $\alpha^*(f_x, f_y)$, a complex conjugate of the object signal spectrum. Correlation, or matched filter type of the localization devices may also be implemented by optical and holographic means; a fact recognized at a very early stage in the history of holography ([1]).

10.2.2 Computer and optical implementations of the matched filter correlator

Two options are available for computer implementation of the matched filter correlator: digital convolution in the signal domain or in the Fourier transform domain as it is described in Ch. 5. The latter is usually more efficient in image processing because using FFT algorithms reduces the computational complexity of matched filtering to, by the order of magnitude, $O(\log N)$ operations per sample, where N is size of the image array. Flow diagram of the matching filtering in the Fourier Transform domain is shown in Fig. 10-2.

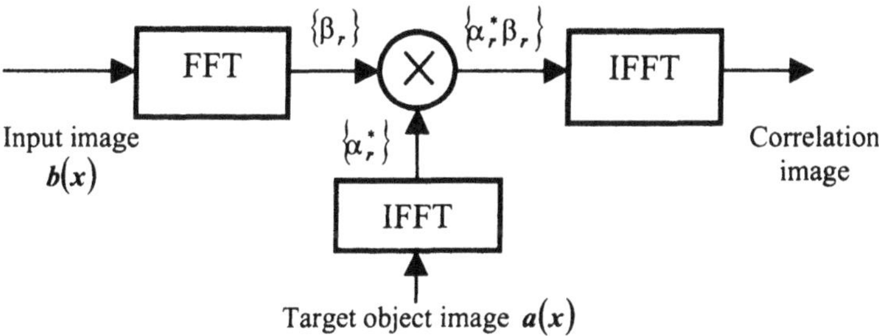

Figure 10-2. Flow diagram of the matched filtering in the Fourier Transform domain

The most straightforward optical implementations of the matched filter correlator is known as the ***4-F correlator***. It is a version of the optical image processing system with two lenses described in Sect. 2.3. Schematic diagram of the $4F$-correlator is shown in Fig. 10-3, a). Using parabolic mirror instead of Fourier lenses makes the correlator to be more compact ([2]). In this correlator shown in Fig. 10-3, b), input image plane, Fourier plane and correlation plane coincide. If a transparency with an input image is placed in the left focus of the mirror and is illuminated by a parallel beam of coherent laser light, its Fourier spectrum appears in the same plane around the mirror optical axis. In this place, a reflective optical spatial modulator with reflectance proportional to the frequency response of the matched filter is positioned. It modifies image spectrum correspondingly. Right part of the mirror performs the second Fourier transform and reconstructs, in the same focal plane, a correlation image centered at the mirror's right focus.

For both devices, the use of prefabricated spatial matched filters in form of an optical or computer generated holograms is assumed. For real time correlation, the Joint Transform Correlator was suggested ([3,4]). It is an opto-electronic device schematically shown in Fig. 10-4.

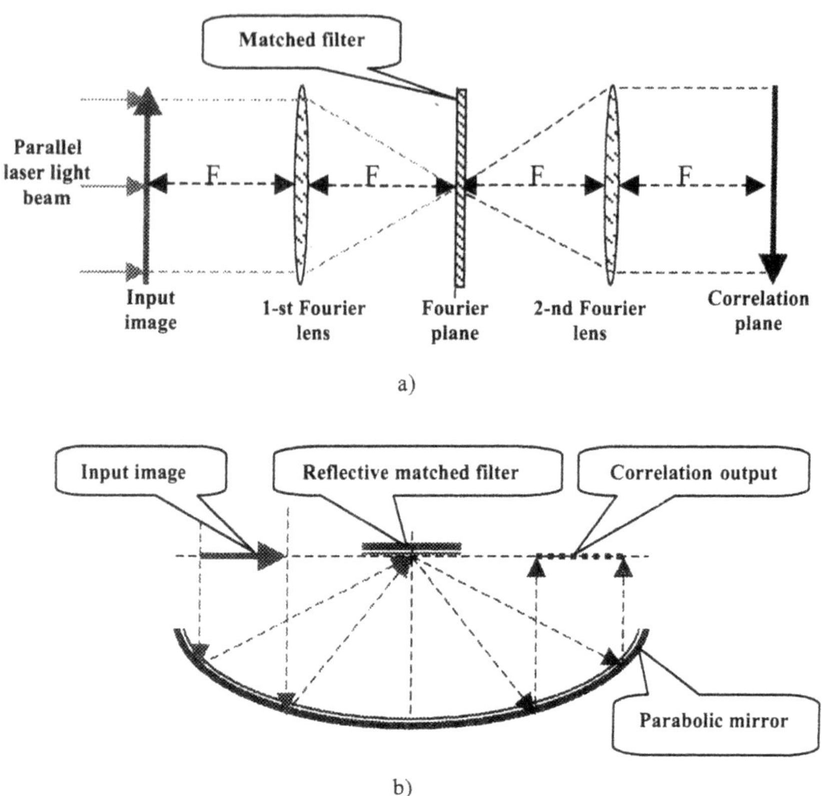

Figure 10-3. 4-F (a) and a parabolic mirror (b) optical correlators

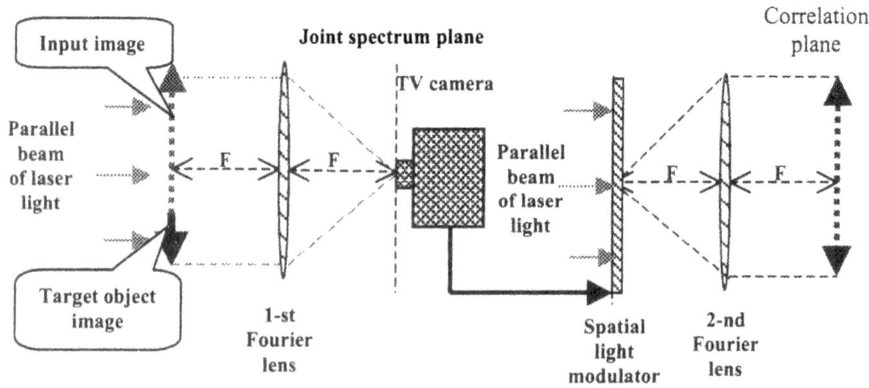

Figure 10-4. Schematic diagram of the Joint Transform Correlator

In the Joint Transform Correlator, input images and target object image are placed aside in the input plane. Squared modulus of Fourier Transform of the sum of these images, or joint spectrum of the images, is converted by TV camera into an electrical signal that controls the transparence of the spatial light modulator. The joint spectrum is described by the equation

$$b(x,y)+a(x+\bar{x},y)\xrightarrow{\text{FourierTransform}}\left|\beta(f_x,f_y)+\alpha(f_x,f_y)\exp(i2\pi f_x\bar{x})\right|^2 =$$
$$\left|\beta(f_x,f_y)\right|^2+\left|\alpha(f_x,f_y)\right|^2+\beta(f_x,f_y)\alpha^*(f_x,f_y)\exp(-i2\pi f_x\bar{x})+$$
$$\beta^*(f_x,f_y)\alpha(f_x,f_y)\exp(i2\pi f_x\bar{x}), \qquad (10.2.12)$$

where $b(x,y)$ and $a(x,y)$ are input and target object images and $\bar{x}$ is a the displacement, along coordinate x, of the target object image with respect the system optical axis. As one can see from this equation, the joint spectrum consists of four components. First two components are power spectra of the input and target object images. The third component is spectrum of their correlation and the fourth component is complex conjugate to the third one.

The second optical part of the Joint Transform Correlator performs Fourier Transform of the joint spectrum. As it follows from Eq. 10.2.12, output image of the Joint Transform Correlator also contains four components. First two components centered at the optical axis are images of autocorrelation functions of input and target object images. The third component is the image of their correlation function. It is reconstructed displaced along x axis by $\bar{x}$. The fourth component is the conjugate correlation image. It is reconstructed in the position $-\bar{x}$ symmetrical to that of the correlation image.

10.2.3 Performance of optimal estimators: normal and anomalous localization errors

Owing to the noise present in the sensor, any measurement device will estimate the object coordinates with a certain error. The optimal ML coordinate estimator of Fig. 10-1 that determines the object coordinates by the position of signal maximum at the output of the correlator (matched filter) provides the most accurate coordinate estimates possible.

The estimation accuracy may be generally characterized by the probability density $p(\varepsilon_x,\varepsilon_y)$ of the estimation errors:

$$\varepsilon_x = x_0-\hat{x}_0, \qquad \varepsilon_y = y_0-\hat{y}_0. \qquad (10.2.13)$$

For the evaluation of $p(\varepsilon_x, \varepsilon_y)$ one should analyze the distribution density of the position of the highest peak of the signal at the output of the matched filter within the area of all possible positions of the target object. This signal is a sum of the target object signal auto-correlation function and of a correlated Gaussian noise resulted from filtering the input white noise by the matched filter. This mixture is a non-stationary random process. This fact complicates the analysis very substantially. An approximate analytical solution of the problem can be obtained using following reasoning.

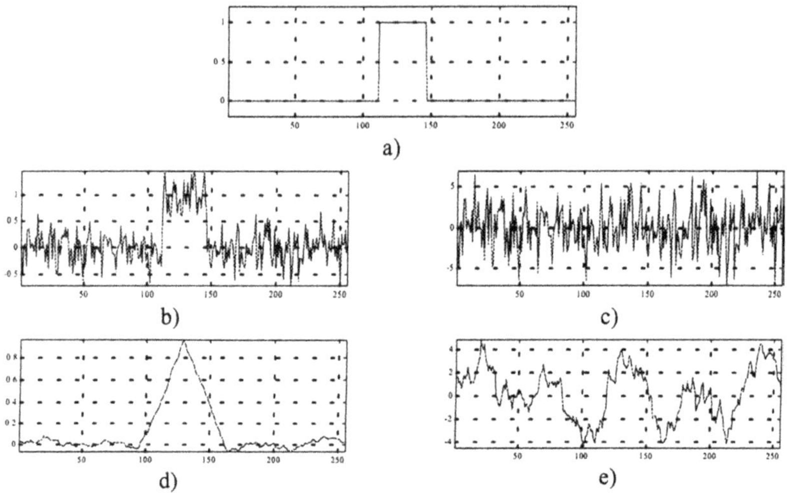

Figure 10-5. Normal and anomalous localization errors: a) - target object signal; b), c) - target object signal hindered by additive noise of different intensity; d) , e) - two corresponding outputs of the filter matched with the target signal that correspond to normal and anomalous localization errors, respectively.

Consider Fig. 10-5. Graph on Fig. 10-5, a) represents a target signal of a rectangular shape. Graphs on Fig. 10-5 b) and c) show mixtures of the target signal and noise with low and high intensity. Corresponding outputs of the correlation filter matched with the target object are shown in Figs. 10-5 d) and e). As one can see from these plots, two essentially different types of possible estimation errors have to be distinguished. One type of the errors occur when noise intensity is relatively low (Fig. 10-5, d)). In such cases the correlator output maxima lie in the vicinity of their correct values, and, therefore, localization errors are also relatively small. The second type of the errors are large errors. They take place when correlator output maxima are found very far away from the actual position of the target object marked by the position of its auto-correlation function (Fig. 10-5, e)). In this case very large localization errors occur.

Therefore small localization errors are caused by distortion of the object's autocorrelation peak shape by the noise and large errors are a result of especially large noise bursts occurring outside the area occupied by the object. These large errors are similar to the so-called "false alarm" errors in signal detection. Following V.A. Kotelnikov [5] we call the first type of error ***normal errors*** because, as we shall see, their density distribution may be approximated by Gaussian (normal) distribution. The errors of the second type we call ***anomalous errors***. The normal errors are responsible for the behavior of the error probability density $p(\varepsilon_x, \varepsilon_y)$ in the vicinity of zero errors $(\varepsilon_x = 0, \varepsilon_y = 0)$. One can say that normal errors characterize the accuracy of coordinate measurements. Anomalous errors, obviously, determine the tails of the error probability density $p(\varepsilon_x, \varepsilon_y)$. One can say that the anomalous errors characterize the measurement reliability. Normal errors are studied in Sects. 10.3. Anomalous errors are discussed in Sect. 10.5.

10.3 PERFORMANCE OF THE OPTIMAL LOCALIZATION DEVICE

10.3.1 Distribution density and variance of normal errors

In order to determine statistical characteristics of normal errors, let us consider the correlator's output signal:

$$R_{ab}(x, y) = \int\int b(\xi, \eta) a(\xi - x, \eta - y) d\xi d\eta =$$

$$\int\int [a(\xi - x_0, \eta - y_0) + n(\xi, \eta)] a(\xi - x, \eta - y) d\xi d\eta =$$

$$R_a(x - x_0, y - y_o) + R_n(x, y), \qquad (10.3.1)$$

where

$$R_a(x, y) = \iint\limits_X a(\xi, \eta) a(\xi - x, \eta - y) d\xi d\eta \qquad (10.3.2)$$

is auto-correlation function of the target object signal, and

$$\tilde{n}(x,y) = \iint_X n(\xi,\eta)\,a(\xi - x,\eta - y)\,d\xi\, d\eta \tag{10.3.3}$$

is a correlated Gaussian random process resulted from filtering input Gaussian noise by the matched filter. It is fully determined statistically by its correlation function

$$\begin{aligned}
&R_{\tilde{n}} = \mathbf{AV}_n\left[\tilde{n}(x_1,y_1)\cdot \tilde{n}(x_2,y_2)\right] = \\
&\iint_X\iint_X \mathbf{AV}_n\{n(\xi_1,\eta_1)n(\xi_2,\eta_2)\}a(\xi_1 - x_1,\eta_1 - y_1)a(\xi_2 - x_2,\eta_2 - y_2)d\xi_1\, d\eta_1\, d\xi_2\, d\eta_2 = \\
&\iint_X\iint_X \mathrm{N}_0\delta(\xi_1 - \xi_2,\eta_1 - \eta_2)a(\xi_1 - x_1,\eta_1 - y_1)a(\xi_2 - x_2,\eta_2 - y_2)d\xi_1\, d\eta_1\, d\xi_2\, d\eta_2 = \\
&\qquad \mathrm{N}_0\iint_X a(\xi_1 - x_1,\eta_1 - y_1)a(\xi_1 - x_2,\eta_1 - y_2)d\xi_1\, d\eta_1 = \\
&\mathrm{N}_0\, R_a\left(x_2 - x_1, y_2 - y_1\right),
\end{aligned} \tag{10.3.4}$$

where $\mathbf{AV}_n(\cdot)$ denotes statistical averaging over the noise $n(x,y)$ ensemble. In the case of normal errors, maxima of the signal $R(x,y)$ at the output of the correlator are located in the close vicinity of the point (x_0, y_0), the position of maximum of $R_a(x,y)$. The following system of equations determines coordinates the maximum of $R(x,y)$:

$$\begin{cases}
\dfrac{\partial}{\partial x} R(x,y) = \dfrac{\partial}{\partial x} R_a\left(x - x_0, y - y_0\right) + \dfrac{\partial}{\partial x} R_n(x,y) = 0 \\
\dfrac{\partial}{\partial y} R(x,y) = \dfrac{\partial}{\partial y} R_a\left(x - x_0, y - y_0\right) + \dfrac{\partial}{\partial y} R_n(x,y) = 0
\end{cases} \tag{10.3.4}$$

Let the solution of this system be:

$$\begin{cases}
\hat{x} = x_0 + n_x; \\
\hat{y} = y_0 + n_y;
\end{cases} \tag{10.3.5}$$

and localization errors n_x and n_y are small. Then we have:

$$\begin{cases}
\dfrac{\partial}{\partial x} R_a\left(x - x_0, y - y_0\right)\Big|_{\substack{x=x_0+n_x \\ y=y_0+n_y}} + \dfrac{\partial}{\partial x} R_n(x,y)\Big|_{\substack{x=x_0+n_x \\ y=y_0+n_y}} = 0 \\
\dfrac{\partial}{\partial y} R_a\left(x - x_0, y - y_0\right)\Big|_{\substack{x=x_0+n_x \\ y=y_0+n_y}} + \dfrac{\partial}{\partial y} R_n(x,y)\Big|_{\substack{x=x_0+n_x \\ y=y_0+n_y}} = 0
\end{cases} \tag{10.3.6}$$

The first terms in Eqs.(10.3.6) may be expressed as:

$$\frac{\partial}{\partial x}R_a(x-x_0,y-y_0)\Big|_{\substack{x=x_0+n_x\\y=y_0+n_y}}=\frac{\partial}{\partial x}R_a(x-x_0,y-y_0)\Big|_{\substack{x=x_0\\y=y_0}}$$
$$+n_x\frac{\partial^2}{\partial x^2}R_a(x-x_0,y-y_0)\Big|_{\substack{x=x_0\\y=y_0}}+n_y\frac{\partial^2}{\partial x\partial y}R_a(x-x_0,y-y_0)\Big|_{\substack{x=x_0\\y=y_0}} \quad (10.3.7)$$
$$=D_x n_x+D_{xy}n_y,$$

where

$$\frac{\partial}{\partial x}R_a(x-x_0,y-y_0)\Big|_{\substack{x=x_0\\y=y_0}}=0 \quad (10.3.8)$$

by definition because ($\{x_0,y_0\}$) are the coordinates of the object, and

$$D_x=\frac{\partial^2}{\partial x^2}R_a(x-x_0,y-y_0)\Big|_{\substack{x=x_0\\y=y_0}};$$
$$D_{xy}=\frac{\partial^2}{\partial x\partial y}R_a(x-x_0,y-y_0)\Big|_{\substack{x=x_0\\y=y_0}}. \quad (10.3.9)$$

Similarly one can obtain :

$$\frac{\partial}{\partial y}R_a(x-x_0,y-y_0)\Big|_{\substack{x=x_0+n_x\\y=y_0+n_y}}=D_{xy}n_x+D_y n_y, \quad (10.3.10)$$

with

$$D_y=\frac{\partial^2}{\partial y^2}R_a(x-x_0,y-y_0)\Big|_{\substack{x=x_0\\y=y_0}}; \quad (10.3.12)$$

Denote

$$-\frac{\partial}{\partial x}R_n(x,y)\Big|_{\substack{x=x_0+n_x\\y=y_0+n_y}}=\nu_x; \quad -\frac{\partial}{\partial y}R_n(x,y)\Big|_{\substack{x=x_0+n_x\\y=y_0+n_y}}=\nu_y. \quad (10.3.13)$$

Then from Eqs.10.3.6 we obtain the equations:

$$\begin{aligned} D_x n_x + D_{xy} n_y &= \nu_x; \\ D_{xy} n_x + D_y n_y &= \nu_y \end{aligned} \tag{10.3.14}$$

from which we find the following relationships for errors n_x and n_y in x and y directions :

$$\begin{aligned} n_x &= \frac{D_y}{D_x D_y - D_{xy}^2} \nu_x - \frac{D_{xy}}{D_x D_y - D_{xy}^2} \nu_y; \\ n_y &= \frac{D_x}{D_x D_y - D_{xy}^2} \nu_y - \frac{D_{xy}}{D_x D_y - D_{xy}^2} \nu_x. \end{aligned} \tag{10.3.15}$$

From Eq. 10.3.15 it follows that small errors $\{n_x, n_y\}$ in the determination of coordinates of the object by the ML-estimator have a Gaussian distribution with zero mean and the following variances :

$$\sigma_x^2 = \mathbf{AV}_n\left(|n_x|^2\right) = \left(\frac{D_y}{D_x D_y - D_{xy}^2}\right)^2 \mathbf{AV}_n\left(|\nu_x|^2\right) + \left(\frac{D_{xy}}{D_x D_y - D_{xy}^2}\right)^2 \mathbf{AV}_n\left(|\nu_y|^2\right)$$

$$-2\frac{D_y D_{xy}}{\left(D_x D_y - D_{xy}^2\right)^2} \mathbf{AV}_n\left(\nu_x \nu_y\right); \tag{10.3.16a}$$

$$\sigma_y^2 = \mathbf{AV}_n\left(n_y\right)^2 = \left(\frac{D_x}{D_x D_y - D_{xy}^2}\right)^2 \mathbf{AV}_n\left(|\nu_y|^2\right) + \left(\frac{D_{xy}}{D_x D_y - D_{xy}^2}\right)^2 \mathbf{AV}_n\left(|\nu_x|^2\right) -$$

$$2\frac{D_x D_{xy}}{\left(D_x D_y - D_{xy}^2\right)^2} \mathbf{AV}_n\left(\nu_x \nu_y\right); \tag{10.3.16b}$$

$$\begin{aligned} \sigma_{xy}^2 &= \mathbf{AV}_n\left(n_x n_y\right) = \frac{D_x D_y - \left(D_{xy}\right)^2}{\left(D_x D_y - D_{xy}^2\right)^2} \mathrm{AV}_n\left(\nu_x \nu_y\right) \\ &- \frac{D_y D_{xy}}{\left(D_x D_y - D_{xy}^2\right)^2} \mathbf{AV}_n\left(|\nu_x|^2\right) + \frac{D_x D_{xy}}{\left(D_x D_y - D_{xy}^2\right)^2} \mathbf{AV}_n\left(|\nu_x|^2\right). \end{aligned} \tag{10.3.16c}$$

Parameters involved in formulas (10.3.16) may be found using the relationship between signal correlation functions and power spectra and properties of the Fourier transform:

$$D_x = \frac{\partial^2}{\partial x^2} R_a(x - x_0, y - y_0)\Big|_{\substack{x=x_0\\y=y_0}}$$

$$= \frac{\partial^2}{\partial x^2} \int_{-\infty}^{\infty}\int_{-\infty}^{\infty} \left|a(f_x, f_y)\right|^2 exp\{-i2p[f_x(x - x_0) + f_y(y - y_0)]\} df_x df_y\Big|_{\substack{x=x_0\\y=y_0}}$$

$$= -4\pi^2 \int_{-\infty}^{\infty}\int_{-\infty}^{\infty} f_x^2 \left|\alpha(f_x, f_y)\right|^2 df_x df_y = -4\pi^2 \bar{f}_x^2 E_a, \qquad (10.3.17)$$

where $\alpha(f_x, f_y)$ is Fourier spectrum of the target object signal :

$$\alpha(f_x, f_y) = \int_{-\infty}^{\infty}\int_{-\infty}^{\infty} a(x, y) \exp\{-i2\pi(f_y x + f_y y)\} df_x df_y \qquad (10.3.18)$$

E_a is its energy:

$$E_a = \int_{-\infty}^{\infty}\int_{-\infty}^{\infty} \left|\alpha(f_x, f_y)\right|^2 df_x df_y \qquad (10.3.19)$$

and $\bar{f}_x^2$ is the inertia moment of its power spectrum along the axis f_x :

$$\bar{f}_x^2 = \frac{\int_{-\infty}^{\infty}\int_{-\infty}^{\infty} f_x^2 \left|\alpha(f_x, f_y)\right|^2 df_x df_y}{\int_{-\infty}^{\infty}\int_{-\infty}^{\infty} \left|\alpha(f_x, f_y)\right|^2 df_x df_y} \qquad (10.3.20)$$

Similarly,

$$D_y = -4\pi^2 \bar{f}_y^2 \int_{-\infty}^{\infty}\int_{-\infty}^{\infty} \left|\alpha(f_x, f_y)\right|^2 df_x df_y = -4\pi^2 \bar{f}_y^2 E_a, \qquad (10.3.21)$$

and

$$D_{xy} = -4\pi^2 \int_{-\infty}^{\infty}\int_{-\infty}^{\infty} f_x f_y \left|\alpha(f_x, f_y)\right|^2 df_x df_y = -4\pi^2 \bar{f}_{xy}^2 E_a, \tag{10.3.22}$$

where

$$\bar{f}_y^2 = \frac{\int_{-\infty}^{\infty}\int_{-\infty}^{\infty} f_y^2 \left|\alpha(f_x, f_y)\right|^2 df_x df_y}{\int_{-\infty}^{\infty}\int_{-\infty}^{\infty} \left|\alpha(f_x, f_y)\right|^2 df_x df_y} \tag{10.3.23}$$

and

$$\bar{f}_{xy}^2 = \frac{\int_{-\infty}^{\infty}\int_{-\infty}^{\infty} f_x f_y \left|\alpha(f_x, f_y)\right|^2 df_x df_y}{\int_{-\infty}^{\infty}\int_{-\infty}^{\infty} \left|\alpha(f_x, f_y)\right|^2 df_x df_y} \tag{10.3.24}$$

Second moments $\mathbf{AV}_n\left(\left|\nu_x\right|^2\right)$, $\mathbf{AV}_n\left(\left|\nu_y\right|^2\right)$ and $\mathbf{AV}_n\left(\nu_x \nu_y\right)$of the noise component in the formulas (10.3.16) may also be found in the spectral domain . For instance, for ν_x we have:

$$\nu_x = -\frac{\partial}{\partial x} R_n(x, y)\Bigg|_{\substack{x = x_0 + n_x \\ y = y_0 + n_y}}. \tag{10.3.25}$$

Hence, power spectrum of ν_x is equal to $4\pi^2 f_x^2$ times power spectrum of $R_n(x, y)$ which, by virtue of Eq.(10.3.3), is equal to $N_0 \left|\alpha(f_x, f_y)\right|^2$. Therefore,

$$\mathbf{AV}_n\left(\nu_x^2\right) = 4\pi^2 \mathrm{N}_0 \int_{-\infty}^{\infty}\int_{-\infty}^{\infty} f_x^2 \left|\alpha(f_x, f_y)\right|^2 df_x df_y = 4\pi^2 \mathrm{N}_0 E_0 \bar{f}_x^2. \tag{10.3.26}$$

In a similar way one can obtain :

$$\mathbf{AV}_n\left(\nu_y^2\right) = 4\pi^2 \mathrm{N}_0 \int_{-\infty}^{\infty}\int_{-\infty}^{\infty} f_y^2 \left|\alpha(f_x, f_y)\right|^2 df_x df_y = 4\pi^2 \mathrm{N}_0 E_0 \bar{f}_y^2 \tag{10.3.27}$$

and

$$\mathbf{AV}_n(\nu_x \nu_y) = 4\pi^2 N_0 \int_{-\infty}^{\infty}\int_{-\infty}^{\infty} f_x f_y |\alpha(f_x, f_y)|^2 df_x df_y = 4\pi^2 N_0 E_0 \bar{f}_{xy}^2 . \quad (10.3.28)$$

After substituting these parameters into Eqs. (10.3.16) and some obvious algebraic transformations we finally arrive at the following relationships for variances of normal errors in object coordinate estimation in the presence of additive white Gaussian noise for the optimal ML-estimator:

$$\sigma_x^2 = \frac{1}{4\pi^2} \frac{\bar{f}_y^2}{\bar{f}_x^2 \bar{f}_y^2 - \left(\bar{f}_{xy}^2\right)^2} \frac{N_0}{E_a}; \quad (10.3.29a)$$

$$\sigma_y^2 = \frac{1}{4\pi^2} \frac{\bar{f}_x^2}{\bar{f}_x^2 \bar{f}_y^2 - \left(\bar{f}_{xy}^2\right)^2} \frac{N_0}{E_a}; \quad (10.3.29b)$$

$$\sigma_{xy}^2 = \frac{1}{4\pi^2} \frac{\bar{f}_{xy}^2}{\bar{f}_x^2 \bar{f}_y^2 - \left(\bar{f}_{xy}^2\right)^2} \frac{N_0}{E_a}. \quad (10.3.29c)$$

Eqs. (10.3.29) demonstrate that variances of normal localization errors are fully determined by the signal-to-noise ratio N_0 / E_a and the inertia moments (10.3.20-24) of power spectrum of the target object signal. These are the only characteristics of the object shape that affect its potential localization accuracy.

Power spectra $|\alpha(f_x, f_y)|^2$ of real valued signals feature the property of central symmetry:

$$|\alpha(f_x, f_y)|^2 = |\alpha(-f_x, -f_y)|^2 . \quad (10.3.30)$$

If the object signal power spectrum is symmetrical with respect to the coordinate axes as well:

$$|\alpha(f_x, f_y)|^2 = |\alpha(f_x, -f_y)|^2 = |\alpha(-f_x, f_y)|^2 , \quad (10.3.31)$$

Eqs.(10.3.29) take the following simpler form :

$$\sigma_x^2 = \frac{1}{\bar{f}_x^2}\frac{N_0}{4\pi^2 E_a}; \tag{10.3.32a}$$

$$\sigma_y^2 = \frac{1}{\bar{f}_y^2}\frac{N_0}{4\pi^2 E_a}; \tag{10.3.32b}$$

$$\sigma_{xy}^2 = 0. \tag{10.3.32c}$$

The last equation shows that if the signal power spectrum is axes-symmetrical, normal localization errors along coordinates x and y are uncorrelated. This situation takes place if the object spectrum $\alpha(f_x, f_y)$, or, which is the same, the object signal $a(x,y)$ is a separable function of the coordinates. Equations (10.3.32) also coincide with the very well known relationship for one-dimensional signals ([5,6]).

It is sometimes more convenient to express variances of normal localization errors in terms of the noise variance $\sigma_{out,n}^2$ at the output of the matched filter rather than in terms of the input noise spectral density N_0. Because this variance is equal to

$$\sigma_{out,n}^2 = N_0 \int_{-\infty}^{\infty}\int_{-\infty}^{\infty} |\alpha(f_x, f_y)|^2 df_x df_y = E_a N_0 , \tag{10.3.33}$$

Eqs.(10.3.29) may be rewritten as

$$\sigma_x^2 = \frac{1}{4\pi^2}\frac{\bar{f}_y^2}{\bar{f}_x^2 \bar{f}_y^2 - \left(\bar{f}_{xy}^2\right)^2}\frac{\sigma_{out,n}^2}{E_a^2}; \tag{10.3.34a}$$

$$\sigma_y^2 = \frac{1}{4\pi^2}\frac{\bar{f}_x^2}{\bar{f}_x^2 \bar{f}_y^2 - \left(\bar{f}_{xy}^2\right)^2}\frac{\sigma_{out,n}^2}{E_a^2}; \tag{10.3.34b}$$

$$\sigma_{xy}^2 = \frac{1}{4\pi^2} \frac{\overline{f_{xy}^2}}{\overline{f_x^2}\,\overline{f_y^2} - \left(\overline{f_{xy}^2}\right)^2} \frac{\sigma_{out,n}^2}{E_a^2}. \tag{10.3.34c}$$

In digital processing, it is natural to represent normal error variances in units of signal discretization intervals $\Delta x, \Delta y$. Denote

$$\sigma_{dx}^2 = \sigma_x^2 / \Delta x^2 ; \tag{10.3.35a}$$

$$\sigma_{dy}^2 = \sigma_y^2 / \Delta y^2 ; \tag{10.3.35b}$$

$$\sigma_{dxy}^2 = \sigma_{xy}^2 / \Delta x \Delta y . \tag{10.3.35b}$$

Then, from Eqs. (10.3.29) obtain:

$$\sigma_{dx}^2 = \frac{1}{4\pi^2} \frac{\overline{f_y^2}\Delta y^2}{\left(\overline{f_x^2}\Delta x^2\right)\left(\overline{f_y^2}\Delta y^2\right) - \left(\overline{f_{xy}^2}\Delta x \Delta y\right)^2} \frac{N_0 / \Delta x \Delta y}{E_a / \Delta x \Delta y}$$

$$= \frac{1}{4\pi^2} \frac{\overline{f_{dy}^2}}{\left(\overline{f_{dx}^2}\right)\left(\overline{f_{dy}^2}\right) - \left(\overline{f_{dxy}^2}\right)^2} \frac{\sigma_n^2}{E_{d,a}} ; \tag{10.3.36a}$$

$$\sigma_{dy}^2 = \frac{1}{4\pi^2} \frac{\overline{f_{dx}^2}}{\left(\overline{f_{dx}^2}\right)\left(\overline{f_{dy}^2}\right) - \left(\overline{f_{dxy}^2}\right)^2} \frac{\sigma_n^2}{E_{d,a}} ; \tag{10.3.36b}$$

$$\sigma_{dxy}^2 = \frac{1}{4\pi^2} \frac{\overline{f_{dxy}^2}}{\overline{f_{dx}^2}\,\overline{f_{dy}^2} - \left(\overline{f_{dxy}^2}\right)^2} \frac{\sigma_n^2}{E_{d,a}} , \tag{10.3.36c}$$

and, for axis symmetrical signal spectrum

$$\sigma_{dx}^2 = \frac{1}{\overline{f_x^2}} \frac{\sigma_n^2}{4\pi^2 E_{d,a}} ; \tag{10.3.37a}$$

$$\sigma_{dy}^2 = \frac{1}{\bar{f}_y^2} \frac{\sigma_n^2}{4\pi^2 E_{d,a}};\tag{10.3.37b}$$

$$\sigma_{dxy}^2 = 0,\tag{10.3.37c}$$

where $\sigma_n^2 = \mathrm{N}_0/\Delta x\Delta y$ is variance of the input noise samples,

$$E_{d,a} = \sum_{k=0}^{K-1} a_k^2\tag{10.3.38}$$

is energy of the discrete signal $\{a_k\}$ of the target object and

$$\bar{f}_{dx}^2 = \bar{f}_x^2 \Delta x^2; \quad \bar{f}_{dy}^2 = \bar{f}_y^2 \Delta y^2; \quad \bar{f}_{dxy}^2 = \bar{f}_{xy}^2 \Delta x \Delta y;\tag{10.3.39}$$

are discrete inertia moments of the target object spectrum. One can obtain them by the discretization of formulas (10.3.20), (10.3.22) and (10.3.23) and using the relationship (10.2.9):

$$\bar{f}_{dx}^2 = \frac{\frac{1}{\Delta x \Delta y} \int\limits_{-1/2\Delta x}^{1/2\Delta x} \int\limits_{-1/2\Delta y}^{1/2\Delta y} \left(f_x^2 \Delta x^2\right) \left|\alpha\left(f_x, f_y\right)\right|^2 df_x df_y}{\frac{1}{\Delta x \Delta y} \int\limits_{-1/2\Delta x}^{1/2\Delta x} \int\limits_{-1/2\Delta y}^{1/2\Delta y} \left|\alpha\left(f_x, f_y\right)\right|^2 df_x df_y} =$$

$$\frac{\sum_{r=1}^{N_x/2} r^2 \left(\Delta f^2 \Delta x^2\right) \sum_{s=1}^{N_y/2} \left|\alpha_{r,s}\right|^2}{\sum_{r=1}^{N_x/2} \sum_{s=1}^{N_y/2} \left|\alpha_{r,s}\right|^2} = \frac{1}{N_x^2} \frac{\sum_{r=1}^{N_x/2} r^2 \sum_{s=1}^{N_y/2} \left|\alpha_{r,s}\right|^2}{\sum_{r=1}^{N_x/2} \sum_{s=1}^{N_y/2} \left|\alpha_{r,s}\right|^2};\tag{10.3.40}$$

$$\bar{f}_{dy}^2 = \frac{1}{N_y^2} \frac{\sum_{r_y=-N_y/2}^{N_y/2-1} r_y^2 \sum_{r_x=-N_x/2}^{N_x/2-1} \left|\alpha_{r_x,r_y}\right|^2}{\sum_{r_y=-N_y/2}^{N_y/2-1} \sum_{r_x=-N_x/2}^{N_x/2-1} \left|\alpha_{r_x,r_y}\right|^2};\tag{10.3.41}$$

$$\bar{f}_{dxy}^2 = \frac{1}{N_x N_y} \frac{\sum_{r=1}^{N_x/2} \sum_{s=1}^{N_y/2} rs|\alpha_{r,s}|^2}{\sum_{r=1}^{N_x/2} \sum_{s=1}^{N_y/2-1} |\alpha_{r,s}|^2}; \qquad (10.3.42)$$

where $N_x = \Delta f_x \Delta x$ is the number of signal samples along coordinate x, $N_y = \Delta f_y \Delta y$ is the number of signal samples along coordinate y, $\{\Delta f_x, \Delta f_y\}$ are discretization intervals in the frequency domain and $\{\alpha_{r,s}\}$is 2-D DFT spectrum of the target object signal $a(x,y)$. It is assumed in Eqs. 10.3.40 - 42 that N_x and N_y are even numbers.

10.3.2 Illustrative examples

In order to gain an intuition concerning the relationship between the signal shape and potential accuracy of its localization in additive Gaussian noise, consider localization of a one-dimensional target object signals in form of Gaussian shaped, rectangular and triangle impulses.

Let

$$a(x) = a_0 \exp\left(-\frac{x^2}{2\sigma_a^2}\right) \qquad (10.3.43)$$

bc a Gaussian-shaped impulse. Its energy is:

$$E_a = \int_{-\infty}^{\infty} |a(x)|^2 dx = a_0^2 \int_{-\infty}^{\infty} \exp\left(-\frac{x^2}{\sigma_a^2}\right) dx = a_0^2 \sqrt{2\pi}\sigma_a / \sqrt{2} = a_0^2 \sqrt{\pi}\sigma_a$$

$$(10.3.44)$$

Introduce an effective impulse width as the width of a rectangular impulse of the same energy:

$$S_a = E_a / a_0^2 . \qquad (10.3.45)$$

From Eq (10.3.44) obtain that for Gaussian impulse:

$$S_a = \sqrt{\pi}\sigma_a . \qquad (10.3.46)$$

This relationship is illustrated in Fig. 10-6.

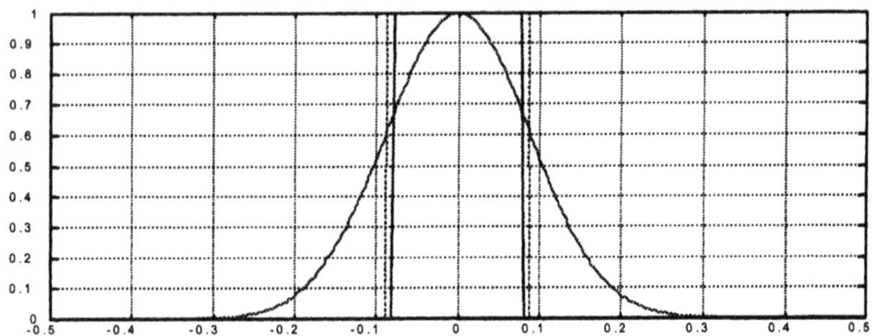

Figure 10-6. Gaussian -shaped impulse and its efficient width shown by solid line. Dashed line shows limits of interval $[-\sigma_a, \sigma_a]$

Spectrum of the Gaussian-shaped impulse is:

$$\alpha(f) = \int_{-\infty}^{\infty} a(x)\exp(i2\pi fx)dx = a_0 \int_{-\infty}^{\infty} \exp\left(-\frac{x^2}{2\sigma_a^2}\right)\exp(i2\pi fx)dx =$$

$$a_0 \int_{-\infty}^{\infty} \exp\left[-\left(\frac{x^2}{2\sigma_a^2} - i2\pi fx + \left(\sqrt{2}\sigma_a i\pi f\right)^2\right)\right]\exp\left(\sqrt{2}\sigma_a i\pi\pi f\right)^2 dx =$$

$$a_0 \exp\left(-2\pi^2\sigma_a^2 f^2\right)\int_{-\infty}^{\infty} \exp\left[-\left(\frac{x^2}{2\sigma_a^2} - i2\pi fx + \left(\sqrt{2}\sigma_a i\pi f\right)^2\right)\right]dx =$$

$$\sqrt{2\pi}\, a_0\sigma_a \exp\left(-2\pi^2\sigma_a^2 f^2\right) \qquad (10.3.47)$$

and its inertia moment is:

$$\overline{f_x^2} = \frac{\int_{-\infty}^{\infty} f^2 |\alpha(f)|^2 df}{E_a} = \frac{2\pi a_0^2\sigma_a^2 \int_{-\infty}^{\infty} f_x^2 \exp\left(-4\pi^2\sigma_a^2 f_x^2\right)df}{a_0^2\sqrt{\pi}\,\sigma_a} = \frac{1}{8\pi^2\sigma_a^2} \qquad (10.3.48)$$

Introduce also an effective spectrum bandwidth:

$$2F_a = E_a / |\alpha(0)|^2 . \qquad (10.3.49)$$

From Eqs.(10.3.44), obtain:

$$2F_a = \frac{1}{2\sqrt{\pi}\sigma_a}. \tag{10.3.50}$$

It follows now from Eqs.10.3.32,a and 10.4.44 through 47 that the potential localization accuracy for the Gaussian-shaped impulse is determined by equations

$$\sigma_x^2 = \frac{1}{4\pi^2 \overline{f_x^2}} \frac{N_0}{E_a} = 2\sigma_a^2 \frac{N_0}{E_a} = \frac{2\sigma_a}{\sqrt{\pi}} \frac{N_0}{a_0^2}, \tag{10.3.51,a}$$

or

$$\frac{\sigma_x^2}{\sigma_a^2} = 2\frac{N_0}{E_a} = \frac{2}{\sqrt{\pi}} \frac{N_0}{a_0^2 \sigma_a}. \tag{10.3.51,b}$$

For discrete signals, Eqs .10.3.51 may be written as:

$$\sigma_{dx}^2 = 2\sigma_{da}^2 \frac{\sigma_n^2}{\sqrt{\pi} a_0^2} = \frac{2\sigma_{da}^2}{\sqrt{\pi}} \frac{1}{PSNR^2} \tag{10.3.52, a}$$

and

$$\frac{\sigma_{dx}^2}{\sigma_{da}^2} = 2\frac{\sigma_n^2}{\sqrt{\pi} a_0^2} = \frac{2}{\sqrt{\pi}} \frac{1}{PSNR^2}. \tag{10.3.52, b}$$

where

$$PSNR = \frac{a_0}{\sigma_n}. \tag{10.3.53}$$

is the ratio of signal peak value to standard deviation of noise.

It is instructive to express, using Eqs. 10.3.46 and 10.3.50, the variance of normal localization errors in terms of the effective impulse width and efficient bandwidth of its spectrum:

$$\sigma_x^2 = \frac{2S_a^2}{\pi} \frac{N_0}{E_a} = \frac{1}{2\pi (2F_a)^2} \frac{N_0}{E_a} = \frac{S_a}{2\pi F_a} \frac{N_0}{E_a} \tag{10.3.54}$$

Thus, given energy of a Gaussian-shaped impulse, potential variance of its localization normal errors is proportional to the squared effective width of the target impulse and inversely proportional to the squared effective bandwidth of its spectrum.

The relationship between impulse width and inertia moment of its spectrum depends on the impulse shape. For rectangular impulses

$$a(x) = a_0 \,\mathrm{rect}\left(\frac{x + S_a/2}{S_a}\right) \tag{10.3.55}$$

one can numerically find that

$$\overline{f_x^2} \cong 0.265/S_a \tag{10.3.56}$$

and, therefore for rectangular impulses

$$\sigma_x^2 \cong 0{,}095 S_a \frac{N_0}{E_a} = 0.095 \frac{N_0}{a_0^2}. \tag{10.3.57}$$

Similarly, for triangle impulses

$$a(x) = \begin{cases} 2a_0(3S_a/2 + x)/3S_a, 3S_a/2 \le x \le 0 \\ 2a_0(3S_a/2 + x)/3S_a, 0 \le x \le 3S_a/2 \end{cases} \tag{10.3.58}$$

the following relationships hold:

$$2F_a = E_a/|\alpha(0)|^2 = 1/S_a \tag{10.3.59}$$

and

$$\left(\overline{f_x^2}\right)^{1/2} \cong 0.183/S_a, \tag{10.3.60}$$

which is very close to the corresponding relationship (Eq. 10.3.48) for Gaussian-shaped impulses for which $\left(\overline{f_x^2}\right)^{1/2} S_a \cong 0.1995$.

Experimental verification of the performance of the described optimal coordinate estimator is illustrated in Figs. 10-7 and 8. Fig. 10-7 shows experimental distribution density of the localization error obtained by computer simulation of localization of Gaussian-shaped impulse of width

$\sigma_a = \mathbf{10}$ for 100000 realizations of noise with $\sqrt{\boldsymbol{E}_a/\mathrm{N}_0} = \mathbf{1.5}$. One can easily distinguish in the plot the contribution to the error distribution density of normal (central peak of the distribution density) and anomalous (uniform tales of the distribution density) errors.

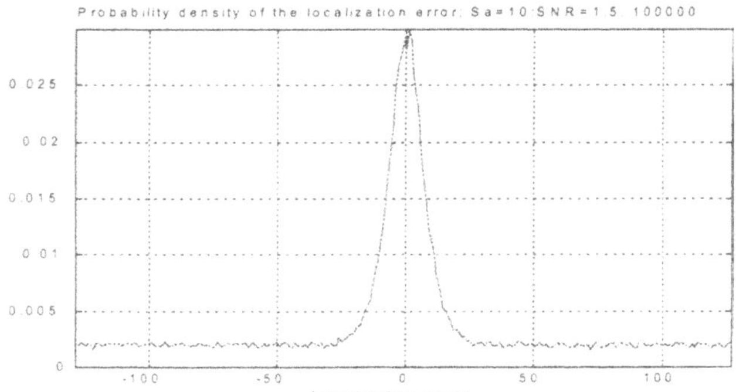

Figure 10-7. Experimental distribution density of the localization error for Gaussian-shaped impulse ($\sigma_a = \mathbf{10}$) and $\sqrt{\boldsymbol{E}_a/\mathrm{N}_0} = \mathbf{1.5}$ obtained for 100000 realizations.

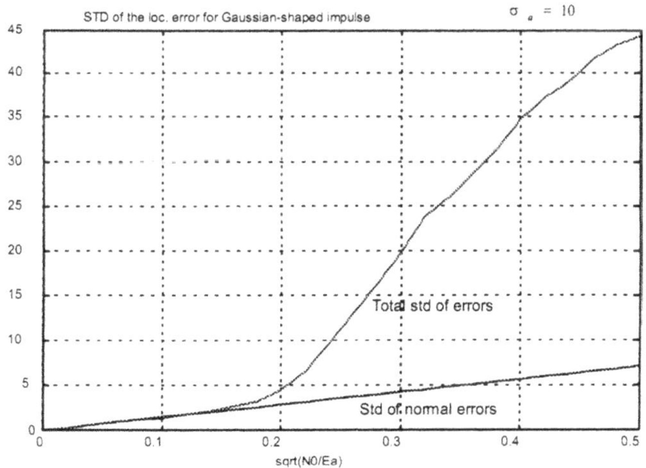

Figure 10-8. Standard deviation of the localization error as a function of noise-to-signal error for Gaussian-shaped impulse with (10000 realizations)

Fig. 10-8 illustrates how standard deviation of the localization error depends on noise-to-signal ratio $\mathrm{N}_0/\boldsymbol{E}_a$. Graphs on the figure are plotted on the base of data obtained by the same computer simulation as that illustrated in Fig. 10-7. Lower curve shows standard deviation of errors smaller then σ_a. As one can see, this curve follows theoretical estimations described by Eqs. 10.3.51 and 52. Upper curve shows standard deviation of the localization error regardless their values. One can see that for low noise-to-signal ratios localization error standard deviation behaves as it is predicted by the theory for normal errors. However, for noise-to-signal ratio larger then about 0.15, error standard deviation starts growing much more rapidly. This behavior is associated with the appearance of anomalous errors. Performance of the optimal localization device in terms of anomalous errors will be discussed in Sect. 10.3.4

10.3.3 Localization accuracy for non-optimal localization devices.

Implementation of the correlator or matched filter assumes exact knowledge of the shape of the target object. In practice, the shape is often not precisely known or the matched filter cannot be precisely implemented. In this section we estimate losses in the localization accuracy owing to the deviation of the real filter in the localization device from the matched filter. We shall analyze these losses for the one-dimensional, or 2-D separable, case.

Let the filter frequency response of the filter in the localization device be $\boldsymbol{H}(f_x)$ instead of $\alpha^*(f_x)$, the matched filter frequency response. The output signal of such a filter will in this case be as follows:

$$\boldsymbol{R}(x) = \int_{-\infty}^{\infty} \alpha(f_x)\boldsymbol{H}(f_x)\exp\{-i2\pi[f_x(x - x_0)]\}df_x + \boldsymbol{R}_{nh}(x), \qquad (10.3.61)$$

where $\boldsymbol{R}_{nh}(x)$ is a result of filtering the white noise component in the observed signal by the filter $\boldsymbol{H}(f_x)$. The location $(x_0 + n_x)$ of the maximum of this signal in the close vicinity of the point x_0 (n_x is small) is defined by the equation:

$$\frac{\partial}{\partial x}\boldsymbol{R}(x)\bigg|_{x=x_0+n_x} =$$
$$(-i2\pi)\int_{-\infty}^{\infty} f_x\alpha(f_x)\boldsymbol{H}(f_x)\exp\{-i2\pi[f_x(x - x_0)]\}df_x\bigg|_{x=x_0+n_x} +$$
$$\frac{\partial}{\partial x}\boldsymbol{R}_{nh}(x)\bigg|_{x=x_0+n_x} = 0 \qquad (10.3.62)$$

As n_x is supposed to be small and, by definition, x_0 is the point of the actual location of the object where the filter response is maximal,

$$\left.(-i2\pi)\int_{-\infty}^{\infty} f_x \alpha^*(f_x)H(f_x)\exp\{-i2\pi[f_x(x-x_0)]\}df_x\right|_{x=x_0+n_x} \propto 4\pi^2 n_x \int_{-\infty}^{\infty} f_x^2 \alpha^*(f_x)H(f_x)df_x \tag{10.3.63}$$

and

$$\int_{-\infty}^{\infty} f_x \alpha^*(f_x)H(f_x)df_x = 0 \tag{10.3.64}$$

Thus, we obtain :

$$\left(4\pi^2 \int_{-\infty}^{\infty} f_x^2 \alpha^*(f_x)H(f_x)df_x\right)n_x = -\left.\frac{\partial}{\partial x}R_{nh}(x)\right|_{x=x_0+n_x}. \tag{10.3.65}$$

Power spectrum of the random process $\left.\frac{\partial}{\partial x}R_{nh}(x)\right|_{x=x_0+n_x}$ is, $4\pi^2 f_x^2 N_0 |H(f_x)|^2$. Then the variance σ_x^2 of a small random error n_x is :

$$\sigma_x^2 = \frac{N_0}{4\pi^2}\frac{\int_{-\infty}^{\infty} f_x^2 |H(f_x)|^2 df_x}{\left(\int_{-\infty}^{\infty} f_x^2 \alpha(f_x)H(f_x)df_x\right)^2}, \tag{10.3.66}$$

or,

$$\sigma_x^2 = \frac{1}{\bar{f}_x^2}\frac{N_0}{4\pi^2}\frac{\int_{-\infty}^{\infty} f_x^2 |H(f_x)|^2 df_x \int_{-\infty}^{\infty} f_x^2 |\alpha(f_x)|^2 df_x}{\left(\int_{-\infty}^{\infty} f_x^2 \alpha(f_x)H(f_x)df_x\right)^2}. \tag{10.3.67}$$

The third factor in this formula is the squared loss factor

$$LSFR = \frac{\left(\int_{-\infty}^{\infty} f_x^2 |H(f_x)|^2 df_x \int_{-\infty}^{\infty} f_x^2 |\alpha(f_x)|^2 df_x \right)^{1/2}}{\int_{-\infty}^{\infty} f_x^2 \alpha(f_x) H(f_x) df_x} \tag{10.3.68}$$

that shows how much the normal error standard deviation for a non-optimal filter exceeds that for the optimal (matched) filter: according to the Schwarz inequality $LSFR \geq 1$. It reaches its minimal value, unity, when the filter is perfectly matched to the object signal, that is when $H(f_x) = \alpha^*(f_x)$.

In order to estimate the order of magnitude of the loss factor we present results of analytical and numerical evaluation of the loss factor for localization of Gaussian-shaped, rectangular and triangle impulses with the help of the filters with, respectively, Gaussian, rectangular and triangle impulse responses of different width.

For Gaussian-shaped impulses, matched

$$a_m(x) = \exp(-\frac{x^2}{2\sigma_m^2}) \tag{10.3.69}$$

and mismatched

$$a_{mm}(x) = \exp(-\frac{x^2}{2\sigma_{mm}^2}) \tag{10.3.70}$$

ones, one can analytically find that

$$LSFR = \left(\frac{1 + \sigma_{mm}^2 / \sigma_m^2}{2\sigma_{mm} / \sigma_m} \right)^{3/2}. \tag{10.3.71}$$

This function is plotted in Fig. 10-9,a under the label “continuous signals”. A curve with the label “discrete signals” in this figure represents results of a numerical evaluation of the loss factor for Gaussian-shaped impulses of 1024 samples with width parameters $\sigma_m = 60$ and $\sigma_{mm} = 11 \div 210$. A certain difference between these curves can be attributed to discretization effects. Fig.10-9, b) represents similar results for localization of a rectangular impulse with $S_a = 21$ samples of 256 with the help of filters with a rectangular impulse response of different width. It is instructive to see that while optimal localization of Gaussian-shaped impulse is relatively tolerant to the inaccuracy of the matched filter width, it is very sensitive to width mismatch between the rectangular impulses.

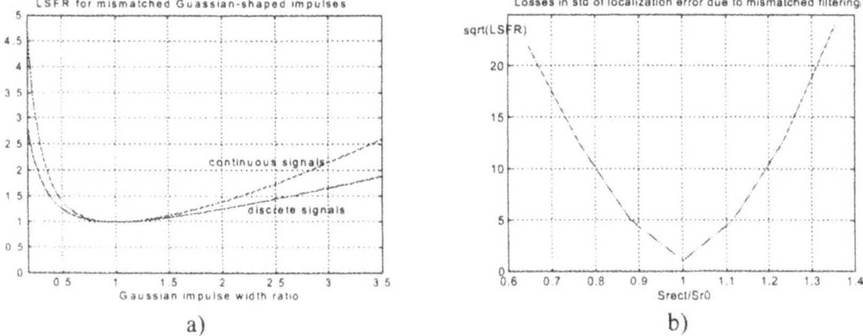

Figure 10-9. Loss factor for mismatched localization of Gaussian-shaped (a) and rectangular impulses

10.3.4 Localization reliability. Probability of anomalous localization errors

In this section we address the phenomenon of anomalous localization errors. We defined anomalous errors as those that occur when the localization device wrongly locates the object at output of the matched filter outside the area occupied by the object signal. Because the noise at the matched filter output results from filtering the white Gaussian noise it is spatially homogeneous. Therefore, large noise outbursts may, with equal probability, be found everywhere in an area outside the target object in is actual location. This implies that the tales of the localization error distribution density are uniform and one can characterize anomalous errors by simply their total probability determined by the area under the distribution density tales. This conclusion is supported by the experimental distribution histogram of the localization error in Fig. 10-4.

Analytical evaluation of the probability of anomalous errors by methods of the theory of random processes requires cumbersome computation and can not be carried out without certain simplifying assumptions ([7]). However its reasonably good estimation may be obtained by the following simple reasoning.

Output of the matched filter outside the area occupied by the object is a correlated Gaussian noise with correlation function defined by Eq. (10.3.4):

$$R_n(x,y) = N_0 R_a(x,y). \tag{10.3.72}$$

Let ΔS be a noise ***correlation interval***, i.e., a minimal distance at which correlation between noise values can be regarded as negligibly small. Because the correlation function of noise at the matched filter output is, according to Eq. 10.3.72, proportional to that of the object signal, ΔS has an order of magnitude of the area occupied by the signal at the matched filter output. Therefore, in the area of the search S there are approximately $Q = S/\Delta S$ uncorrelated samples of Gaussian noise with variance $\sigma_n^2 = R_n(0,0) = N_0 R_a(0,0) = N_0 E_a$.

Let us take one of this samples in the point of actual location of the object. According to Eq. 10.3.1, its value is equal to $R_a(0,0) + R_n = E_a + R_n$, where R_n is a Gaussian zero mean random value with variance σ_n^2. One can then find the probability of anomalous errors P_{ae} as a value complementary to the probability that none of the rest of $(Q-1)$ uncorrelated samples of Gaussian noise outside the object location exceeds this value:

$$P_{ae} = 1 - \frac{1}{\sqrt{2\pi\sigma_n^2}} \int_{-\infty}^{\infty} \exp\left(-\frac{n^2}{2\sigma_n^2}\right) dn \left[\frac{1}{\sqrt{2\pi\sigma_n^2}} \int_{-\infty}^{E_a+R_n} \exp\left(-\frac{t^2}{2\sigma_n^2}\right) dt\right]^{Q-1} =$$

$$\frac{1}{\sqrt{2\pi}} \int_{-\infty}^{\infty} \exp\left(-\frac{n^2}{2}\right) \left\{1 - \left[\Phi\left(\sqrt{\frac{E_a}{N_0}} + n\right)\right]^{Q-1}\right\} dn =$$

$$\frac{1}{\sqrt{2\pi}} \int_{-\infty}^{\infty} \exp\left(-\frac{n^2}{2}\right) \left\{1 - \left[\Phi\left(\frac{E_a}{\sigma_n} + n\right)\right]^{Q-1}\right\} dn, \qquad (10.3.73.)$$

where

$$\Phi(x) = \frac{1}{\sqrt{2\pi}} \int_{-\infty}^{x} \exp\left(-\frac{n^2}{2}\right) dn \qquad (10.3.74)$$

is the "error integral".

Formula (10.3,73) known in communication theory as Kotelnikov's integral ([2]) determines the probability of errors in communication channels with Q orthogonal signal and additive white Gaussian noise ([5,6]). The remarkable feature of this integral is its threshold behavior for large Q. Consider

$$\lim_{Q\to\infty} \left(\ln \left[\Phi\left(\sqrt{\frac{E_a}{N_0}} + n\right)\right]^{Q-1} \right) \qquad (10.3.75)$$

for an auxiliary variable $\mathrm{E}=\sqrt{E_a/N_0 \ln Q}$. By L'Hospital's rule we have:

$$\lim_{Q\to\infty}\left(\ln\left[\Phi\left(\sqrt{\frac{E_a}{N_0}}+n\right)\right]^{Q-1}\right)=\lim_{Q\to\infty}\left[\frac{d}{dQ}\ln\Phi\left(\mathrm{E}\sqrt{\ln Q}\right)\Big/\frac{d}{dQ}\frac{1}{Q}\right]=$$

$$\lim_{Q\to\infty}\frac{-\mathrm{E}}{\Phi\left(\mathrm{E}\sqrt{\ln Q}\right)}Q^{\left(1-\varepsilon^2/2\right)}=\begin{cases}-\infty, & \text{if } \mathrm{E}^2/2>1\\ 0, & \text{if } \mathrm{E}^2/2\le 1\end{cases}. \quad (10.3.76)$$

Therefore,

$$\lim_{Q\to\infty}P_{ae}=\begin{cases}1, & \text{if } E_a/2N_0\le \ln Q\\ 0, & \text{if } E_a/2N_0>\ln Q\end{cases}. \quad (10.3.77)$$

This feature implies that if the area of search is large enough with respect to the size of the object, the probability of anomalous errors may become enormously high when input signal-to-noise ratio is lower than the fundamental threshold defined by Eq. 10.3.77. This also means that when area of search is increased the signal-to-noise ratio must also be increased in order to keep the probability of anomalous errors low. Moreover, for any given intensity of noise, there exists a trade off between localization accuracy defined by the variance of normal errors and localization reliability described by the probability of anomalous errors. Increasing the accuracy achieved by widening the object signal spectrum with signal energy fixed results in increasing the probability of anomalous errors: widening spectrum width is equivalent to narrowing the signal, and, consequently, to increasing the ratio Q of the area of search to the object signal area.

Results of a numerical verification of the above relationships by computer simulation of the optimal localization device are shown in Fig.10-10.

In conclusion it should be noted that the formula (10.3.77) has a general fundamental meaning. Note that $\ln Q$ may be regarded as the entropy, in bits, of the results of measurements of object coordinate in which Q different object positions can be distinguished. Therefore, Eq.(10.3.77) gives the absolute lower bound for the object signal energy per bit of measurement information required for reliable parameter estimation in the presence of white Gaussian noise:

$$E_a/\log_2 Q>2N_0\ln 2. \quad (10.3.78)$$

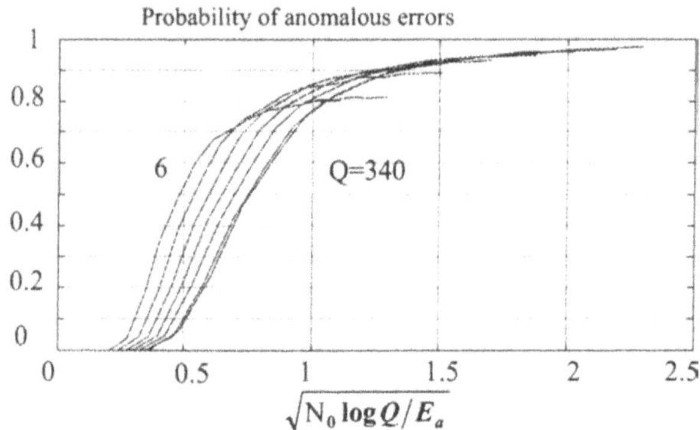

Figure 10-10. Probability of anomalous errors as a function of normalized noise-to-signal ratio $\sqrt{N_0 \ln Q / E_a}$ for localization of rectangular impulses of 2;5;11;21;41;81;161 samples within an interval of 1024 samples (10000 realizations). Note that, according to Eq. (10.3.77), the threshold value of the normalized noise-to-signal ratio is $\sqrt{2}/2 \approx 0.707$

10.4 LOCALIZATION OF AN OBJECT IN THE PRESENCE OF ADDITIVE CORRELATED GAUSSIAN NOISE

10.4.1 Localization of a target object in the presence of non-white (correlated) additive Gaussian noise

In this section we extend the additive signal/noise model of Eq. (10.2.1) to the case of non-white, or correlated, noise. Let the noise power spectrum be $N_0 |H_n(f_x, f_y)|^2$ where $|H_n(f_x, f_y)|^2$ is a normalized spectrum shaping function

$$\int_{-\infty}^{\infty} |H_n(f_x)|^2 df_x = 1 \tag{10.4.1}$$

such that noise variance is the same as that of the white noise with spectral density N_0.

The problem of target localization in the presence of correlated noise may be reduced to the white noise case if we pass the observed signal plus noise mixture through a filter with frequency response

$$H_{wht}(f_x, f_y) = \frac{1}{H_n(f_x, f_y)}. \tag{10.4.2}$$

We will refer to this filter as to ***whitening filter***. At the output of the whitening filter, noise power spectrum becomes uniform with spectral density N_0 while the target object signal spectrum becomes equal to $\alpha(f_x, f_y) / H_n(f_x, f_y)$. Hence, in this case, ML-optimal localization device should consist of the whitening filter followed by a filter matched to the target object signal at the whitening filter output and by the device for localizing the signal maximum.

The whitening and the matched filters may be combined into one filter with a frequency response

$$H_{opt}(f_x, f_y) = \frac{\alpha^*(f_x, f_y)}{|H_n(f_x, f_y)|^2} \tag{10.4.3}$$

We will refer to this filter as to the ***optimal filter***.

In this way we arrive at the optimal localization device shown in Fig.10-11 that provides maximum likelihood (ML) estimation of the target object coordinates.

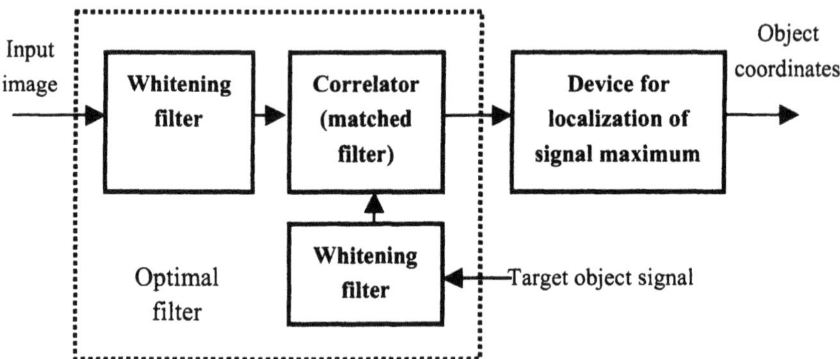

Figure 10-11. Schematic diagram of the optimal device for localizing objects in the presence of additive correlated Gaussian noise.

From the above reasoning it follows that, for a target signal observed in a mixture with additive correlated Gaussian noise, the optimal filter generates a signal which is proportional to a posteriori probability distribution function of target coordinates. Now we will prove that the optimal filter has yet another important feature. It also provides the highest possible for all linear filters ratio of its response to the target object signal to standard deviation of noise at its output (signal-to-noise ratio). This feature enables designing optimal localization devices with linear filtering in case of non-Gaussian signal observation model. We will use this feature in Ch. 11 for the design of optimal devices for target location in clutter.

Consider response of an arbitrary filter with a frequency response $H(f_x, f_y)$ to the target signal $a(x - x_0, y - y_0)$ and noise $n(x)$ with spectral density $N_0|H_n(f_x, f_y)|^2$. Filter output to the target signal at the point (x_0, y_0) is equal to

$$b_0 = \int_{-\infty}^{\infty}\int_{-\infty}^{\infty} \alpha(f_x, f_y) H(f_x, f_y) \exp\{i2\pi [f_x(x - x_0) + f_y(y - y_0)]\} df_x df_y \Big|_{\substack{x=x_0 \\ y=y_0}} =$$

$$= \int_{-\infty}^{\infty}\int_{-\infty}^{\infty} \alpha(f_x, f_y) H(f_x, f_y) df_x df_y . \qquad (10.4.4)$$

Standard deviation of noise at the filter output noise may be found as:

$$\sigma_n = \left(N_0 \int_{-\infty}^{\infty}\int_{-\infty}^{\infty} |H(f_x,f_y)|^2 |H_n(f_x,f_y)|^2 df_x df_y \right)^{1/2}. \tag{10.4.5}$$

Their ratio, called signal-to-noise ratio, is then equal:

$$SNR = \frac{b_0}{\sigma_n} = \frac{\int_{-\infty}^{\infty}\int_{-\infty}^{\infty} \alpha(f_x,f_y) H(f_x,f_y) df_x df_y}{\left(N_0 \int_{-\infty}^{\infty}\int_{-\infty}^{\infty} |H(f_x,f_y)|^2 |H_n(f_x,f_y)|^2 df_x df_y \right)^{1/2}}. \tag{10.4.6}$$

It is follows from Schwarz's inequality ([8]) that

$$\int_{-\infty}^{\infty}\int_{-\infty}^{\infty} \alpha(f_x,f_y) H(f_x,f_y) df_x df_y =$$
$$\int_{-\infty}^{\infty}\int_{-\infty}^{\infty} \frac{\alpha(f_x,f_y)}{|H_n(f_x,f_y)|} \left[H(f_x,f_y) |H_n(f_x,f_y)| \right] df_x df_y \le$$
$$\le \left(\int_{-\infty}^{\infty}\int_{-\infty}^{\infty} \frac{|\alpha(f_x,f_y)|^2}{|H_n(f_x,f_y)|^2} df_x df_y \right)^{1/2} \left(\int_{-\infty}^{\infty}\int_{-\infty}^{\infty} |H(f_x,f_y)|^2 |H_n(f_x,f_y)|^2 df_x df_y \right)^{1/2} \tag{10.4.7}$$

with equality taking place for the optimal filter of Eq.10.4.3. Therefore

$$H_{opt}(f_x,f_y) = \frac{\alpha^*(f_x,f_y)}{|H_n(f_x,f_y)|^2} = \arg\max_{H(f_x,f_y)} (SNR). \tag{10.4.8}$$

10.4.2 Localization accuracy of the optimal filter

Because the optimal filter is reduced to the matched filter by the whitening filter operation, the potential localization accuracy of the optimal filter may be found from Eqs. 10.3.36 a-c for the localization accuracy in the presence white noise appropriately modified to account for the whiting filter:

$$\sigma^2_{x,NW} = \frac{\bar{f}^2_{y,NW}}{\bar{f}^2_{y,NW}\bar{f}^2_{y,NW} - \left(\bar{f}^2_{xy,NW}\right)^2} \frac{N_0}{4\pi^2 E_{a,NW}}; \tag{10.4.9, a}$$

$$\sigma^2_{y,NW} = \frac{\bar{f}^2_{x,NW}}{\bar{f}^2_{y,NW}\bar{f}^2_{y,NW} - \left(\bar{f}^2_{xy,NW}\right)^2} \frac{N_0}{4\pi^2 E_{a,NW}}; \qquad (10.4.9, b)$$

$$\sigma^2_{xy,NW} = \frac{\bar{f}^2_{xy,NW}}{\bar{f}^2_{y,NW}\bar{f}^2_{y,NW} - \left(\bar{f}^2_{xy,NW}\right)^2} \frac{N_0}{4\pi^2 E_{a,NW}}. \qquad (10.4.9, c)$$

where:

$$E_{a,NW} = \int_{-\infty}^{\infty}\int_{-\infty}^{\infty} \frac{\left|\alpha\left(f_x, f_y\right)\right|^2}{\left|H_n\left(f_x, f_y\right)\right|^2} df_x df_y; \qquad (10.4.10)$$

$$\bar{f}^2_{x,NW} = \frac{\int_{-\infty}^{\infty}\int_{-\infty}^{\infty} f_x^2 \frac{\left|\alpha\left(f_x, f_y\right)\right|^2}{\left|H_n\left(f_x, f_y\right)\right|^2} df_x df_y}{\int_{-\infty}^{\infty}\int_{-\infty}^{\infty} \frac{\left|\alpha\left(f_x, f_y\right)\right|^2}{\left|H_n\left(f_x, f_y\right)\right|^2} df_x df_y}; \qquad (10.4.11)$$

$$\bar{f}^2_{y,NW} = \frac{\int_{-\infty}^{\infty}\int_{-\infty}^{\infty} f_y^2 \frac{\left|\alpha\left(f_x, f_y\right)\right|^2}{\left|H_n\left(f_x, f_y\right)\right|^2} df_x df_y}{\int_{-\infty}^{\infty}\int_{-\infty}^{\infty} \frac{\left|\alpha\left(f_x, f_y\right)\right|^2}{\left|H_n\left(f_x, f_y\right)\right|^2} df_x df_y}; \qquad (10.4.12)$$

$$\bar{f}^2_{xy,NW} = \frac{\int_{-\infty}^{\infty}\int_{-\infty}^{\infty} f_x f_y \frac{\left|\alpha\left(f_x, f_y\right)\right|^2}{\left|H_n\left(f_x, f_y\right)\right|^2} df_x df_y}{\int_{-\infty}^{\infty}\int_{-\infty}^{\infty} \frac{\left|\alpha\left(f_x, f_y\right)\right|^2}{\left|H_n\left(f_x, f_y\right)\right|^2} df_x df_y}. \qquad (10.4.13)$$

It is very instructive to compare the potential localization accuracy in the cases of white and non-white noise. Naturally, the comparison must be done under certain normalization conditions. Let us compare the accuracy for the same input noise variance. For the sake of simplicity we shall consider the one-dimensional case. In this case we have from Eq.(10.4.9a) for white noise that

$$\sigma^2_{x,NW} = \frac{1}{\bar{f}^2_{x,NW}} \frac{N_0}{4\pi^2 E_{a,NW}} . \tag{10.4.14}$$

Let us give this formula the form, which explicitly contains the error variance σ^2_x for the case of white noise (Eqs. 10.3.37, a):

$$\sigma^2_{x,NW} = \frac{1}{\bar{f}^2_{x,NW}} \frac{N_0}{4\pi^2 E_{a,NW}} = \sigma^2_x \frac{E_a \bar{f}^2_x}{E_{a,NW} \bar{f}^2_{x,NW}} . \tag{10.4.15}$$

The last factor in this formula shows the change of error variance owing to non-whiteness of noise. Because all functions involved in this factor are nonnegative, the following inequality takes place:

$$G = \frac{E_a}{E_{a,NW}} \frac{\bar{f}^2_x}{\bar{f}^2_{x,NW}} = \frac{\int_{-\infty}^{\infty} f_x^2 |\alpha(f_x)|^2 df_x}{\int_{-\infty}^{\infty} f_x^2 \frac{|\alpha(f_x)|^2}{|H_n(f_x)|^2} df_x} \le$$

$$\frac{\int_{-\infty}^{\infty} f_x^2 \frac{|\alpha(f_x)|^2}{|H_n(f_x)|^2} df_x \int_{-\infty}^{\infty} |H_n(f_x)|^2 df_x}{\int_{-\infty}^{\infty} f_x^2 \frac{|\alpha(f_x)|^2}{|H_n(f_x)|^2} df_x} = \int_{-\infty}^{\infty} |H_n(f_x)|^2 df_x = 1 \tag{10.4.16}$$

Therefore, the potential localization accuracy in the presence of non-white noise is always better then that for white noise with the same variance. Note that in the trivial special case when noise is band limited and its bandwidth is less that that of the signal, optimal filter is a band pass filter which lets through all input signal frequencies where the noise spectrum vanishes to zero. The resulting signal is therefore noise free and the potential localization error variance is equal to zero.

In order to make a numerical evaluation of the decrease of the localization error variance for non-white noise, consider a one-dimensional Gaussian-shaped signal

$$a(x) = a_0 \exp(-\frac{1}{2\sigma_a^2}). \tag{10.4.17}$$

observed in a mixture with Gaussian noise with power spectrum:

$$|H_n(f_x)|^2 = 2F_n \exp(-\frac{|f_x|}{F_n}). \tag{10.4.18}$$

In this case optimal filter frequency response is

$$H_{opt}(f_x) \propto \exp\left[-2\pi^2\sigma_a^2\left(|f_x|^2 - \frac{|f_x|}{2\pi^2\sigma_a^2 F_n}\right)\right] \propto$$

$$\exp\left[-2\pi^2\sigma_a^2\left(|f_x| - \frac{1}{4\pi^2\sigma_a^2 F_n}\right)^2\right] = \exp\left[-2\pi\left(S_a|f_x| - \frac{1}{4\pi S_a F_n}\right)^2\right], \tag{10.4.19}$$

where $S_a = \sqrt{\pi}\sigma_a$ is the impulse efficient width (Eq. 10.3.46). One can see from Eq. 10.4.19 that optimal filter is in this case a sort of a band-pass filter with central frequency $1/4\pi S_a^2 F_n$. A family of its frequency responses is illustrated in Fig. 10-12.

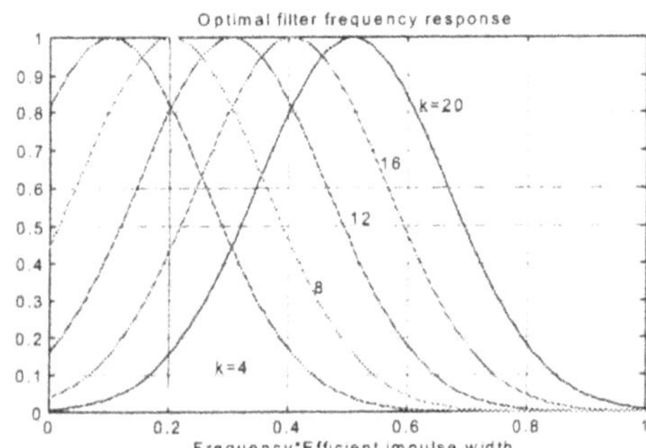

Figure 10-12. Frequency responses of the optimal filter for Gaussian-shaped impulse $a(x) = \exp(-\pi x^2 / 2S_a^2)$ and non-white noise with exponentially decaying power spectrum $|N_n(f)|^2 = \exp(-|f|/F_n), k = S_a F_n$;

10.5 OPTIMAL LOCALIZATION IN COLOR AND MULTI COMPONENT IMAGES

10.5.1 Optimal localization device

The results presented above for optimal localization in monochrome images may be extended to localization in color, or, generally, multi-component images, observed in the presence of additive Gaussian noise in each component.

Let $\{a_{k,m}(x_0, y_0)\}$, $k = 0,1,...,N-1$, be m-th component of the target object signal, M be the number of components $m = 1,2,...,M$ (for color images M=3). Let also $\{n_{k,m}\}$ and $\{b_{k,m}\}$ be the samples of the corresponding components of the additive Gaussian noise and of the observed signal, such that

$$b_{k,m} = a_{k,m}(x_0, y_0) + n_{k,m}. \tag{10.5.1}$$

Let also assume that components of the additive noise as well as samples of noise within each component are all mutually uncorrelated. This assumption directly implies, that probability of observing signal $\{b_{k,m}\}$ provided target object is located in coordinates (x_0, y_0) is

$$P(\{n_{k,m} = b_{k,m} - a_{k,m}(x_0, y_0)\}) \propto \prod_{m=1}^{M}\prod_{k=0}^{N-1} \exp\left\{-\frac{1}{2\sigma_{n,m}^2}\left[b_{k,m} - a_{k,m}(x_0, y_0)\right]^2\right\}, \tag{10.5.2}$$

where $\sigma_{n,m}^2$ denotes the variance of the m-th noise component. Therefore, optimal MAP- and ML-localization devices are defined, respectively, by equations

$$\{(\hat{x}_0, \hat{y}_0)\} = \underset{(x_0, y_0)}{\arg\max}\left\{\sum_{m=1}^{M}\sum_{k=0}^{N-1} b_{k,m}a_{k,m}(x_0, y_0) - \sum_{m=1}^{M}\sigma_{n,m}^2 \ln P(x_0, y_0)\right\} \tag{10.5.3}$$

and

$$\{(\hat{x}_0, \hat{y}_0)\} = \underset{(x_0, y_0)}{\arg\max} \left\{ \sum_{m=1}^{M} \sum_{k=0}^{N-1} b_{k,m} a_{k,m}(x_0, y_0) \right\}, \tag{10.5.4}$$

For continuous signals, MAP-estimation is

$$\{(\hat{x}_0, \hat{y}_0)\} = \underset{(x_0, y_0)}{\arg\max} \left\{ \sum_{m=1}^{M} \frac{1}{\mathrm{N}_{0,m}} \int_{-\infty}^{\infty}\int_{-\infty}^{\infty} b_m(x, y) a_m(x - x_0, y - y_0) dx dy - \ln P(x_0, y_0) \right\} \tag{10.5.5}$$

where $\mathrm{N}_{0,m}$ is the spectral density of the m-th noise component, and ML-estimation is

$$\{(\hat{x}_0, \hat{y}_0)\} = \underset{(x_0, y_0)}{\arg\max} \left\{ \sum_{m=1}^{M} \frac{1}{\mathrm{N}_{0,m}} \int_{-\infty}^{\infty}\int_{-\infty}^{\infty} b_m(x, y) a_m(x - x_0, y - y_0) dx dy \right\} \tag{10.5.6}$$

Therefore optimal localization device should consist of M parallel component wise correlators, or matched filters, of an adder for a weighted summation correlator outputs and of a unit for determining coordinates of the signal maximum at the adder's output. Its schematic diagram is shown in Fig. 10-13.

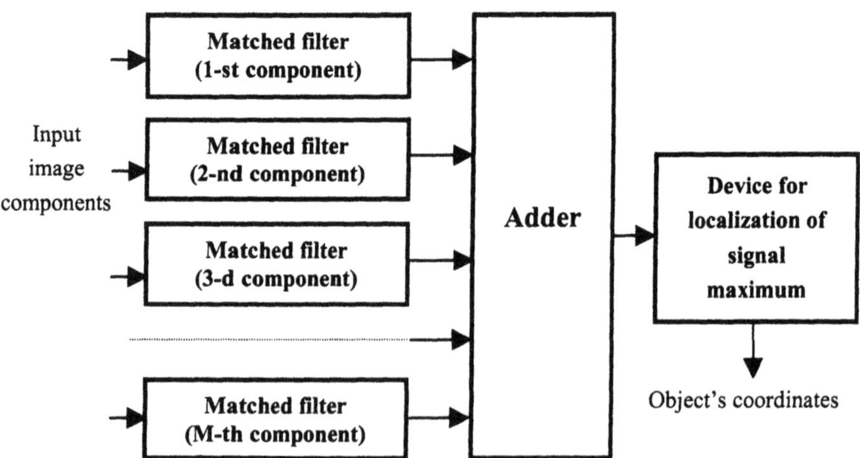

Figure 10-13. Schematic diagram of the optimal device for localization of a target object in multi component images

10.5.2 Localization accuracy and reliability

Variances of normal localization errors, that characterize the accuracy of optimal localization, may be found with just the same technique as it was done for single-component images in Sect. 10.3.1. In this way one can obtain ([7]):

$$\sigma^2_{x,M} = \bar{f}^2_{y,a}\frac{\bar{f}^2_{x,a}\bar{f}^2_{y,a}\gamma_x + (\gamma_y - 2\gamma_{xy})\bar{f}^2_{xy,a}}{\left(\bar{f}^2_{x,a}\bar{f}^2_{y,a} - \left(\bar{f}^2_{xy,a}\right)^2\right)^2}\frac{\sigma^2_{n,M}}{4\pi^2 E_{a,M}}; \tag{10.5.7}$$

$$\sigma^2_{y,M} = \bar{f}^2_{x,a}\frac{\bar{f}^2_{x,a}\bar{f}^2_{y,a}\gamma_y + (\gamma_x - 2\gamma_{xy})\bar{f}^2_{xy,a}}{\left(\bar{f}^2_{x,a}\bar{f}^2_{y,a} - \left(\bar{f}^2_{xy,a}\right)^2\right)^2}\frac{\sigma^2_{n,M}}{4\pi^2 E_{a,M}}; \tag{10.5.8}$$

$$\sigma^2_{xy,M} = \bar{f}^2_{xy,a}\frac{\bar{f}^2_{x,a}\bar{f}^2_{y,a}(\gamma_{xy} - \gamma_x - \gamma_y) + \gamma_{xy}\left(\bar{f}^2_{xy,M}\right)^2}{\left(\bar{f}^2_{x,a}\bar{f}^2_{y,a} - \left(\bar{f}^2_{xy,a}\right)^2\right)^2}\bar{f}^2_{xy,M}\frac{\sigma^2_{n,M}}{4\pi^2 E_{a,M}}; \tag{10.5.9}$$

where

$$\gamma_x = \bar{f}^2_{x,n} / \bar{f}^2_{x,a};\ \gamma_y = \bar{f}^2_{y,n} / \bar{f}^2_{y,a};\ \gamma_{xy} = \bar{f}^2_{xy,n} / \bar{f}^2_{xy,a}; \tag{10.5.10}$$

$$\bar{f}^2_{x,a} = \frac{\sum_{m=1}^{M}\frac{1}{N^2_{0,m}}\int_{-\infty}^{\infty}\int_{-\infty}^{\infty} f_x^2 \left|\alpha(f_x, f_y)\right|^2 df_x df_y}{\sum_{m=1}^{M}\frac{1}{N^2_{0,m}}\int_{-\infty}^{\infty}\int_{-\infty}^{\infty} \left|\alpha(f_x, f_y)\right|^2 df_x df_y}; \tag{10.5.11}$$

$$\bar{f}^2_{y,a} = \frac{\sum_{m=1}^{M}\frac{1}{N^2_{0,m}}\int_{-\infty}^{\infty}\int_{-\infty}^{\infty} f_y^2 \left|\alpha(f_x, f_y)\right|^2 df_x df_y}{\sum_{m=1}^{M}\frac{1}{N^2_{0,m}}\int_{-\infty}^{\infty}\int_{-\infty}^{\infty} \left|\alpha(f_x, f_y)\right|^2 df_x df_y}; \tag{10.5.12}$$

$$\bar{f}^2_{xy,a} = \frac{\sum_{m=1}^{M}\frac{1}{N^2_{0,m}}\int_{-\infty}^{\infty}\int_{-\infty}^{\infty} f_x f_y \left|\alpha(f_x, f_y)\right|^2 df_x df_y}{\sum_{m=1}^{M}\frac{1}{N^2_{0,m}}\int_{-\infty}^{\infty}\int_{-\infty}^{\infty} \left|\alpha(f_x, f_y)\right|^2 df_x df_y}; \tag{10.5.13}$$

$$\bar{f}^2_{x,a} = \frac{\sum_{m=1}^{M} \frac{1}{N_{0,m}} \int_{-\infty}^{\infty}\int_{-\infty}^{\infty} f_x^2 \left|\alpha(f_x, f_y)\right|^2 df_x df_y}{\sum_{m=1}^{M} \frac{1}{N_{0,m}} \int_{-\infty}^{\infty}\int_{-\infty}^{\infty} \left|\alpha(f_x, f_y)\right|^2 df_x df_y}; \tag{1.6.14}$$

$$\bar{f}^2_{y,a} = \frac{\sum_{m=1}^{M} \frac{1}{N_{0,m}} \int_{-\infty}^{\infty}\int_{-\infty}^{\infty} f_y^2 \left|\alpha(f_x, f_y)\right|^2 df_x df_y}{\sum_{m=1}^{M} \frac{1}{N_{0,m}} \int_{-\infty}^{\infty}\int_{-\infty}^{\infty} \left|\alpha(f_x, f_y)\right|^2 df_x df_y}; \tag{10.5.15}$$

$$\bar{f}^2_{xy,a} = \frac{\sum_{m=1}^{M} \frac{1}{N_{0,m}} \int_{-\infty}^{\infty}\int_{-\infty}^{\infty} f_x f_y \left|\alpha(f_x, f_y)\right|^2 df_x df_y}{\sum_{m=1}^{M} \frac{1}{N_{0,m}} \int_{-\infty}^{\infty}\int_{-\infty}^{\infty} \left|\alpha(f_x, f_y)\right|^2 df_x df_y} \tag{10.5.16}$$

where

$$E_{a,M} = \sum_{m=1}^{M} \frac{1}{N^2_{0,m}} \int_{-\infty}^{\infty}\int_{-\infty}^{\infty} \left|\alpha(f_x, f_y)\right|^2 df_x df_y; \tag{10.5.17}$$

$$\sigma^2_{n,M} = \sum_{m=1}^{M} N_{0,m} \int_{-\infty}^{\infty}\int_{-\infty}^{\infty} \left|\alpha(f_x, f_y)\right|^2 df_x df_y. \tag{10.5.18}$$

In the case of the signal symmetry, when $\bar{f}^2_{xy,M} = 0$,

$$\sigma^2_{x,M} = \frac{\gamma_x}{\bar{f}^2_{x,M}} \frac{\sigma^2_{n,M}}{4\pi^2 E^2_{a,M}}; \quad \sigma^2_{y,M} = \frac{\gamma_y}{\bar{f}^2_{y,M}} \frac{\sigma^2_{n,M}}{4\pi^2 E^2_{a,M}}; \quad \sigma^2_{xy,M} = 0. \tag{10.5.19}$$

As for the probability of anomalous errors, it can be found from Eq. (10.3.73) with substitution of the involved values E_a and σ_n by $E_{a,M}$ and $\sigma_{n,M}$, respectively.

10.5.3 Optimal localization in multi component images with correlated noise.

As in the case of single component images, for localization in multi component images, when noise components are correlated, input image and target object image pre-whitening may be used in order to reduce the problem to the previous one. The optimal localization device of Fig. 10-3 should then be modified as it is shown in Fig. 10-14.

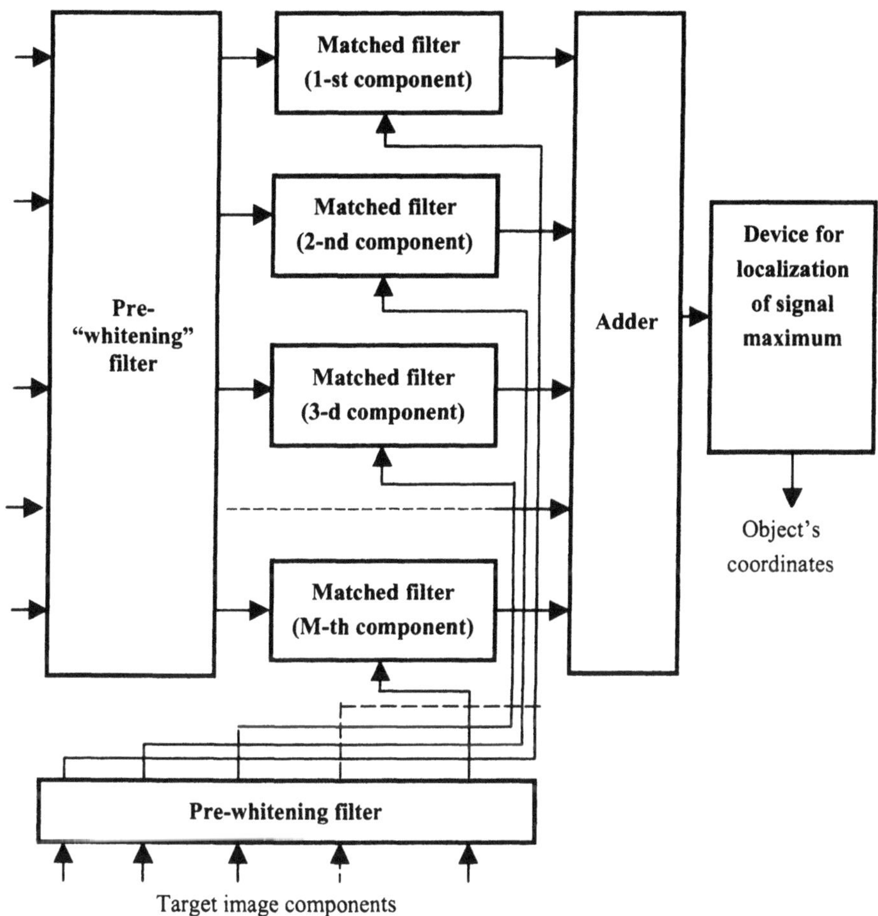

Figure 10-14. Schematic diagram of the optimal device for localizing objects in multi-component images with additive correlated noise in the channels.

Frequency response of the pre-whitening filter is defined in the 3-D frequency domain as

$$H_{pw}(\sigma, f_x, f_y) = \frac{1}{\left|N_0(\sigma, f_x, f_y)\right|^{1/2}}, \qquad (10.5.20)$$

where $\{f_x, f_y\}$ are spatial frequencies, σ is an index of component-wise Fourier Transform (for multi-component images, the latter is the component wise Discrete Fourier Transform) and $N(\sigma, f_x, f_y)$ is a spectral density of noise defined in the 3-D frequency domain.

10.6 OBJECT LOCALIZATION IN THE PRESENCE OF MULTIPLE NONOVERLAPPNING NON-TARGET OBJECTS

In this section, we discuss an image model in which observed image signal $b(x,y)$ contains, along with the target object signal $a(x-x_0,y-y_0)$ in coordinates (x_0,y_0) and additive white Gaussian sensor noise $n(x)$, some number Q_e of non-target objects $\{a_q(x-x_q,y-y_q)\}$ with coordinates $\{(x_q,y_q)\}$:

$$b(x,y)=a(x-x_0,y-y_0)+\sum_{q=1}^{Q_e}a_q(x-x_q,y-y_q)+n(x,y). \qquad (10.6.1)$$

Suppose also that the non-target objects do not overlap one another and the target object. The most immediate illustration of such a situation would be the task of locating a specific character in a a printed text.

As non-target objects do not overlap the target object under, it is clear that the design of the filter ensuring the highest localization accuracy, i.e. the lowest variance of normal errors, is governed by the same reasonings as those for above described additive noise model with no non-target objects. Therefore optimal filter in this case will also be the matched filter and the variance of normal errors will be defined by the same formulas as those in the cases of no non-target objects (Eqs. 10.3.34 and 10.3.36). Presence of non-target objects has an effect only on the probability of anomalous errors.

In order to estimate the probability of anomalous errors in target localization in the presence of multiple non-target objects let us introduce some quantitative characteristics of the non-target objects. Let

- $P(Q_e)$ be the probability of appearance of exactly Q_e non-target objects in the area of search.
- Non-target objects form some C classes, such as, for instance, classes of characters in optical character recognition, according to the maximal values $R_{0,q}^{(c)}$ of their cross correlation function with the target object

$$R_{0,q}^{(c)}=\max\left(R_{a,q}^{(c)}(x,y)=\int_{-\infty}^{\infty}\int_{-\infty}^{\infty}a(x-\xi,y-\eta)a_q^{(c)}(\xi,\eta)d\xi d\eta\right), \qquad (10.6.2)$$

$c=1,2,...,C$

- $P(Q_{e,1}, Q_{e,2}, \ldots Q_{e,C})$ be the joint probability of appearance of $Q_{e,1}$ objects of the first class, $Q_{e,2}$ objects of the second class and so on, given the total number of false objects is Q_e.

Obviously, in the presence of non-target objects anomalous errors will occur mainly as a result of false localization of the target object at the location of one of the non-target objects. Signal values at these are equal to maxima of the matched filter responses to the non-target objects, i.e., to cross-correlations (Eq. 10.6.2) between the target object and the non-target objects, plus correlated noise resulting from the additive noise at the matched filter input. Then the probability of anomalous errors may be evaluated as being the complement to the probability that all signal values at the output of the matched filter at locations of non-target objects do not exceed the signal value at the location of the actual location of the target object. As non-target objects do not overlap one another, the values of additive noise at the points of location of non-target objects may be regarded uncorrelated. Therefore, one can, by analogy with Eq.(10.3.73), evaluate probability of anomalous localization errors (probability of false detection) as

$$P_a = 1 - \frac{1}{\sqrt{2\pi}} \int_{-\infty}^{\infty} \exp\left(-\frac{n^2}{2}\right) dn \left\{ \sum_{Q_e} \sum_{Q_{e,1}, \ldots Q_{e,C}} P(Q_{e,1}, Q_{e,2}, \ldots Q_{e,C}) \prod_{c=1}^{C} \left[\Phi\left(\frac{E_a - R_{0,q}^{(c)}}{\sqrt{E_a N_0}} + n \right) \right]^{Q_{e,c}} \right\} \tag{10.6.3}$$

where as before

$$\Phi(x) = \frac{1}{\sqrt{2\pi}} \int_{-\infty}^{x} \exp\left(-\frac{n^2}{2}\right) dn \tag{10.6.4}$$

It is convenient to represent Eq.10.6.3 through the appearance probabilities $\{P_e(c)\}$ of non-target objects of the class c making use the fact that the probability distribution of the appearances of exactly $\{Q_{e,1}, Q_{e,2}, \ldots Q_{e,C}\}$ of each class non-target objects from Q_e is polynomial and is connected with the probabilities $\{P_e(c)\}$ by the following equation:

$$P(Q_{e,1}, Q_{e,2}, \ldots Q_{e,C}) = Q_e! \prod_{c=1}^{C} \frac{[P_e(c)]^{Q_{e,c}}}{Q_{e,c}!}. \tag{10.6.5}$$

After substitution of Eq. 10.6.5 into Eq. 10.6.4 and with the use of the identity

$$\left(\sum_{c=1}^{C} P_e(c)\right)^{Q_e} = \sum_{Q_{e,c}} Q_e! \prod_{c=1}^{C} \frac{[P_e(c)]^{Q_{e,c}}}{Q_{e,c}!} \tag{10.6.6}$$

we obtain:

$$P_a = \sum_{Q_e} P(Q_e) \left\{ \frac{1}{\sqrt{2\pi}} \int_{-\infty}^{\infty} \exp\left(-\frac{n^2}{2}\right) \left\{ 1 - \left[\sum_{c=1}^{C} P_e(c) \Phi\left(\frac{E_a - R_{0,q}^{(c)}}{\sqrt{E_a N_0}} + n \right) \right]^{Q_e} \right\} dn \right\}. \tag{10.6.7}$$

Thus, for estimating the probability of anomalous errors, it is necessary to know noise spectral density N_0, (or, for signals in discrete representation, noise variance) at the input of the matched filter, target object signal energy E_a, differences $\left\{E_a - R_{0,q}^{(c)}\right\}$ between auto and cross-correlation peaks, probabilities $\{P_e(c)\}$ of non-target objects of different classes, and probability $P(Q_e)$ of observing in the area of search exactly Q_e non-target objects.

One can obtain more tractable formulas and practical recommendations by considering the following special cases.

1. Let probabilities $P(Q_e)$ are concentrated within a relatively narrow range $\left[Q_{e,\min}, Q_{e,\max}\right]$ of values of Q_e: $\left(Q_{e,\max} - Q_{e,\min}\right)/Q_{e,\min} << 1$. For instance, in case of optical character recognition, this assumption is equivalent to the assumption that variations of characters that belong to the same class are small. Then for $Q_{e,\min} \le Q_0 \le Q_{e,\max}$ we obtain an estimate:

$$\left(1 - \frac{Q_{e,\max} - Q_{e,\min}}{Q_{e,\min}}\right) P_a(Q_{e,0}) \le P_a \le \left(1 + \frac{Q_{e,\max} - Q_{e,\min}}{Q_{e,\min}}\right) P_a(Q_{e,0}), \tag{10.6.8}$$

where

$$P_a(Q_{e,0}) = \frac{1}{\sqrt{2\pi}} \int_{-\infty}^{\infty} \exp\left(-\frac{n^2}{2}\right) \left\{ 1 - \left[\sum_{c=1}^{C} P_e(c) \Phi\left(\frac{E_a - R_{0,q}^{(c)}}{\sqrt{E_a N_0}} + n \right) \right]^{Q_{e,0}} \right\} dn. \tag{10.6.9}$$

Therefore a good estimate of the probability P_a of anomalous errors in this case is the value $P_a(Q_{e,0})$ at some intermediate point $Q_{e,0}$.

2. The probability distribution $P(Q_e)$ is concentrated around some point $Q_{e,0}$ and there is only one class of the non-target objects ($C = 1$). In this case

$$P_a(Q_{e,0}) = \frac{1}{\sqrt{2\pi}} \int_{-\infty}^{\infty} \exp\left(-\frac{n^2}{2}\right)\left\{1 - \left[\Phi\left(\frac{E_a - R_{0,q}^{(c)}}{\sqrt{E_a N_0}} + n\right)\right]^{Q_{e,0}}\right\} dn. \quad (10.6.10)$$

This expression is similar to Eq. 10.3.73 for the probability of anomalous errors in the absence of non-target objects. The only difference is that now signal-to-noise ratio is substantially reduced by the cross-correlation peak of the non-target and target objects.

Obviously, the probability of anomalous error also features threshold behavior:

$$\lim_{Q\to\infty} P_{ae} = \begin{cases} 1, & if \dfrac{E_a - R_{0,q}^{(c)}}{\sqrt{E_a N_0}} \le \sqrt{2\ln Q} \\ 0, & if \dfrac{E_a - R_{0,q}^{(c)}}{\sqrt{E_a N_0}} > \sqrt{2\ln Q} \end{cases} \quad (10.6.11)$$

For instance, for Q = 16 to 2048 and

$$\frac{E_a - R_{0,q}^{(c)}}{\sqrt{E_a N_0}} \cong (1.1 - 1.6)\sqrt{\ln Q_0} \cong 1.8 - 4.4. \quad (10.6.12)$$

$P_{a,e} < 10^{-2}$ and it increases rapidly with decreasing signal-to-noise ratio $\frac{E_a - R_{0,q}^{(c)}}{\sqrt{E_a N_0}}$. Therefore the probability of anomalous errors in the presence of multiple false objects may be very high. Note, for example, that the number of characters on a standard printed page is about 1600. Fig. 10-15 illustrates the phenomenon of false detection on an example of character recognition.

In order to decrease $P_{a,e}$, it is necessary to increase signal- to-noise ratio by suppressing cross-correlation peaks for non-target objects with respect to the auto-correlation peak of the target object. This requires an appropriate modification of the matched filter. Therefore, for localization of a target in

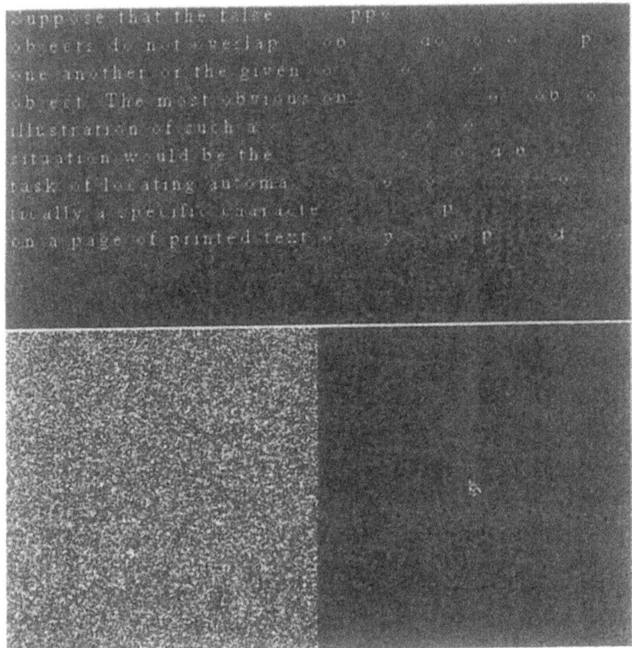

Figure 10-15. Detection, by means of matched filtering, of a character with and without non-target characters in the area of search. Upper row, left image: noisy image of a printed text with standard deviation of additive noise 15 (within signal range 0-255). Upper row, right image: results of detection of character "o" (right); one can see quite a number of false detections. Bottom row, left: a noisy image with a single character "o" in its center; standard deviation of the additive noise is 175. Highlighted center of the right image shows that the character is correctly localized by the matched filter.

the presence of non-target objects, it is not possible to simultaneously secure the minimum of normal error variance and the minimum of the probability of anomalous errors with the same estimator and optimal localization must be carried out into two steps. The first step is target detection with minimal probability of anomalous errors. A coarse but reliable estimate of the object coordinates to the accuracy of the object size is obtained at this stage. For the accurate estimation of target coordinates, the second step is needed in which localization with minimal variance of normal errors is obtained in the area of search in the vicinity of the location found at the first step. The optimal localization device for the second step is the matched filter correlator. The problem of the design of optimal device for reliable detection of targets on the background of a clutter of non-target objects is addressed in Ch. 11.

References

1. A.Van der Lugt, Signal Detection by Complex Spatial Filtering, IEEE Trans., IT-10, 1964, No. 2, p. 139
2. L.P. Yaroslavsky, M.G. Ilieva, V.N. Karnaukhov, I.D. Nikolov, G.I. Sokolinov, A Device for Pattern Recognition, Certificate of Authorship No 1414170, of 24.4.1986, Reg. 01.04.1988
3. C. S. Weaver and J. W. Goodman, " A technique for optically convoluting two functions," Applied Optics, **5**, pp. 1248-1249 (1966).
4. F. T. S. Yu and X. J. Lu, "A real-time programmable Joint Transform Correlator", Opt. Commun., Vol. 52, 1984, p. 47.
5. V. Kotelnikov, The Theory of Potential Noise Immunity, Moscow, Energia, 1956 (In Russian)
6. J. M. Wozencraft, I. M. Jacobs, Principles of Communication Engineering, N.Y.,Wiley, 1965
7. L.P. Yaroslavsky, The Theory of Optimal Methods for Localization of Objects in Pictures, In: Progress in Optics, Ed. E. Wolf, v.XXXII, Elsevier Science Publishers, Amsterdam, 1993
8. L. Schwartz, "Analyse Mathematique", Hermann, Paris, 1967

Chapter 11

TARGET LOCATING IN CLUTTER

11.1 PROBLEM FORMULATION

In this chapter we discuss the problem of locating targets in images that contain, besides the target object, a clutter of non-target objects that obscure the target object. As it follows from the discussion in Sect. 10.7, background non-target objects represent the main obstacle for reliable object localization in this case. Our purpose therefore is to find out how can one design a localization device that minimizes the danger of false identification of the target object with one of the non-target objects.

A general solution of this problem similar to that described in Ch. 10 for the model of a single target object observed in additive signal independent Gaussian noise does not seem to be feasible in this case for two reasons. First, the least bit constructive statistical or whatever description of image ensemble is very problematic owing to the great variability of real life images. Second, and this is even more important, in applications where target location in images with cluttered background is required one is interested in the most reliable localization not on average over a hypothetical image ensemble but on each individual image. Therefore, optimization and adaptation of the localization algorithm for individual images are desired.

This requirement of adaptivity will be imperative in our discussion. The second imperative requirement will be that of low computational complexity of localization algorithms. Bearing this in mind we shall restrict the discussion with the same type of localization devices as those described in Ch. 10, i.e., the devices that consist of a linear filter followed by a unit for locating signal maximum at the filter output (Fig. 11-1). Owing to the existence of fast and recursive algorithms of digital linear filtering (Chs. 5, 6), such devices have low computational complexity in their computer implementation. They also have promising optical and electro-optical implementations. In such type of the devices, it is the linear filter that has to be optimized and is the subject of the adaptation.

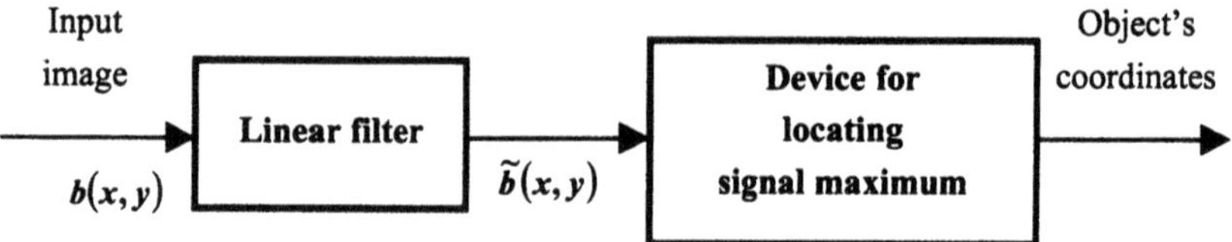

Figure 11-1. Schematic diagram of the localization device

We will begin the discussion with a formulation of the notion of optimality using an illustrative diagram of Fig. 11-2. Let us assume that input image $b(x,y)$ with a target object in coordinates (x_0,y_0) may be decomposed into S fragments of areas $\{S_s\}$, $s=1,...,S$ that may be regarded homogeneous in terms of the price for false localization errors within these areas. Let also $h_s(\tilde{b}_{bg}/(x_0,y_0))$ be a histogram of signal values $\tilde{b}(x,y)$ at the filter output as measured for the s-th fragment over background points not contained within the target object and $\tilde{b}_0$ is the filter output at the object location.

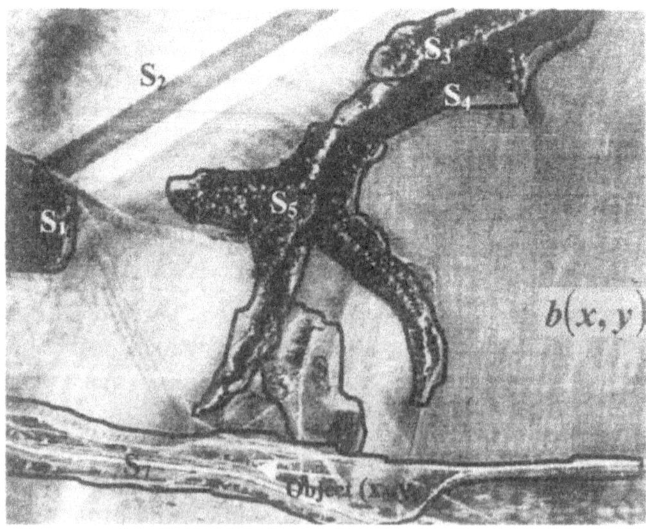

Figure 11-2. Illustrative diagram of basic notations

The localization device under consideration makes its decision regarding the coordinates of the desired object upon the position of the highest maximum at the signal output. Consequently, the integral

$$P_{a,s} = \mathbf{AV}_{\mathrm{bg}} \int_{\tilde{b}_0}^{\infty} h_s\left(\tilde{b}_{bg} / (x_0, y_0)\right) d\tilde{b}_{bg}, \tag{11.1.1}$$

where $\mathbf{AV}_{\mathrm{bg}}$ symbolizes averaging over an ensemble of the background component of the image, represents the rate of the s-th fragment points that, on average (over sensor's noise and different possible background objects), can be erroneously assigned by the decision unit to the target object.

Generally speaking, linear filter output at the target object location $\tilde{b}_0$ should also be regarded a random variable because it may depend on signal sensor noise, photographic environment, illumination, object orientation, neighboring objects, and numerous other factors unknown for the filter design. In order to take this uncertainty into the consideration, introduce a function $p(\tilde{b}_0)$ that describes a priori probability density of $\tilde{b}_0$. Unknown object coordinates (x_0, y_0) should also be regarded random. Moreover, the price of the measurement errors may differ over different images fragments. To allow for these factors, we introduce weighting functions $w_s(x_0, y_0)$ and W_s characterizing the a priori significance of the localization within the s-th fragment and for each s-th fragment, respectively, as :

$$\iint_{S_s} w_s(x_0, y_0) dx_0 dy_0 = 1; \tag{11.1.2}$$

$$\sum_{s=1}^{S} W_s = 1. \tag{11.1.3}$$

Then the quality of estimating coordinates by the device of Fig. 11-1 may be described by a weighted mean with respect to $p(\tilde{b}_0)$, $\{w_s(x_0, y_0)\}$, and $\{W_s\}$ of the integral of Eq.(11.1.1):

$$\overline{P}_a = \int_{-\infty}^{\infty} p(\tilde{b}_0) d\tilde{b}_0 \sum_{s=1}^{S} W_s \iint_{S_s} w_s(x_0, y_0) \left(\mathbf{AV}_{\mathrm{bg}} \int_{b_0}^{\infty} h_s\left(\tilde{b}_{bg} / (x_0, v_0)\right) db \right) dx_0 dy_0. \tag{11.1.4}$$

A device that provides the minimum for the $\overline{\boldsymbol{P}}_a$ will be regarded as optimal on average with respect to sensor's noise and background components of the image. For the design of the adaptive device optimal for a fixed image, one should drop out from the above formula averaging $\mathbf{AV}_{\mathrm{bg}}$ over the background.

11.2 LOCALIZATION OF PRECISELY KNOWN OBJECTS: SPATIALLY HOMOGENEOUS OPTIMALITY CRITERION

11.2.1 Optimal adaptive correlator

Assume that the target object is defined precisely. In this context, this means that the response of any filter to this object may be determined exactly. Therefore, probability density $p(\tilde{b}_0)$ is the delta-function:

$$p(b_0) = \delta\left(\tilde{b}_0 - \bar{b}_0\right). \tag{11.2.1}$$

where $\bar{b}_0$ is the known filter response at the point of the target object location. Then Eq. (11.1.4) that defines the localization quality becomes:

$$\bar{P}_a = \sum_{s=1}^{S} W_s \iint_{S_s} w_s(x_0, y_0)\left(\mathrm{AV}_{bg} \int_{\bar{b}_0}^{\infty} h_s\left(\tilde{b}_{bg} / (x_0, y_0)\right) d\tilde{b}_{bg} \right) dx_0 dy_0 \; . \tag{11.2.2}$$

Denote the histogram averaged within each fragment over (x_0, y_0) by

$$\bar{h}_s\left(\tilde{b}_{bg}\right) = \iint_{S_s} w_s(x_0, y_0) h_s\left(\tilde{b}_{bg} / (x_0, y_0)\right) dx_0 dy_0 \; . \tag{11.2.3}$$

Then obtain:

$$\bar{P}_a = \sum_{s=1}^{S} W_s \left(\mathrm{AV}_{bg} \int_{\bar{b}_0}^{\infty} \bar{h}_s\left(\tilde{b}_{bg}\right) d\tilde{b}_{bg} \right). \tag{11.2.4}$$

Suppose that weights $\{W_s\}$ do not depend on s and are equal to $1/S$. We refer to this case as to the case of the spatially homogeneous optimality criterion. Then

$$\bar{h}\left(\tilde{b}_{bg}\right) = \frac{1}{S} \sum_{s=1}^{S} \bar{h}_s\left(\tilde{b}_{bg}\right) \tag{11.2.5}$$

is the histogram of the filter output signal amplitudes as measured over the background part of the whole image and averaged with respect to the unknown coordinates of the target object. By substituting Eq.(11.2.5) into Eq.(11.2.4), we obtain

$$\overline{P}_a = \int_{\overline{b}_0}^{\infty} \mathbf{AV}_{bg}\left(\overline{h}(b)\right) db . \qquad (11.2.6)$$

Let us now determine the frequency response $H(f_x, f_y)$ of a filter that minimizes $\overline{P}_a$. The choice of $H(f_x, f_y)$ affects both $\overline{b}_0$ and histogram $\overline{h}(\widetilde{b}_{bg})$. As $\overline{b}_0$ is the filter response at the target object position, it may be determined through the target object Fourier spectrum $\alpha(f_x, f_y)$ as

$$\overline{b}_0 = \int_{-\infty}^{\infty} \alpha(f_x, f_y) H(f_x, f_y) df_x df_y . \qquad (11.2.7)$$

As for the relationship between $\overline{h}(\widetilde{b}_{bg})$and $H(f_x, f_y)$, it is, generally speaking, of a more involved nature. The explicit dependence of $\overline{h}(\widetilde{b}_{bg})$ on $H(f_x, f_y)$ may only be written for the second moment m_2^2 of the histogram $\overline{h}(b)$ using Parseval's relation for the Fourier transform:

$$m_2^2 = \iint_S w(x_0, y_0) dx_0 dy_0 \int_{-\infty}^{\infty} \widetilde{b}_{bg}^2 h(\widetilde{b}_{bg} / (x_0, y_0)) d\widetilde{b}_{bg} =$$
$$\frac{1}{S_{bg}} \iint_S w(x_0, y_0) dx_0 dy_0 \int_{S_{bg}} \widetilde{b}_{bg}^2 (x, y) dx dy =$$
$$\frac{1}{S_{bg}} \iint_S w(x_0, y_0) dx_0 dy_0 \int_{-\infty}^{\infty}\int_{-\infty}^{\infty} \left|\beta_{bg}^{(0,0)}(f_x, f_y)\right|^2 \left|H(f_x, f_y)\right|^2 df_x df_y =$$
$$\int_{-\infty}^{\infty}\int_{-\infty}^{\infty} \overline{\left|\beta_{bg}^{(0,0)}(f_x, f_y)\right|^2} \left|H(f_x, f_y)\right|^2 df_x df_y , \qquad (11.2.8)$$

where S_{bg} is the area of the background part of the images, or the image search area minus the area occupied by the target object located at coordinates (x_0, y_0), $\beta_{bg}^{(0,0)}(f_x, f_y)$ is Fourier spectrum of this background part, and

$$\overline{\left|\beta_{bg}^{(0,0)}(f_x, f_y)\right|^2} = \frac{1}{S_{bg}} \iint_S w(x_0, y_0) \left|\beta_{bg}^{(0,0)}(f_x, f_y)\right|^2 dx_0 dy_0 . \tag{11.2.9}$$

Therefore, for the optimization of the localization device filter we will rely upon the Tchebyshev's inequality

$$\text{Probability}\left(x \geq \overline{x} + b_0\right) \leq \sigma^2 / b_0^2 . \tag{11.2.10}$$

that connects the probability that a random variable x exceeds some threshold b_0 and the variable's mean value $\overline{x}$ and standard deviation σ .

Applying this relationship to Eq.(11.2.6), we obtain:

$$\overline{P}_a = \int_{\overline{b}_0}^{\infty} \text{AV}_{\text{bg}}\left(\overline{h}(b)\right) db \leq \text{AV}_{\text{bg}} \frac{m_2^2 - \overline{b}^2}{\left(\overline{b}_0 - \overline{b}\right)^2} , \tag{11.2.11}$$

where $\overline{b}$ is mean value of the histogram $\overline{h}(b)$. By virtue of the properties of the Fourier Transform the latter can be found as:

$$\overline{b} = \frac{1}{S_{bg}} \iint_S w(x_0, y_0) \alpha_{bg}^{0,0}(0,0) H(0,0) dx_0 dy_0 . \tag{11.2.12}$$

It follows from this equation that the mean value of the histogram over the background part of the picture, $\overline{b}$, is determined by the filter frequency response $H(0,0)$ at the point $(f_x = 0, f_y = 0)$. This value defines a constant bias of the signal at the filter output which is irrelevant for the device that localizes the signal maximum. Therefore, one can choose $H(0,0) = 0$ and disregard $\overline{b}$ in (11.2.11). Then we can conclude that, in order to minimize the rate $\overline{P}_a$ of anomalous errors, a filter should be found that maximizes ratio of its response $\overline{b}_0$ to the target object to standard deviation $\left(m_2^2\right)^{1/2}$ of its response to the background image component

$$SCR = \frac{\overline{b}_o}{\left(m_2^2\right)^{1/2}} = \frac{\int_{-\infty}^{\infty}\int_{-\infty}^{\infty} \alpha(f_x, f_y) H(f_x, f_y) df_x df_y}{\left(\int_{-\infty}^{\infty}\int_{-\infty}^{\infty} \text{AV}_{\text{bg}}\left(\left|\alpha_{bg}^{0,0}(f_x, f_y)\right|^2\right) \left|H(f_x, f_y)\right|^2 df_x df_y\right)^{1/2}} \tag{11.2.13}$$

We will refer to this ratio as to ***signal-to-clutter ratio***.

It follows from Schwarz's inequality ([1]) :

$$\left(\int_{-\infty}^{\infty}\int_{-\infty}^{\infty}\alpha(f_x,f_y)H(f_x,f_y)df_x df_y\right)^2 =$$

$$\left(\int_{-\infty}^{\infty}\int_{-\infty}^{\infty}\frac{\alpha(f_x,f_y)H(f_x,f_y)}{\left(\mathbf{AV}_{\mathrm{bg}}\overline{\left|\alpha_{bg}^{0,0}(f_x,f_y)\right|^2}\right)^{1/2}}\left(\mathbf{AV}_{\mathrm{bg}}\overline{\left|\alpha_{bg}^{0,0}(f_x,f_y)\right|^2}\right)^{1/2}df_x df_y\right)^2 \le$$

$$\int_{-\infty}^{\infty}\int_{-\infty}^{\infty}\frac{\left|\alpha(f_x,f_y)\right|^2}{AV_{\mathrm{bg}}\overline{\left|\alpha_{bg}^{0,0}(f_x,f_y)\right|^2}}\left|H(f_x,f_y)\right|^2\left(\mathbf{AV}_{\mathrm{bg}}\overline{\left|\alpha_{bg}^{0,0}(f_x,f_y)\right|^2}\right)df_x df_y\ . \tag{11.2.13}$$

that frequency response of such an optimal filter is defined by the equation:

$$H_{opt}(f_x,f_y)=\frac{\alpha^*(f_x,f_y)}{\mathbf{AV}_{\mathrm{bg}}\overline{\left|\alpha_{bg}^{0,0}(f_x,f_y)\right|^2}}, \tag{11.2.14}$$

where asterisk * denotes complex conjugation.

One can see that filter defined by Eq. 11.2.14 is analogous to the optimal filter (Eq. 10.4.3) introduced in Sect.10.4 for object localization in correlated Gaussian noise. The numerator of its frequency response is the frequency response of the filter matched to the target object. The denominator of its frequency response is power spectrum $\mathbf{AV}_{\mathrm{bg}}\left(\overline{\left|\alpha_{bg}^{0,0}(f_x,f_y)\right|^2}\right)$ of the image background component averaged over an ensemble of images that may contain the same target. It plays here the role additive Gaussian noise power spectrum plays in the filter of Eq. 10.4.3. There is, however, an important difference. When averaging is applied to $\overline{\left|\alpha_{bg}^{0,0}(f_x,f_y)\right|^2}$ the filter will be optimal (in terms of minimization of the misdetection rate) also on average over the ensemble of possible images. However, as we mentioned in Sect. 11.1, in applications, one frequently needs a filter optimal for each particular image rather then on the average. This means that in this case the averaging over the background image component $\mathbf{AV}_{\mathrm{bg}}$ must be dropped out from the optimality equation 11.2.13. In this way we arrive at the following filter:

$$H_{opt}(f_x, f_y) = \frac{\alpha^*(f_x, f_y)}{\left|\alpha_{bg}^{0,0}(f_x, f_y)\right|^2}. \tag{11.2.15}$$

This filter will be optimal for the particular observed image. Because its frequency response depends on the image background component the filter is adaptive. We will call this filter "***optimal adaptive correlator***".

To be implemented, optimal adaptive correlator needs knowledge of power spectrum of the background component of the image. However coordinates of the target object are not known. Therefore one cannot separate the target object and the background in the observed image and can not exactly determine the background component power spectrum and implement the exact optimal adaptive correlator. Yet, one can attempt to approximate it by means of an appropriate estimation of the background component power spectrum from the observed image.

11.2.2 Implementation issues, practical recommendations and illustrative examples

There might be different approaches to substantiating spectrum estimation methods using additive and implant models for the representation of target and background objects within images. ([2,3]). In the additive model, input image $b(x,y)$ is regarded as an additive mixture of the target object $a(x-x_0, y-y_0)$ and image background component $a_{bg}(x,y)$:

$$b(x,y) = a(x-x_0, y-y_0) + a_{bg}(x,y), \tag{11.2.16}$$

where (x_0, y_0) are unknown coordinates of the target object. Therefore power spectrum of the image background component averaged over the set of possible target locations may be estimated as

$$\begin{aligned}\overline{\left|\alpha_{bg}^{(0,0)}(f_x, f_y)\right|^2} = {}& \left|\beta(f_x, f_y)\right|^2 + \left|\alpha(f_x, f_y)\right|^2 + \\ & \beta^*(f_x, f_y)\alpha(f_x, f_y)\overline{\exp\left[i2\pi(f_x x_0 + f_y y_0)\right]} + \\ & \beta(f_x, f_y)\alpha^*(f_x, f_y)\overline{\exp\left[-i2\pi(f_x x_0 + f_y y_0)\right]},\end{aligned} \tag{11.2.17}$$

where the horizontal bar denotes averaging over target coordinates (x_0, y_0).

Functions $\overline{\exp\left[-i2\pi\left(f_x x_0 + f_y y_0\right)\right]}$ and $\overline{\exp\left[i2\pi\left(f_x x_0 + f_y y_0\right)\right]}$ are Fourier transforms of the distribution density $p(x_0, y_0)$ of the target coordinates:

$$\overline{\exp\left[\pm i2\pi\left(f_x x_0 + f_y y_0\right)\right]} = \int_X \int_Y p(x_0, y_0) \exp\left[\pm i2\pi\left(f_x x_0 + f_y y_0\right)\right] dx_0 dy_0 \tag{11.2.18}$$

In the assumption that the target object coordinates are uniformly distributed within the input image area, functions $\overline{\exp\left[-i2\pi\left(f_x x_0 + f_y y_0\right)\right]}$ and $\overline{\exp\left[i2\pi\left(f_x x_0 + f_y y_0\right)\right]}$ are sharp peaks with maximum at $f_x = f_y = 0$ and negligible values for all other frequencies. Therefore, for the additive model of the target object and image background component, averaged power spectrum of the background component may be estimated as:

$$\overline{\left|\alpha_{bg}^{(0,0)}\left(f_x, f_y\right)\right|^2} \cong \left|\beta\left(f_x, f_y\right)\right|^2 + \left|\alpha\left(f_x, f_y\right)\right|^2; f_x, f_y \neq 0. \tag{11.2.19}$$

The implant model assumes that

$$a_{bg}(x, y) = b(x, y) w(x - x_0, y - y_0), \tag{11.2.20}$$

where $w(x - x_0, y - y_0)$ is a target object window function:

$$w(x, y) = \begin{cases} 0 & \text{within the target object} \\ 1 & \text{elsewhere} \end{cases}. \tag{11.2.21}$$

In a similar way as it was done for the additive model and in the same assumption of uniform distribution of the coordinates over the image area, one can show that in this case power spectrum of the background image component averaged over all possible target coordinates may be estimated as

$$\overline{\left|\alpha_{bg}^{0,0}\left(f_x, f_y\right)\right|^2} \cong \int\int \left|W\left(p_x, p_y\right)\right|^2 \left|\beta\left(f_x - p_x, f_y - p_y\right)\right|^2 dp_x dp_y, \tag{11.2.22}$$

where $W\left(f_x, f_y\right)$ is Fourier transform of the window function $w(x, y)$. This method of estimating background component power spectrum resembles the conventional methods of signal spectral estimation that assume convolving power spectrum of the signal with a certain smoothing spectral window function $\overline{W}(p_x, p_y)$ ([4]):

$$\overline{\left|\alpha_{bg}^{0,0}(f_x, f_y)\right|^2} \cong \int\int \overline{W}(p_x, p_y)\left|\beta(f_x - p_x, f_y - p_y)\right|^2 dp_x dp_y\,, \qquad (11.2.23)$$

or, in the discrete representation,

$$\overline{\left|\alpha_{bg}^{0,0}(r, s)\right|^2} \cong \sum_{r_W}\sum_{r_W} \overline{W}(r_W, s_{W_y})\left|\beta(r - r_W, s - s_W)\right|^2 \qquad (11.2.24)$$

Both models imply that, as a zero order approximation, input image power spectrum may be used as an estimate of the background component power spectrum:

$$\overline{\left|\alpha_{bg}^{0,0}(r, s)\right|^2} \cong \left|\beta(r, s)\right|^2. \qquad (11.2.25)$$

This approximation is justified by the natural assumption that the target object size is substantially smaller then the size of the input image and that its contribution to the entire image power spectrum is small with respect to that of the background image component.

Experimental comparison of these approaches showed that the spectrum smoothing method provides better and more stable and image independent results in terms of the achievable signal-to-clutter ratio. Experimental data show also that a rectangular window may be used for image spectrum smoothing and that the window size is not very critical. Practical recommendations are rectangular window functions of 5x5 to 15x15 pixels.

In the conclusion of this section we illustrate advantages of the optimal adaptive correlator over the conventional matched filter correlator that is optimal for target location in Gaussian noise in the absence of non-target objects and clutter.

Fig.11-3 compares matched filter correlator and optimal adaptive correlator for localization of the corresponding image fragments in stereoscopic images. This comparison shows that the matched filter is incapable of reliable locating small image fragments of one of the stereo images in the second image while the optimal adaptive correlator does substantially suppress the background image component and therefore has much better discrimination capability.

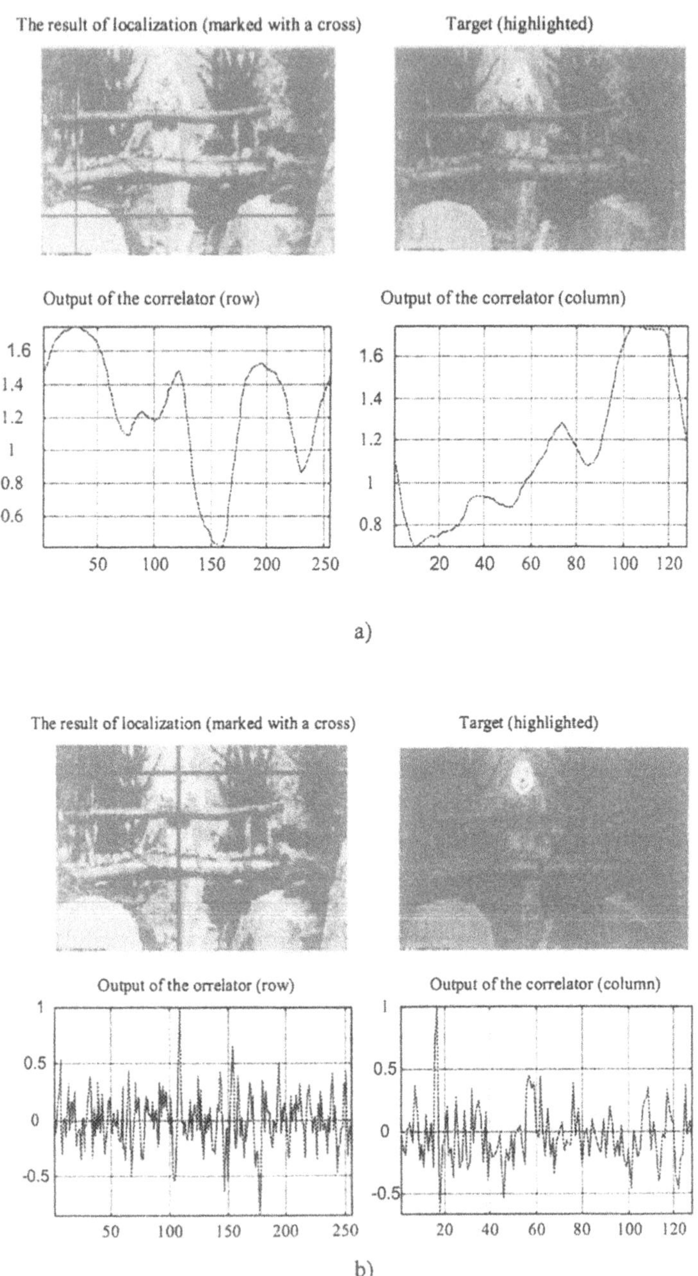

Figure 11-3. Localization of correspondent points in stereoscopic images with the matched filter (a) and the optimal adaptive correlators (b)

In Fig. 11-4 data are presented of an experiment in which a round spot of about 32 pixels in diameter was randomly placed within the image of chest X-ray of 512x512 pixels. The initial X-ray and outputs of the matched filter and optimal adaptive correlators are shown in Fig. 11-4, a) in form of gray scale images and graphs of signal rows that go through the spot center in one of its positions. The shape of the spot is illustrated in 3-D plot of Fig. 11-4, b) in coordinates gray scale value versus spatial coordinates.

Average rates of false localization found for different contrast of the spot over the ensemble of the spot random positions uniformly distributed within the image are plotted in graphs of Fig. 11-4, c) in coordinates rate of misdetection versus ratio of the spot maximum to standard deviation of the image gray scale values.. These results clearly show that, with the optimal adaptive filter correlator, the spot can be reliably detected for much lower contrast that with the matched filter correlator.

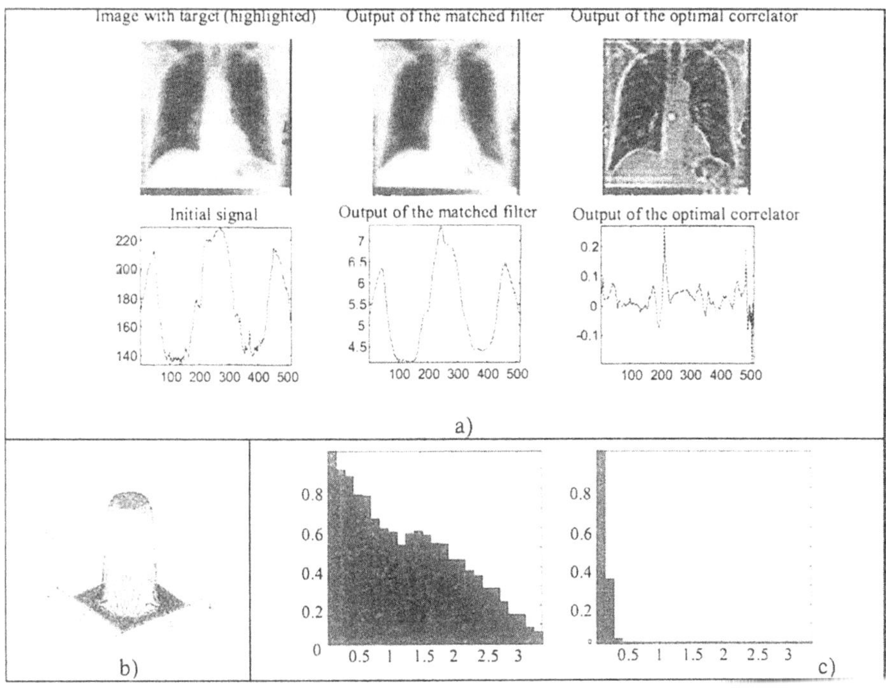

Figure 11-4. Comparison of localization reliability of the matched filter and optimal adaptive correlators: a) a test X-ray image and outputs of the matched filter and of the optimal adaptive correlators; b) - 3-D plot of the target test spot; c) - plots of false detection rates versus test spot contrast ratio to the image standard deviation for the matched filter correlator (left) and for the optimal adaptive correlator (right).

11.3 LOCALIZATION OF INEXACTLY KNOWN OBJECT: SPATIALLY HOMOGENEOUS CRITERION

Optimal adaptive correlator introduced in Sect. 11.2.1 is tuned to a target object that is exactly known. However in reality there might be some uncertainty about target object parameters such as orientation, scale, other geometrical distortions, distortions caused by sensor's noise, etc. In these cases the probability density $p(\widetilde{b}_0)$ of the filter response to the target object in its coordinates can no longer be regarded to be a delta function. Therefore the filter of the optimal estimator must secure the minimum of the integral:

$$\overline{P}_a = \int_{-\infty}^{\infty} p(\widetilde{b}_0) db_0 \int_{\widetilde{b}_0}^{\infty} \mathrm{AV}_{\mathrm{bg}}\left(\overline{h}\left(\widetilde{b}_{bg}\right)\right) d\widetilde{b}_{bg}, \qquad (11.3.1)$$

where $\overline{h}\left(\widetilde{b}_{bg}\right)$ is defined by Eq.(11.2.3).

Two options to reach this goal may be considered, depending on implementation restrictions.

a. Localization device with multiple filters. Split the interval of possible values of $\widetilde{b}_0$ into subintervals within which $p(\widetilde{b}_0)$ may be regarded to be constant. Then

$$\overline{P}_a = \sum_{q=1}^{Q} p_q \int_{\widetilde{b}_0^{(q)}}^{\infty} \mathrm{AV}_{\mathrm{bg}}\left[\overline{h}\left(\widetilde{b}_{bg}\right)\right] d\widetilde{b}_{bg}, \qquad (11.3.2)$$

where $\widetilde{b}_o^{(q)}$ is a representative of q-th interval, p_q is the area under $p(\widetilde{b}_0)$ over the q-th interval and Q is the number of the intervals. Because $p_k \geq 0$, $\overline{P}_a$ is minimal, if

$$\overline{P}_a^{(k)} = \int_{\widetilde{b}_0^{(k)}}^{\infty} \mathrm{AV}_{\mathrm{bg}}\left[\overline{h}\left(\widetilde{b}_{bg}\right)\right] d\widetilde{b}_{bg} \qquad (11.3.3)$$

is minimal and the problem reduces to the above problem of localizing a precisely known object. The only difference is that now one should consider Q different versions of exactly known target object. Therefore, optimal filters

$$H_{opt}^{(q)}(f_x, f_y) = \frac{\alpha^{*(q)}(f_x, f_y)}{\mathbf{AV}_{bg}\left|\alpha_{bg}^{0,0}(f_x, f_y)\right|^2} \tag{11.3.4}$$

need to be generated separately for each of Q "representative" of all possible object variations of the target object. This solution is, of coarse, more expensive in terms of the computational or hardware implementation complexity than that with a single filter.

b . *Localization device with optimal-on-average filter.* If variances of the target object variable parameters are not too large, one may, at the expense of worsening the localization reliability, solve the problem with a single filter as though the target object were exactly known. The optimal filter in this case has to be corrected with due regard to the object parameter variations. In order to find the frequency response of this filter, change the variables $\tilde{b}_{bg} \to \tilde{\tilde{b}} = \tilde{b}_{bg} - \tilde{b}_0$ and change the order of integration in Eq. 11.3.1:

$$\overline{P}_a = \int_0^\infty d\tilde{\tilde{b}}_1 \int_{-\infty}^\infty p(\tilde{b}_0)\mathbf{AV}_{bg}\left[\overline{h}\left(\tilde{\tilde{b}} + \tilde{b}_0\right)\right] d\tilde{\tilde{b}} . \tag{11.3.5}$$

The internal integral in Eq. 11.3.5, is a convolution of distributions, or distribution of the difference of two variables $\tilde{\tilde{b}}$ and $\tilde{b}_0$. Denote this distribution by $\overline{h}_p\left(\tilde{\tilde{b}} - \tilde{b}_0\right)$. Its mean value is equal to the difference of mean values of $\tilde{b}_0$ and b_{av} of distributions $p(\cdot)$ and $\overline{h}(\cdot)$, respectively and its variance is equal to the sum $\left[m_2^2 - (b_{av})^2\right] + \sigma_b^2$ of variances of these distributions, where σ_b^2 is the variance of the distribution $p(\cdot)$. Therefore,

$$\overline{P}_a = \int_0^\infty \overline{h}_p\left(\tilde{\tilde{b}}\right) d\tilde{\tilde{b}} = \int_{\tilde{b}_0}^\infty \mathbf{AV}_{bg}\left[\overline{h}_p\left(\tilde{\tilde{b}} - \tilde{b}_0\right)\right] d\tilde{\tilde{b}} . \tag{11.3.6}$$

Thus, the problem has again been boiled down to that of Sect.11.2, and one can use Eq. 11.2.14, with corresponding modifications of its numerator and denominator, to determine the optimal filter frequency response. As $\tilde{b}_0$ is the mean value of the filter output at the target object location over the distribution $p(\tilde{b}_0)$, a complex conjugate spectrum $\alpha^*(f_x, f_y)$ of the object in the numerator of the formula (11.2.14) should be replaced by a complex conjugate of object spectrum $\overline{\alpha}^*(f_x, f_y)$

averaged over variations of the target object. As for the denominator of Eq. 11.2.14, it should be modified by adding to it the variance of the object spectrum

$$\left|\alpha_{ef}(f_x,f_y)\right|^2 = \mathbf{AV}_{\mathrm{obj}}\left[\left|\alpha(f_x,f_y)-\overline{\alpha}(f_x,f_y)\right|^2\right], \qquad (11.3.7)$$

where $\mathbf{AV}_{\mathrm{obj}}(\cdot)$ denotes the operator of averaging over variations of the target object uncertain parameters. As a result, we obtain the following equation for the "optimal-on-average filter" frequency response:

$$\overline{H}_{opt}(f_x,f_y)=\frac{\overline{\alpha}^*(f_x,f_y)}{\mathrm{AV}_{\mathrm{bg}}\overline{\left|\alpha_{bg}^{0,0}(f_x,f_y)\right|^2}+\overline{\left|\alpha_{ef}(f_x,f_y)\right|^2}}. \qquad (11.3.8)$$

This filter will be optimal-on-average over all variations of the target object and over all possible backgrounds. By eliminating averaging over the background we obtain the optimal-on-average adaptive filter:

$$\overline{H}_{opt}(f_x,f_y)=\frac{\overline{\alpha}^*(f_x,f_y)}{\overline{\left|\alpha_{bg}^{0,0}(f_x,f_y)\right|^2}+\overline{\left|\alpha_{ef}(f_x,f_y)\right|^2}} \qquad (11.3.9)$$

which will be optimal-on-average over variations of the object for a fixed background; that is for the given observed image. For the design of this filter, methods of estimating background image component power spectrum $\overline{\left|\alpha_{bg}^{0,0}(f_x,f_y)\right|^2}$ discussed in the previous section may be used. In a special case when variations of the target object are caused by additive sensor's noise of the imaging system, the variance of the object spectrum is equal to the spectral density of the noise, and optimal filter becomes:

$$\overline{H}_{opt}(f_x,f_y)=\frac{\overline{\alpha}^*(f_x,f_y)}{\overline{\left|\alpha_{bg}^{0,0}(f_x,f_y)\right|^2}+\overline{\left|\nu(f_x,f_y)\right|^2}}. \qquad (11.3.10)$$

where $\overline{\left|\nu(f_x,f_y)\right|^2}$ denotes the noise spectral density. As it was already mentioned, such an averaged filter cannot provide as large a "signal-to-clutter" ratio as the filter designed for a precisely known target object. Indeed, for each specific target object with spectrum $\alpha(f_x,f_y)$ the following inequality for "signal-to-clutter ratio" $\overline{\boldsymbol{SCR}}$ for the filter of Eq. 11.3.9 holds:

$$\overline{\boldsymbol{SCR}} = \int\limits_{-\infty}^{\infty}\int\limits_{-\infty}^{\infty} \frac{\alpha^*(f_x, f_y)\overline{\alpha}^*(f_x, f_y)}{\left|\alpha_{bg}^{0,0}(f_x, f_y)\right|^2 + \left|\alpha_{ef}(f_x, f_y)\right|^2} df_x df_y \leq$$

$$\int\limits_{-\infty}^{\infty}\int\limits_{-\infty}^{\infty} \frac{\left|\alpha(f_x, f_y)\right|^2}{\left|\alpha_{bg}^{0,0}(f_x, f_y)\right|^2 + \left|\alpha_{ef}(f_x, f_y)\right|^2} df_x df_y \, . \qquad (11.3.11)$$

The right hand part of the inequality is the signal-to-clutter ratio for the optimal filter, exactly matched to the target object.

11.4 LOCALIZATION METHODS FOR SPATIALLY INHOMOGENEOUS CRITERIA

Let us return back to the general formula of Eq. 11.1.4 that accounts for possible inhomogeneity of the localization criterion by means of weight coefficients $\{W_s\}$. Depending on the implementation constraint, one of three ways to attain the minimum of $\overline{P}_a$ may be suggested in this case.

a. *Re-adjustable localization device with fragment-wise optimal adaptive correlator.* Given nonnegative weights $\{W_s\}$, the minimum of $\overline{P}_a$ is attained at the minima of integrals

$$\overline{P}_{a,s} = \int_{b_0} p(b_0)db_0 \iint_{S_s} w_s(x_0, y_0)\left(\mathbf{AV}_{bg} \int_{b_0}^{\infty} h_s(b/(x_0, y_0))db \right) dx_0 dy_0 . \quad (11.4.1)$$

This implies that the filter should be re-adjustable for each s-th fragment. Frequency responses of fragment-wise optimal filters are determined as it was shown in Sect. 11.2 for the homogeneous criterion on the basis of measurements of spectra of individual fragments.

According to Eq. 11.1.4, the fragments are assumed not to overlap. However, from the very sense of Eq. 11.1.4, it is obvious that it gives rise to a sliding window processing. In the sliding window processing, localization is carried out by applying to image fragments within the sliding window at each window position the optimal adaptive correlator that is designed upon estimates of local power spectra within the window. Then the device for locating signal maximum analyzes obtained output image and generates coordinates of the highest correlation peak. We will refer to sliding window optimal adaptive correlators as to ***local adaptive correlators***.

Experimental verification of the local adaptive correlators confirms that they substantially outperform global optimal adaptive correlators in terms of signal-to-cluster ration they provide. Fig. 11-5 illustrates application of the local adaptive correlator to localization of a small fragment of one of the stereoscopic images in the second image. Size of the fragment is 8x8 pixels, image size is 128x256 pixels, size of the sliding window is 32x32 pixels. Graphs on the figure illustrate difference between filter response to the target and that to the rest of the image. The same filter applied globally rather than locally fails to properly localize the fragment (compare with the example shown in Fig. 11-3, when global optimal adaptive correlator provides approximately the same signal-to-clatter ration for a target fragment of 16x16 pixels).

In conclusion note the similarity between local adaptive correlator and local adaptive filters for image denoising described in Sect.8.3.

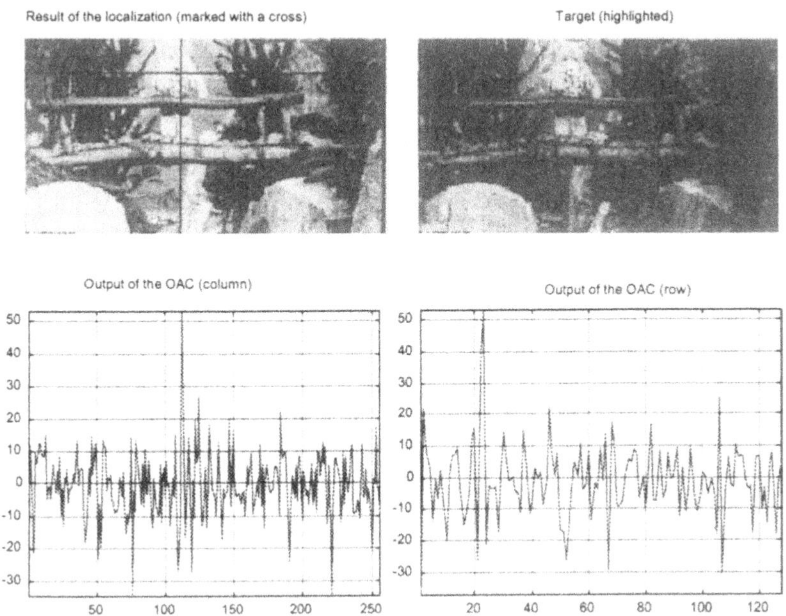

Figure 11-5. Localization of a small (8x8 pixels, highlighted) fragment of the right of two stereoscopic images on the left image by the local adaptive correlator.

b. *Non-re-adjustable localization device.* When there is no possibility of implementing the local adaptive localization device with fragment-wise or sliding processing , the alternative is to design a localization device adjusted to the power spectra of image segments averaged over the set of segments with weights $\{W_s\}$. Indeed, it follows from Eq.(11.1.4), that

$$\overline{P}_a = \int_{-\infty}^{\infty} p(\widetilde{b}_0)db_0 \sum_{s=1}^{S} W_s \iint_{S_s} w_s(x_0, y_0) \cdot \left[\mathbf{AV}_{\text{bg}} \int_{\widetilde{b}_0}^{\infty} h_s\left(\widetilde{b} / (x_0, y_0)\right) d\widetilde{b} \right] dx_0 dy_0 =$$

$$\int_{-\infty}^{\infty} p(\widetilde{b}_0)d\widetilde{b}_0 \, \mathbf{AV}_{\text{bg}} \int_{b_0}^{\infty} \overline{\overline{h}}_s\left(\widetilde{b}\right) d\widetilde{b}, \qquad (11.4.2)$$

where $\overline{\overline{h}}_s(b)$ is a histogram averaged over both $\{W_s\}$ and $\{w_s(x_0, y_0)\}$,

$$\overline{\overline{h}}_s(b)=\sum_{s=1}^{S}W_s\iint_{S_s}w_s(x_0,y_0)\left(\mathbf{AV}_{bg}\int_{b_0}^{\infty}h_s(b/(x_0,y_0))db\right)dx_0dy_0 . \qquad (11.4.3)$$

Now one can, by analogy with Eqs.11.3.9, conclude that the optimal filter may be designed upon averaged power spectra of background components of image segments as

$$\overline{\overline{H}}_{opt}(f_x,f_y)=\frac{\alpha^*(f_x,f_y)}{\mathbf{AV}_S\overline{\left|\alpha_{bg}^{(s)}(f_x,f_y)\right|^2}} . \qquad (11.4.4)$$

where $\mathbf{AV}_S$ denotes averaging power spectra $\left|\alpha_{bg}^{(s)}(f_x,f_y)\right|^2$ of background components of image fragments over the image area:

$$\mathbf{AV}_S\overline{\left|\alpha_{bg}^{(s)}(f_x,f_y)\right|^2}=\sum_{s=1}^{S}W_s\iint_{S_s}w_s(x_0,y_0)\left|\alpha_{bg}^{(s)}(f_x,f_y)\right|^2dx_0dy_0 . \qquad (11.4.5)$$

Thus, the optimal filter frequency response in this case depends on the weights $\{W_s\}$ and is determined by the averaged power spectrum of all segments. Of course, such an "averaged" filter may not be as efficient for individual image fragments as the filter adapted to these fragments but it is definitely less expensive for computation and hardware implementation. One can also notice that if a set of images is given for which the filter should be optimized, the averaging should be extended to this set. In the limit, it reduces to the estimation of the image statistical power spectrum, and we arrive at the optimal filter of Eq. 10.4.2 for additive non-white Gaussian noise.

The above reasoning may also be applied to pattern recognition tasks. When it is required to recognize an image from a given set of images. It follows from Eq. (11.4.4) that, prior to recognition, images in the set should be normalized and the normalization can be implemented in the Fourier domain by the filter:

$$H_{norm}(f_x,f_y)=\frac{1}{\sum_{k=1}^{N}\left|\alpha_k(f_x,f_y)\right|^2} , \qquad (11.4.6)$$

where k is image index and N is the number of images in the set. The recognition can then be carried out by filtering the normalized images with filters

$$H_k(f_x, f_y) = \frac{\alpha_k^*(f_x, f_y)}{\sum_{k=1}^{N} |\alpha_k(f_x, f_y)|^2}. \tag{11.4.7}$$

and by subsequent determining the filter that provides the highest output.

c. *Image homogenization.* In cases when non re-adjustable localization device does not provide admissible localization reliability while one cannot afford the computational expenses for the implementation of the local adaptive correlator, an intermediate solution, an image preprocessing that improves image homogeneity may be used ([5]). After the preprocessing, optimal adaptive correlator should be built for the preprocessed image. We will refer to such a preprocessing as to ***image homogenization***.

The idea of the image homogenization arises from the following reasoning. In terms of the design of the optimal adaptive correlator, image may be regarded homogeneous if spectra of all image fragments involved in the averaging procedure in Eq. (11.4.5) are identical. While this is, in general, not the case, one can attempt, before applying the optimal adaptive correlator, to reduce the diversity of image local spectra by means of an appropriate image preprocessing.

One of the computation-wise simplest preprocessing homogenization methods is the preprocessing that standardizes image local mean and local variance. In discrete signal representation, it is defined for image fragments of $(2N_1+1)(2N_2+1)$ pixels as

$$\tilde{b}_{k,l} = \sqrt{\frac{\sigma_{st}^2}{\overline{b_{k,l}^2}}}\left[b_{k,l} - \bar{b}_{k,l}\right]. \tag{11.4.8}$$

where (k,l) are running coordinates of the window central pixel, $\{b_{k,l}\}$ are image samples,

$$\bar{b}_{k,l} = \frac{1}{(2N_1+1)(2N_2+1)} \sum_{n=-N_1}^{N_1} \sum_{m=-N_2}^{N_2} b_{k-n,l-m}. \tag{11.4.9}$$

is image local mean over the window and

$$\overline{b_{k,l}^2} = \frac{1}{(2N_1+1)(2N_2+1)} \sum_{n=-N_1}^{N_1} \sum_{m=-N_2}^{N_2} b_{k-n,l-m}^2 - \left[\bar{b}_{k,l}\right]^2. \tag{11.4.10}$$

is its local variance.

Such an image homogenization procedure equalizes average of local power spectra over all frequencies rather then entire spectra and may be regarded as a zero order approximation to making spectra identical on each particular frequency. As it was described in Sect. 7.1, computation of image local mean and variance may be very efficiently carried out recursively such that the computational complexity may not depend on the window size.

Selection of the window size is governed by the image inhomogeneity. A good practical recommendation is to select the window size commensurable with the size of the target object. Localization with the use of the homogenization preprocessing is illustrated in Fig. 11-6 on an example of localization of corresponding points on the same stereoscopic images as those used in Figs. 11-3 and 11-5. Comparing images in these figures, one can see that the homogenization may visually be interpreted as an enhancement of local contrast of image details. Note that such a preprocessing may be also used for image enhancement aimed at easing image visual analysis (see Sect. 8.6.4). Its relation to the problem of localization of objects in images provides a good explanation why it does help to visually analyze images.

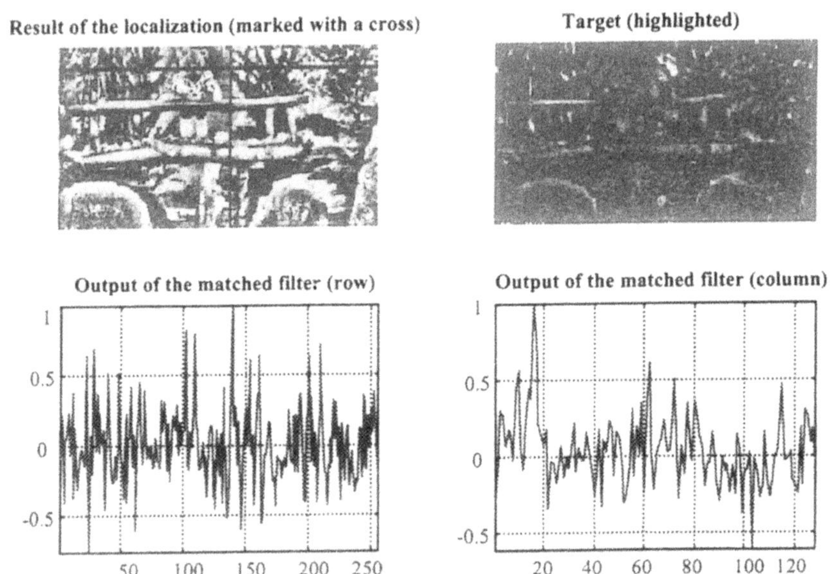

Figure 11-6. Illustration of improvement of the localization with the use of image "homogenization"

11.5 OBJECT LOCALIZATION AND IMAGE BLUR

In this section we analyze how such common image distortion as blur affects the localization relaibility. We will assume that the optimal adaptive correlator of Eq. 11.2.15 is employed in the localization device.

As it was described in Sect. 8.2.3, image blur can usually be described in terms of distortion of image spectra in imaging systems. Yet another image distortion factor that has to be taken into account is imaging system noise. Let frequency response of the imaging system be $H_{ims}(f_x, f_y)$ and let the noise be additive with spectral density $N(f_x, f_y)$. Then, if the target object has spectrum $\alpha(f_x, f_y)$, its spectrum at the output of the imaging system will be equal to $\alpha(f_x, f_y)H_{ims}(f_x, f_y)$. The image background component spectrum $\overline{\left|\alpha_{bg}^{(0,0)}(f_x, f_y)\right|^2}$ will also be modified by the imaging system to become $\overline{\left|\alpha_{bg}^{(0,0)}(f_x, f_y)\right|^2}\left|H_{ims}(f_x, f_y)\right|^2 + N(f_x, f_y)$. In this estimate we assume averaging of the background component spectrum over an ensemble of its different possible realizations generated by the noise. The optimal adaptive correlator should then have the following frequency response:

$$H_{opt}^{(bl)}(f_x, f_y) = \frac{\alpha^*(f_x, f_y)H_{ims}(f_x, f_y)}{\overline{\left|\alpha_{bg}^{(0,0)}(f_x, f_y)\right|^2}\left|H_{ims}(f_x, f_y)\right|^2 + N(f_x, f_y)} \qquad (11.5.1)$$

In order to give to this equation a clear physical interpretation, modify it in the following way:

$$H_{opt}^{(bl)}(f_x, f_y) =$$

$$\frac{\alpha^*(f_x, f_y)}{\overline{\left|\alpha_{bg}^{(0,0)}(f_x, f_y)\right|^2}\left|H_{ims}(f_x, f_y)\right|^2} \frac{H_{ims}^*(f_x, f_y)\overline{\left|\alpha_{bg}^{(0,0)}(f_x, f_y)\right|^2}}{\overline{\left|\alpha_{bg}^{(0,0)}(f_x, f_y)\right|^2}\left|H_{ims}(f_x, f_y)\right|^2 + N(f_x, f_y)} =$$

$$= \frac{\alpha^*(f_x, f_y)}{\overline{\left|\alpha_{bg}^{(0,0)}(f_x, f_y)\right|^2}}\left[\frac{1}{H_{ims}^*(f_x, f_y)} \frac{\left|H_{ims}(f_x, f_y)\right|^2\overline{\left|\alpha_{bg}^{(0,0)}(f_x, f_y)\right|^2}}{\overline{\left|\alpha_{bg}^{(0,0)}(f_x, f_y)\right|^2}\left|H_{ims}(f_x, f_y)\right|^2 + N_{ims}(f_x, f_y)}\right] \qquad (11.5.2)$$

The first factor in this formula

$$H_{opt}(f_x,f_y)=\frac{\alpha^*(f_x,f_y)}{\overline{\left|\alpha_{bg}^{(0,0)}(f_x,f_y)\right|^2}} \qquad (11.5.3)$$

is obviously the optimal adaptive filter designed for the undistorted image. As for the second factor,

$$H(f_x,f_y)=\frac{1}{H_{ims}^*(f_x,f_y)}\frac{\left|H_{ims}(f_x,f_y)\right|^2\overline{\left|\alpha_{bg}^{(0,0)}(f_x,f_y)\right|^2}}{\overline{\left|\alpha_{bg}^{(0,0)}(f_x,f_y)\right|^2}\left|H_{ims}(f_x,f_y)\right|^2+N_{ims}(f_x,f_y)}, \qquad (11.5.4)$$

it may be regarded as the frequency response of an image deblurring filter. As one can see, it is an empirical Wiener deblurring filter introduced in Sect. 8.2.3. This means, that the optimal adaptive correlator defined by Eq. 11.5.2 may be treated as a device that performs a deblurring operation and then optimal adaptive correlation for the deblurred image.

It will be instructive now to estimate "signal-to-clutter" ratio that this filter can provide. By analogy with Eqs. 11.2.12 and 13, one can obtain in this case that

$$SCR_{blr}=\int_{-\infty}^{\infty}\int_{-\infty}^{\infty}\frac{\left|\alpha(f_x,f_y)\right|^2\left|H_{ims}(f_x,f_y)\right|^2}{\overline{\left|\alpha_{bg}^{(0,0)}(f_x,f_y)\right|^2}\left|H_{ims}(f_x,f_y)\right|^2+N_{ims}(f_x,f_y)}df_x df_y=$$

$$\int_{-\infty}^{\infty}\int_{-\infty}^{\infty}\frac{\left|\alpha(f_x,f_y)\right|^2}{\overline{\left|\alpha_{bg}^{(0,0)}(f_x,f_y)\right|^2}}\frac{\overline{\left|\alpha_{bg}^{(0,0)}(f_x,f_y)\right|^2}\left|H_{ims}(f_x,f_y)\right|^2}{\overline{\left|\alpha_{bg}^{(0,0)}(f_x,f_y)\right|^2}\left|H_{ims}(f_x,f_y)\right|^2+N_{ims}(f_x,f_y)}df_x df_y\le$$

$$\int_{-\infty}^{\infty}\int_{-\infty}^{\infty}\frac{\left|\alpha(f_x,f_y)\right|^2}{\overline{\left|\alpha_{bg}^{(0,0)}(f_x,f_y)\right|^2}}df_x df_y=SCR_{id}, \qquad (11.5.5)$$

where SCR_{id} is signal-to-clutter ratio for the nonblurred image. One can evaluate from Eq. (11.5.5) how much image blur worsen the localization reliability. In particular, Eq. (11.5.5) shows that so long the noise level in the imaging system is not very high, image blur may have marginal effect on the performance of the optimal filter.

11.6 OBJECT LOCALIZATION AND EDGE DETECTION. SELECTION OF REFERENCE OBJECTS FOR TARGET TRACKING

It is widely believed in the image processing community that the information conveyed by images is contained mostly in image high frequency components and in edges. The theory of the optimal adaptive correlator provides a rational explanation of this belief. Represent, for instance, Eq. (11.2.13) for the optimal adaptive correlator in the following way:

$$H_{opt}^{0}(f_x,f_y)=\frac{\alpha^*(f_x,f_y)}{\left|\alpha_{bg}^{0,0}(f_x,f_y)\right|^2}=\frac{\alpha^*(f_x,f_y)}{\left[\left|\alpha_{bg}^{0,0}(f_x,f_y)\right|^2\right]^{1/2}}\frac{1}{\left[\left|\alpha_{bg}^{0,0}(f_x,f_y)\right|^2\right]^{1/2}}. \quad (11.6.1)$$

In this representation, the optimal adaptive correlator is regarded as consisting of two filters in cascade. The filter represented by the second factor in Eq. (11.6.1), may be regarded as an analog of the whitening filter introduced in Sect. 10.4 for optimal target localization in correlated Gaussian noise. This filter being applied to the input image makes the image background component spectrum almost uniform. Practically, the same may be said with respect to the input image spectrum for all reasonable estimates of image background component power spectrum from the power spectrum of the entire input image. The filter represented by the first factor in Eq. 11.6.1 is, obviously, nothing but a matched filter for the target object, modified by the same whitening operator.

This implies two conclusions. First of all, power spectra of images usually (though not always) tend to decay on high frequencies. Therefore spectrum whitening usually results in emphasizing high frequency image components with respect to its low frequency components. It is in this sense one can tell that high frequency image components are more important for image object recognition. Second, if one visually compares image before and after whitening such as, for instance, image shown in Figs. 11-7 and 8, one can see that image whitening results in what is visually interpreted as edge enhancement.

An important feature of optimal adaptive correlators is their adaptivity. Their frequency responses are determined by power spectra of the input images. Therefore the whitening operation is, in principle, also adaptive. In general, the whitening tends to suppress those image frequency components that have high energy and that are, therefore, responsible for features common to the majority of objects represented in image. Low energy frequency components that are responsible for "uncommon", or rare and to objects and object features are, on contrary, emphasized. In a sense, one can say that whitening is an operation that automatically enhances dissimilarities and suppresses similarities of objects in images. This is well illustrated by whitening of a test image of geometrical figures and printed characters shown in Fig. 11-8 and of a test texture image shown in Fig. 11-9. In Fig. 11-8 one can see that the whitening does enhance edges but the enhancement is very selective. Vertical and horizontal edges that are common to all figures and characters are enhanced much less then circumferences, slanted edges and corners in geometrical figures and characters and those are also enhanced very selectively according to their rate of occurrences in the image. Whitening of texture image in Fig. 11-9 may also be interpreted as edge enhancement but the nature of this enhancement is substantially different from that in Figs. 11-7 and 8. In this example, the whitening automatically detects boundary between different textures that differ in their spatial statistical properties rather then in their gray level distribution.

Initial image

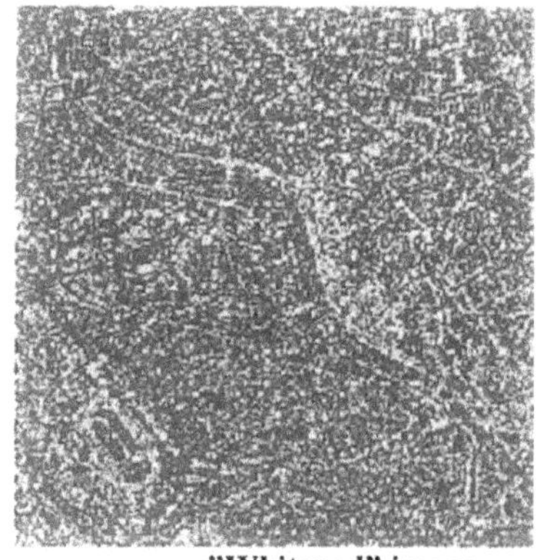

"Whitened" image

Figure 11-7. Image whitening as edge detection

Initial test image

Whitened image and its magnified fragments

Figure 11-8. Image whitening: edges and corners in geometrical figures

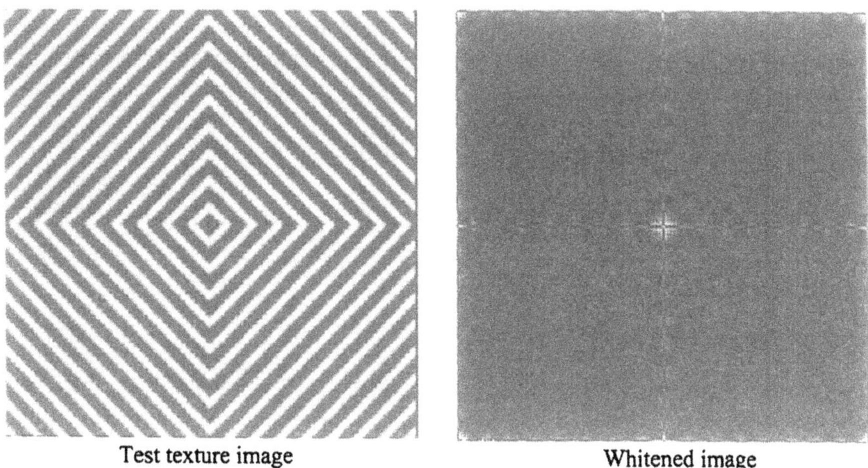

Figure 11-9. Whitening of a test texture image

It is very instructive to analyze impulse response of the whitening filter. Diagrams on Fig. 11-9 show several most valuable central samples of 1-D sections of impulse responses of the whitening filters for the air photograph and for the test image of geometrical figures shown in Figs. 11-7 and 8.

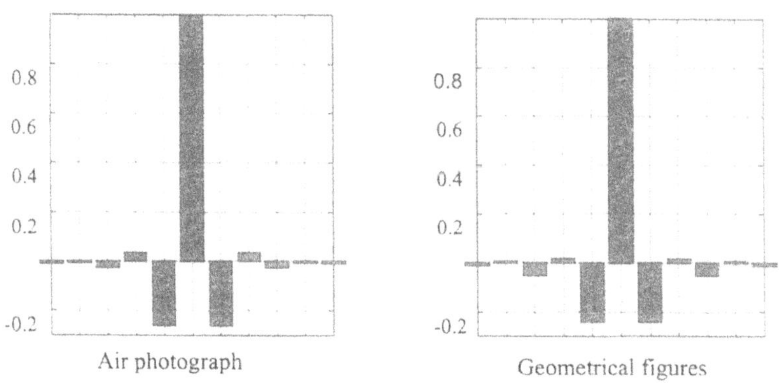

Figure 11-10. 1-D sections of impulse responses of whitening filters for two images of Figs. 11-7 and 8

One can see that those impulse responses are not identical though they are rather similar. The remarkable fact is that the dissimilarity between the whitening filter impulse responses is not commensurable with the dissimilarity between the two images. This may be attributed to the statistical nature of the image power spectra measured for the entire image with many objects in it. One can also note that whitening filter impulse responses resemble very much the impulse response of the filter that computes image spatial ***Laplacian*** defined by the matrix:

$$\mathbf{L} = \begin{bmatrix} -c_1 & -c_2 & -c_1 \\ -c_2 & 1 & -c_2 \\ -c_1 & -c_2 & -c_1 \end{bmatrix}, \qquad (11.6.2)$$

where c_1 and c_2 are non-negative constants such that $c_1 < c_2$ and $c_1 + c_2 = \mathbf{0.25}$. Most frequently settings $c_1 = \mathbf{0}$ and $c_2 = \mathbf{0.25}$ are used.

This observation suggests an explanation and a justification to the common belief in the importance of the Laplacian operator in image processing (see, for instance, [6]).

One can get an additional insight into the importance of different frequency bands in images from evaluating the potential "signal-to-clutter' ratio at the output of the optimal adaptive correlator. The "signal-to-clutter" ratio ***SCR*** is a natural figure of merit of target objects that determines the potential reliability of their localization. One can obtain from Eqs. (11.2.13 – 11.2.15) that for the optimal adaptive correlator

$$SCIR = \int_{-\infty}^{\infty}\int_{-\infty}^{\infty} \frac{\left|\alpha(f_x, f_y)\right|^2}{\left|\alpha_{bg}(f_x, f_y)\right|^2} df_x df_y \,. \qquad (11.6.3)$$

It follows from this formula that the signal-to-clutter ratio at the output of the optimal adaptive correlator is equal to the energy of the whitened spectrum of the target object. Therefore the higher the energy of the whitened spectrum, the higher is the signal-to-clutter ratio that one can expect at the optimal adaptive correlator output for this object. It is this relationship that quantitatively measures the importance of different signal frequency components.

This relationship provides also a useful quantitative measure for the selection of target objects in cases when target objects have to be chosen from the image. This is the case, for example, in image registration and matching with a map. Similar problems occur in stereo image analysis, in robot vision navigation, in multi modal medical imaging. In such application, the question is how to make the choice of the target objects to the best advantage. As it follows from Eq. (11.6.2.) those image fragments will be the best target objects that have maximal energy of their "whitened" power spectrum. Therefore, the reference objects with intensive high frequency components, that is, image fragments that are visually interpreted as containing the most intensive edges or texture, will be the best candidates.

This conclusion is illustrated in Fig.11-11 where the "goodness" of the fragments of 7x7 pixels in terms of the ***SCR***-factor is represented by the pixel gray levels for test images of Figs. 11-8 and 9. One can see from these images that edge rich areas are indeed most appropriate potential candidates for reference object. However, the remarkable observation is that, for rectangles in the image of geometrical figure and for the image of periodical textures, corners have much higher "goodness". This result is very natural. Edges in all rectangles are similar and cannot be regarded as their specific features. For the texture image, corners of areas of textures with different orientation are certainly the more specific objects.

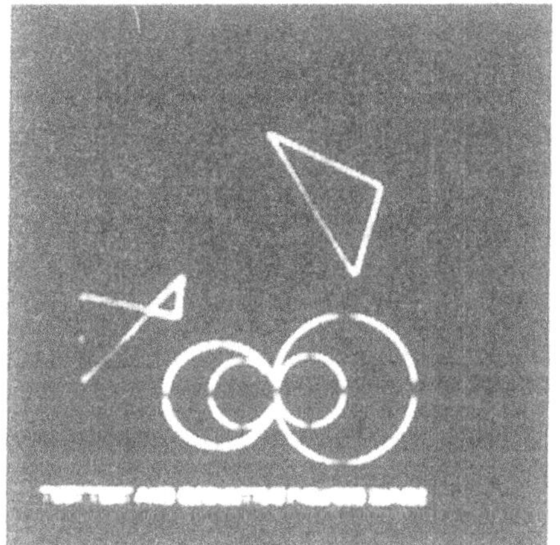

Figure 11-11. Potential reference objects for test images of Figs. 11-8 and 9: object "goodness" is displayed proportional to pixel gray levels

11.7 OPTIMAL ADAPTIVE CORRELATOR AND OPTICAL CORRELATORS

11.7.1 Optical correlators with improved discrimination capability

After the introduction of holographic matched spatial filters and optical correlators ([7]) it was recognized very soon that the ability of matched filters to discriminate effectively between an object of one class and objects belonging to other classes is not nearly so good as its ability to combat additive Gaussian noise. Accordingly, attempts were undertaken to find filters with better discriminatory capability.

One of the oldest ideas is that by S.Lowenthal and Y.Belvaux ([8]) of preprocessing the picture as well as the reference object by an appropriate differential operator before they are correlated. This can be accomplished by spatial filtering with coherent light; for example, by using opaque stops or slits, centered on the optical axis, or by introducing into the coherent optical correlator an additional spatial filter.

Another simple method for increasing discrimination by differentiation is to use the nonlinearity of the medium on which the holographic matched filter is recorded. It was noted that overexposing the spectral information from the target material of which the filter is made can result in high pass filtering similar to the spatial differentiation of ([9,10]). More recently it was noted that Joint Transform Correlators appear to have better discrimination capability than do the matched filters if the nonlinearity of the medium used in their Fourier plane to record the joint spectrum is properly chosen.

Fabrication of the matched filter requires, in general, recording both amplitude and phase information. This is not a trivial process that requires using special recording media and means, especially, if the filter is to be a synthetic one, generated by a computer (some associated problems are discussed in Sect. 13.3). In order to simplify fabrication of the filter J.L. Horner and J.R. Leger proposed using, instead of the matched filter, a ***phase-only filter*** (POF), a filter with constant amplitude transmittance and phase transmittance equal to that of the matched filter ([11,12]). If the frequency response of the matched filter is $\alpha^*(f_x, f_y)$, frequency response of the phase-only filter is

$$POF(f_x, f_y) = \frac{\alpha^*(f_x, f_y)}{|\alpha(f_x, f_y)|}. \tag{11.7.1}$$

An additional simplification is the so called binary phase-only filter (BPOF) [11,12] in which only the signs of the real and imaginary parts of the matched filter complex transmittance are recorded. Having been introduced mainly to avoid the need for complex filters and to increase the filter efficiency in using light energy in the optical correlators, POF and BPOF appeared to have better discrimination capability than the conventional matched filter. This fact along with their simplicity of implementation contributed greatly to the popularity of the POF.

Many different improvements of the POF have been proposed. Here are some of them, to name a few:

- amplitude compensated matched filters (ACMF, [13]):

$$ACMF(f)=\frac{\alpha^*(f_x,f_y)}{|\alpha(f_x,f_y)|^2}; \tag{11.7.2}$$

- amplitude-modulated POF (AMPOF, [14])

$$AMPOF(f)=\frac{\alpha^*(f_x,f_y)}{|\alpha(f_x,f_y)|^2+\varepsilon^2}, \tag{11.7.3}$$

  where ε^2 is a small constant introduced to prevent dividing by zero;

- POF with an improved signal-to-noise ratio (POFISNR, [15]):

$$POFISNR(f)=\begin{cases}\dfrac{\alpha^*(f_x,f_y)}{|\alpha(f_x,f_y)|^2}, & for\ (f_x,f_y)\in F\\ 0, & otherwise\end{cases}, \tag{11.7.4}$$

  where F is an area in Fourier domain chosen by some optimization procedure to improve signal-to-noise ratio;

- A family of ternary matched filters TMF ([16]), a natural generalizations of the BPOF, in which the real and imaginary parts of the matched filter transmittance are quantized to 3 levels (1, -1, 0) instead of two, and the location of the areas where transmittance of the filter must be zeroed, is chosen by an optimization procedure designed to improve signal-to-noise ratio;

- Optimal BPOF (OPOF) [17]:

$$OPOF(f_x,f_y)=\exp\{i[e(f_x,f_y)\varphi_1+(1-e(f_x,f_y))\varphi_2]\}, \tag{11.7.5}$$

where binary function $e(f_x, f_y)$ that takes values 0 and 1, and constants $\{\phi_1, \phi_2\}$ have to be chosen to optimize the filter figures of merits.

- "Minimum Average Correlation Energy" filters (MACE- filters, [18]). They maximize the ratio of the squared peak value of the correlation function to the average correlation plane energy. The MACE-filters are, in principle, designed to identify (or detect) multiple targets (in the presence of virtually different types of uncertainty in their a priori description), but for precisely known objects they coincide with the ACMF-filters.
- Entropy Optimized Filters (EOF [19]) that provide a strong and narrow peak for a match between the input and filter function as contrasted with a uniform signal distribution for any pattern to be rejected. It was shown by computer simulation, that the discriminatory power of EOF is much better than that of POF and is comparable with that of MACE-filters. It was also noted, that a dominant feature of the EOF, observed experimentally, is substantial enhancement of the high frequency components of images. Another important feature of the EOF is its adaptivity, because the filter optimality criterion takes into account the patterns to be rejected.
- Phase-only correlators (POC[20]):

$$POC(f_x, f_y) = \frac{1}{|\beta(f_x, f_y)|} \frac{\alpha^*(f_x, f_y)}{|\alpha(f_x, f_y)|}, \tag{11.7.6}$$

where $\beta(f_x, f_y)$ is spectrum of the image in which the target object has to be detected.

As one can see, the design of only two latter filters takes, in one or another way, into account objects that have to be discriminated from the target object by the filter. All other filters are optimized only in terms of the sharpness of the peak of their response to the target object. Obviously, the latter does not necessarily secures the high discrimination capability of the filters with respect to non-target objects. Nevertheless, all these filters may be regarded as non-adaptive approximates to the optimal adaptive correlator defined by Eq.(11.2.15) and its implementations described in Sect. 11.2.2. This explains their improved discrimination capability as compared to the matched filter.

For instance, frequency response of the optimal adaptive correlator may be approximated as

$$H(f_x, f_y) \cong \frac{\alpha^*(f_x, f_y)}{|\beta(f_x, f_y)|^2} = \frac{1}{|\beta(f_x, f_y)|} \frac{\alpha^*(f_x, f_y)}{|\beta(f_x, f_y)|}. \tag{11.7.7}$$

when power spectrum of the input image is used as a zero order approximation for the power spectrum of its background component. If the input image contains only the target object without any non-target objects, the second filter in the formula (11.7.7) will just be the phase-only-filter, and the total filter (11.7.7) will simply be the amplitude-compensated -phase-only filter, as well as a MACE-filter designed for a single object. Of course, at the filter output we shall have a delta-function. Furthermore, in pattern recognition the main danger of false recognition is associated with non-target objects most nearly similar to the given target object. Hence, in this case, the phase-only-filter represents a more or less good approximation to the second filter in (11.7.7), and the other filters mentioned above approximate the optimal filter itself. But it is precisely the extent of the dissimilarity between spectra of target object and non-target objects that is the most important for discriminating objects by optimal adaptive correlators. Optimal adaptive correlators take advantage of these dissimilarities.

As it was shown in Sect. 11.6, the inherent part of the optimal adaptive correlation is the whitening operation. Owing to the fact that usually (but not always!) image power spectra decay at higher spatial frequencies, whitening results in enhancement of high frequencies. This why the above mentioned recommendations about enhancement of high frequencies improve the object detectibility. But, in distinction to these ad hoc recommendations, Eqs. 11.6.1 and 11.7.7 explicitly indicate to what degree and in which specific way this enhancement must be carried out for each specific image with its specific set of non-target objects.

Above mentioned improvements in discrimination capability of Joint Transform Correlators achieved by proper selection of nonlinearity of the medium used for recording joint spectrum may also be explained in a similar way as it will be shown in the next section.

11.7.2 Computer and optical implementations of the optimal adaptive correlator

Computer implementation of the optimal adaptive correlator is most naturally based on convolution in Fourier domain because its design is based on measuring image power spectra as it is described by Eqs. 11.2.15, 11.2.16 and 11.2. 17. Flow diagram of the corresponding computer algorithm is shown in Fig. 11-11. In local adaptive correlation in sliding window, this algorithm is applied locally at every window position, spectra of image fragments within the filter window are computed recursively using the algorithm described in Sect. 5.4. Note the similarity between local adaptive correlation and local adaptive image restoration described in Sect. 8.3.

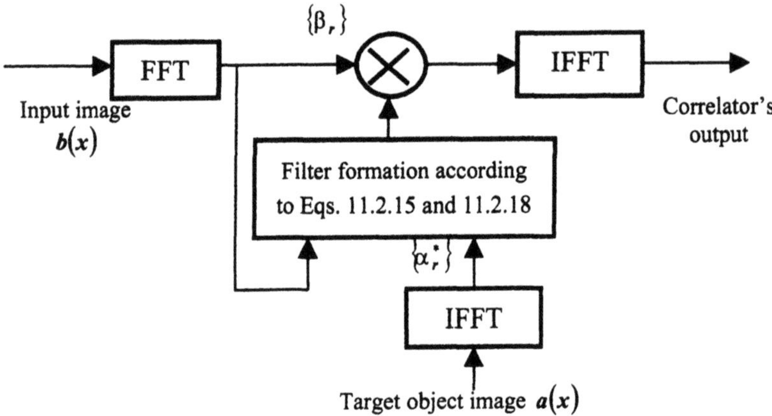

Figure 11-12. Flow diagram of optimal adaptive correlation

Optimal adaptive correlators may also be implemented in nonlinear optical and electro-optical correlators. Nonlinear optical correlators are optical correlators with a nonlinearity in their Fourier plane. Two types of nonlinear optical correlators are most feasible for the implementation of optimal adaptive correlators: optical correlators with k-th low nonlinearity ([21]) and Joint Transform correlators ([22]).

Consider a 4-F coherent optical correlators with a non-linear light sensitive transparent medium installed in its Fourier plane (Fig. 11-12, a). Let's assume that transparency T of the medium is inversely proportional to k-th power of the intensity I of the incident light:

$$T \propto \frac{1}{I^k}. \tag{11.7.8}$$

We ill refer to this type of the nonlinearity as to **k-th law nonlinearity**. The correlator of Fig. 11-12 with a matched filter $\alpha^*(f_x, f_y)$ installed in its Fourier plane after the nonlinear medium will implement input image filtering with frequency response

$$H(f_x, f_y) = \frac{\alpha^*(f_x, f_y)}{\left(\left|\beta(f_x, f_y)\right|^2\right)^k}. \tag{11.7.9}$$

because light intensity in the Fourier plane of the correlator is proportional to power spectrum $\left|\beta\left(f_x, f_y\right)\right|^2$ of the input image. Selecting nonlinear medium with the nonlinearity index $k \approx 1$ and placing it in the Fourier plane slightly out of focus or/and using low resolution nonlinear medium, one can implement the optimal adaptive correlator of Eq. 11.2.15 with estimation of the background image component power spectrum described by Eqs. 11.2.23, 24.

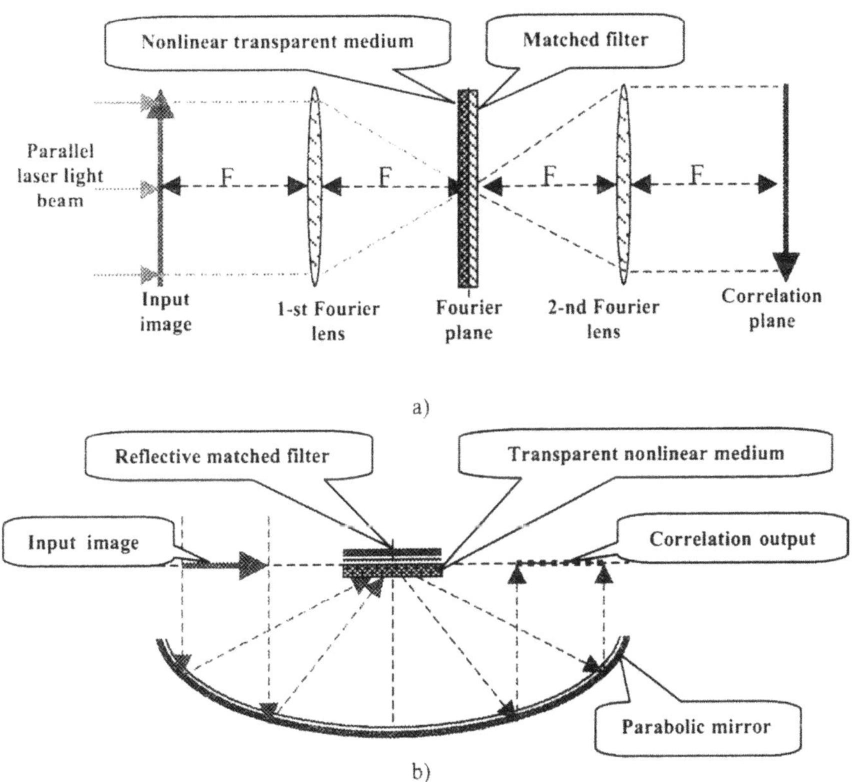

Figure 11-13. Schematic diagram of optical correlators with a nonlinearity in the Fourier plane: a) - 4-F nonlinear optical correlator; b) Nonlinear optical correlator with a parabolic mirror

Important practical limitation of such an implementation is high dynamic range of image power spectra which requires the corresponding high dynamic range of the nonlinear medium. This requirement is substantially weakened for the optical correlator with a parabolic mirror as Fourier transformer shown in Fig. 11.-12, b). In this correlator, light passes through the nonlinear medium twice which means that the required dynamic range of the nonlinear medium is roughly a square root of that required for the medium in 4-F correlators.

Computer simulation of nonlinear optical correlator with ***k***-th law nonlinearity and its comparison with matched filter, phase-only filter and phase-only correlators demonstrated that ([21])

- Optimal value of the nonlinearity index ***k*** is close to 1.
- Input image power spectrum moderate smoothing for the implementation of the correlator significantly improves the discrimination capability of the correlator evaluated in terms of signal-to-clutter ratio.
- ***k***-th law nonlinear correlator substantially outperforms phase-only correlator, phase-only and filter matched filter in discriminating target objects against non-target objects. In the experiments, signal-to-clutter ratio for these four types of correlators related, by the order of magnitude, as 7:3:2:1. At the same time, two to five times gap was observed between signal-to-clutter ratio for the optimized ***k***-th law nonlinear correlator and the ideal optimal adaptive correlator designed, in the simulation, for the exactly known power spectrum of the background image component.
- A trade-off exists between the optimal nonlinearity index ***k*** and the degree of the dynamic range limitation of the nonlinear medium; a higher degree of limitation requires a higher value of ***k***.
- Combination of input image power spectrum smoothing, high nonlinearity index k and a high degree of the dynamic range limitation promises considerable improvement in the nonlinear correlator's light efficiency while preserving their high discrimination capability.

Joint Transform correlators offer yet another possibility for implementing optimal adaptive correlator. In nonlinear JTCs, input image and target object joint spectrum is non-linearly transformed in an electronic amplifier before modulating the output spatial light modulator (Fig. 11-14). If the amplifier is the logarithmic one, signal recorded on the spatial light modulator is proportional to

$$\begin{aligned}
&\ln\left[\left|\beta(f_x,f_y)+\alpha(f_x,f_y)\exp(i2f_x\bar{x})\right|^2\right]=\\
&\ln\Big\{\left[\left|\beta(f_x,f_y)\right|^2+\left|\alpha(f_x,f_y)\right|^2\right]+\\
&\beta^*(f_x,f_y)\alpha(f_x,f_y)\exp(i2f_x\bar{x})+\beta(f_x,f_y)\alpha^*(f_x,f_y)\exp(-i2f_x\bar{x})\Big\}\cong\\
&\ln\left[\left|\beta(f_x,f_y)\right|^2+\left|\alpha(f_x,f_y)\right|^2\right]+
\end{aligned}$$

$$\frac{\beta^*(f_x,f_y)\alpha(f_x,f_y)}{\left|\beta(f_x,f_y)\right|^2+\left|\alpha(f_x,f_y)\right|^2}\exp(i2f_x\bar{x})+$$
$$\frac{\beta(f_x,f_y)\alpha^*(f_x,f_y)}{\left|\beta(f_x,f_y)\right|^2+\left|\alpha(f_x,f_y)\right|^2}\exp(-i2f_x\bar{x}), \qquad (11.7.10)$$

where, as in Eq. 10.2.12, $\beta(f_x,f_y)$ is spectrum of the input image, $\alpha(f_x,f_y)$ is spectrum of the target object, $\bar{x}$ is displacement between input image and target object template in the correlator's input plane, and where a natural assumption that

$$\left|\beta(f_x,f_y)\right|^2 >> \left|\alpha(f_x,f_y)\right|^2 \qquad (11.7.11)$$

is made on the base that the size of the target object is much smaller then the input image size.

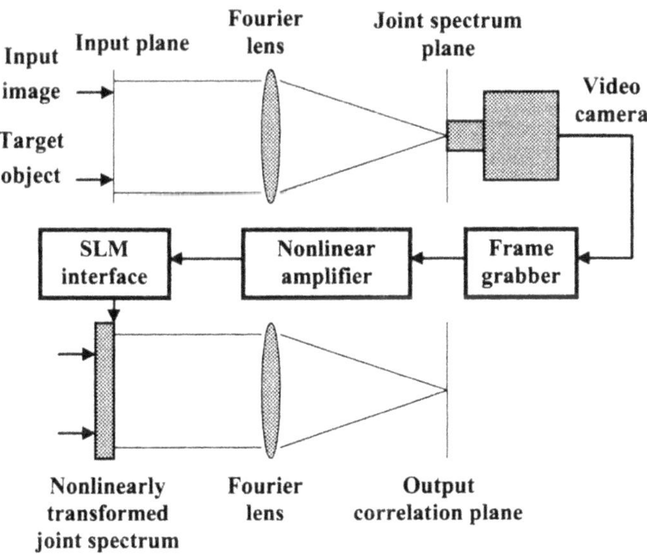

Figure 11-14. Schematic diagram of the Joint Transform correlator with a nonlinear transformation of the joint spectrum

As one can see, the last term in Eq. 11.7.10 approximates the output signal of the optimal adaptive correlator in which power spectrum of the background image component is estimated as it is described by Eq. 11.2.19 for the additive model of target objects and input images.

Logarithmic nonlinearity is not the only option for nonlinear joint transform correlators. Nonlinearity optimization in nonlinear joint transform correlators was addressed in Ref. [22]. It was shown there that

- different types of nonlinear transformation of the joint spectrum may be used for substantial improvement of the correlator's discrimination capability, provided an appropriate choice of the nonlinearity parameters: logarithmic transformation, the $\boldsymbol{k}$-th law transformation, with $\boldsymbol{k} < \mathbf{1}$, in combination with limitation of the dynamic range and thresholding the joint spectrum
- A moderate blur of the joint spectrum before its nonlinear transformation is admissible and may even improve the correlator's discrimination capability.
- All types of the nonlinear joint transform correlators are tolerant to reasonable deviations of nonlinearity parameters from their optimal values.

11.8 TARGET LOCATING IN COLOR AND MULTI COMPONENT IMAGES

11.8.1 Theoretical framework

The theory of the optimal adaptive correlator may be extended to the problem of reliable target localization in color and multi-component images with a localization device composed of a multi component linear filter, a summation unit and a unit that determines coordinates of the highest signal peak at its input (Fig. 11-15)

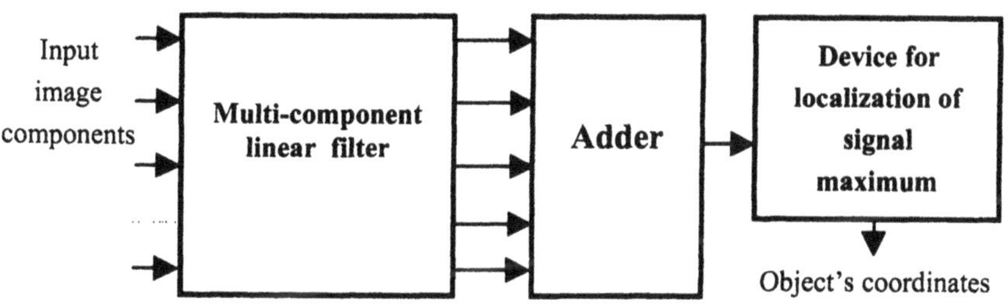

Figure 11-15. Schematic diagram of the localization device for multi-component images

We will consider here only the problem of localization of an exactly known object for spatially homogeneous criterion. Following the reasoning that resulted in Eq. 11.2.13 for the filter that maximizes signal-to-clutter ratio for single component images, one can conclude that, for multi-component images, multi-component filter will be optimal that maximizes signal-to-clutter ratio at the output of the summation unit:

$$H(f_x, f_y, f_m) = \underset{H(f_x, f_y, f_m)}{\arg\max} \left(SCR = \frac{\sum_{f_x}\sum_{f_y}\sum_{f_m} \alpha(f_x, f_y, f_m) H(f_x, f_y, f_m)}{\left\{ \mathrm{AV}_{bg} \sum_{f_x}\sum_{f_y} \left| \sum_{f_m} \alpha_{bg}^{(0,0)}(f_x, f_y, f_m) H(f_x, f_y, f_m) \right|^2 \right\}^{1/2}} \right),$$

(Eq.11.8.1)

where (f_x, f_y) are indices of spatial frequencies, f_m is the frequency index of the component-wise Discrete Fourier Transform, $\alpha_{bg}^{(0,0)}(f_x, f_y, f_m)$ is 3-D spectrum of the target object, $\alpha_{bg}^{(0,0)}(f_x, f_y, f_m)$ is 3-D spectrum of the background clutter (note that here we consider discrete Fourier spectra), horizontal bar denotes average over unknown coordinates of the target object. Reformulate Eq.11.8.1 in terms of power spectrum of the clutter $\overline{\left|\alpha_{bg}^{(0,0)}(f_x, f_y, f_m)\right|^2}$ that may be estimated from the observed image. To this goal, one can rely on the inequality:

$$\left|\sum_{f_m} \nu_{cl}(f_x, f_y, f_m) \cdot H(f_x, f_y, f_m)\right|^2 \leq M \sum_{f_m} \left|\nu_{cl}(f_x, f_y, f_m)\right|^2 \cdot \left|H(f_x, f_y, f_m)\right|^2, \qquad \text{(Eq. 11.8.2)}$$

where M is the number of image components, which implies that

$$SCR \geq \frac{\left(\sum_{f_x}\sum_{f_y}\sum_{f_m} \alpha(f_x, f_y, f_m) \cdot H(f_x, f_y, f_m)\right)^2}{M \sum_{f_x}\sum_{f_y}\sum_{f_m} \mathbf{AV}_{bg} \overline{\left|\alpha_{bg}^{(0,0)}(f_x, f_y, f_m)\right|^2} \cdot \left|H(f_x, f_y, f_m)\right|^2}, \qquad \text{(Eq. 11.8.3)}$$

Right part of Eq. 11.8.3 defines the lower bound of ***SCR*** for any spectrum of clutter and any filter frequency response. It follows now from Cauchy-Schwarz-Buniakowsky inequality that choosing

$$H_{opt}(f_x, f_y, f_m) = \frac{\alpha^*(f_x, f_y, f_m)}{\mathbf{AV}_{bg} \overline{\left|\alpha_{bg}^{(0,0)}(f_x, f_y, f_m)\right|^2}} \qquad \text{(Eq. 11.8.4)}$$

guaranties that this lower bound is made the highest possible:

$$SNR \geq \frac{1}{M} \sum_{f_x}\sum_{f_y}\sum_{f_m} \frac{\left|\alpha(f_x, f_y, f_m)\right|^2}{\mathbf{AV}_{bg} \overline{\left|\alpha_{bg}^{(0,0)}(f_x, f_y, f_m)\right|^2}}. \qquad \text{(Eq. 11.8.5)}$$

Dropping out averaging $\mathbf{AV}_{bg}$ over background clutter ensemble, obtain frequency response of the optimal adaptive multi-component correlator:

$$H_{opt}(f_x, f_y, f_m) = \frac{\alpha^*(f_x, f_y, f_m)}{\overline{\left|\alpha_{bg}^{(0,0)}(f_x, f_y, f_m)\right|^2}} \qquad (11.8.6)$$

For the implementation of the optimal adaptive multi-component correlator, one can assume, similarly to the case of single component images, smoothing power spectrum $\left|\beta(f_x, f_y, f_m)\right|^2$ of the image with a spatial and component-wise spectral window:

$$H_{opt}(f_x, f_y, f_m) = \frac{\alpha^*(f_x, f_y, f_m)}{\left|\beta(f_x, f_y, f_m)\right|^2 \bullet W(f_x, f_y, f_m)}, \qquad (11.8.7)$$

where $W(f_x, f_y, f_m)$ is a 3-D spectral window function and symbol $\bullet$ denotes the convolution operation.

In the conclusion note that all solutions that were obtained in Sects. 11.3 and 11.4 for inexactly known target objects and for spatially inhomogeneous criterion may be extended to the multi-component case in a straightforward way.

11.8.2 Separable component-wise implementations of the optimal adaptive multi-component correlator ([23])

3-D image processing makes serious demands to the computing resources in digital image processing. Electro-optical implementation is also possible only in a form of a separable, component-wise processing. Therefore we will consider the implementation that is schematically represented in Fig. 11-16. It assumes separable space and component-wise processing where primary image components are first transformed before their use as inputs of multiple channel correlators. Outputs of the channel correlators are then combined to form a single output signal in which target is localized by locating position of the highest peak or peaks that exceed a certain threshold.

Such a channel decomposition can be based on an assumption that power spectrum of the background clutter $\overline{\left|\alpha_{bg}^{(0,0)}(f_x, f_y, f_m)\right|^2}$ may be approximated as a separable function:

$$\overline{\left|\alpha_{bg}^{(0,0)}(f_x, f_y, f_m)\right|^2} = \overline{\left|\alpha_{bg}^{(m)}(f_m)\right|^2} \cdot \overline{\left|\alpha_{bg}^{(xy)}(f_x, f_y)\right|^2} \qquad (11.8.9)$$

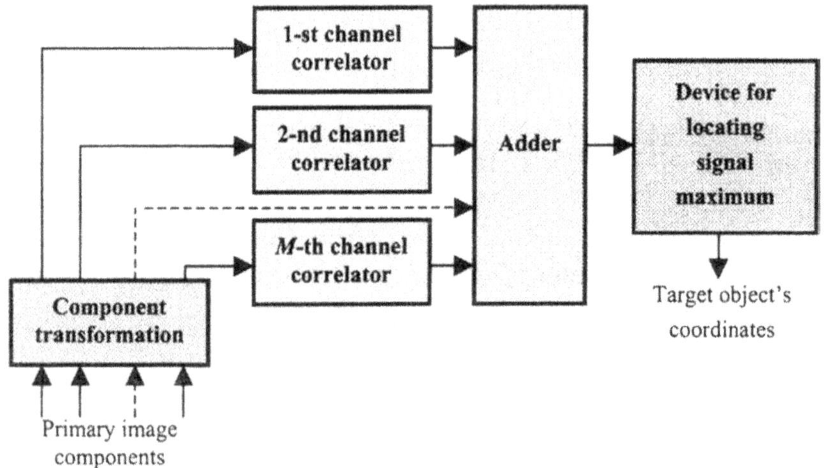

Figure 11-16. A separable implementation of the multi-component localization device

In this case, the localization is implemented with separate component-wise and spatial-wise "whitening filters"

$$H_{1w}(f_m) = \frac{1}{\left(\overline{\left|\alpha_{cl}^{(m)}(f_m)\right|^2}\right)^{1/2}} \tag{11.8.10}$$

and

$$H_{2w}(f_x, f_y) = \frac{1}{\left(\overline{\left|\alpha_{bg}^{(xy)}(f_x, f_y)\right|^2}\right)^{1/2}}, \tag{11.8.11}$$

respectively. The target object should also be subjected to the same transformations. Therefore channel correlators in Fig. 11-16 should be optimal adaptive correlators for image components transformed by component-wise "whitening" filters defined by Eq. 11.8.10. For the involved separated spectral factors of the background clutter power spectrum, the following empirical estimations from input image power spectrum $\left|\beta(f_x, f_y, f_m)\right|^2$ may be suggested:

$$\overline{\left|\alpha_{bg}^{(m)}(f_m)\right|^2} \cong \sum_{f_x} \sum_{f_y} \left|\beta(f_x, f_y, f_m)\right|^2 \tag{11.8.12}$$

and

$$\overline{\left|\alpha_{bg}^{(xy)}(f_x, f_y)\right|^2} \cong \sum_{f_m} \left|\beta(f_x, f_y, f_m)\right|^2 . \qquad (11.8.13)$$

Flow diagram of a computer implementation of the separable optimal adaptive multi-component correlator is shown in Fig. 11-17.

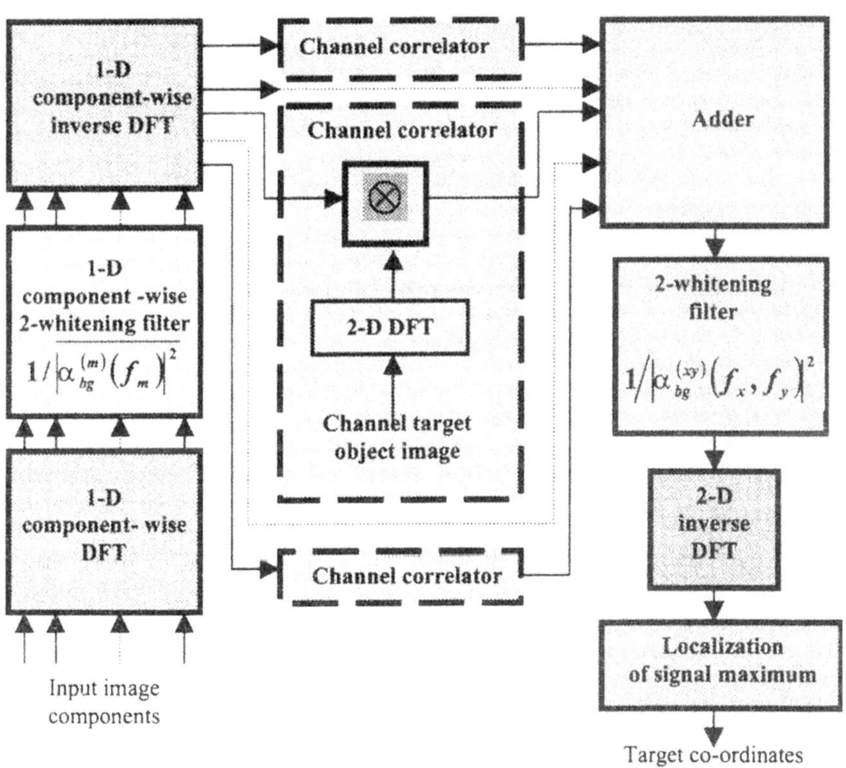

Figure 11-17. Flow diagram of computer implementation of the separable optimal adaptive multi-component correlator

Component transformation for the general case of multi component images has not as yet been discussed in the literature. However, component transformation in the context of object recognition in color images represented by their RGB components has obtained considerable attention. One of the approaches involves preprocessing of RGB components on the base of human perception ([24]). In this case an element-wise linear preprocessing based on the opponent-color theory proposed in Ref. [25] is performed over the RGB components. Another example is a projection of color images on a certain generalized color plane suggested Ref. [26,27]. This approach is aimed also at reducing the number of components used in correlation process and therefore at simplification of multi channel optical setup. Of course, the option of reducing the number of channels after transformation of the image components is open for any transformation method. Other examples of component wise color image preprocessing aimed at improvement of the correlator's discrimination capability are suggested in [28] on the base of the above discussed theory and proved their efficiency. They are modifications of the following two methods:

- Component "circular whitening":

$$\tilde{b}_{k,l}^{(m)} = \frac{b_{k,l}^{(m)} - \left(b_{k,l}^{(m+1)\bmod 3} + b_{k,l}^{(m+2)\bmod 3}\right)/2}{\left[\left(\sum_{m=0}^{2} b_{k,l}^{(m)}\right)^2 - 2\sum_{k=0}^{2} b_{k,l}^{(m)} b_{k,l}^{(m+1)\bmod 3}\right]^{1/2}} + 1; \quad m = 0,1,2; \qquad (11.8.14)$$

- Component centering (component-wise Laplacian):

$$\tilde{b}_{k,l}^{(m)} = b_m - \frac{b_{k,l}^{(m+1)\bmod 3} + b_{k,l}^{(m+2)\bmod 3}}{2}; \quad m = 0,1,2, \qquad (11.8.15)$$

where $\left\{b_{k,l}^{(m)}\right\}$ are samples of color image components, m is component index and (k,l) are coordinate indices.

As one can see, both methods involve component wise discrete Laplacian. The first method assumes independent component wise whitening of all pixels in 3 image components without any spectral estimation by averaging as suggested by Eq. 11.8.12 and 13.

Component wise whitening according Eq. 11.8.10 of a color image is illustrated in Figs. 11- 18. Fig. 11-19 shows diagram of samples of the color whitening filter impulse response for this particular image. One can see that the whitening filter impulse response is not exactly the Laplacian's operator samples (-0.5,1,-0.5) and that it may change from image to image. But the difference is not very substantial and experiments show that variations of the whitening filter impulse response are also not very substantial.

Figure 11-18. Color image RGB components (upper row) and the result of component wise whitening (bottom row)

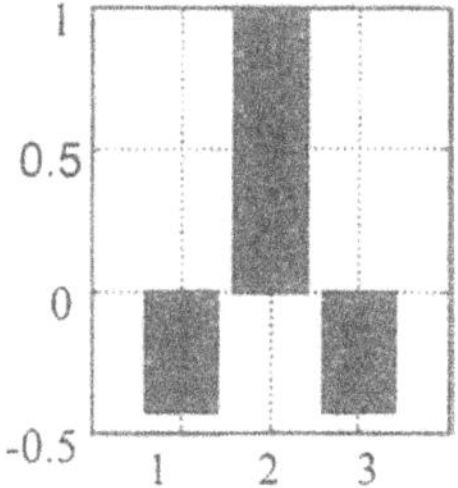

Figure 11-19. Impulse response of the whitening filter.

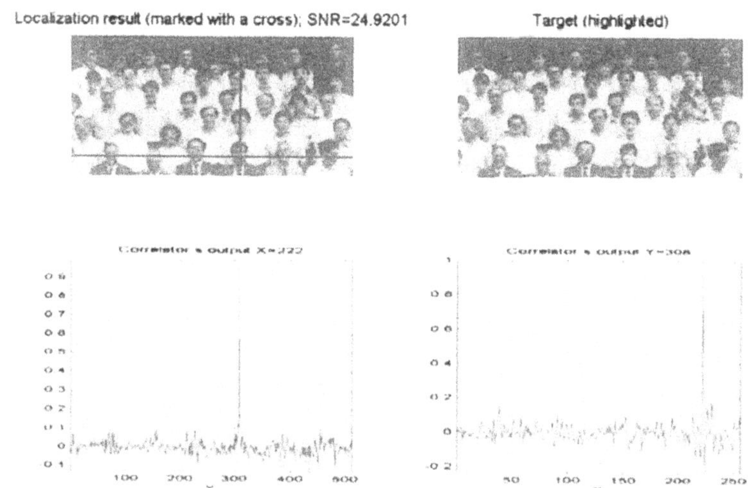

Figure 13-20. Localization of a target in a color image: separable optimal correlator (images are printed in gray scale)

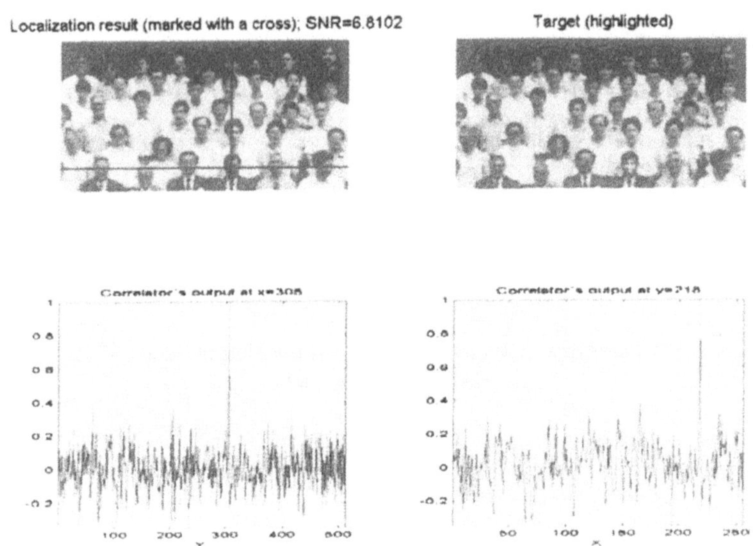

Figure 11-21. Localization of a target in the single component image (Green component of the image in Fig. 11-20)

Figs. 11-20 and 21 illustrate that the use of color information substantially improves reliability of object detection and localization. Fig. 11-20 shows result of localization of a small target (highlighted in figures) on a color image. Fig. 11-21 demonstrates the localization result obtained using a single component (Green component in this example) of the image. They show that "signal-to-clutter" ratios for these two cases are 24.9 versus 6.8 which is more then 3 times better for the color correlator.

References

1. L. Schwartz, "Analyse Mathematique", Hermann, Paris, 1967
2. L. Yaroslavsky, "*Adaptive Nonlinear Correlators for Object location and Recognition*", In:Advanced Optical Correlators for Pattern Recognition and Association, Ed. Kazuyoshi Itoh,Research Signpost, Trivandrum, Kerala, State, India, 1997, pp. 47-61.
3. L. Yaroslavsky, "*Nonlinear Optical Correlators with improved discrimination Capability for Object Location and Pattern Recognition*", In : Optical Patern Recognition, Ed. F.S.T. Yu and S. Jutamulia, Cambridge University Press, 1998, 141-170
4. R.B. Blackman and J.W. Tukey, The Measurement of Power Spectra, New York: Dover, 1958
5. L. Yaroslavsky, O. Lopez-Coronado, Juan Campos, Input image "homogenization" method for improvement the correlator's discrimination capability, Optics Letters, 15 July, 1998
6. D. Marr, Vision. A Computational Investigation into the Human Representation and Processing of Visual Information, W.H. Freeman and Co., N.Y. 1982
7. A. Vander Lugt, Signal Detection by Complex Spatial Filtering, IEEE Trans., IT-10, 1964, No. 2, p. 139
8. S. Lowenthal, Y. Belvaux, Reconnaissance des Formes par Filtrage des Frequences Spaciales, Optica Acta, v. 14, p. 245-258, 1967
9. A. Vander Lugt, A Review of Optical Data-Processing Techniques, Optica Acta, v. 15, 1-33, 1968
10. R. A. Binns, A. Dickinson, B. M. Watrasiewicz, Method of Increasing Discrimination of Optical Filtering, Appl. Optics, v. 7, 1047-1051, 1968
11. J. L. Horner, P.D. Gianino, Phase-only Matched Filtering, Appl. Opt., v. 23, p. 812-816, 1984
12. J.L Horner, J. R. Leger, Pattern Recognition with Binary Phase-only Filters, Appl. Opt., v. 24, 609-611, 1985
13. G.-G. Mu, X.-M. Wang, Z.-O. Wang, Appl. Opt., v. 27, 3461, 1988
14. A.A. S. Awwal, M. A. Karim, S. R. Jahan, Improved Correlation Discrimination Using an Amplitude-modulated Phase-only Filter, Appl. Opt., v. 29, 233-236, 1990
15. B. V. K. Vijaya Kumar, Z. Bahri, Phase-only Filters with Improved Signal to Noise Ratio, Appl. Opt., 28, pp. 250-257, 1989
16. F. M. Dickey, B. V. K. Vijaya Kumar, L. A. Romero, J. M. Connelly, Complex Ternary Matched Filters Yielding High Signal to Noise Ration, Opt. Eng. , v. 29, p. 994-1001, 1990
17. W. W. Farn, J. W. Goodman, Optimal Binary Phase-only Matched Filters, Appl. Opt., v. 27, pp. 4431-4437, 1988
18. A. Mahalanobis, B. V. K. Vijaya Kumar, D. Casasent, Minimum Average Correlation Energy Filters, Appl. Opt., v. 26, pp. 3633-3640, 1987
19. M. Fleisher, U. Mahlab, J. Shamir, Entropy Optimized Filter for Pattern Recognition, Appl. Opt. V. 29, pp. 2091-2098, 1990
20. K. Chalasinska-Macukow, "Generalized Matched Spatial Filters with Optimum Light Efficiency", in Optical Processing and Computing, H. H. Arsenault, T. Szoplik and B. Macukow, Ed. (Academic Press, Boston, 1989), p. 31-45.
21. L.P. Yaroslavsky, "Optical Correlators with (-k)th Law Nonlinearity: Optimal and Suboptimal Solutions", Applied Optics, **34**, pp. 3924-3932 (1995)
22. L. Yaroslavsky, E. Marom, Nonlinearity Optimization in Nonlinear Joint Transform Correlators , Applied Optics , vol. 36, No. 20, 10 July, 1997, pp. 4816-4822
23. L. Yaroslavsky, Optimal Target Location in Color and Multi-component Images, Asian Journal of Physics, Vol 8, No 3 , 1999, pp. 100-113

24. K. Yamaba and Y.Miyake, "Color character recognition method based on human perception," Opt. Eng. **32**, 33-40 (1993).
25. S.L.Guth, "Model for color vision and light adaptation," J. Opt. Soc. Am. **8**, 976-993 (1991).
26. E.Badique, Y.Komiya, N.Ohyama, J.Tsujiuchi and T.Honda, "*Color image correlation*," Optics Comm. **62**, 181-186, (1987).
27. E.Badique, N.Ohyama, T.Honda and J.Tsujiuchi, "*Color image correlation for spatial/spectral recognition and increased selectivity*," Optics Comm. **68**, 91-96, (1988).
28. V. Kober, V. Lashin, L. Moreno, J. Campos, L. Yaroslavsky, and M.J. Yzuel, Color component transformations for optical pattern recognition, JOSA, A/vol. 14, No.10, October 1997, pp.2656-2669

Chapter 12

NONLINEAR FILTERS IN IMAGE PROCESSING

Since J.W. Tukey introduced median filters in signal processing ([1]), a vast variety of nonlinear filters and families of nonlinear filters for image processing has been suggested. In order to ease navigating in this ocean of filters we will provide in this chapter a classification of the filters described in the literature and describe some most useful filters for image denoising, enhancement and segmentation. The classification is aimed at revealing general common principles in the filter design and their efficient implementation in serial computers and parallel computational networks.

In Sect. 12.1 main assumptions and fundamental notions of pixel neighborhoods and their attributes and estimation operations that constitute the base of the classification are introduced. Then, in Sects. 12.2-4, respectively, listed and explained are: typical pixel and neighborhood attributes, typical estimation operations involved in the filter design and typical neighborhood building operations that were found in the result of analysis of a large variety of nonlinear filters known from literature ([2-10]). In Sect. 12.5 classification tables of the filters are provided in which filters are arranged according to the order and the type of neighborhood they use. Iterative, cascade and recursive implementation of filters are reviewed in Sects. 7 and 8. Sect. 9 illustrates some new filters that naturally follow from the classification and in Sect.10 filter implementation in parallel neuromorphic structures is briefly discussed.

12.1 CLASIFICATION PRINCIPLES

12.1.1 Main assumptions and definitions

We will assume that images are single component signals with scalar and quantized values.

Main common properties of the filters that constitute the base of thei classification are:

- Filtering is performed within a filter window. For processing signals and images filter window scans input image sample by sample.
- In each position $\boldsymbol{k}$ of the window, with $\boldsymbol{k}$ being a coordinate in the signal domain, filters generate, from input signal samples $\left\{b_n^{(k)}\right\}$ within the window, an output value $\hat{a}_k$ for this position by means of a certain estimation operation **ESTM** applied to a certain subset $NBH\left\{b_n^{(k)}\right\}$ of window samples:

$$\left\{b_n^{(k)}\right\} \rightarrow \hat{a}_k : \hat{a}_k = \mathbf{ESTM}\left(NBH\left\{b_n^{(k)}\right\}\right). \tag{12.1}$$

We will associate this estimate of the filter output with window central sample and will call subset $NBH\left\{b_n^{(k)}\right\}$ of samples used as inputs to estimation operations ***neighborhood*** of the window central sample.

- The neighborhood is formed on the base of window sample attributes (to be discussed in Sect.12.1.2). The process of forming neighborhood may, in general, be multi stage one beginning from the initial W-neighborhood (***Wnbh***) formed from all filter window samples through a series of intermediate neighborhoods. Intermediate neighborhoods may, in addition to its pixel attributes, have attributes of their own associated with the neighborhood as a whole and obtained through an estimation operation over the neighborhoods formed at the previous stage.

Nonlinear filters will be specified in terms of neighborhood forming and estimation operations they use. Such an approach enables unified treatment of the filters and their structurization that can be directly translated into their algorithmic implementation on serial computers and hardware implementation on parallel computational networks.

This concept is schematically illustrated in a flow diagram of Fig. 12-1. Filter window samples with their attributes form a primary "window neighborhood" ***Wnbh***. On the next level, this first level neighborhood $NBH^1 \equiv Wbnh$ is used to form, through a number of neighborhood building operations, a second level neighborhood NBH^2. In this illustrative example, it is used for generating, by means of an estimation operation, filter output pixel for the particular position of the window.

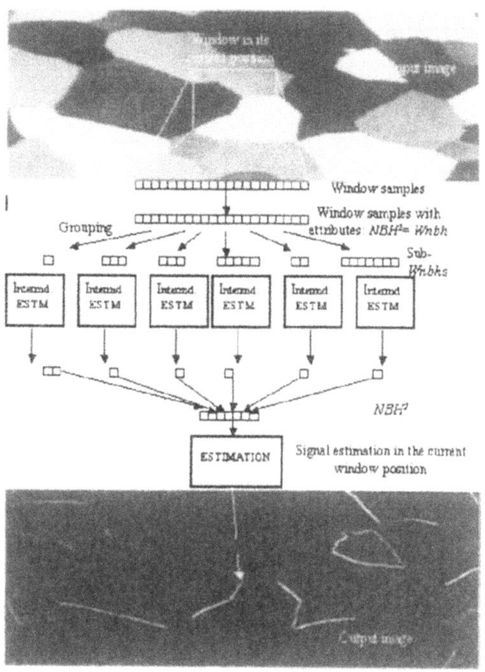

Figure 12 -1. The concept of image nonlinear filtering

12.1.2 Typical signal attributes

Primary signal sample attributes that determine filtering operations are sample (pixel) gray values and their co-ordinates. It turns out, however, that a number of attributes other then only these primary ones are essential for nonlinear filtering. Table 12-1 lists typical digital signal sample attributes that are involved in the design of nonlinear filters known from the literature. As one can see from the table, these "secondary" attributes reflect features of pixels as members of their neighborhood.

Attributes **Rank** and **Cardinality** describe statistical properties of pixels in neighborhoods. They are interrelated and may actually be regarded as two faces of the same quality. While **Rank,** as it is described in Sect. 7.1.1, is associated with the variational row, i.e., ordered in ascending order sequence of neighborhood pixel values, **Cardinality** is associated with the histogram over the neighborhood. "Geometrical" attributes describe properties of images as surfaces in 3-D spaces, respectively. **Membership in**

neighborhood and **Spatial connectedness** are binary attributes that take values 0 and 1 and specify topological relationship between signal sample and a neighborhood. Neighborhood elements are regarded spatially connected if one can connect them by a line that passes through the samples that all belong to the neighborhood.

Table12-1. Typical attributes of digital signals

Attribute		Definition
	Primary attributes	
Value	a_k	Sample (pixel) gray value
Co-ordinate	$k(a)$	Sample (pixel) coordinate in the filter window
	Secondary attributes	
Cardinality	$H(a)$=HIST(NBH,a)	Number of neighborhood elements with the same value as that of element **a** (defined for quantized signals): $H(a)=\sum_{k\in NBH}\delta(a-a_k)$
Rank	R_a=RANK(NBH,a)	- Number of neighborhood elements with values lower than a - Position of value a in *variational row* (ordered, in ascending values order, sequence of the neighborhood elements) - $R_a=\sum_{v=0}^{a}H(v)$
Geometrical attributes	COORD(NBH,R)	Co-ordinate of the element with rank R (R -th rank order statistics)
	GRDNT(NBH,k)	Signal gradient in position k
	CURV(NBH,a_r)	Signal curvature in position k
Membership in the neighborhood	MEMB(NBH,a).	A binary attribute that evaluates by 0 and 1 membership of element **a** in the neighborhood.
Spatial connectedness	CONCTD(NBH,a)	A binary attribute that evaluates by 0 and 1 spatial connectedness of element **a** with other elements of the neighborhood

12.1.3 Estimation operations

Typical estimation operations used in known nonlinear filters are listed in Table 12-2. Two lasses of estimation operations may be distinguished: data smoothing operations, and operations that evaluate data spread in the neighborhood. In the filter design, selection of estimation operation is, in general, governed by requirements of statistical or other optimality of the estimate. For instance, arithmetic **MEAN** is an optimal MAP- (Maximal A Posteriori Probability) estimation of a location parameter of data in the assumption that data are observations of a single value distorted by an additive uncorrelated Gaussian random values (noise). It is also an estimate that minimizes mean squared deviation of the estimate from the data. **PROD** is an operation homomorphic to the addition involved in **MEAN** operation: sum of logarithms of a set of values is logarithm of their product.

Table12-2. Estimation operations

Operation	Denotation	Definition
SMTH (*NBH*): Data smoothing operations		
Arithmetic and geometric mean	**MEAN(*NBH*)**	Arithmetic mean of samples of the neighborhood
	PROD(*NBH*)	Product of samples of the neighborhood
Order statistics	**K_ROS(*NBH*)**	Value that occupies *K*-th place (has *rank* ***K***) in the variational row over the neighborhood. Special cases:
	MIN(*NBH*)	Minimum over the neighborhood (the first term of the variational row)
	MEDN(*NBH*)	Central element (median) of the variational row
	MAX(*NBH*)	Maximum over the neighborhood (the last term of the variational row);
Histogram mode	**MODE(*NBH*)**	Value of the neighborhood element with the highest cardinality: $\mathbf{MODE}(NBH) = \arg\max\left(H\left(NBH\right)\right)$
RAND(*NBH*)	A random (pseudo-random) number taken from an ensemble with the same gray level distribution density as that of elements of the neighborhood	
SPRD(*NBH*): Operations that evaluate spread of data within the neighborhood		
STDEV(*NBH*)	Standard deviation over the neighborhood	
IQDIST(*NBH*)	Interquantil distance $\mathbf{R\_ROS}(NBH) - \mathbf{L\_ROS}(NBH)$, *where* $1 \le L < R \le \mathbf{SIZE}(NBH)$.	
RNG(*NBH*)	Range $\mathbf{MAX}(NBH) - \mathbf{MIN}(NBH)$	
SIZE(*NBH*)	Number of elements of the neighborhood	

ROS operations may be optimal MAP estimations for other then additive Gaussian noise models. For instance, if neighborhood elements are observations of a constant distorted by addition to it independent random values with exponential distribution density. **MEDN** is known to be optimal MAP estimation of the constant. It is also an estimate that provides minimum to average modulus of its deviation from the data. If additive noise samples have one-sided distribution and affect not all data, **MIN** or **MAX** might be optimal estimations. **MODE** can be regarded an operation of obtaining MAP estimation if distribution histogram is considered a posteriori distribution of signal gray level.

RAND is a "stochastic" estimation operation. It generates an estimate that, statistically, is equivalent to all above "deterministic" estimates. This quite unusual operation is illustrated in Fig. 12-2. Images on this figure clearly show that replacement of image samples with pseudo-random numbers by **RAND**-operation provides a reasonably good estimate of pixel gray levels from pixel neighborhoods. It is especially true for higher order neighborhoods. For instance, one can hardly distinguish visually image (c) from initial image (a). Difference (e) between these two images looks almost chaotic which means that no substantial visual information is lost in image (c).

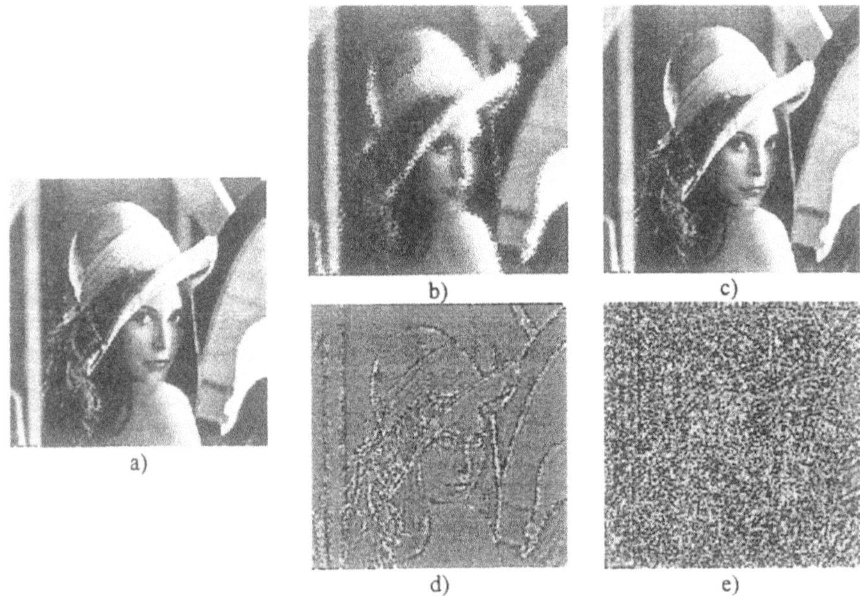

Figure 12-2. Illustration of **RAND**(*NBH*) as a data smoothing operation: a) - initial test image with 256 quantization levels; b) - result of applying operation RAND(Wnbh) in the window of 7×7 pixels; c) - result of applying operation RAND(EV-nbh) operation in the filter window of 7×7 pixels and $\varepsilon_v^+=\varepsilon_v^-=10$ (for the definition of EV-neighborhood see Sect. 12.1.4); d) - difference between images (a) and (d); standard deviation of the difference is 26; t) - difference between images and (e); standard deviation of the difference is 5.6.

All above-mentioned operations belong to a class of data smoothing operations. We will use for them a common denotation **SMTH**. The rest of operations generate numerical measures of neighborhood data spread. We will use for them a common denotation **SPRD.** Their two modifications, "interquantil distance" **IQDIST** and "range" **RNG** ones, are recommended as a replacement for standard deviation for the evaluation of spread of data.. **SIZE** operation computes number of samples in the neighborhood, when it does not directly follows from the neighborhood definition.

12.1.4 Neighborhood building operations

Neighborhood building operations may be unified in two groups: operations that generate a scalar attribute of the neighborhood as a whole ("scalar" operations) and those (vectorial) that are used in multi stage process of forming neighborhood. The latter generate, for neighborhood elements, a new set of elements and their attributes. This set of elements forms a neighborhood of the next stage. "Scalar" neighborhood building operations are basically the same as estimation operations listed in Table 12-2. Typical vectorial operations are listed in Table 12-3.

Table 12 -3. Vectorial neighborhood building operations

FUNC(*NBH*): Element wise functional transformation of neighborhood elements	
MULT_Attr(*NBH*): Multiplying elements of the neighborhood by some weights	
MULT_C(*NBH*)	weighting coefficients are defined by element co-ordinates
MULT_V(*NBH*)	weight coefficients are defined by element values
MULT_R(*NBH*)	weight coefficients are defined by element ranks
MULT_H(*NBH*)	weight coefficients are defined by the cardinality of the neighborhood elements
MULT_G(*NBH*)	weight coefficients are defined by certain geometrical attributes of the neighborhood elements,
MULT_AA(*NBH*) **MULT_CR(*NBH*)**	weight coefficients depend on combination of attributes, for instance, on both co-ordinates and ranks of neighborhood elements
REPL_Attr(*NBH*): Replicating elements of the neighborhood certain number of times according to elements' attributes	
SELECT_Attr(*NBH*)	Attribute controlled selection of one sub-neighborhood from a set: **SELECT_A(*NBH*)** = ***Subnbh***
MIN_Std($SubWnbh_1$, $SubWnbh_2$,..., $SubWnbh_n$)	Standard deviation over the neighborhood as the attribute
MIN_RNG($SubWnbh_1$, $SubWnbh_2$, ..., $SubWnbh_n$)	Neighborhood range as the attribute

Table 12 -3 (contd.) Vectorial neighborhood building operations

***C*-neighborhoods: pixel co-ordinates as attributes**	
SHnbh Shape-neighborhoods	Selection of neighborhood elements according to their co-ordinates. In 2-D and multi-dimensional cases: neighborhoods of a certain spatial shape.
***V*-neighborhoods:** pixel values as attributes	
"*Epsilon-V*"-neighborhood of element a_k: $EVnbh(NBH;a_k;\varepsilon_v^+;\varepsilon_v^-)$	A subset of elements with values $\{a_n\}$ that satisfy inequality: $a_k-\varepsilon_v^-\le a_n\le a_k+\varepsilon_v^+$.
"***K*** *nearest by value*"-neighborhood of element a_k $KNVnbh(NBH;a_k,K)$:	A subset of K elements with values $\{a_n\}$ closest to that of element a_k.
Range-neighborhood: $RNGnbh(NBH,V_{mn},V_{mx})$-	A subset of elements with values $\{V_k\}$ within a specified range $\{V_{mn}<V_k<V_{mx})$
***R*-neighborhoods:** pixel ranks as attributes	
$ERnbh(NBH;a_k;\varepsilon_R^+;\varepsilon_R^-)$ "*epsilon-R*"- neighborhood	A subset of elements with ranks $\{R_n\}$ that satisfy inequality: $R_k-\varepsilon_R^-\le R_n\le R_k+\varepsilon_R^+$.
$KNRnbh(NBH;a_k,K)$ - "***K****-nearest by rank*" neighborhood of element a_k	A subset of K elements with ranks closest to that of element a_k.
$Qnbh(NBH,R_{left},R_{right})$ Quantil-neighborhood	Elements (order statistics) whose ranks $\{R_r\}$ satisfy inequality $1<R_{left}<R_r<R_{right}<SIZE(Wnbh)$
***H*-neighborhoods:** pixel cardinalities as attributes	
$CLnbh(NBH;a_k)$ - "Cluster" neighborhood of element a_k.	Neighborhood elements that belong to the same cluster of the histogram over the neighborhood as that of element a_k.
***G*-neighborhoods:** Geometrical attributes	
FLAT(NBH) – "Flat"-neighborhood	Neighborhood elements with values of Laplacian (or module of gradient) lower than a certain threshold
Linear combination of elements of neighborhood	
T(*NBH*)	Orthogonal transform **T** of neighborhood elements
DEV(*NBH,a*)	Differences between elements of the neighborhood and certain value ***a***

One can distinguish the following groups of vectorial neighborhood building operations:

- functional element-wise transformations;
- replication operations;
- selection operations;
- operations that use sample co-ordinates as attributes (*C*-neighborhoods);
- operations that use sample gray values as attributes (*V*-neighborhoods);

- operations that use sample ranks as attributes (***R***-neighborhoods)
- operations that use sample cardinalities as attributes (***H***-neighborhoods);
- operations that use geometrical attributes (***G***-neighborhoods);
- linear combination operations.

By functional transformations we assume applying, element-wise, nonlinear functions such as, for instance, logarithmic one, to the neighborhood elements. A special case of the functional transformations, **MULT**-operations, is multiplying neighborhood elements by scalar "weight" coefficients that are selected according certain attributes (co-ordinates, value, rank, cardinality), or according to a combination of attributes. Replication **REPL**-operations can be regarded as a version of weighting with integer weights and are used in data sorting.

Selection **SELECT_A** operations select from the neighborhood some sub-groups of elements, or sub-neighborhoods according to certain attributes of the sub-neighborhoods such as, for instance, standard deviation od signal values in the sub-neighborhoods or sub-neighborhood range.

Neighborhoods with elements selected according to their coordinates in the filter window are referred to as shape-neighborhoods as they correspond to certain spatial shapes, such as, for instance, circle, cross, diagonal, etc. Shape neighborhoods are key notions in morphological filters ([2]).

Signal gray value is one of the most important signal attributes. Two types of neighborhood are based on it: ***EV***- and ***KNV***-neighborhoods. The former is composed of signal samples with gray values that deviate from the gray value of the window central sample not more then by certain predefined values $\{\varepsilon_v^+, \varepsilon_v^-\}$. The latter unify in the neighborhood certain predefined number ***K*** of window samples with gray values closest to that of the window central sample.

For ranks as signal attributes, one can introduce such analogs of ***EV***- and ***KNV***-neighborhoods as ***ER***-, ***KNR***- neighborhoods. Quantil ***Q***-neighborhood is also a variant of the rank based ***R***-neighborhoods.

EV-, ***KNV***-, ***ER***-neighborhoods are used in a number of very efficient noise cleaning algorithms (see Sect. 12.4). Fig. 12-3 illustrates the principle feature controlled selection of neighborhood elements on an example of ***EV***-, ***KNV***-, ***ER***-neighborhoods.

Cluster ***CL***- neighborhood is an example of neighborhoods built using pixel cardinality as the attribute. Neighborhoods built using filter window geometrical attributes are exemplified by **FLAT**-neighborhood composed of pixels with low gradient.

Neighborhoods built using linear combination of widow samples operations are exemplified by spectra of window samples obtained using orthogonal transforms such as DCT, DFT, Haar or alike and by differences between window samples and certain value, such as, for instance, that of the window central pixel.

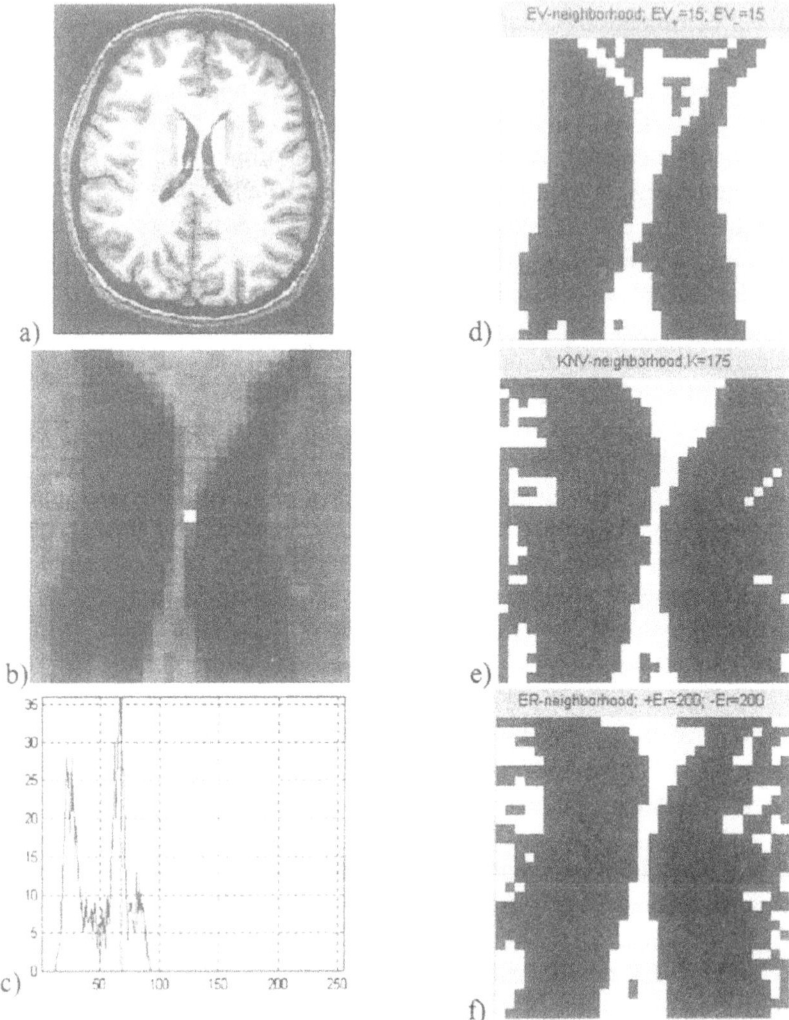

Figure-12.3. Examples of formation of EV-, KNV- and ER-neighborhoods: a) - input image; b) - image fragment in a window 31x31 pixels (highlighted in (a)) with window central pixel marked by the white square; c - histogram of the fragment with central pixel gray value indicated by a line; d-f - results of the fragment segmentation to EV, KNV and ER - neighborhoods (shown in white on black background).

12.2 FILTER CLASSIFICATION TABLES

12.2.1 Transferential filters

In this section we provide, in tables 12-4 through 12-7, a list of a number of the most nonlinear filters classified according to the number of stages in building neighborhood they use for generating final estimation of the filter output. We call them transferential in order to distinguish them from filters with feedback described in Sect. 12.2.2. The list is mostly based on data collected in Ref.[9] as well as in Refs. [3, 4, 8, 9, 10] to which readers can refer for detailed information regarding properties and applications of the filters. Most of the filters are intended for cleaning additive noise in data. In other cases we provide an additional indication of the filtering purpose.

Table 12-4 lists the simplest nonlinear filters that use one-stage $\boldsymbol{NBH}^1$ neighborhood, the primary window ***Wnbh***-neighborhood. In particular, one can find in the table such popular in signal and image processing filters as moving average, median and local histogram equalization filters. We will illustrate capabilities of these filters in Sect. 12.4.

It appears that the majority of known nonlinear filters belongs to the family of two-stage $\boldsymbol{NBH}^2$ neighborhood filters listed in Table 12-5. According to the type of the $\boldsymbol{NBH}^2$-neighborhood used, the filters form four groups: **MULT_A**-, **REPL_A**-, **V**-, and **R**-neighborhood filters. In Sect. 12.4 we will compare capabilities of some of these filters with corresponding one stage filters.

Among three-stage neighborhood filters listed in Table 12-6 one can find two large families of filters: transform domain filters and stack filters. Transform domain filters nonlinearly modify transform coefficients of filter window samples to generate filter output by means of applying to them operation **MEAN** which is an implementation of the inverse transform for the window central sample. Two the most advanced modifications of these filters are sliding window DCT ([8,11]) and wavelet shrinkage filters ([12]). They are described in details in Sect. 8.3. A popular in image processing community Local Linear Minimum Mean Square Error (LLMMSE-) filter is a special case of transform domain filters in which signal squared transform coefficients $|\mathbf{T}(\boldsymbol{Wnbh})|^2$ (signal spectral estimations) are replaced by their mean values $(\mathbf{STD}(\boldsymbol{Wnbh}))^2$.

Stack filters belong to yet another large family of filters. They originate from the idea of threshold decomposition of multilevel signals to binary signals to which Boolean functions are then applied ([13]).

Four stage neighborhood filters are exemplified in Table 12-7 by a family of polynomial filters and by Weighted Majority of ***m*** Values with Minimum

Range (WMVMS-, or Shorth-) filter that implements an idea of data smoothing by averaging over data subset that has minimal spread.

Table 12-4. One stage (***Wnbh***-based) filters

Signal "smoothing" filters		
Moving average filter		$\hat{a}_k = \mathrm{MEAN}(\boldsymbol{Wnbh})$
"Ranked order" ("percentile") filters		$\hat{a}_k = \mathrm{K\_ROS}(\boldsymbol{Wnbh})$
	Median filter	$\hat{a}_k = \mathrm{MEDN}(\boldsymbol{Wnbh})$
	MAX-filters	$\hat{a}_k = \mathrm{MAX}(\boldsymbol{Wnbh})$
	MIN- filters	$\hat{a}_k = \mathrm{MIN}(\boldsymbol{Wnbh})$
Adaptive Mode Quantization filter		$\hat{a}_k = \mathrm{MODE}(\boldsymbol{Wnbh})$
Signal "enhancement" filters		
Local histogram equalization		$\hat{a}_k = \mathrm{RANK}(\boldsymbol{Wnbh})$
Quasi-range filter		$\hat{a}_k = \mathrm{QSRNG}(\boldsymbol{Wnbh}) = \mathrm{R\_ROS}(\boldsymbol{Wnbh}) - \mathrm{L\_ROS}(\boldsymbol{Wnbh})$
Local variance filter		$\hat{a}_k = \mathrm{STDEV}(\boldsymbol{Wnbh})$

Table 12-5. Two stage (***NBH***^{2}-based) filters

FUNC(*NBH*)		
General		$\hat{a}_k = \mathrm{MEAN}(\mathrm{FUNC}(\boldsymbol{NBH}))$
MULT-A- neighborhood filters		
RMSE optimal linear filters		$\hat{a}_k = \mathrm{MEAN}(\mathrm{MULT\_C}(\boldsymbol{Wnbh}))$
L-filters, Rank Selection filters; C-filters (Ll-filters)		$\hat{a}_k = \mathrm{MEAN}(\mathrm{MULT\_R}(\boldsymbol{Wnbh}))$ $\hat{a}_k = \mathrm{MEAN}(\mathrm{MULT\_RC}(\boldsymbol{Wnbh}))$
REPL-A - neighborhood filters		
Weighted median filters		$\hat{a}_k = \mathrm{MEDN}(\mathrm{REPL\_C}(\boldsymbol{Wnbh}))$;
Weighted K-ROS - filters		$\hat{a}_k = \mathrm{K\_ROS}(\mathrm{REPL\_C}(\boldsymbol{Wnbh}))$
Morphological filters	Dilation filter:	$\hat{a}_k = \mathrm{MAX}(\boldsymbol{SHnbh})$
	Erosion filter	$\hat{a}_k = \mathrm{MIN}(\boldsymbol{SHnbh})$
	Soft Morph. filters	$\hat{a}_k = \mathrm{ROS}(\boldsymbol{SHnbh})$
V-neighborhood filters		
K-Nearest Neighbor filter		$\hat{a}_k = \mathrm{MEAN}(\boldsymbol{KNV}(\boldsymbol{Wnbh}; \boldsymbol{a}_k; \boldsymbol{K}))$
"Sigma"- filter		$\hat{a}_k = \mathrm{MEAN}(\boldsymbol{EVnbh}(\boldsymbol{Wnbh}; \boldsymbol{a}_k; \varepsilon_V^+; \varepsilon_V^-))$
Modified Trimmed Mean filters		$\hat{a}_k = \mathrm{MEAN}(\mathrm{EV}(\boldsymbol{Wnbh}; \mathrm{MEDN}(\boldsymbol{Wnbh}); \varepsilon_V^+; \varepsilon_V^-))$

Table 12-5 (Contd.) Two stage (***NBH***^{2}-based) filters

R-neighborhoods		
Alpha-trimmed mean, median		$\hat{a}_k = \mathbf{MEAN}(Qnbh(Wnbh, R_{left}, R_{right}))$; $\hat{a}_k = \mathbf{MEDN}(Qnbh(Wnbh, R_{left}, R_{right}))$
Impulse noise filtering filters	General	$\hat{a}_k = \mathbf{MEMB}(Qnbh(Wnbh, R_{left}, R_{right}), a_k) \cdot a_k + [1 - \mathbf{MEMB}(Qnbh(Wnbh, R_{left}, R_{right}), a_k)] \times \mathbf{SMTH}(Qnbh(Wnbh, R_{left}, R_{right}))$
	Rank Conditioned Median filter	$\hat{a}_k = \mathbf{MEMB}(Qnbh(Wnbh, R_{left}, R_{right}), a_k) \cdot a_k + [1 - \mathbf{MEMB}(Qnbh(Wnbh, R_{left}, R_{right}), a_k)] \times \mathbf{MEDN}(Qnbh(Wnbh, R_{left}, R_{right}))$

Table -6. Tree stage (***NBH***^{3}-based) filters

Transform domain filters	"Soft" thresholding	$\hat{a}_k = \mathbf{MEAN}(\mathbf{H} \cdot \mathbf{T}(Wnbh))$; $\mathbf{H} = diag\left\lfloor \max\left(\lvert \mathbf{T}(Wnbh)\rvert^2 - \sigma^2 / \lvert \mathbf{T}(Wnbh)\rvert^2, 0\right)\right\rfloor$
	"Hard" thresholding	$\hat{a}_k = \mathbf{MEAN}(\mathbf{STEP}\{\lvert\mathbf{T}(Wnbh)\rvert - \sigma\} \cdot \mathbf{T}(Wnbh))$, where σ is a filter parameter, $\mathbf{STEP}(x) = \begin{cases} 0, x \le 0 \\ 1, x > 0 \end{cases}$
LLMMSE-filter	$\hat{a}_k = \left(1 - \frac{\sigma^2}{(\mathbf{STD}(Wnbh))^2}\right) a_k + \frac{\sigma^2}{(\mathbf{STD}(Wnbh))^2} \mathbf{MEAN}(Wnbh)$, where σ^2 is a filter parameter	
Double Window Modified Trimmed Mean filter	$\hat{a}_k = \mathbf{MEAN}(EVnbh(Wnbh; \mathbf{MEDN}(SHnbh); \varepsilon_V^+; \varepsilon_V^-))$.	
Stack filters	$\hat{a}_k = \mathbf{MAX}(\mathbf{MIN}(SubWnbh_1), \mathbf{MIN}(SubWnbh_2), ..., \mathbf{MIN}(SubWnbh_n))$.	

Table -7. Four stage (***NBH***^{4}-based) filters

Polynomial filters	$\hat{a}_k = \mathbf{MEAN}(MULT\_C(\mathbf{PROD}(SubWnbh_1), ..., \mathbf{PROD}(SubWnbh_n)))$
WMVMS - filters	$\hat{a}_k = \mathbf{MEAN}(\mathbf{MULT\_R}(\mathbf{MIN\_RNG}(\{SubRnbh_i^{(m)}\})))$, where $\{SubRnbh_i^{(m)}\}$ are rank based sub-neighborhoods of m elements.

12.2.2 Iterative filtering.

An important common feature of the nonlinear filters is their local adaptivity: the way how filter output is computed depends, in each filter window position, on window sample attributes. As it was already mentioned in Sect. 8. 3, this filtering method may mathematically be explained with the use of local criteria of processing introduced in Sect. 2.1.1. According to Eq. 2.1.4, optimal processing filter is the filter that, from window samples $\{b_n^{(k)}\}$ in the window position k, generates estimates $\{\hat{a}_n^{(k)}\}$ of the window samples that minimize average deviation $\{LOSS(a_n^{(k)},\hat{a}_n^{(k)})\}$ of the estimates from samples $\{a_n^{(k)}\}$ of a hypothetical true signal

$$\{\hat{a}_n^{(k)}\} = \underset{\{b_n^{(k)}\}\to\hat{a}_k}{\arg\min}\, AV\left\{\sum_n LOC(m;a_k)LOSS(a_n^{(k)},\hat{a}_n^{(k)})\right\}. \tag{12.2.1}$$

The averaging in the criterion is two-fold: statistical averaging over an ensemble of images for which the filter is intended and a spatial averaging. The spatial averaging over window sample index n is, in general, a weighted summation carried out over a subset of signal samples associated with central sample a_k of the window that form its neighborhood NBH and are defined by a locality function $LOC(m;a_k)$. As one can see from Eq. 12.2.1, the locality function, and therefore the neighborhood, in principle, depends on "true" value a_k of the central sample. Therefore optimal processing algorithm depends on signal true values that are required to specify the locality function. For these values are not known and are the goal of the processing, this implies that optimal estimation algorithm should, in principle, be iterative:

$$\hat{a}_k^{(t)} = \mathbf{ESTM}(NBH^{(t-1)}), \tag{12.2.2}$$

where t is iteration index. In iterative filtering, filters are supposed to converge to "true" values. Therefore, in particular applications, filters should be selected according to their "root" signals (fixed points).

Experimental experience shows that iterative nonlinear noise cleaning filters substantially outperform non-iterative ones. We will compare iterative and non iterative filtering in Sect. 12.3.

12.2.3 Multiple branch parallel, cascade and recursive filters

An important problem of iterative filtering is that of adjustment of neighborhood building and estimation operations according to changing statistics of data that takes place in course of iterations. This may require iterative wise change of the filter parameters. One of the possible solutions of the problem is to combine in one filter several filters acting in parallel branches and switch between them under control of a certain auxiliary filter that evaluates the changing statistics.

Iterative filtering and modification of the estimation and neighborhood building operations may also be implemented in cascade filtering when each filter in cascade operates with its own neighborhood and estimation operation. Several examples of cascade filters are listed in Table 8.

Table -8. Cascade filters

<table>
<tr><td colspan="3">Multistage Median Filters: cascaded median filters</td></tr>
<tr><td colspan="3">Median Hybrid Filters: cascaded alternating median and linear filters</td></tr>
<tr><td rowspan="4">Alternative sequential morphological filters</td><td>Closing</td><td>$\mathbf{MIN}(SHbnh(\mathbf{MAX}(SHnbh)))$</td></tr>
<tr><td>Opening</td><td>$\mathbf{MAX}(SHnbh(\mathbf{MIN}(SHnbh)))$</td></tr>
<tr><td>Close-opening</td><td>$\mathbf{MAX}(SHnbh(\mathbf{MIN}(SHnbh(\mathbf{MIN}(SHnbh(\mathbf{MAX}(SHnbh)))))))$</td></tr>
<tr><td>Open-closing</td><td>$\mathbf{MIN}(SHnbh(\mathbf{MAX}(SHnbh(\mathbf{MAX}(SHnbh(\mathbf{MIN}(SHnbh)))))))$</td></tr>
<tr><td>Quasi-spread filter</td><td colspan="2">$\hat{a}_k = \mathbf{QSPREAD}(Wnbh) =$ $\mathrm{R\_ROS}(Wnbh(\mathbf{SMTH}(NBH))) - \mathrm{L\_ROS}(Wnbh(\mathbf{SMTH}(NBH)))$</td></tr>
<tr><td>“Wilcoxon test”-filter</td><td colspan="2">$\hat{a}_k = \mathbf{MEAN}(Wnbh(\mathbf{RANK}(Wnbh)))$</td></tr>
<tr><td>“Tamura’s test”-filter</td><td colspan="2">$\hat{a}_k = \mathbf{MEAN}\left(Wnbh\left((\mathbf{RANK}(Wnbh))^P\right)\right)$</td></tr>
<tr><td>“Median test “filter</td><td colspan="2">$\hat{a}_k = \mathbf{MEAN}(Wnbh(sign(\mathbf{RANK}(Wnbh) - \mathbf{SIZE}(Wnbh)/2)))$</td></tr>
</table>

Computational expenses associated with iterative and cascade filtering in conventional sequential computers may be reduced by using, as window samples in the process of scanning signal by the filtering window, those that are already estimated in previous positions of the window. Two examples of recursive filters are shown in Table 9. Recursive filters are not relevant for parallel implementation.

Table -9. Two examples of recursive ***NBH*** filters

Recursive Median filters:	$\hat{a}_k = \mathbf{MEDN}(\boldsymbol{RecWnbh})$
Recursive Algorithm for filtering impulse noise	$\hat{a}_k = a_k \cdot \mathbf{RECT}(\Delta) + [\mathbf{MEAN}(\boldsymbol{RecWnbh}) + \delta_2 \boldsymbol{sign}(\Delta)] \cdot \mathbf{RECT}(-\Delta)$ where $\Delta = \delta_1 - \lvert a_k - \mathbf{MEAN}(\boldsymbol{RecWnbh}) \rvert$ and δ_1 and δ_2 are detection and correction thresholds,

12.2.4 Miscellaneous new filters

The presented classification eases analysis of structure of nonlinear filters by reducing it to the analysis of what type of neighborhood, neighborhood building and estimation operations they use. This analysis may also lead to new filters that "fill in" logical niches in the classification. Several examples of such filters are given In Table 12-10. Examples of filter applications will be illustrated in Sect. 12.3.

SizeEV-controlled Sigma-filter improves image noise cleaning capability of the "sigma"-filter (Table 12.5). Original "Sigma" filter tends to leave untouched isolated noisy pixels whose ***EV***-neighborhood is very small in size if they substantially deviate from their true values. In **SizeEV-controlled Sigma-filter,** size of ***EV***-neighborhood is evaluated as an additional attribute of EV-neighborhood and, if it is lower then a certain threshold ***Thr***, median over the window (or, in general any other of **SMTH** operations) instead of **MEAN(*EV*)** is used to estimate window central pixel.

Size(Evnbh) is a useful attribute of ***EV***-neighborhood that, by itself, can be used to characterize image gray level local inhomogeneity, for example, as "an edge detector", as it will be illustrated in Sect. 12.3

"***Cardnl-filter***" that generates image of local cardinality of its pixels can be regarded as a special case of **Size(*Evnbh*)** filter for $\varepsilon_V^+ = \varepsilon_V^+ = 0$. It can be used for enhancement of small gray level inhomogeneity of images that are composed of almost uniform patches.

P-histogram equalization generalizes local histogram equalization which is its special case for ***P***=1. While histogram equalization enhances local contrasts on a gray level $\boldsymbol{q}$ proportionally to histogram $\boldsymbol{h}(\boldsymbol{q})$, contrast amplification coefficient for ***P***-histogram equalization is proportional to $[\boldsymbol{h}(\boldsymbol{q})]^P$, ***P***-th power of $\boldsymbol{h}(\boldsymbol{q})$. When ***P***=0, ***P***-histogram equalization results in automatic local gray level normalization by local minimum and maximum. For 256 quantization levels, it transforms image gray levels according to the equation:

$$\hat{a}_k = 255\frac{a_k - \mathrm{MIN}(NBH)}{\mathrm{MAX}(NBH) - \mathrm{MIN}(NBH)}. \qquad (12.2.3)$$

Table -10. Some new filters

Size controlled (SC-) "Sigma"-filter	$\hat{a}_k = sw \cdot \mathrm{MEAN}(EVnbh(a_k)) + (1 - sw)\mathrm{MEDN}(SHnbh)$ $sw = \mathrm{RECT}(\mathrm{SIZE}(EVnbh(Wnbh; a_k; \varepsilon_V^+; \varepsilon_V^-)) - Thr)$
Size_EV -filter	$\hat{a}_k = \mathrm{SIZE}(EVnbh(Wnbh; \varepsilon Vpl; \varepsilon Vmmn))$
P- histogram equalization	$\hat{a}_k = \sum_{q=0}^{a}(h(q))^P / \sum_{q=0}^{a_{max}}(h(q))^P$
"Cardnl-filter"	$\hat{a}_k = \mathrm{HIST}(Wnbh, a_k)$
NBH²-histogram equalization	
EVnbh-histogram equalization	$\hat{a}_k = \mathrm{RANK}(EVnbh(Wnbh; a_k; \varepsilon Vpl; \varepsilon Vmn))$
KNVnbh-histogram equalization	$\hat{a}_k = \mathrm{RANK}(KNV(Wnbh; a_k; K))$
"***SHnbh***-histogram equalization	$\hat{a}_k = \mathrm{RANK}(SHnbh)$
Spatially connected (SC-) EV- and R-neighborhood filters	
SC-*K*-Nearest Neighbor filter	$\hat{a}_k = \mathrm{MEAN}(\mathrm{CONCTD}(KNV(Wnbh; a_k; K), a_k))$
SC-"Sigma"-filter	$\hat{a}_k = \mathrm{MEAN}(\mathrm{CONCTD}(EVnbh(Wnbh; a_k; \varepsilon Vpl; \varepsilon Vmn), a_k))$
SC-Modified Trimmed Mean filters	$\hat{a}_k =$ $\mathrm{MEAN}(\mathrm{CONCTD}(EVnbh(Wnbh; \mathrm{MEDN}(Wnbh); \varepsilon Vpl; \varepsilon Vmn), \mathrm{MEDN}(Wnbh)))$
SC-Alpha-trimmed mean, median	$\hat{a}_k = \mathrm{MEAN}(\mathrm{CONCTD}(Qnbh(Wnbh, R_{left}, R_{right}), \mathrm{MEDN}(Wnbh)))$; $\hat{a}_k = \mathrm{MEDN}(\mathrm{CONCTD}(Qnbh(Wnbh, R_{left}, R_{right}), \mathrm{MEDN}(Wnbh)))$

Intermediate values of ***P*** enable flexible, with only one parameter ***P***, control of image local contrast enhancement. One of the immediate applications of ***P***-histogram equalization is image dynamic range blind calibration.

EV- , ***KNV***- and ***SH***-neighborhood equalizations *represent* ***NBH²-histogram equalization.*** It is yet another generalization of the local histogram equalization algorithm when it is carried out over neighborhoods other then initial window neighborhood. Conventional local histogram equalization and ***EV***-neighborhood equalization are compared in Fig. 12-11.

V- and ***R***- neighborhood based filters listed in Table 5 (***K***-Nearest Neighbor, Sigma, Trimmed mean filters and alike) select from samples of the filter window those that are "useful" for subsequent estimation operation by evaluation of their "nearness" to the selected sample in terms of their gray levels and ranks. As one can see from Fig. 12-3, d - f, it certainly may happen that the resulting neighborhood will contain samples that are not spatially connected to the window central pixel. One can refine such a selection by involving an additional check-up for spatial connectivity of neighborhood elements. This is of especial importance in image denoising applications.

12.3 PRACTICAL EXAMPLES

In this section we provide illustrations of application and comparison of nonlinear filters for image denoising, enhancement and segmentation.

12.3.1 Image denoising

In the design of filters for image denoising one should distinguish cases of additive noise, impulse noise and of mixture of both types of noise. The following equation describes the principle of nonlinear filtering for cleaning additive noise:

$$\hat{a}_{k,l}^{(t)} = \mathbf{SMTH}\left(NBH\left(\hat{a}_{k,l}^{(t-1)}\right)\right), \tag{12.3.1}$$

where $\mathbf{SMTH}(\cdot)$ is one of the data smoothing estimation operations listed in Tab. 12.2.

Because additive noise distorts signal gray values, iterative smoothing over *EV*-neighborhoods is most natural. The simplest *EV*-smoothing is "sigma"-filter (Table 12-5):

$$\hat{a}_k = \mathbf{MEAN}\left(EVnbh\left(Wnbh; a_k; \varepsilon_V^+; \varepsilon_V^-\right)\right) \tag{12.3.2}$$

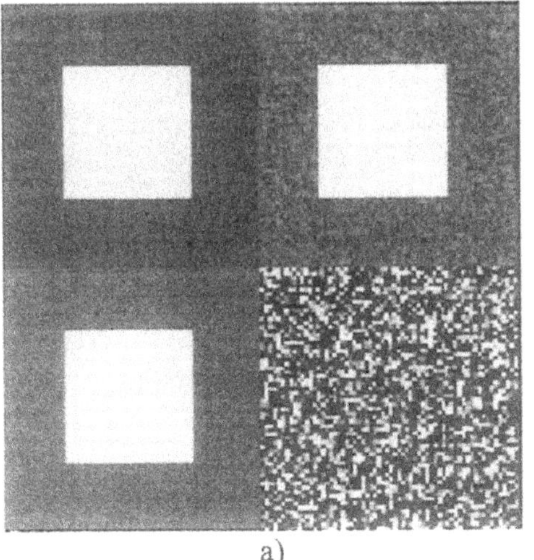

a)

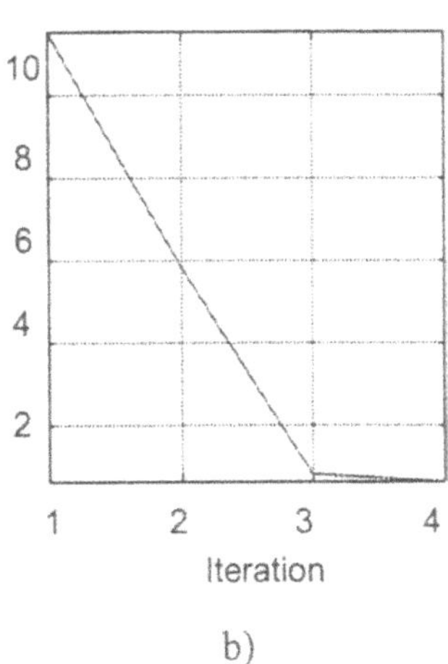

b)

Figure 12-4. Iterative EV-neighborhood filtering additive noise: a) - test piece-wise constant image (1), image distorted by additive noise with a limited dynamic range (2), filtered image after 4 iterations (3), difference between noisy (2) and filtered (3) images amplified for better visibility; b) - standard deviation of difference image as a function of the iteration index

Similar noise smoothing capability has a modification of "Sigma"-filter with ***MEDN*** as a smoothing operation:

$$\hat{a}_k = \mathbf{MEDN}\left(\boldsymbol{EVnbh}\left(\boldsymbol{Wnbh}; \boldsymbol{a}_k; \varepsilon_V^+; \varepsilon_V^-\right)\right). \tag{12.3.3}$$

Being applied iteratively, these filters converge to a piece-wise constant image, which is their root (fixed point) signal. This property enables very efficient suppressing additive noise with values within a limited dynamic range $\left[\varepsilon_v^-, \varepsilon_v^+\right]$ as it is illustrated in Fig. 12-4.

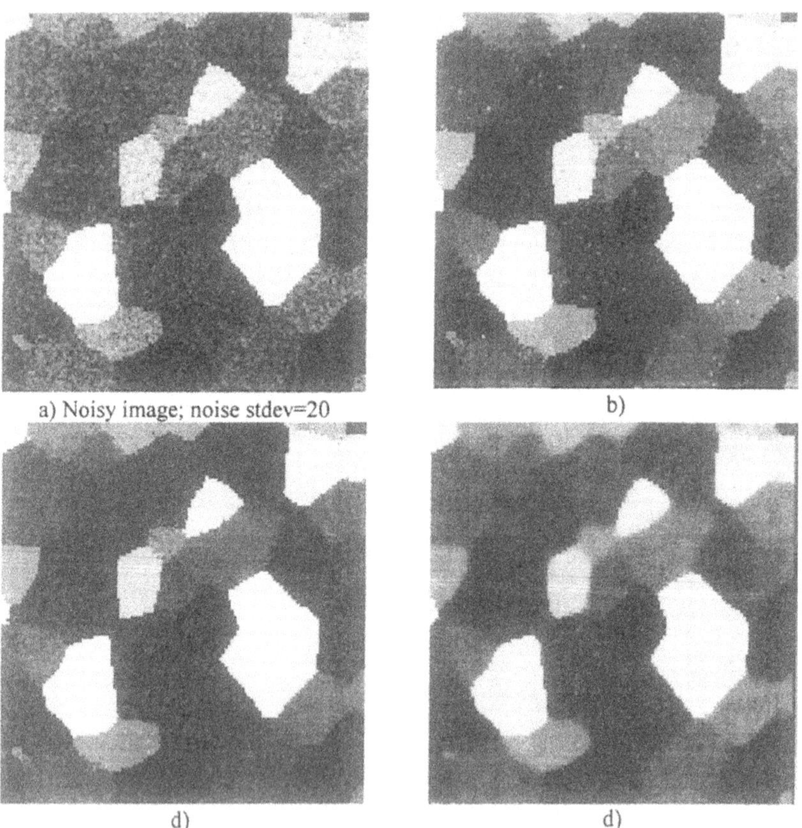

Figure 12-5. Filtering additive noise with normal distribution a) - initial test piece-wise image distorted by additive Gaussian noise with standard deviation 20 within image dynamic range 0-255; b) - result of 5 iterations of "Sigma" filter, with ***Wnbh***=5x5, $\varepsilon_v^- = \varepsilon_v^+ = \mathbf{20}$; c) - result of 5 iterations of size-controlled "Sigma" filter (Eq. 12.3.4) with the same parameters and threshold size $\boldsymbol{Thr} = \mathbf{5}$; d) filtering same image with median filter in the window of 5x5 pixels.

Range $\left[\varepsilon_v^-,\varepsilon_v^+\right]$ defines resolving power of ***EV***-neighborhood smoothing filters in image gray values. When additive noise has a distribution with substantial tails outside this range, "Sigma"-filter and its median modification tends to leave isolated noise outbursts outside the filter range. For such outbursts, EV-neighborhood is very small and no averaging is applied to them. This phenomenon is illustrated in Fig. 12-5, a) and b). As one can see in Fig. 12-5, c), size-controlled "sigma"-filter described in Sect. 12.2.4 (Table 12-10)

$$\hat{a}_k = sw\cdot \mathbf{MEAN}\left(\boldsymbol{EVnbh}\left(a_k\right)\right)+\left(1-sw\right)\mathbf{MEDN}\left(\boldsymbol{SHnbh}\right);$$
$$sw = \mathbf{RECT}\left(\mathbf{SIZE}\left(\boldsymbol{EVnbh}\left(\boldsymbol{Wnbh};a_k;\varepsilon_V^+;\varepsilon_V^-\right)\right)-\boldsymbol{Thr}\right) \qquad (12.3.4)$$

efficiently fixes this drawback of the "Sigma-filter". For comparison, Fig. 12-5, d) shows result of filtering same noisy image with a popular median filter (Table 12-4):

$$\hat{a}_k = \mathbf{MEDN}(\boldsymbol{Wnbh}) \qquad (12.3.5)$$

As one can see, "Sigma-" and SC-"Sigma"-filters outperform median filter very substantially in terms of both noise suppressing capability and preserving image details.

The capability of SC-"sigma" filter to detect isolated noise outbursts may also be effectively utilized for filtering mixture of additive and impulse noise (see Fig. 12-6)

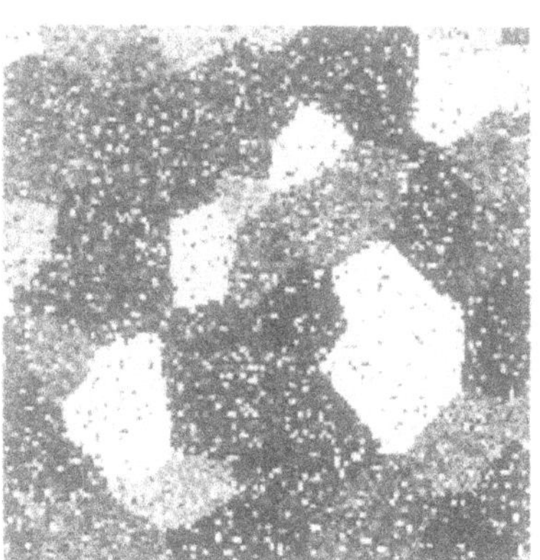

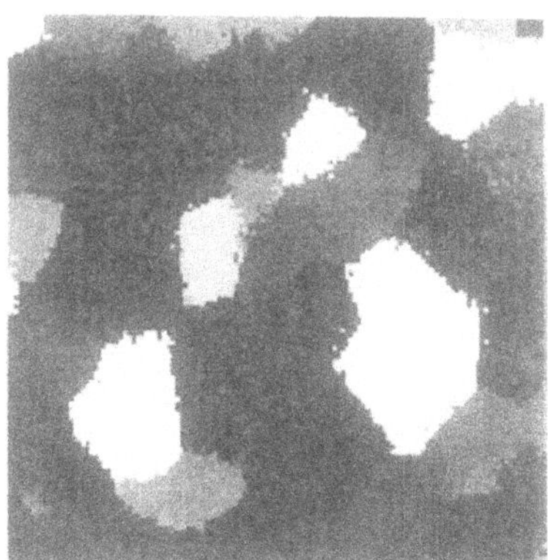

Figure 12-6. Iterative SC-"Sigma" filtering mixture of additive and impulse noise: a) a piece-wise test image distorted by additive Gaussian noise with standard deviation 20 (in image range 0-255) and by impulse noise with probability 0.15; b) SC "Sigma"-filtering result with the same filter parameters as in Fig. 12-6.

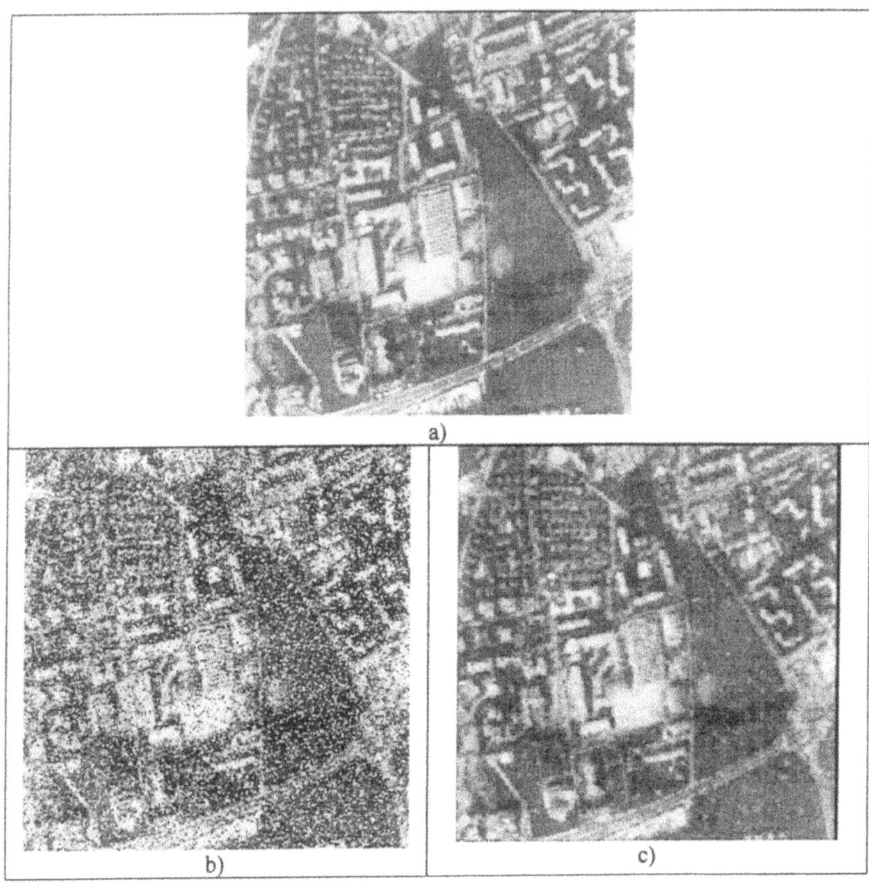

Figure 12-7. Filtering impulse noise: a) initial image; b) - image distorted by impulse noise with probability0.3; c) -filtering result; d)

Detection of noise impulses is a key stage in filtering impulse noise. For filtering impulse noise, one should first detect pixels distorted by noise and then replace defective pixels by its estimate obtained by smoothing over a pixel's neighborhood. This general filtering principle is described by the equation

$$\hat{a}_{k,l}^{(t)} = sw \cdot \hat{a}_{k,l}^{(t-1)} + (1 - sw) \cdot \mathbf{SMTH}_{est}\left(NBH\left(\hat{a}_{k,l}^{(t-1)}\right)\right), \qquad (12.3.6)$$

where sw is a binary switch variable

$$sw = \mathbf{MEMB}\left\{\mathbf{SMTH}_{det}\left(NBH\left(\hat{a}_{k,l}^{(t-1)}\right)\right), \hat{a}_{k,l}^{(t-1)}\right\} \qquad (12.3.7)$$

and $\mathbf{SMTH}_{est}(NBH)$ and $\mathbf{SMTH}_{det}(NBH)$ are smoothing operation for estimation and detection, respectively.

Performance of impulse noise filtering is characterized by the rate of false detecting noise impulses (probability of false alarms), by the rate of missing noise impulses (probability of missing) and by measures of image distortions owing to false alarms, owing to noise impulse missing and of distortions associated with replacement of properly detected image samples by their estimation. Fig. 12-7 illustrates these principles of impulse noise filtering on an example of filter that uses MEAN operation over simple spatial neighborhood ***Wnbh*** of 3x3 pixels:

$$\mathbf{SMTH}_{est}\left(NBH\left(\hat{a}_{k,l}^{(t-1)}\right)\right)=\mathbf{SMTH}_{det}\left(NBH\left(\hat{a}_{k,l}^{(t-1)}\right)\right)=MEAN(Wnbh)\,. \tag{12.3.8}$$

12.3.2 Image enhancement.

Image enhancement is an image processing aimed at assisting visual image analysis. Basically, two types operations are carried out for image enhancement: amplification of local contrasts (***local contrast enhancement***)and detection and enhancement of object boundaries, which is conventionally referred to as ***edge detection***.

Local contrast enhancement is amplification of a deviation of imaged objects from background that eases visual object detection and analysis. The deviation of objects from background is determined in terms of different pixel attributes. One of the most natural attributes is pixel gray value. Local contrast enhancement methods based on pixel gray values belong to a family of methods that may be called "unsharp masking" methods. These methods amplify difference between images and their smoothed copies obtained by one or another image smoothing operation:

$$\hat{a}_k = a_k + g\left[a_k - \mathbf{SMTH}(NBH, a_k)\right]. \tag{12.3.9}$$

When $\mathbf{MEAN}(Wnbh)$ is used as a smoothing operation, the method is similar to aperture correction methods described in Sect. 8.2.3. Using nonlinear smoothing operations reduces artifacts such as object negative framing associated with mutual influence of objects and image background onto the smoothing result that characteristic for linear smoothing. Unsharp masking with linear and nonlinear smoothing as a local contrast enhancement method is illustrated in Fig. 12-8.

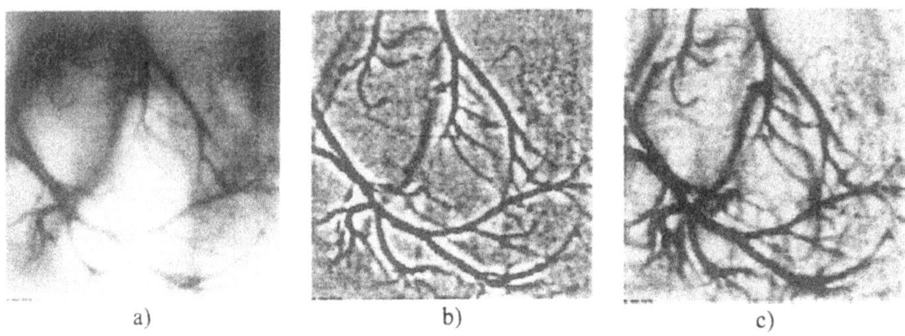

Figure 12-8. Image unsharp masking using **MEAN** (***Wnbh***) and **MEDN**(***Wnbh***) in the window of 15x15 pixels as smoothing operations

Another useful attribute that may be used for emphasizing objects with respect to their background is rank. Local contrast enhancement methods based on pixel ranks are referred to as ***local histogram equalization.*** The family of local histogram equalization algorithms is described by the equation:

$$\hat{\boldsymbol{a}}_k = \mathbf{RANK}(\boldsymbol{NBH}, \boldsymbol{a}_k). \tag{12.3.10}$$

P-histogram equalization described in Sect. 12.2.4 adds flexibility to these algorithms. Fig. 12-9 illustrates image local contrast enhancement by local histogram equalization and ***P***-histogram equalization over spatial neighborhood ***Wnbh***. ***P***-histogram equalization provides softer contrast enhancement and results in more naturally looking images.

Local histogram equalization over other types of neighborhoods may also be useful. Fig. 12-10 illustrates local histogram equalization carried out in windows of different spatial shapes.

Ordinary local histogram equalization over a spatial window and over ***KNV***- and ***EV***-neighborhoods are illustrated in Fig. 12-11. One can observe that latter two methods reveal some details that are invisible in the original image and in the image obtained after local histogram equalization. It is instructive to compare local histogram equalization with unsharp masking shown in Fig. 12-8

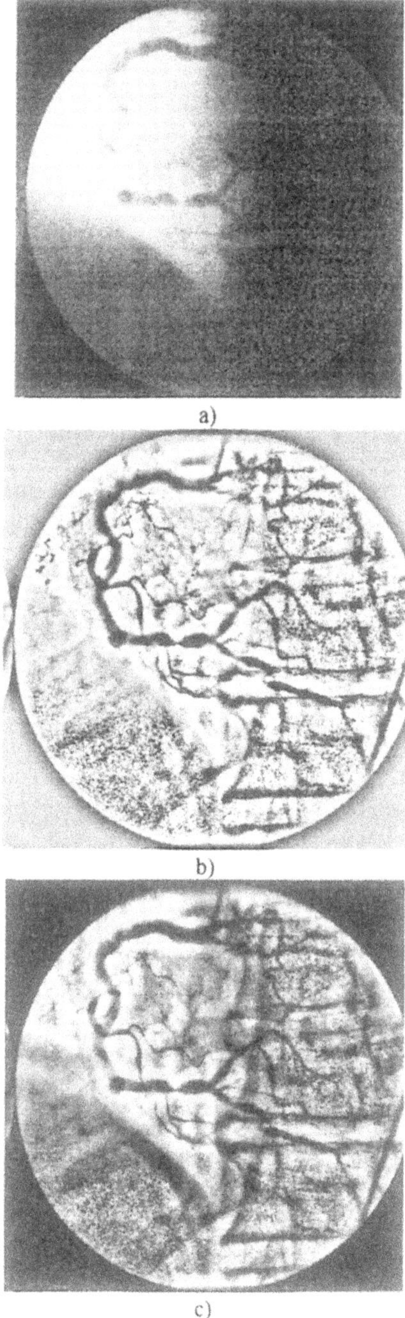

Figure 12-9. Local histogram equalization and *P*-histogram equalization: a) - initial image (256x256 pixels) of eye blood vessels; b) - local histogram equalization over spatial window of 35x35 pixels; c) - P-histogram equalization with ***P***=0.5.

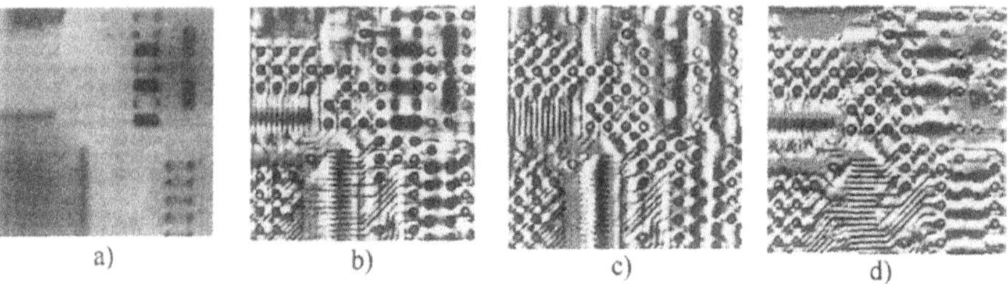

Figure 12-10. Local histogram equalization over Shape-neighborhoods: a) - initial image of a printed circuit board; b) - d) - local histogram equalization over spatial windows of 35x35, 1x35 and 35x1 pixels, respectively.

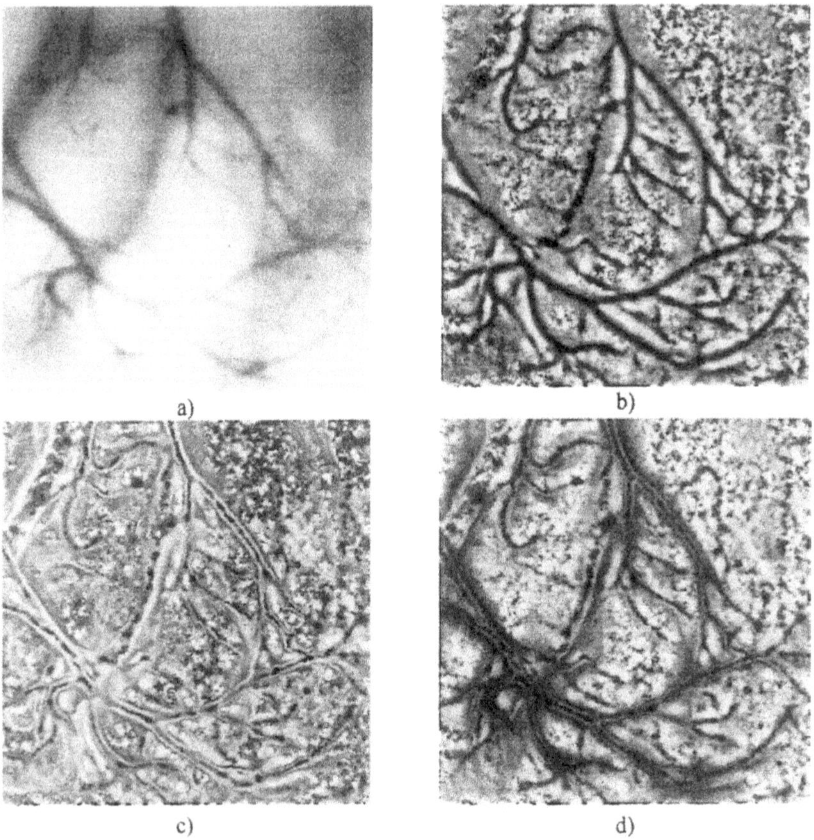

a) b)

c) d)

Figure 12-11. Comparison of local histogram equalization over spatial neighborhood and over EV- and KNV-neighborhoods applied to an angiogram. A) - initial image (256 quantization levels, 256x256 pixels); b) local histogram equalization over spatial window of 15x15 pixels; c) - local histogram equalization over KNV-neighborhood in the same spatial window with K=113 (half of the window size); d) local histogram equalization over EV-neighborhood in the same spatial window with $\varepsilon_v^+ = \varepsilon_v^- = \mathbf{10}$.

Other pixel attributes that may be useful for visual image analysis are size of EV-neighborhood and pixels cardinality. Image enhancement by means of measuring and displaying these attributes is illustrated in Fig. 12-12 on an example of a microscopic image of crystals.

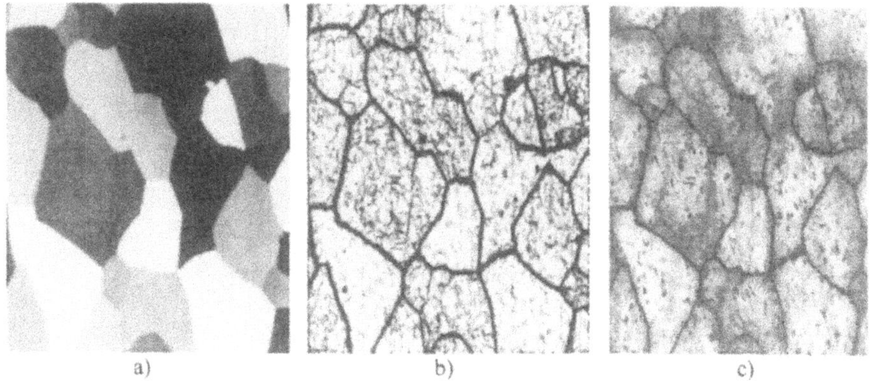

Figure 12-12. Local contrast enhancement using Size-EV and cardinality as pixel attributes: a) - original image (256 gray levels); b) -c): ***Size-EV*** ($\varepsilon_v^+ = \varepsilon_v^- = \mathbf{3}$) and pixel cardinality displayed as images.

12.3.3 Image segmentation and edge detection

Image segmentation and edge detection are interconnected types of image processing. Image segmentation is aimed on segmenting image into non-overlapping areas according to their certain features. Provided appropriately selected features, segmentation separates different objects in images. Segmentation results in piece-wise constant images in which pixel values are indices of segmented areas. In this sense it is an operation akin to image smoothing for piece-wise constant model of images. This smoothing eliminates small variations of object attributes, selected as segmentation features. As a result of smoothing for segmentation, image histogram degenerates. Its modes that correspond to different segments are becoming sharper and narrower in the process of iterative smoothing and tend to converge to a histogram that consist of a set of delta-functions. Fig. 12-13 illustrates image segmentation by iterative smoothing and the phenomenon of degenerating image histograms. Smoothing was carried out by the algorithm:

$$\hat{a}_k = \mathbf{MEAN}\left(EVnbh\left(Wnbh; a_k; \varepsilon_V^+; \varepsilon_V^-\right)\right) \tag{12.3.11}$$

in the window of 5x5 pixels with $\varepsilon_v^+ = \varepsilon_v^- = \mathbf{3}$ (image for 256 quantization levels).

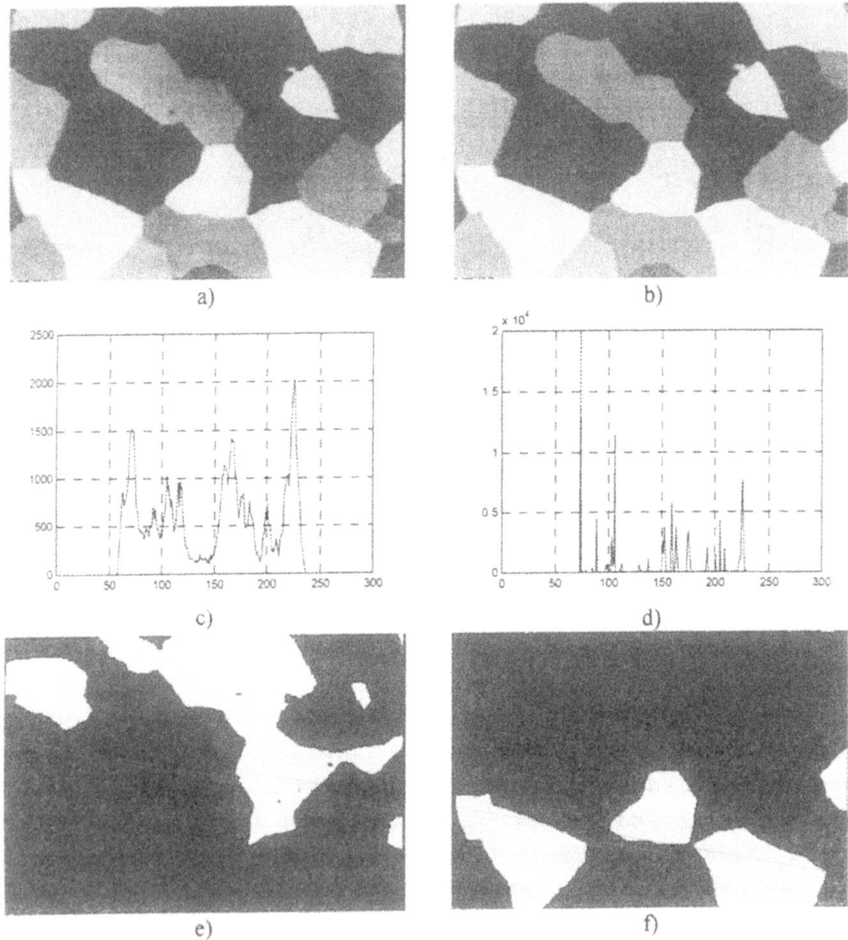

Figure 12-13. Image segmentation by means of iterative smoothing: a) - original image; b) - smoothed image after 10 iterations of the algorithm of Eq. 12.3.11; c) - gray level histogram of the original image; d) - gray level histogram of the smoothed image; e) and f) - image segments that correspond to two extreme modes of histogram d)

Edge detection is outlining object boundaries. In principle, edge detection should be preceded by image segmentation. In fact, edge detection in perfectly segmented image is a trivial task of marking boundaries between image segments. However in practice of image processing edge detection is frequently considered as a primary task and is carried out by means of measuring deviations in one or another pixel attribute from those of neighboring pixels. In this respect edge detection methods are akin to the above described methods of local contrast enhancement. Fig. 12-14 illustrates edge detection algorithms that use three types of neighborhood attributes: standard deviation within spatial window **STDEV(*Wnbh*)**, **SEIZE(*EV*(*Wnbh*))** and **MAX(*Wnbh*)-MIN(*NBH*)** .

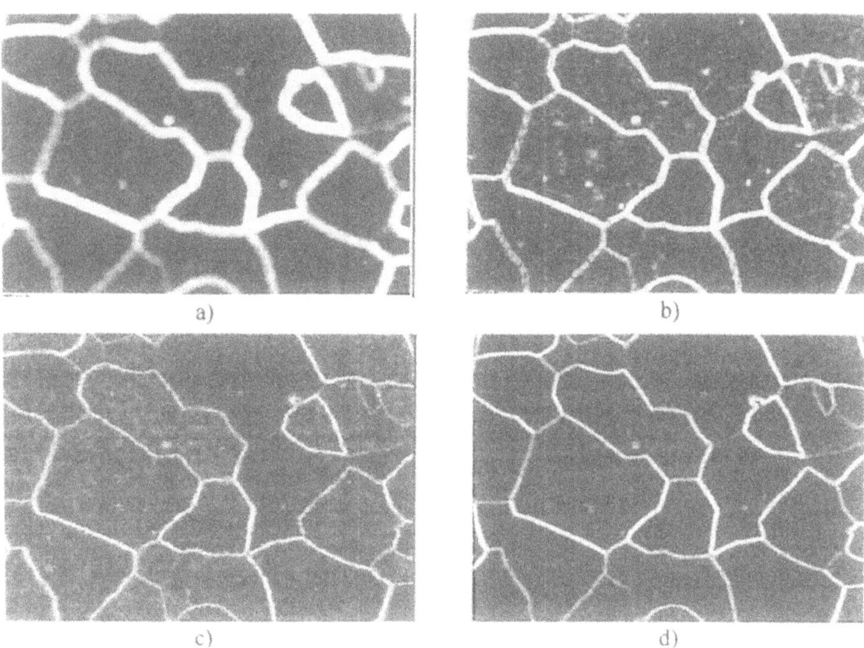

Figure 12-14. Edge detection using different image attributes: a) - local standard deviation in the window of 7x7 pixels; b) - local standard deviation in the window of 3x3 pixels; c) - **SIZE**(*EV(Wnbh)*) in the window of 7x7 pixels, $\varepsilon_V^+ = \varepsilon_V^- = \mathbf{3}$ (256 level of quantization); d) - difference between local maxima and minima over spatial window of 3x3 pixels. Image c) is identical to the image of Fig. 12-12 b) shown in negative.

Fig. 12-15 illustrates the importance of image smoothing before edge detection. Compare magnified image fragments to see that smoothing image before applying edge detection processing enables preserving fine image details that otherwise are destroyed by noise in the original image.

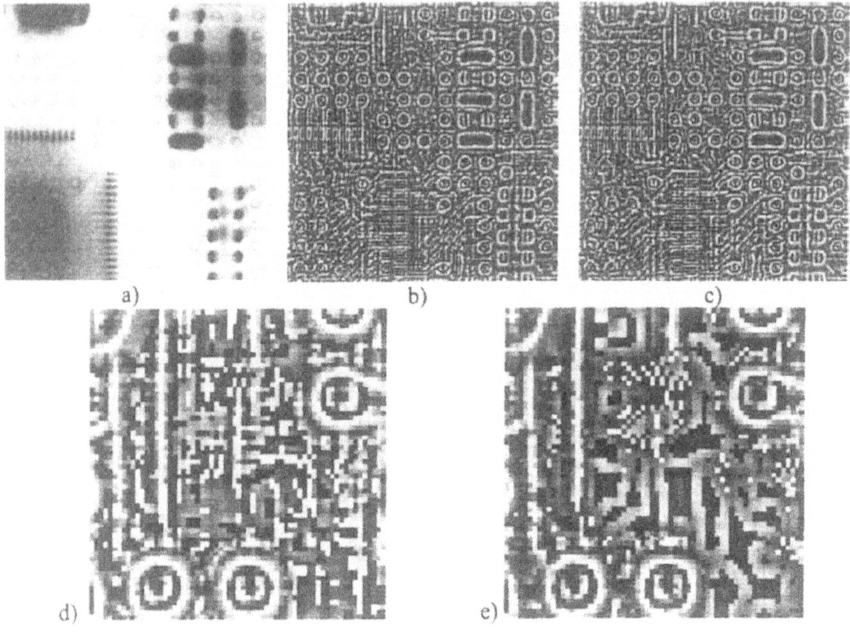

Figure 12-15. The importance of image smoothing before edge detection: a) initial image of a printed circuit board; b) - edge image generated by the algorithm MAX(Wnbh3x3)-MIN(Wnbh3x3; c) - edge image generated by the same algorithm after image smoothing by Sigma-filter (MEAN(EV(*Wnbh*5x5,3,3))); d), e) - corresponding magnified fragments of the edge images.

12.3.4 Implementation issues

Described nonlinear filters are inherently parallel. Therefore, the most natural way to implement them is to use multi layer parallel networks with neuro-morphic structure. Each level of such a network contains an array of elementary processors that implement estimation operations described in Sect. 12.1.3. The processors in each layer have to be are connected with corresponding sub-arrays of processors on the previous layer that form their neighborhood according to the principles described in Sect. 12.1.4. For N stage filters, there should be N such layers. The last layer in the structure produces filter output. In each layer, processors work in parallel and process "neighborhood" pixels formed on the previous layer to produce output for the next level or finally, the filter output. One can not avoid mentioning that such a multi layer structure much resembles the structure of visual retina in animal and human eyes.

Modern advances in "smart pixel arrays" promise a possible electronic implementation of such an artificial retina. Another option is associated with opto-electronic implementations that are based on the natural parallelism of optical processors ([14]).

In serial computers, software implementation of the filters may be based of efficient recursive algorithms for computing local histograms and order statistics described in Sect. 7.1.1.

References

1. J. W. Tukey, *Exploratory Data Analysis*, Addison Wesley, 1971
2. J. Serra, *Image Analysis and Mathematical Morphology*, Academic Press, 1983, 1988
3. V. Kim, L. Yaroslavsky, "Rank algorithms for picture processing", *Computer Vision, Graphics and Image Processing*, v. 35, 1986, p. 234-258
4. I. Pitas, A. N. Venetsanopoulos, *Nonlinear Digital Filters. Principles and Applications.* Kluwer, 1990
5. E. R. Dougherty, *An Introduction to Morphological Image Processing*, SPIE Press, 1992
6. H. Heijmans, *Morphological Operators*, Academic Press, 1994
7. E. R. Dougherty, J. Astola, *An Introduction to Nonlinear Image Processing*, SPIE Optical Engineering Press, 1994
8. L. Yaroslavsky, M. Eden, *Fundamentals of Digital Optics*, Birkhauser, Boston, 1996
9. J. Astola, P. Kuosmanen, *Fundamentals of Nonlinear Digital Filtering*, CRC Press, Roca Baton, New York, 1997
10. E. Dougherty, J. Astola, *Nonlinear Filters for Image Processing*, Eds., IEEE publ., 1999
11. L. P. Yaroslavsky, K. O. Egiazarian, J. T. Astola, "Transform domain image restoration methods: review, comparison and interpretation", *Photonics West, Conference 4304, Nonlinear Processing and Pattern Analysis*, 22-23 January, 2001, *Proceedings of SPIE*, v. 4304.
12. D. L. Donoho, I.. M. Johnstone, "Ideal Spatial Adaptation by Wavelet Shrinkage", *Biometrica*, 81(3), pp. 425-455, 1994
13. P. D. Wendt, E. J. Coyle, and N. C. Gallagher, Jr., "Stack Filters", *IEEE Trans. On Acoust., Speech and Signal Processing*, vol. ASSP-34, pp. 898-911, Aug. 1986.
14. T. Szoplik, *Selected Papers On Morphological Image Processing: Principles and Optoelectronic Implementations*, SPIE, Bellingham, Wa., 1996 (MS 127).

Chapter 13

COMPUTER GENERATED HOLOGRAMS

13.1 MATHEMATICAL MODELS

Basic stages in the synthesis of computer generated holograms are: (i) formulating mathematical models of the object and of the usage of the hologram; (ii) computing the ***mathematical hologram***, or distribution of the wave front amplitude and phase in the hologram plane; (iii) encoding samples of the mathematical hologram for recording them on the physical medium; (iiii) fabrication of the hologram (Fig. 13-1).

Mathematical model of the object is intended to specify the amplitude and phase distribution of the object wave front. Two types of the models can be distinguished: 2-D or 3-D arrays of points (object samples) and models that represent objects as compositions of elementary diffracting elements such as point scatters, segments of 2-D or 3-D curves, slits, etc. The former is more general. It requires specifying amplitudes and phases of the object wave front in all points of the array.

Mathematical models of usage of to be generated holograms describe geometry of wave propagation from objects to the hologram plane and specify criteria for performance evaluation of the hologram.

Computing the mathematical hologram simulates numerically wave propagation for the selected models of the object and of the hologram usage. Mathematical holograms are arrays of complex numbers that represent amplitudes and phases of hologram samples. For fabrication of holograms these numbers should be converted into an array of numbers that control optical properties of the physical recording medium used for recording the hologram. We will refer to this conversion as to ***hologram encoding***.

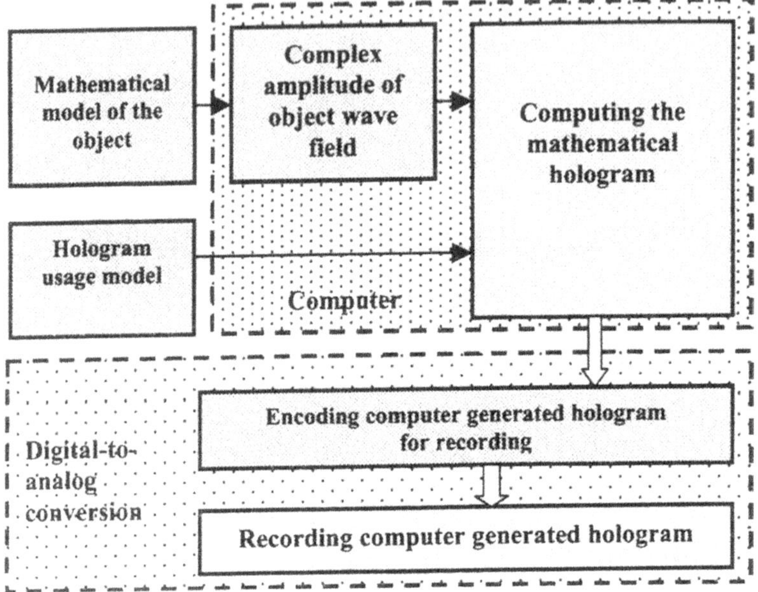

Figure 13-1. Basic stages in synthesis of computer generated holograms

Major applications of computer generated holograms are spatial filters for optical information processing, computer generated optical elements such as beam formers, deflectors and special diffraction gratings, and information display. Mathematical models for computer generated spatial filters such as, for instance, filters for implementing optimal adaptive correlators discussed in Chapt 11 are defined by the filter frequency response. Mathematical models for computer generated optical elements are directly formulated by the requirements to the specific optical elements. In this chapter we will describe a mathematical model for computer generated display holograms.

Consider a schematic diagram for the visual observation of objects shown in Fig.13-2. The observer's position with respect to the observed object is defined by the observation surface where the observer's eyes are situated. The set of observation foreshortenings is defined by the object observation angle. In order that the observer may see the object at the given observation angle, it suffices to reproduce, by means of a hologram, the distribution of intensities and phases of the light waves scattered by the object on the observation surface.

Consider monochromatic object illumination, which enables one to describe light wave transformations in terms of the wave amplitude and phase. Object characteristics defining its ability to reflect and scatter incident radiation may be described for our purposes by a radiation reflection factor with respect to the light intensity $O(x,y,z)$ or amplitude $o(x,y,z)$

which are functions of the object's surface coordinates (x, y, z) and are connected with the relationship: $O(x, y, z) = |o(x, y, z)|^2$.

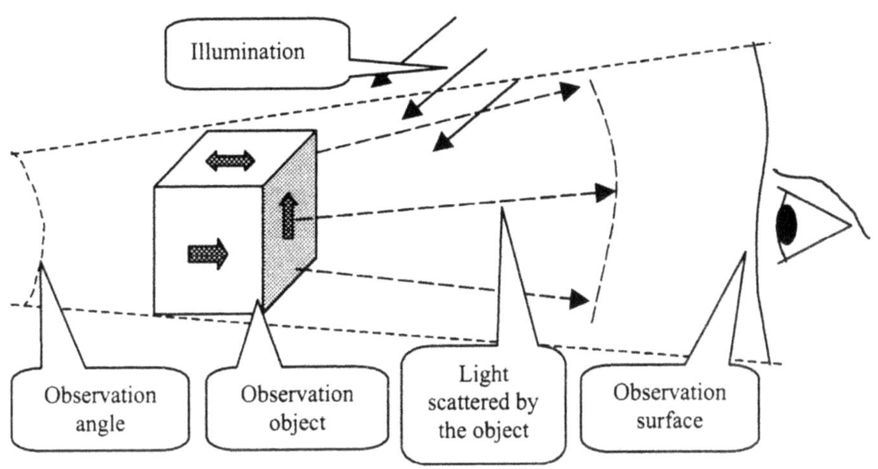

Figure 13-2. Scheme of observation of holograms

The intensity of the scattered wave $I_o(x, y, z)$ and its complex amplitude $A_o(x, y, z)$ at the point (x, y, z) are related to the intensity $I(x, y, z)$ and complex amplitude $A(x, y, z)$ of thc incident wave as follows:

$$\begin{aligned} I_o(x, y, z) &= O(x, y, z) I(x, y, z) \\ A_o(x, y, z) &= o(x, y, z) A(x, y, z) \end{aligned} . \qquad (13.1.1)$$

The amplitude reflection factor $o(x, y, z)$ may be regarded as a complex function of the spatial coordinates represented as

$$o(x, y, z) = |o(x, y, z)| \exp[i\theta_o(x, y, z)]. \qquad (13.1.2)$$

Its modulus $|o(x, y, z)|$ and phase $\theta_o(x, y, z)$ show how the modulus $|A(x, y, z)|$ and the phase $\omega(x, y, z)$ of the complex amplitude $|A(x, y, z)| \exp[i\omega(x, y, z)]$ of incident light are changed after reflection by the object surfacc at thc point (x, y, z):

$$
\begin{aligned}
|A_o(x,y,z)| &= |A(x,y,z)||o(x,y,z)| \\
\omega_o(x,y,z) &= \omega(x,y,z) + \theta_o(x,y,z)
\end{aligned}, \tag{13.1.3}
$$

where $|A_o(x,y,z)|$ and $\omega_o(x,y,z)$ are, respectively, the modulus and the phase of the reflected wave front.

The relation between the complex amplitude $\Gamma(\xi,\eta,\zeta)$ of the light wave field over an arbitrary surface defined at the coordinates (ξ,η,ζ) and the complex amplitude $A_o(x,y,z)$ of the wave front at the object surface can by described by a wave propagation integral over the object surface S_o

$$
\Gamma(\xi,\eta,\zeta) = \iiint_{S_o} A_o(x,y,z) T(x,y,z;\xi,\eta,\zeta) dxdydz \,, \tag{13.1.4}
$$

whose kernel $T(x,y,z;\xi,\eta,\zeta)$ depends on the spatial disposition of the object and the observation surface. Reconstruction of the object can be described by a back propagation integral over the observation surface S_s :

$$
\tilde{A}_o(x,y,z) = \iiint_{S_s} \Gamma(\xi,\eta,\zeta) \tilde{T}(\xi,\eta,\zeta;x,y,z) d\xi d\eta d\zeta \,, \tag{13.1.5}
$$

where $\tilde{A}_o(x,y,z)$ is complex amplitude of the object as it is seen from the observation surface and $\tilde{T}(\xi,\eta,\zeta;x,y,z)$ is a kernel reciprocal to $T(x,y,z;\xi,\eta,\zeta)$.

Thus, the hologram synthesis lies in the computation of $\Gamma(\xi,\eta,\zeta)$ through $A_o(x,y,z)$ which is to be defined by the object description and illumination conditions, and in recording the results on a physical medium in a form allowing interaction with radiation for visualizing or reconstructing $\tilde{A}_o(x,y,z)$ according to Eq. (13.1.5).

3-D integration required by Eq. 13.1.4 can be reduced to much more simple 2-D integration by taking into consideration the following natural limitations of visual observation:

- The pupil of the observer's eye is usually much smaller than the distance to the observation surface.
- The area of the observation surface is large as compared to the inter-pupil distance and the surface may be regarded as flat.
- The depth of objects situated at a distance convenient for observation usually are small as compared with the distance.

In view of these limitations, one may break down the observation surface into areas approximated by planes, and, using the laws of geometrical optics, replace the wave amplitude and phase distributions over the object surface by those of the wave in a plane tangent to the object (or sufficiently close to it, so as to make diffraction effects negligible) and parallel to the given plane area of the observation surface. In this way the problem of hologram synthesis over the whole observation surface boils down to the synthesis of fragmentary holograms for the plane areas of this surface, with the complete hologram being composed as a mosaic of fragmentary ones. Instead of Eq.(13.1.4) one then obtains for fragmentary holograms:

$$\overline{\Gamma}(\xi,\eta)=\iint_{\overline{S}_o}\overline{A}_o(x,y)T_D(x,y;\xi,\eta)dxdy\ , \tag{13.1.6}$$

where $\overline{A}_o(x,y)=\left|\overline{A}_o(x,y)\right|\exp\left|i\overline{\omega}_o(x,y)\right|$ is a complex function resulting from recalculation of the amplitude and phase of the field reflected by the object onto the plane (x,y) tangent to it and parallel to the observation plane (ξ,η); $T_D(x,y;\xi,\eta)$ is a reduced 2-D transformation kernel for the distance between these planes D.

If the geometrical dimensions of the object under observation and of the observation surface fragment are small as compared with the distance D from the object to the observation plane, the integral relationship of Eq. 13.1.7 is, as it was shown in Sect. 2.3.2, reduced to integral Fresnel transform:

$$\overline{\Gamma}_{Fr}(\xi,\eta)=\int_{-\infty}^{\infty}\int_{-\infty}^{\infty}\overline{A}_o(x,y)\exp\left\{i\pi\left[(x-\xi)^2+(y-\eta)^2\right]/\lambda D\right\}dxdy, \tag{13.1.7}$$

where λ is the illumination wave length. Discrete representation of the integral Fresnel transform suitable for its numerical computing is discussed in Sect. 4.4. The holograms synthesized by means of this relationship will be referred to as synthesized ***Fresnel holograms***.

Further simplification is possible if

$$\pi\left(x^2+y^2\right)/\lambda D<<1 \tag{13.1.8}$$

in which case the integral of Eq.(13.1.8) can be approximated as

$$\overline{\Gamma}_{Fr}(\xi,\eta)\cong\exp\left[i\pi\left(\xi^2+\eta^2\right)\right]\int_{-\infty}^{\infty}\int_{-\infty}^{\infty}\overline{A}_o(x,y)\exp\left[i2\pi(x\xi+y\eta)/\lambda D\right]dxdy \tag{13.1.9}$$

through the integral Fourier transform:

$$\overline{\Gamma}_F(\xi,\eta) = \int_{-\infty}^{\infty}\int_{-\infty}^{\infty} \overline{A}_o(x,y)\exp[i2\pi(x\xi + y\eta)/\lambda D]dxdy \ . \qquad (13.1.10)$$

The holograms synthesized according to Eq. 13.1.10 by means of Fourier transformation will be referred to as synthesized ***Fourier holograms***. Fourier and Fresnel holograms can be computed through Discrete Fourier transforms discussed in Sect. 4.3 and 4.4.

Schemes of using computer generated display Fresnel and Fourier holograms for visual observation of virtual objects are shown in Fig. 13-3, a) and b), respectively.

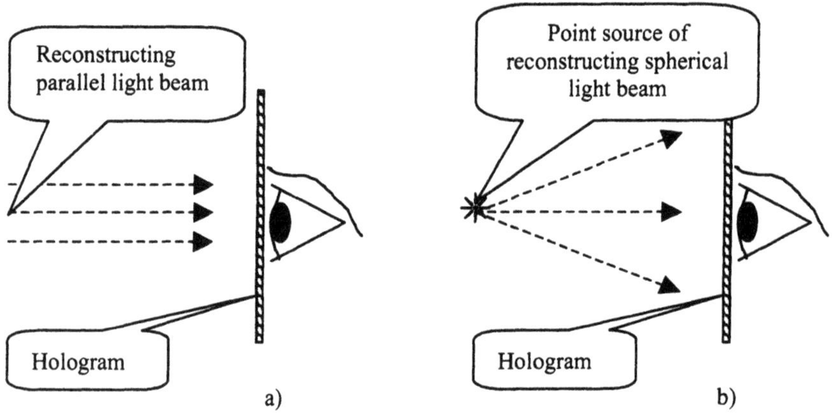

Figure 13-3. Scheme of visual observation of virtual objects using computer generated Fresnel (a) and Fourier (b) holograms

From the standpoint of object wave front reconstruction, Fresnel holograms differ from Fourier holograms in that they have focusing properties and are capable of reproducing the finite distance from the observation surface to the object. For reconstruction of Fourier holograms, spherical reconstruction light beam is needed and reconstructed image is observed in the plane of the reconstructing beam point source.

Mathematically, reconstructing an object by Fresnel and Fourier holograms is described by the inverse Fresnel and Fourier transformations, respectively. When holograms are observed visually, these transformations are performed by the optical system of the eye.

Having passed from the spatial 3D-problem to the problem of treating its 2D approximation, we, strictly speaking, lose the possibility of taking into account the precise effect of object depth relief on the wave front. Even Fresnel holograms use only a single distance value from the object to the observation plane rather than object relief depth. Nevertheless, it is still possible to synthesize holograms which are capable of reconstructing 3-D images for visual observation and retain the most important property of holographic visualization - naturalness of object visual observation. At least two options exist for this: composite computer generated stereo- and macro-holograms and programmed diffuser holograms,

Stereo-holograms are pairs of holograms synthesized to be viewed by left and right eye and to reconstruct the corresponding stereo views of the object ([1]). Composite stereo holograms are composed of multiple holograms that reproduce different aspects of the object as observed from different positions. Macro-holograms are large composite stereo-holograms that cover wide observation angle and can be used as a wide observation window. An example of a composite stereo-hologram that cover full 360° of observation angle with horizontal parallax is shown in Fig. 13-4.

Programmed diffuser holograms are computer generated Fourier holograms that imitate properties of diffuse surfaces to scatter irradiation non-uniformly in the space and in this way provide visual clue about the object surface shape (See Sect. 7.2.4).

Figure 13-4. Composite stereo macro hologram, a holographic "movie", composed of 800 elementary holograms (adopted from [1]). 1 - holograms recorded on a photographic film, 2 - annular holder, 3 - conic reflector. Holograms are illuminated from above through the conic reflector and are viewed as through a window.

13.2 METHODS FOR ENCODING AND RECORDING COMPUTER GENERATED HOLOGRAMS

Optical media for recording computer generated holograms may be classified into three categories: ***amplitude-only media***, ***phase-only media***, and ***combined amplitude/phase media***.

In amplitude-only media, the controlled optical parameter is the light intensity transmission or reflection factor. This is the most common and available class, whose typical representatives are the standard silver galid photographic emulsions used in photography and optical holography ([2]).

In phase-only media, the optical thickness of the medium can be controllable, for example, by varying the medium refractive index, or physical thickness, or both. Phase-only media include thermoplastic materials, photo resists, bleached photographic materials, media based on photo polymers, etc. Recently, micro-lens array and micro-mirror array technology has emerged ([3,4]) that offer new opportunities for recording computer generated holograms.

Combined media allow independent control of both the light intensity transmission factor and of the optical thickness. Currently, these are photographic materials with two or more layers sensitive to radiation of different wavelengths. This permits the user to control of the transparency of certain layers and the optical thickness of others by exposing each layer to its wavelength independently.

Special digitally controlled hologram recording devices are required for modulating optical parameters of these media according to the mathematical hologram. No such special purpose devices exist as yet, and computer printer and display devices designed for the output of characters, graphs, and gray-scale and color images are used instead. The distinctive feature of many of such devices is that they perform only binary or two level modulation of the medium's optical parameters. Amplitude-only and phase-only media whose controlled optical parameters may assume only two values will be referred to as ***binary media***. The use of amplitude and phase media in the binary mode is quite inefficient in terms of the medium information capacity because the possibility of writing information on them is defined only by their spatial degrees of freedom (their resolution power), whereas in amplitude and phase media, in principle, the degrees of freedom related to a transmission (reflection, refraction) factor may be used as well. The major advantage of the binary mode is simplicity of media exposure, photo-chemical processing when it is needed and copying.

The most important characteristics of recorders are their sampling step; that is, the distances $\Delta\xi$ and $\Delta\eta$ between neighboring, separately exposed, resolution cells, and the total number of cells which may be exposed. The sampling step defines the angular dimensions of the reconstructed image. For instance, in order to make the image's angular dimensions about $\mathbf{10^o}$ or more with a reconstruction light wavelength of about 0.5 μm, $\Delta\xi$ and $\Delta\eta$ should be 3 μm at most. The existing recording devices have a sampling step of about 1 through 10 μm, and the total number of resolution cells, by the order of magnitude, from $\mathbf{10^3 \times 10^3}$ to $\mathbf{10^4 \times 10^4}$. At the early stages of digital holography, holograms were recorded by means of standard computer plotters and printers. The obtained hologram hard copies were then photographically reduced in order to achieve acceptable values of the sampling step ([1, 5,6]).

The existing methods for recording synthesized holograms on amplitude-only, phase-only, binary, and combined amplitude/phase media may be classified with respect to different features. Methods of representing complex numbers describing samples of the mathematical hologram will be chosen here as the classification key (Fig. 13-5). For other possible classifications the reader is referred to the reviews [5-7].

Complex numbers may be represented in two ways: in the exponential form $A\exp(i\phi)$ where A and ϕ are, respectively, the amplitude and phase of the numbers or additively as a sum of two or more basis complex numbers. Obviously, for hologram recording on the combined amplitude/phase media, the exponential form is the most suitable. It does not require any special encoding. Computer generated holograms recorded on combined media are capable of wave front reconstruction on optical axis of the reconstruction set-up at the zero-order diffraction order and are called ***on-axis holograms***.

For recording on phase-only or amplitude-only media, special hologram encoding methods are required. One of the most straightforward encoding methods oriented on using phase-only media is that of ***kinoform*** ([8,9]). Kinoform is a computer generated hologram in which amplitude data of the mathematical hologram intended for recording are disregarded and only the hologram sample phases are recorded on a phase-only medium. Although disregarding hologram sample amplitudes results in substantial distortions such as appearance of speckle noise in the reconstructed wave front (see Sect. 7.3), kinoforms are advantageous in terms of the energy of reconstruction light because the total energy of the reconstruction light is transformed into the energy of the reconstructed wave field without being absorbed in the hologram. Moreover, the distortions may to some degree be reduced by an appropriate choice of the diffusion component of the wave field phase on the object using an iterative optimization procedure similar to that described in Sect. 7.2.4 ([6]).

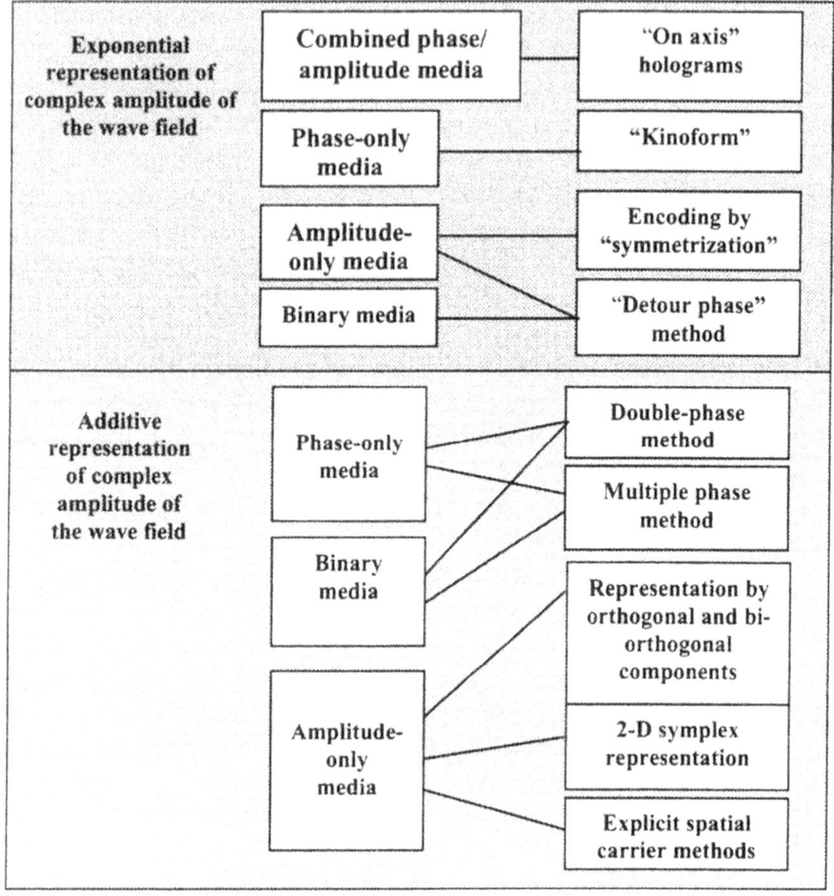

Figure 13-5. Classification of methods for recording computer generated holograms

In hologram recording on amplitude media, the main problem is recording the phase component of hologram samples. The ***symmetrization method*** ([1]) offers a straightforward solution of this problem for recording Fourier holograms. The method assumes that prior to hologram synthesis, the object be symmetrized, so that, according to properties of Fourier transform, its Fourier hologram contains only real valued samples and, thus, may be recorded on amplitude-only media. As real numbers may be both positive and negative, holograms should be recorded in the amplitude-only medium with a constant positive bias, making all the recorded values positive.

For SDFT as a discrete representation of the integral Fourier transform, symmetry properties (see Sect. 4.3.2) require symmetrization through a rule depending on the shift parameters. For example, with integer $2u$ and $2v$ the symmetrization rule for the object wave front specified by the array of its samples $\{\overline{A}_{k,l}\}$, $k = 0,1,\ldots,N_1 - 1$, $l = 0,1,\ldots,N_2 - 1$ becomes

$$\tilde{\overline{A}}_{k,l} = \begin{cases} \overline{A}_{k,l}, 0 \le k \le N_1 - 1; 0 \le l \le N_2 - 1 \\ \overline{A}_{2N-k,l}, N_1 \le k \le 2N_1 - 1; 0 \le l \le N_2 - 1 \end{cases} \qquad (13.2.1)$$

which implies symmetrization by object duplication. In doing so, the number of samples of the object and, correspondingly, its Fourier hologram, is twice that of the original object. It is this two-fold redundancy that enables one to avoid recording the phase component. Symmetrization by quadruplicating is possible as well. This consists in symmetrizing the object according to the rule of Eq.(13.2.1) with respect to both indices k and l. Hologram redundancy becomes in this case four-fold. Holograms of symmetrized objects are also symmetrical and reconstruct duplicated or quadruplicated objects depending on the particular symmetrization.

Notably, duplication and quadruplication do not imply a corresponding increase in computer time for execution of SDFT at the hologram calculation step because, for computation of SDFT, one may use combined algorithms making use of signal redundancy for accelerating computations (see Sect.5.3).

Historically, one of the first methods for recording synthesized holograms was that proposed for amplitude-only binary media by A.Lohmann and his collaborators ([10,11]). In this method, an elementary cell of the medium is allocated for reproducing the amplitude and phase of each sample of a mathematical hologram. The modulus of the complex number is represented by the size of the opening (aperture) in the cell and the phase - by the position of the opening within the cell.

All the cells corresponding to mathematical hologram samples are arranged over a regular (usually, rectangular) raster. A shift of the aperture by $\Delta\tilde{\xi}$ in a given cell with respect to its raster node corresponds to a phase detour for this cell equal to $(2\pi\Delta\tilde{\xi}\cos\theta)/\lambda$ for hologram reconstruction at an angle θ to the system's optical axis perpendicular to the hologram plane (Fig.13-6), where λ is wavelength of the illumination light used for hologram reconstruction. The use of the aperture spatial shift for representing complex number phase was named the ***detour phase method*** ([6])

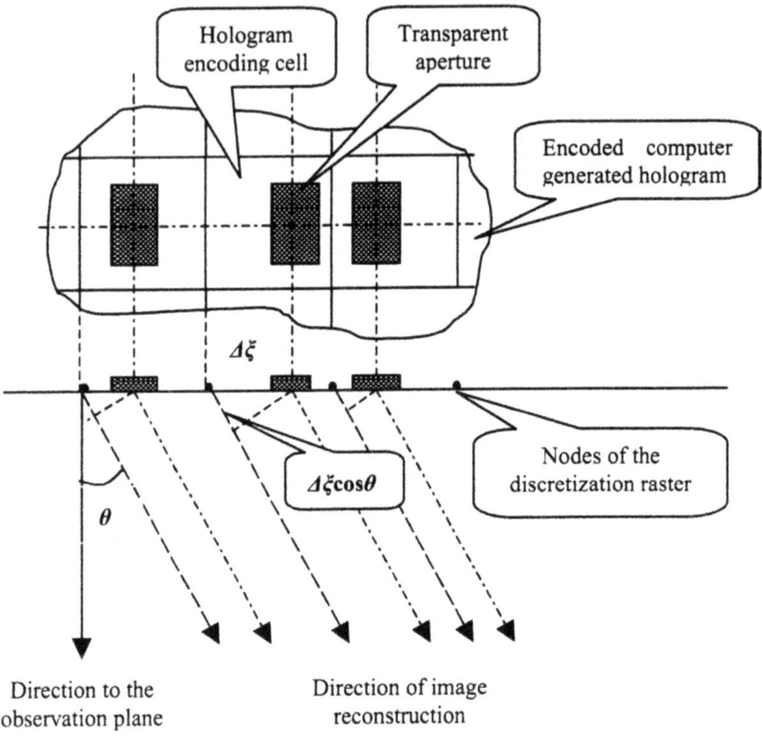

Figure 13-6. Detour phase method for coding phase shift by spatial shift of the transparent aperture

As was already noted, only spatial degrees of freedom are used for recording in binary media; therefore, the number of binary medium cells should exceed the number of hologram samples by a factor equal to the product of the amplitude and phase quantization levels. This product may run into several tens or even hundreds. The low effectiveness of using the degrees of freedom of the hologram carrier is the major drawback of the binary hologram method. However many modern technologies such as lithography and laser and electron beam pattern generators ([12]) provide much higher number of spatial degrees of freedom that the required number of hologram samples. Therefore this drawback is usually disregarded and such merits of binary recording as simpler recording and copying technology of synthesized holograms and the possibility of using commercially available devices made it the most widespread. Numerous modifications of the method are known, oriented to different types of recording devices.

With an additive representation, the complex number (regarded as a vector in the complex plane) is a sum of several components. For recording on amplitude-only media these components should be assigned standard directions on a complex plane (phase angle) and controllable length (amplitude).

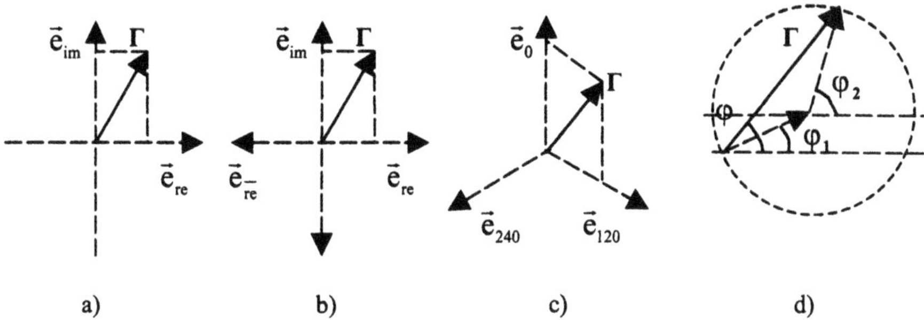

Figure 13-7. Additive representation of complex numbers over orthogonal basis (a), over biorthogonal basis (b), over 2-D symplex (c) and as a sum of two equal length vectors (d).

The simplest case is the representation of a vector $\mathbf{\Gamma}$ by its orthogonal components, say, real Γ_{re} and imaginary Γ_{im} parts:

$$\mathbf{\Gamma} = \Gamma_{re}\mathbf{e}_{re} + \Gamma_{im}\mathbf{e}_{im}, \tag{13.2.2}$$

where $\mathbf{e}_{re}$ and $\mathbf{e}_{im}$ are orthogonal unit vectors. When recording a hologram, the phase angle between the orthogonal components may be encoded by the detour phase method, Γ_{re} and Γ_{im} being recorded into neighboring hologram resolution cells neighboring in a raster rows as shown in Fig.13-7, a. The reconstructed image will be then observed at an angle θ_ξ to the axis defined as follows:

$$\Delta\xi \cos\theta_\xi = \lambda / 4. \tag{13.2.3}$$

For recording negative values of Γ_{re} and Γ_{im} a constant positive bias may be added to recorded values. We will refer to this method as to ***orthogonal encoding*** method.

Since two resolution elements are used here for recording one hologram sample, such a hologram has double redundancy, as in the symmetrization with duplication.

With such encoding and recording of a hologram, one must take into account that the optical path differences $\lambda / 2$ and $3\lambda / 4$ will correspond to the next pair of hologram resolution cells (see Fig.13-8 a); that is, values of Γ_{re} and Γ_{im} for each odd sample of the mathematical hologram should be recorded with opposite signs.

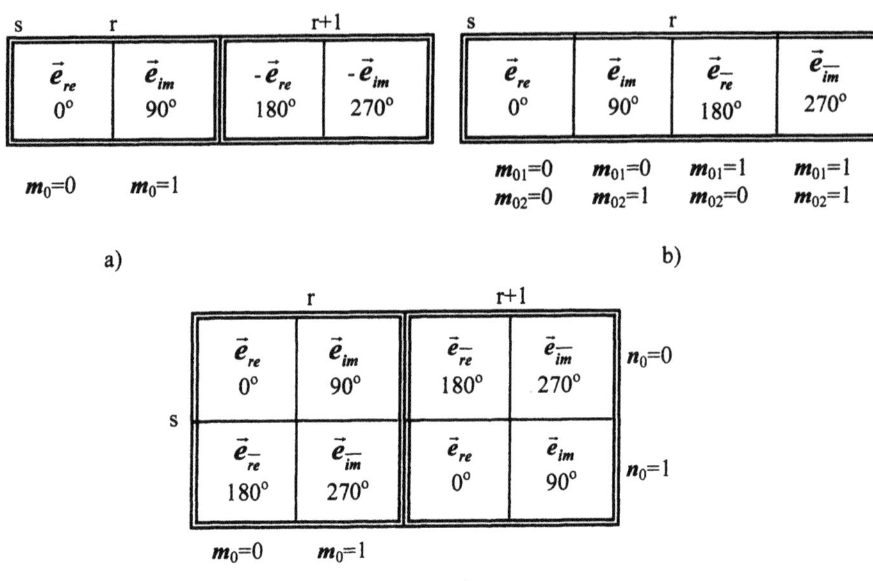

Figure 13-8. Hologram encoding by decomposition of complex numbers in orthogonal (a) and biorthogonal (b, c) bases. Hologram encoding cells corresponding to one hologram sample is outlined with a double line.

This encoding technique may be described formally as follows. Let $\{r,s\}$ be indices of mathematical hologram samples $\{\Gamma_{r,s}\}$, $r = 0,1,...,M-1$, $\{m,n\}$ be indices of the recording medium resolution cells, $\{\widetilde{\Gamma}_{r,s}\}$ be samples of the encoded hologram ready for recording and let index m is counted as a two digit number:

$$m = 2r + m_0,\ m_0 = 0,1, \qquad (13.2.4)$$

and index n is counted as $n = s$. The encoded hologram can then be written as

$$\widetilde{\Gamma}_{m,n} = \frac{1}{2}(-1)^r (i)^{m_0}\left[(-1)^{m_0}\Gamma_{r,s} + \Gamma^*_{r,s}\right] + b, \qquad (13.2.5)$$

where b is a positive bias constant needed to avoid recording negative values.

In reconstruction of computer generated holograms recorded with a constant bias, substantial part of energy of the reconstructing light beam is not used for image reconstruction and goes to the zero-order diffraction spot.

One can avoid the constant biasing for recording Γ_{re} and Γ_{im} by allocating, for recording one hologram sample, four neighboring-in-raster medium resolution cells rather then two ([5]). This arrangement is shown in Fig.13-8, b. With a reconstruction angle as defined by Eq.(13.2.3), phase detours $\mathbf{0}$, $\pi/\mathbf{2}$, π and $3\pi/\mathbf{2}$ correspond to them. Therefore, cells in each group of four cells allocated for recording of one sample of the mathematical hologram should be written in the following order: $\left(\Gamma_{re}+\left|\Gamma_{re}\right|\right)/2$; $\left(\Gamma_{im}+\left|\Gamma_{im}\right|\right)/2$; $\left(\left|\Gamma_{re}\right|-\Gamma_{re}\right)/2$; $\left(\left|\Gamma_{im}\right|-\Gamma_{im}\right)/2$. One can see that in this case all the recorded values are nonnegative. The method represents a vector in the complex plane in a ***biorthogonal basis*** $(\vec{\mathbf{e}}_{re}, \vec{\mathbf{e}}_{\overline{re}}, \vec{\mathbf{e}}_{im}, \vec{\mathbf{e}}_{\overline{im}})$:

$$\mathbf{\Gamma}=\frac{1}{2}\left(\Gamma_{re}+\left|\Gamma_{re}\right|\right)\vec{\mathbf{e}}_{re}+\frac{1}{2}\left(\Gamma_{re}-\left|\Gamma_{re}\right|\right)\vec{\mathbf{e}}_{\overline{re}}+$$
$$+\frac{1}{2}\left(\Gamma_{im}+\left|\Gamma_{im}\right|\right)\vec{\mathbf{e}}_{im}+\frac{1}{2}\left(\Gamma_{im}-\left|\Gamma_{im}\right|\right)\vec{\mathbf{e}}_{\overline{im}}. \qquad (13.2.6)$$

In terms of indices $(\boldsymbol{m},\boldsymbol{n})$ of recording medium resolution cells counted as

$$m=4r+2m_{01}+m_{02};\ r=0,1,\dots,M-1;\ m_{01}=0,1;\ m_{02}=0,1;\ n=s, \qquad (13.2.7)$$

this encoding method may be described as follows:

$$\tilde{\Gamma}_{m,n}=\frac{1}{4}\left\{i^{m_{02}}\left[(-1)^{m_{02}}\Gamma_{r,s}+\Gamma^{*}_{r,s}\right]+(-1)^{m_{01}}i^{m_{02}}\left|(-1)^{m_{02}}\Gamma_{r,s}+\Gamma^{*}_{r,s}\right|\right. . \qquad (13.2.8)$$

In hologram encoding by this method, it is required, that the size of the medium resolution cell in one direction be four times smaller than that in the perpendicular one. In this way the proportions of the reconstructed picture can be preserved. A natural way to avoid this anisotropy is to allocate pairs of resolution cells in two neighboring raster rows for each sample of the mathematical hologram as it is shown in Fig.13-7, c); that is, to record the hologram samples according to the following relationship:

$$\tilde{\Gamma}_{m,n}=\frac{1}{4}\left\{i^{m_0}\left[(-1)^{m_0}\Gamma_{r,s}+\Gamma^{*}_{r,s}\right]+(-1)^{n_0}i^{m_{00}}\left|(-1)^{m_0}\Gamma_{r,s}+\Gamma^{*}_{r,s}\right|\right., \qquad (13.2.9)$$

where indices (m, n) are counted as two digit numbers:

$$m = 2r + m_0;\ n = 2s + n_0;\ m_0 = 0,1;\ n_0 = 0,1. \tag{13.2.10}$$

With this encoding method, images are reconstructed along a direction making angles θ_ξ and θ_η with axes (ξ, η) in the hologram plane defined by the equations:

$$\Delta\xi \cos\theta_\xi = \lambda / 2;\ \Delta\eta \cos\theta_\eta = \lambda / 2. \tag{13.2.11}$$

Representation of complex numbers in the biorthogonal basis of Eq.(13.2.6) is redundant because two of the four components are always zero. This redundancy is reduced in the representation of complex numbers with respect to a ***two-dimensional simplex*** ($\vec{\mathbf{e}}_0, \vec{\mathbf{e}}_{120}, \vec{\mathbf{e}}_{240}$) (Fig.13.6, c):

$$\mathbf{\Gamma} = \Gamma_0 \vec{\mathbf{e}}_0 + \Gamma_1 \vec{\mathbf{e}}_{120} + \Gamma_2 \vec{\mathbf{e}}_{240}. \tag{13.2.12}$$

Similarly to the biorthogonal basis, this basis is also not linearly independent because

$$\vec{\mathbf{e}}_0 + \vec{\mathbf{e}}_{120} + \vec{\mathbf{e}}_{240} = \mathbf{0}, \tag{13.2.13}$$

and its redundancy is exploited to insure that components $\{\Gamma_0, \Gamma_1, \Gamma_2\}$ of complex vectors be non-negative.
Two versions of hologram coding by the two-dimensional simplex are known. In the first version, complex vectors in the complex plane are represented as the sum of two components directed along those of three vectors ($\vec{\mathbf{e}}_0, \vec{\mathbf{e}}_{120}, \vec{\mathbf{e}}_{240}$) which confine the third part of the plane where this vector is situated ([13]). It follows that of the three numbers $\{\Gamma_0, \Gamma_1, \Gamma_2\}$ defining vector $\mathbf{\Gamma}$ through Eq.(13,.2.2), two are always non-negative and are projections of the vector $\mathbf{\Gamma}$ on the corresponding basis vectors, and the third one is zero. The following relations may be obtained for components $\{\Gamma_0, \Gamma_1, \Gamma_2\}$ from this condition

$$\Gamma_0 = \left[(1 + \mathit{sign}A)(B - |B|) + (1 - \mathit{sign}A)(C + |C|)\right] / 2\sqrt{3};$$
$$\Gamma_1 = \left[(1 + \mathit{sign}C)(A - |A|) + (1 - \mathit{sign}C)(B + |B|)\right] / 2\sqrt{3};$$
$$\Gamma_0 = \left[(1 + \mathit{sign}B)(C - |C|) + (1 - \mathit{sign}B)(A + |A|)\right] / 2\sqrt{3}, \tag{13.2.14}$$

where

$$A = (\Gamma + \Gamma^*)/2;$$

$$B = [\Gamma \exp(i2\pi/3) + \Gamma^* \exp(-i2\pi/3)]/2;$$

$$C = [\Gamma \exp(-i2\pi/3) + \Gamma^* \exp(i2\pi/3)]/2, \quad (13.2.15)$$

where $signX = X/|X|$.

The second version of the method makes use of the fact that, by virtue of Eq. 13.2.13, the addition of an arbitrary constant to $\{\Gamma_0, \Gamma_1, \Gamma_2\}$ leaves Eq. 13.2.12 unchanged. It represents $\{\Gamma_0, \Gamma_1, \Gamma_2\}$as

$$\{\Gamma_t = \Gamma_t^{(0)} + V\};\ t = 0,1,2 \quad (13.2.16)$$

The presence of an arbitrary constant V allows to impose on $\{\Gamma_t^{(0)}\};\ t = 0,1,2$ an additional constraint:

$$\Gamma_0^{(0)} + \Gamma_1^{(0)} + \Gamma_2^{(0)} = 0. \quad (13.2.17)$$

In this case $\{\Gamma_t^{(0)}\}$ are computed as

$$\Gamma_0^{(0)} = (\Gamma + \Gamma^*)/3 = 2A/3;$$

$$\Gamma_1^{(0)} = [\Gamma \exp(-i2\pi/3) + \Gamma^* \exp(i2\pi/3)]/3 = 2C/3;$$

$$\Gamma_2^{(0)} = [\Gamma \exp(i2\pi/3) + \Gamma^* \exp(-i2\pi/3)]/3 = 2B/3 \quad (13.2.18)$$

and constant V is chosen so as to make $\{\Gamma_0, \Gamma_1, \Gamma_2\}$ nonnegative. The bias V, evidently, may differ between different hologram samples. The set of these biases for hologram samples may be optimized so as to improve the quality of reconstructed images ([14]).

When recording holograms with this method, one may encode the phase angle corresponding to unit vectors ($\vec{\mathbf{e}}_0, \vec{\mathbf{e}}_{120}, \vec{\mathbf{e}}_{240}$) by means of the mentioned above detour phase method by writing the components $\{\Gamma_0, \Gamma_1, \Gamma_2\}$ of the simplex-decomposed vector into groups of threc neighboring hologram resolution cells in the raster as it is shown in Fig.13-9 , a, b and c. The latter two arrangements are more isotropic then the first one. Coordinate scale ratio for arrangement b) is 2:1.5 and for arrangement c) is $3\sqrt{3}:4$ whereas for arrangement a) it is3:1.

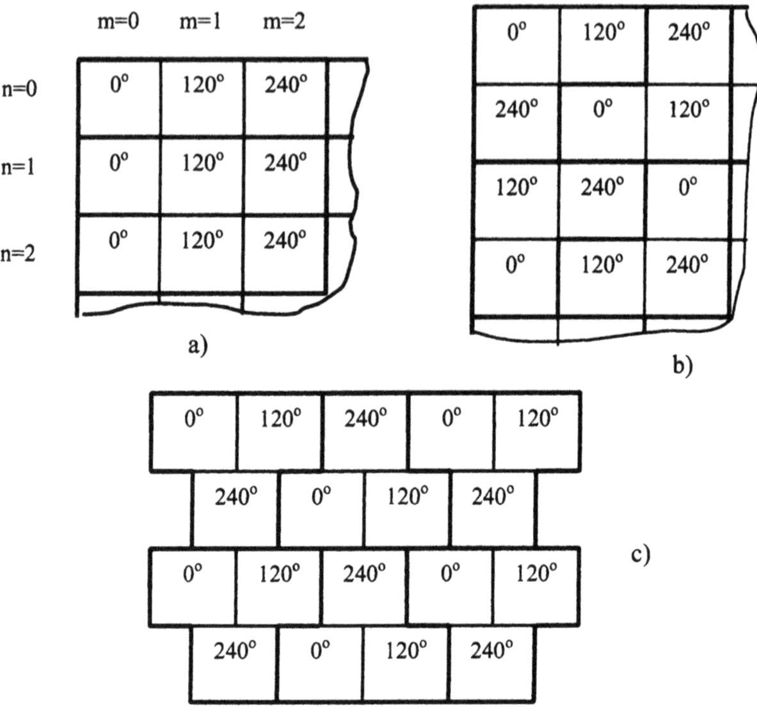

Figure 13-9. Three methods of arrangement of recording medium resolution cells for hologram encoding by decomposition of complex numbers with respect to a 2-D simplex. Triples of medium resolution cells used for recording one hologram sample are outlined with a bold line.

For holograms recorded by this method according to the arrangement of Fig. 13-8, a), images are reconstructed at angles $\theta_\xi = \mathbf{arccos}(\lambda / 3\Delta\xi)$ to the axis ξ coinciding with the direction of the hologram raster rows and $\theta_\eta = \mathbf{0}$ to the perpendicular axis. For the arrangements of Fig. 13-8, b) and c) the reconstruction angles are, respectively, $\theta_\xi = \mathbf{arccos}(\lambda / 3\Delta\xi)$, $\theta_\eta = \mathbf{arccos}(2\lambda / 3)$ and $\theta_\xi = \mathbf{arccos}(4\lambda / 3\Delta\xi)$, $\theta_\eta = \mathbf{arccos}(2\lambda / 3)$.

One can show by using Eq.(13.2.14) that the following formula relating values written in recording medium resolution cells with indices $(\boldsymbol{m},\boldsymbol{n})$ with samples $\{\Gamma(\boldsymbol{r},\boldsymbol{s})\}$ of the mathematical hologram corresponds to the first version of the coding method:

$$\widetilde{\Gamma}_{m,n} = \frac{\mathbf{1}}{\mathbf{2}\sqrt{\mathbf{3}}}[(\mathbf{1} + \boldsymbol{sign}\{\mathbf{Re}[\Gamma_{r,s}\,\mathbf{exp}(-\,\boldsymbol{i2\pi m_0}\,/\,\mathbf{3})]\})\times$$

$$\left\{\mathbf{Re}\left[\Gamma_{r,s}\exp(-i2\pi m_0/3)\right]-\left|\mathbf{Re}\left[\Gamma_{r,s}\exp(-i2\pi m_0/3)\right]\right|\right\}+$$

$$\left(1-sign\left\{\mathbf{Re}\left[\Gamma_{r,s}\exp(-i2\pi m_0/3)\right]\right\}\right)\times$$

$$\left\{\mathbf{Re}\left\{\Gamma_{r,s}\exp\left[i2\pi(m_0+1)/3\right]\right\}+\left|\mathbf{Re}\left\{\Gamma_{r,s}\exp\left[i2\pi(m_0+1)/3\right]\right\}\right|\right\}. \tag{13.2.19}$$

where horizontal and vertical indices $\{m,n\}$ (Fig. 13-8, a)) are counted as: $m=3r+m_0$, $m_0=0,1,2$, $n=s$ and $\mathbf{Re}(X)$ is real part of X.

The above described encoding methods based on additive representation of complex vectors are applicable for recording on amplitude-only media, both continuos tone and binary. In the latter case, projections of the complex number on basic vectors are represented by varying the size of the transparent aperture in each of the appropriate resolution cells.

For recording on phase media, additive representation of complex vectors assumes complex vector representation as a sum of complex vectors of a standard length. Two versions of such encoding are known: double phase method and multiple phase method.

In the ***double-phase method*** ([10,15]), hologram samples are encoded as

$$\mathbf{\Gamma}=|\Gamma|\exp(i\varphi)=A_0\exp(i\varphi_+)+A_0\exp(i\varphi_-), \tag{13.2.20}$$

where (φ_1,φ_2) are component phase angles defined by the following equations:

$$\varphi_+=\varphi+\mathbf{arccos}\left(|\Gamma|/2A_0\right);\ \varphi_-=\varphi-\mathbf{arccos}\left(|\Gamma|/2A_0\right), \tag{13.2.21}$$

that can be easily derived from the geometry of the method illustrated in Fig. 13-6, d.

Two neighboring medium resolution cells should be allocated for representing two vector components. Formally, the encoded in this way hologram may be written as

$$\Gamma_{m,n}=A_0\exp\left\{i\left[\varphi_{r,s}+(-1)^{m_0}\mathbf{arccos}\left(\left|\Gamma_{r,s}\right|/2A_0\right)\right]\right\}, \tag{13.2.22}$$

where $\{\mathbf{\Gamma}_{r,s}=|\mathbf{\Gamma}_{r,s}|\exp(i\varphi_{r,s})\}$ are samples of the mathematical hologram, indices of the recording medium resolution cells (m,n) are counted as $m=2r+m_0$, $m_0=0,1$, $n=s$ and A_0 is found as half of maximal value of $\{|\mathbf{\Gamma}_{r,s}|\}$.

For hologram encoding with the double-phase method, images are reconstructed in a direction normal to the hologram plane, because the optical path difference of rays passing along this direction through the neighboring resolution cells is zero. This holds, however, only for the central area of the image through which the optical axis of the reconstruction system passes. In the peripheral areas of the image, some phase shift between the rays appears, thus leading to distortions in the peripheral image. This will be discussed in more details in Sect. 13.3.

The method can also be used for recording on amplitude binary media with phase-detour encoding of phases. Two versions of such an implementation of the double-phase method were suggested ([16]). In the first version, two elementary cells of the binary medium are allocated to each of the two component vectors, their phases being coded by a shift of the transparent (or completely reflecting) aperture along a direction perpendicular to the line connecting cell centers as it is illustrated in Fig.13-9, a. This technique exhibits above mentioned distortions due to mutual spatial shift of elementary cells. The second version reduces these distortions by means of decomposing each elementary cell into sub-cells alternating as shown in Fig.12-9, b. With the decomposition into K sub-cells, the relative shift of elementary cells is K times less then without decomposition and the peripheral image distortions are accordingly lower.

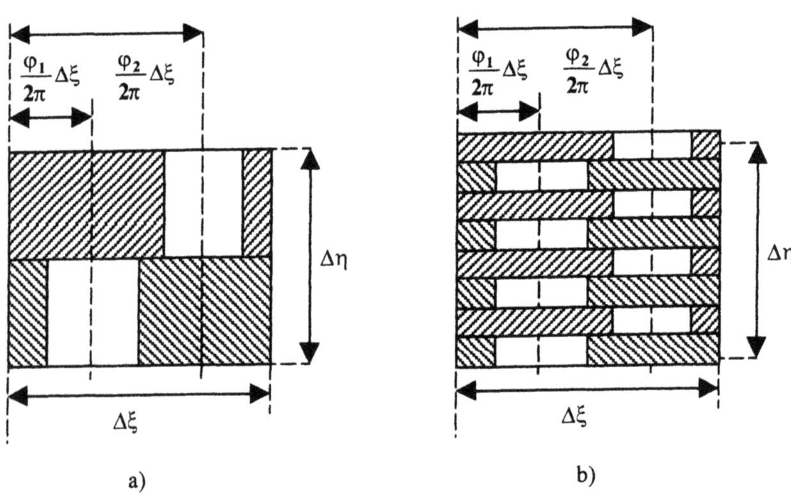

Figure 13-10. Double-phase encoding method for hologram recording on a binary phase medium: two separate cells (a); decomposition of cells into alternating sub-cells

The double-phase method is readily generalized to ***multi-phase encoding*** by vector decomposition into an arbitrary number Q of equal-length vector components:

$$\boldsymbol{\Gamma} = \sum_{q=1}^{Q} A_0 \exp(i\varphi_q). \tag{13.2.23}$$

As Γ is a complex number, Eq.(13.2.23) represents two equations for Q unknown values of $\{\varphi_q\}$. They have a unique solution defined by Eq. 13.2.21 only for $Q = 2$. When $Q > 2$, $\{\varphi_q\}$ may be chosen in a rather arbitrary manner. For example, for odd Q one may choose phases $\{\varphi_q\}$ so as to make of them an arithmetic progression

$$\varphi_{q+1} - \varphi_q = \varphi_q - \varphi_{q-1} = \Delta\varphi . \tag{13.2.24}$$

In this case, the following equations can be derived for the increment $\Delta\varphi$:

$$\frac{\sin(Q\Delta\varphi/2)}{\sin(\Delta\varphi/2)} = \frac{|\Gamma|}{A_0}. \tag{13.2.25}$$

and for phase angles $\{\varphi_q\}$:

$$\varphi_q = \varphi + [q - (Q+1)/2]\Delta\varphi , \quad q = 1,2,\dots,Q. \tag{13.2.26}$$

For odd Q, Eq.(13.2.25) boils down to an algebraic equation of power $(Q-1)/2$ with respect to $\sin^2(\Delta\varphi/2)$. Thus, for $Q = 3$,

$$\Delta\varphi = 2\arcsin\left(\frac{\sqrt{3}}{2}\sqrt{1 - |\Gamma|/3}\right). \tag{13.2.27}$$

For even Q, it is more expedient to separate all the component vectors into two groups having the same phase angles φ_+ and φ_- that are defined by analogy with Eq.(13.2.21) as

$$\varphi_+ = \varphi + \arccos(|\Gamma|/QA_0); \ \varphi_- = \varphi - \arccos(|\Gamma|/QA_0). \tag{13.2.28}$$

With multi-phase encoding, the dynamic range of possible hologram values may be extended because the maximal reproducible amplitude in this case is QA_0. Most interesting of the $Q > 2$ cases are those with $Q = 3$ and $Q = 4$ because the two-dimensional spatial degrees of freedom of the medium and hologram recorder may be used more efficiently through allocation of the component vectors according to Figs.13-8, c) and 13-9, b) and c).

All hologram encoding methods can be treated in a unified way as methods that explicitly or implicitly introduce to recorded holograms some form of a spatial carrier similarly to how optical holograms are recorded. This can be shown by writing Eqs.13.2.5, 13.2.8, 13.2.9, 13.2.19 and 13.2.22 in the following equivalent form explicitly containing hologram samples multiplied by those of the spatial carrier with respect to one or both coordinates:

$$\tilde{\Gamma}_{m,n} = \mathrm{Re}\{\Gamma_{r,s} \exp(-i2\pi m/2)\}/4 + b,$$
$$m = 2r + m_0;\ m_0 = 0,1;\ n = s; \qquad (13.2.5')$$

$$\tilde{\Gamma}_{m,n} = \frac{1}{2}\mathrm{rctf}\{\mathrm{Re}[\Gamma_{r,s} \exp(-i2\pi m/2)]\};$$
$$m = 4r + 2m_{01} + m_{02};\ m_{01}, m_{02} = 0,1;\ n = s; \qquad (13.2.8')$$

$$\tilde{\Gamma}_{m,n} = \frac{1}{2}\mathrm{rctf}\{\mathrm{Re}[\Gamma_{r,s} \exp(-i2\pi (m+2n)/4)]\};$$
$$m = 2r + m_0;\ n = 2s + n_0, m_0, n_0 = 0,1; \qquad (13.2.9')$$

$$\tilde{\Gamma}_{m,n} = \frac{1}{\sqrt{3}}\sum_{p=0}^{1}\left(\mathrm{hlim}\{\mathrm{Re}[\Gamma_{r,s} \exp(-i2\pi (m+p)/3)]\}\right)\times$$
$$\mathrm{rctf}\{\mathrm{Re}[\Gamma_{r,s} \exp(-i2\pi (m+p+1/2)/3)]\};$$
$$m = 3r + m_0; m_0 = 0,1,3;\ n = s; \qquad (13.2.19')$$

$$\tilde{\Gamma}_{m,n} = A_0 \exp\{i[\varphi_{r,s} - \cos(2\pi m/2)\arccos(|\Gamma_{r,s}|/2A_0)]\};$$
$$m = 2r + m_0; m_0 = 0,1;\ n = s, \qquad (13.2.22')$$

where $\mathrm{rctf}(X)$ is the "rectifier" function

$$\mathrm{rctf}(X) = \begin{cases} X, & X \geq 0 \\ 0, & X < 0 \end{cases}, \qquad (13.2.29)$$

and $\mathbf{hlim}(X)$ is the "hard-limiter" function:

$$\mathbf{hlim}(X)=\begin{cases}\mathbf{1}, & X \geq \mathbf{0} \\ \mathbf{0}, & X < \mathbf{0}\end{cases}. \qquad (13.2.30)$$

As one may see from these expressions, the spatial carrier has a period no greater than one half that of hologram sampling. At least two samples of the spatial carrier have to correspond to one hologram sample in order to enable reconstruction of amplitude and phase of each hologram sample through the modulated signal of the spatial carrier. This redundancy implies that, in order to modulate the spatial carrier by a hologram, one or more intermediate samples are required between the basic ones. They may be determined by any of variety of interpolation methods.

The simplest interpolation method, zero order interpolation, simply. repeats samples. It is this interpolation that is implied in the recording methods described above. For instance, according to Eq.(13.2.5') each hologram sample is repeated twice for two samples of the spatial carrier, in Eq.(13.2.8') it is repeated four times for four samples, etc. Of course such an interpolation found in all modifications of the detour phase method yields a very rough approximation of hologram intermediate samples. In Sect.13.3 it will be shown that the distortions of the reconstructed image caused by improper interpolation of hologram samples manifest themselves in aliasing effects.

As we have shown in Ch. 9 aliasing free interpolation can be achieved with discrete sinc-interpolation. Discrete sinc-interpolated values of the desired intermediate samples can be obtained by using, at the hologram synthesis, Shifted DFT with appropriately found shift parameters that correspond to the position of required intermediate samples of the hologram. It should be also noted that the symmetrization method may be regarded as an analog of the method of Eq.(13.2.5) with discrete sinc-interpolation of intermediate samples. In the symmetrization method, such an interpolation is secured automatically and, as will be seen in Sect.13.5, the restored images are free of aliasing images.

Along with methods that introduce the spatial carriers implicitly, explicit introduction of spatial carriers is also possible that directly simulates optical recording of holograms and interferograms.

Of the methods oriented to amplitude-only media, one can mention those of Burch ([17]):

$$\tilde{\Gamma}_{m,n} = \mathbf{1} + \left|\Gamma_{m,n}\right| \mathbf{cos}\left(2\pi \boldsymbol{fm} + \varphi_{m,n}\right) \qquad (13.2.31)$$

and of Huang and Prasada ([18])

$$\tilde{\Gamma}_{m,n} = \left|\Gamma_{m,n}\right|\left[1 + \cos\left(2\pi fm + \varphi_{m,n}\right)\right] \tag{13.2.32}$$

where f is frequency of the spatial carrier.

Of binary-media-oriented methods, one may mention that described in the overview [5]:

$$\tilde{\Gamma}_{m,n} = \mathbf{hlim}\left\{\cos\left[\arcsin\left(\left|\Gamma_{m,n}\right|/A_0\right)\right] + \cos\left(2\pi fm + \varphi_{m,n}\right)\right\}, \tag{13.2.33}$$

where A_0 is maximal value of $\left|\Gamma_{m,n}\right|$.

Of the phase-media-oriented methods, we may mention that of Kirk and Jones ([19]):

$$\tilde{\Gamma}_{m,n} = A_o \exp\left\{i\left[\varphi_{m,n} - h_{m,n}\cos\left(2\pi fm\right)\right]\right\}$$

where $h_{m,n}$ depends in a certain way on $\left|\Gamma_{m,n}\right|$ and on the diffraction order where the reconstructed image should be obtained. In a sense, this method is equivalent to the multi phase encoding method. When $f = 1/2$ and

$$h_{m,n} = \arccos\left(\left|\Gamma_{m,n}\right|/2A_0\right) \tag{13.2.34}$$

it coincides with the double phase encoding method according to Eq.(13.2.22).

13.3 RECONSTRUCTION OF COMPUTER GENERATED HOLOGRAMS

In the reconstruction stage, computer generated holograms synthesized with the use of discrete representations of wave propagation transformations are subjected, to analog optical transformations. The discrepancy between them and their discrete representation affects in a certain way the hologram reconstruction result. Transformations of a digital signal into a physical hologram according to the method of hologram recording its influence as well. In order to illustrate techniques for analysis of the hologram reconstruction that account for these factors, we will consider in this section the reconstruction, in an analog set-up performing the optical Fourier transform, of computer generated Fourier holograms for three methods of hologram encoding: for symmetrization method, for orthogonal encoding method (Eq.(13.2.5)), and for double phase recording on a phase medium (Eq.(13.2.22)).

The following characteristics of the hologram recording device affect the reconstruction result: type of the discretization raster, discretization intervals, recording aperture and physical size of recorded holograms. Let (ξ,η) be physical coordinates on the hologram recording medium, $(\Delta\xi,\Delta\eta)$ be discretization intervals of the rectangular sampling grid along coordinates (ξ,η), (ξ_0,η_0) be shift parameters that depend on the geometry of positioning the hologram in the reconstruction set up, $\boldsymbol{h}_{rec}(\xi,\eta)$ be recording aperture function and $w(\xi,\eta)$ be a window function that defines physical size of the recorded hologram: $w(\xi,\eta)\geq 0$, when (ξ,η) belongs to the hologram area and $w(\xi,\eta)=0$, otherwise (see Fig. 13-11).

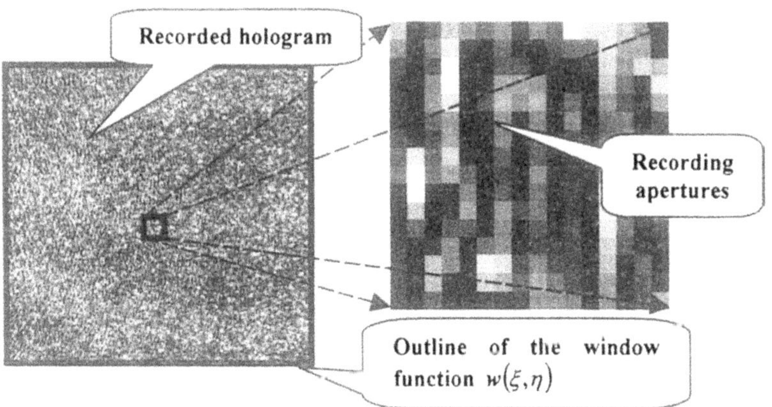

Figure 13-11. Definitions related to recorded physical computer generated holograms

For the symmetrization method, the following equation describes conversion of the numerical matrix of the mathematical hologram $\{\Gamma_{r,s}\}$ into the physical hologram $\tilde{\Gamma}(\xi,\eta)$ recorded on an amplitude-only medium:

$$\tilde{\Gamma}(\xi,\eta) = w(\xi,\eta)\sum_r \sum_s (\Gamma_{r,s} + b) h_{rec}(\xi + \xi - r\Delta\xi_0, \eta + \eta_0 - s\Delta\eta), \quad (13.3.1)$$

where b is a constant bias required for eliminating negative values in recording $\{\Gamma_{r,s}\}$ and indices (r,s) run over all available hologram samples. Eq. (13.3.1) may be rewritten in a form of the convolution:

$$\tilde{\Gamma}(\xi,\eta) = w(\xi,\eta)\cdot\Big\{h_{rec}(\xi,\eta)\otimes \\ \sum_{r=-\infty}^{\infty}\sum_{s=\infty}^{\infty}(\Gamma_{r,s} + b)\delta(\xi + \xi_0 - r\Delta\xi)\delta(\eta + \eta_0 - s\Delta\eta)\Big\}, \quad (13.3.2)$$

where $\otimes$ stands for the convolution and summation limits are changed to $[-\infty,\infty]$ having in mind that hologram window function $w(\xi,\eta)$ selects available hologram samples from virtual infinite number of samples.

Let now (x,y) be coordinates in reconstructed image plane situated at a distance D from the hologram plane and λ be wavelength of the reconstructing light beam. According to the convolution theorem of the integral Fourier transform, the result of the optical Fourier transform performed at the reconstruction of the hologram $\tilde{\Gamma}(\xi,\eta)$ can then be represented as:

$$A_{rec}(x,y) = \int_{-\infty}^{\infty}\int_{-\infty}^{\infty}\Gamma(\xi,\eta)\exp[i2\pi(x\xi + y\eta)/\lambda D]d\xi d\eta =$$

$$W(x,y)\otimes\Big\{H_{rec}(x,y)\int_{-\infty}^{\infty}\int_{-\infty}^{\infty}\sum_{r=-\infty}^{\infty}\sum_{s=-\infty}^{\infty}(\Gamma_{r,s} + b)\delta(\xi + \xi_0 - r\Delta\xi)\delta(\eta + \eta_0 - s\Delta\eta)\times$$

$$\exp[i2\pi(x\xi + y\eta)/\lambda D]d\xi d\eta\Big\} =$$

$$W(x,y)\otimes\{H_{rec}(x,y)\exp[-i2\pi(x\xi_0 + y\eta_0)/\lambda D]\times$$

$$\sum_r \sum_s (\Gamma_{r,s} + b)\exp[-i2\pi(r\Delta\xi x + s\Delta\eta y)/\lambda D], \quad (13.3.3)$$

where

$$W(x,y)=\int_{-\infty}^{\infty}\int_{-\infty}^{\infty} w(\xi,\eta)\exp[i2\pi(x\xi+y\eta)/\lambda D]d\xi d\eta \tag{13.3.4}$$

and

$$H_{rec}(x,y)=\int_{-\infty}^{\infty}\int_{-\infty}^{\infty} h_{rec}(\xi,\eta)\exp[i2\pi(x\xi+y\eta)/\lambda D]d\xi d\eta \tag{13.3.5}$$

are Fourier transforms of the hologram window function and of the aperture function of the hologram recording device.

Introduce now an array $\overline{A}_{k,l}^{(o)}$, $k=0,...,2N_1-1$, $l==0,1,...,N_2-1$, of samples of the object wave front. In the symmetrization method, the array is symmetrical as defined by (Eq. 13.2.1). For Fourier holograms, mathematical hologram $\Gamma_{r,s}$ is computed as Shifted DFT of this array:

$$\Gamma_{r,s}\propto\sum_{k=0}^{2N_1-1}\sum_{l=0}^{N_2-1}\overline{A}_{k,l}^{(o)}\exp\left\{i2\pi\left[\frac{(k+u)(r+p)}{2N_1}+\frac{(l+v)(s+q)}{N_2}\right]\right\}, \tag{13.3.6}$$

where (u,v) and (p,q) are shift parameters. Substitute Eq. 13.3.6 into Eq. 13.3.3 and obtain:

$$A_{rec}(x,y)\propto W(x,y)\otimes\{H_{rec}(x,y)\exp[-i2\pi(x\xi_0+y\eta_0)/\lambda D]\times$$

$$\left\{\sum_{r=-\infty}^{\infty}\sum_{s=-\infty}^{\infty}\sum_{k=0}^{2N_1-1}\sum_{l=0}^{N_2-1}\overline{A}_{k,l}^{(o)}\exp\left\{i2\pi\left[\frac{(k+u)(r+p)}{2N_1}+\frac{(l+v)(s+q)}{N_2}\right]\right\}+b\right\}\times$$

$$\exp[-i2\pi(r\Delta\xi x+s\Delta\eta y)/\lambda D], \tag{13.3.7}$$

or, using Poisson's summation formula (Eq. 3.3.16),

$$A_{rec}(x,y)\propto W(x,y)\otimes\{H_{rec}(x,y)\exp[-i2\pi(x\xi_0+y\eta_0)/\lambda D]\times$$

$$\left\{\sum_{m=-\infty}^{\infty}\sum_{n=-\infty}^{\infty}\sum_{k=0}^{2N_1-1}\sum_{l=0}^{N_2-1}\overline{A}_{k,l}^{(o)}\exp\left\{i2\pi\left[\frac{(k+u)p}{2N_1}+\frac{(l+v)q}{N_2}\right]\right\}\times\right.$$

$$\delta\left(\frac{k+u}{2N_1}-\frac{\Delta\xi x}{\lambda D}+m\right)\delta\left(\frac{l+v}{N_2}-\frac{\Delta\eta y}{\lambda D}+n\right)\Bigg\}+$$

$$b\sum_{m=-\infty}^{\infty}\sum_{n=-\infty}^{\infty}\delta\left(\frac{k+u}{2N_1}-\frac{\Delta\xi x}{\lambda D}+m\right)\delta\left(\frac{l+v}{N_2}-\frac{\Delta\eta y}{\lambda D}+n\right). \tag{13.3.8}$$

As may be seen from Eq.(13.3.8), at synthesis and recording of Fourier holograms one can take $p=q=0$. One can also neglect the phase term $\exp[-i2\pi(x\xi_0+y\eta_0)/\lambda D]$ that is irrelevant for display holograms. With these settings, obtain finally:

$$A_{rec}(x,y)\propto$$

$$H_{rec}(x,y)\cdot\left\{\sum_{m=-\infty}^{\infty}\sum_{n=-\infty}^{\infty}\sum_{k=0}^{2N_1-1}\sum_{l=0}^{N_2-1}\overline{A}_{k,l}^{(o)}W\left(\frac{k+u}{2N_1}-\frac{\Delta\xi x}{\lambda D}+m,\frac{l+v}{N_2}-\frac{\Delta\eta y}{\lambda D}+n\right)\right.$$

$$\left.+b\sum_{m=-\infty}^{\infty}\sum_{n=-\infty}^{\infty}W\left(\frac{k+u}{2N_1}-\frac{\Delta\xi\, x}{\lambda D}+m,\frac{l+v}{N_2}-\frac{\Delta\eta\, y}{\lambda D}+n\right)\right\}. \tag{13.3.9}$$

It follows from this formula that

- the hologram placed into the optical Fourier system reconstructs the original object wave front in several diffraction orders defined by indices m and n;
- the object wave front is reconstructed by interpolation of its samples with the interpolation function equal to Fourier transform of the hologram window function;
- the pattern of the reconstructed images samples in the diffraction orders is masked by a function $H_{rec}(x,y)$, spatial frequency response of the hologram recording device.
- constant bias in hologram recording results in appearance in the reconstructed image of bright spots in the center of each diffraction order (last term in Eq. 13.3.9)

Arrangement of diffraction orders in the reconstructed image plane under shift parameters $u=-N_1, v=-N_2/2$ is shown in Fig. 13-12.

In case of orthogonal encoding according to Eq.(13.2.5), one obtains, in the same denotations that were adopted above:

$$\widetilde{\Gamma}(\xi,\eta)=\frac{1}{2}w(\xi,\eta)\left\{\sum_{r=-\infty}^{\infty}\sum_{s=-\infty}^{\infty}\sum_{m_0=0}^{1}\left\{(-1)^r(i)^{m_0}\left[(-1)^{m_0}\Gamma_{r,s}+\Gamma_{r,s}^{*}\right]+b\right\}\times\right.$$

$$h_{rec}(\xi+\xi_0-r\Delta\xi,\eta+\eta_0-s\Delta\eta)\}=$$

$$w(\xi,\eta)\left\{\sum_{r=-\infty}^{\infty}\sum_{s=-\infty}^{\infty}\sum_{m_0=0}^{1}\left\{\mathbf{Re}\left[(-1)^r(-i)^{m_0}\Gamma_{r,s}\right]+b\right\}\times\right.$$

$$h_{rec}(\xi+\xi_0-r\Delta\xi,\eta+\eta_0-s\Delta\eta)\}=$$

$$w(\xi,\eta)\left\{h_{rec}(\xi,\eta)\otimes\sum_{r=-\infty}^{\infty}\sum_{s=-\infty}^{\infty}\sum_{m_0=0}^{1}\left\{\mathbf{Re}\left[(-1)^r(-i)^{m_0}\Gamma_{r,s}\right]+b\right\}\times\right.$$

$$\delta(\xi+\xi_0-r\Delta\xi)\delta(,\eta+\eta_0-s\Delta\eta)\}. \qquad (13.3.10)$$

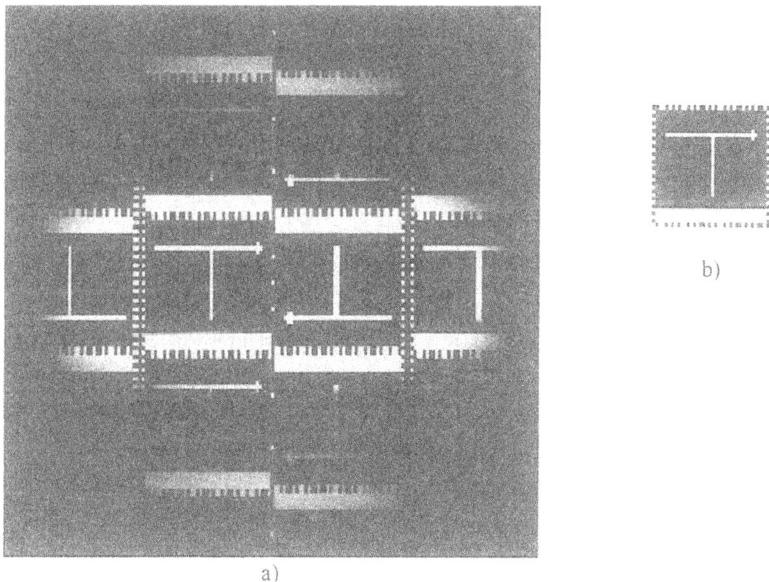

Figure 13-12. Reconstruction of a hologram synthesized using image symmetrization by duplication (a) for a test object shown in b).

Following the reasoning and settings used in deriving Eq.(13.3.8), one can obtain that, in the Fourier transform reconstruction scheme, such a hologram reconstructs the following wave front:

$$A_{rec}(x,y) \propto H_{rec}(x,y)\left\{\cos\left[\pi\left(\frac{1}{4}+\frac{\Delta\xi\, x}{\lambda D}\right)\right]\times\right.$$

$$\sum_{m=-\infty}^{\infty}\sum_{n=-\infty}^{\infty}\sum_{k=0}^{N_1}\sum_{l=0}^{N_2}\overline{A}_{k,l}^{(o)} W\left(\frac{k+u}{N_1}-\frac{\Delta\xi\, x}{\lambda D}+m+\frac{1}{2},\frac{l+v}{N_2}-\frac{\Delta\eta\, y}{\lambda D}+n\right)+$$

$$\sin\left[\pi\left(\frac{1}{4}-\frac{\Delta\xi\, x}{\lambda D}\right)\right]\times$$

$$\sum_{m=-\infty}^{\infty}\sum_{n=-\infty}^{\infty}\sum_{k=0}^{N_1}\sum_{l=0}^{N_2}\overline{A}_{N_1-k,N_2-l}^{(o)\,*} W\left(\frac{k+u}{N_1}-\frac{\Delta\xi\, x}{\lambda D}+m+\frac{1}{2},\frac{l+v}{N_2}-\frac{\Delta\eta\, y}{\lambda D}+n\right)+$$

$$\left. b\cos\left(\pi\frac{\Delta\xi\, x}{\lambda D}\right)\sum_{m=-\infty}^{\infty}\sum_{n=-\infty}^{\infty} W\left(\frac{k+u}{N_1}+\frac{\Delta\xi\, x}{\lambda D}+m+\frac{1}{2},\frac{l+v}{N_2}+\frac{\Delta\eta\, y}{\lambda D}+n\right)\right\},$$

(13.3.11)

where, as above, (u,v) and (p,q) are shift parameters of SDFT used in computing the mathematical hologram, N_1 and N_2 are dimensions of the array $\{\overline{A}_{k,l}^{(o)}\}$. As in the above case of hologram encoding by symmetrization, one should take $p=q=0$ and $\xi_0=\eta_0=0$ at hologram synthesis and recording.

One can see from Eq. (13.3.11) that, similarly to the above case, the reconstructed image contains a number of diffraction orders, masked by the function $H_{rec}(x,y)$, spatial frequency response of the hologram recording device, and a central spot in several diffraction orders due to the constant bias in the hologram. But here, in contrast to the above case, each diffraction order contains two superimposed images of the object, - a direct image and its conjugate rotated by 180° with respect the former. Each of them is additionally masked by functions $\left\{\cos\left[\pi\left(\frac{1}{4}+\frac{\Delta\xi\, x}{\lambda D}\right)\right]\right.$ and $\left\{\sin\left[\pi\left(\frac{1}{4}-\frac{\Delta\xi\, x}{\lambda D}\right)\right]\right.$, respectively. Therefore, in the central region of the direct image the conjugate image is suppressed, but at its periphery the aliasing conjugate image has an intensity comparable with that of the direct image. This is accounted for by the fact, that here additional hologram samples necessary for representing the spatial carrier are obtained by zero order interpolation of samples of the mathematical hologram.

The pattern of diffraction orders of the direct and conjugate images and their corresponding masking functions are shown in Fig.11-13 for $u = -N_1/2$ and $v = -N_2/2$. One may easily see the direct image and its conjugate, an aliasing image whose contrast increases along the horizontal axis from the center to the periphery.

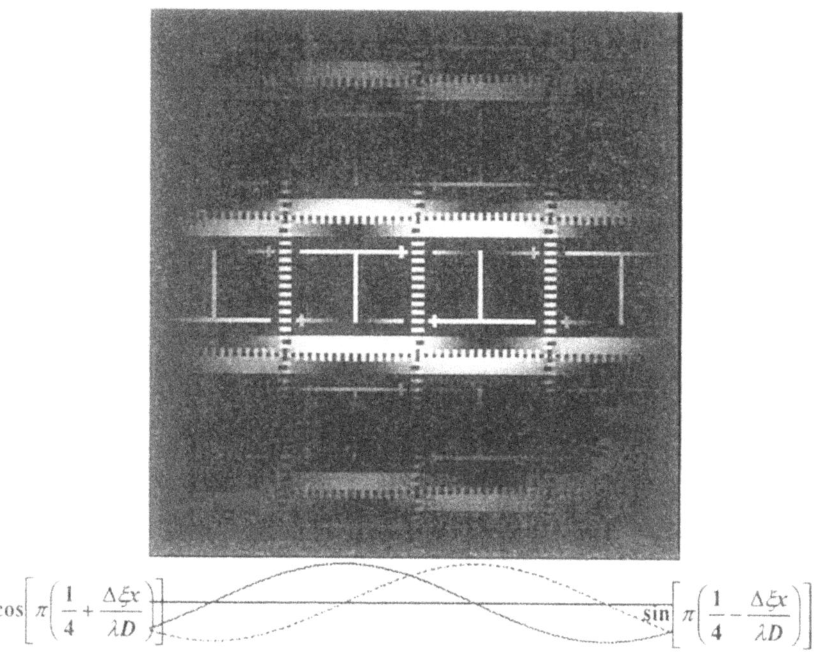

Figure 13-13. Image reconstruction for the orthogonal encoding method. At the bottom, spatial weighting functions of the direct and conjugate images are depicted.

The phase recording of the hologram on phase-only media according Eq.(13.2.22) is described, in the same denotations, as

$$\tilde{\Gamma}(\xi,\eta) = w(\xi,\eta)\left\{\sum_{r=-\infty}^{\infty}\sum_{s=-\infty}^{\infty}\sum_{m_0=0}^{1}\exp\left\{i\left[\varphi_{r,s} + (-1)^{m_0}\arccos\left(\left|\Gamma_{r,s}\right|/2A_0\right)\right]\right\}\times\right.$$

$$h_{rec}\left[\xi+\xi_0-(2r+m_0)\Delta\xi, \eta+\eta_0-s\Delta\eta\right] =$$

$$w(\xi,\eta)\left\{h_{rec}(\xi,\eta)\otimes\sum_{r=-\infty}^{\infty}\sum_{s=-\infty}^{\infty}\sum_{m_0=0}^{1}\exp\left\{i\left[\varphi_{r,s} + (-1)^{m_0}\arccos\left(\left|\Gamma_{r,s}\right|/2A_0\right)\right]\right\}\times\right.$$

$$\delta\left[\xi+\xi_0-(2r+m_0)\Delta\xi\right]\delta(\eta+\eta_0-s\Delta\eta). \tag{13.3.12}$$

In a Fourier hologram reconstruction setup, this hologram is Fourier transformed and reconstructs the wave front described as

$$A_{rec}(x,y)=\int_{-\infty}^{\infty}\int_{-\infty}^{\infty}\widetilde{\Gamma}(\xi,\eta)\exp[-i2\pi(x\xi+y\eta)/\lambda D]d\xi d\eta =$$

$$W(x,y)\otimes\left[H_{rec}(x,y)\left(\sum_{s=-\infty}^{\infty}\sum_{r=-\infty}^{\infty}\sum_{m_0=0}^{1}\exp\{i[\varphi_{r,s}+(-1)^{m_0}\arccos(|\Gamma_{r,s}|/2A_0)]\}\times\right.\right.$$

$$\left.\exp(-i2\pi\{[(2r+m_0)\Delta\xi-\xi_0]x+(s\Delta\eta-\eta_0)y\}/\lambda D)\right]=$$

$$W(x,y)\otimes\left\{2H_{rec}(x,y)\exp(-i2\pi\{[(\xi_0-\Delta\xi/2)x+\eta_0 y]\}/\lambda D)\times\right.$$

$$\left[\cos\left(\pi\frac{\Delta\xi\, x}{\lambda D}\right)\sum_{r=-\infty}^{\infty}\sum_{s=-\infty}^{\infty}(|\Gamma_{r,s}|/2A_0)\exp(i\varphi_{r,s})\exp\left[-i2\pi\frac{(2r\Delta\xi x+s\Delta\eta y)}{\lambda D}\right]+\right.$$

$$\left.\left.\sin\left(\pi\frac{\Delta\xi\, x}{\lambda D}\right)\sum_{r=-\infty}^{\infty}\sum_{s=-\infty}^{\infty}\sqrt{1-\frac{|\Gamma_{r,s}|^2}{4A_0^2}}\exp(i\varphi_{r,s})\exp\left[-i2\pi\frac{(2r\Delta\xi x+s\Delta\eta y)}{\lambda D}\right]\right]\right\}. \tag{13.3.13}$$

By substituting $\Gamma_{r,s}$ with its expression through SDFT($u,v;p,q$):

$$|\Gamma_{r,s}|\exp(i\varphi_{r,s})=\sum_{k=0}^{N_1-1}\sum_{l=0}^{N_2-1}\overline{A}_{k,l}^{(o)}\exp\left\{i2\pi\left[\frac{(k+u)(r+p)}{N_1}+\frac{(l+v)(s+q)}{N_2}\right]\right\} \tag{13.3.14}$$

and introducing an auxiliary function $\widetilde{A}_{k,l}^{(0)}$ defined through equation

$$\sqrt{4A_0^2-\left|\Gamma_{r,s}\right|^2}\exp\left(i\varphi_{r,s}\right)=$$
$$\sum_{k=0}^{N_1-1}\sum_{l=0}^{N_2-1}\widetilde{A}_{k,l}^{(o)}\exp\left\{i2\pi\left[\frac{(k+u)(r+p)}{N_1}+\frac{(l+v)(s+q)}{N_2}\right]\right\} \tag{13.3.15}$$

obtain after some transformations and setting $p=q=0$:

$$A_{rec}(x,y)\propto H_{rec}(x,y)\times$$

$$\left\{\cos\left(\pi\frac{\Delta\xi\, x}{\lambda D}\right)\sum_{m=-\infty}^{\infty}\sum_{n=-\infty}^{\infty}\sum_{k=0}^{N_1}\sum_{l=0}^{N_2}\overline{A}_{k,l}^{(o)}W\left(\frac{k+u}{N_1}-\frac{2\Delta\xi\, x}{\lambda D}+m,\frac{l+v}{N_2}-\frac{\Delta\eta\, y}{\lambda D}+n\right)\times\right.$$

$$\left.\sin\left(\pi\frac{\Delta\xi\, x}{\lambda D}\right)\sum_{m=-\infty}^{\infty}\sum_{n=-\infty}^{\infty}\sum_{k=0}^{N_1}\sum_{l=0}^{N_2}\widetilde{A}_{k,l}^{(o)}W\left(\frac{k+u}{N_1}-\frac{2\Delta\xi\, x}{\lambda D}+m,\frac{l+v}{N_2}-\frac{\Delta\eta\, y}{\lambda D}+n\right)\right\}$$

(13.3.16)

The result of reconstruction of a hologram recorded on a phase medium by the two phase method is, thus, similar to the reconstruction of an hologram with orthogonal encoding (see Eq. 13.3.11). Image is also reconstructed in a number of diffraction orders masked by the function $H_{rec}(x,y)$. There is also superposition of the aliasing image described by the function $\widetilde{A}_{k,l}^{(o)}$ over the original image $\overline{A}_{k,l}^{(o)}$ and the original and aliasing images are additionally masked by the functions $\cos\left(\pi\frac{\Delta\xi\, x}{\lambda D}\right)$ and $\sin\left(\pi\frac{\Delta\xi\, x}{\lambda D}\right)$, respectively. In the center of the proper image aliasing image is fully attenuated but over the peripheral area it may be of the same intensity as the proper image. In contrast to the orthogonal coding method, the aliasing image here is not conjugate to the original one, but is similar to it, in a sense, because, according to Eq. 13.3.15, it has the same phase spectrum and a distorted amplitude spectrum. Unlike the holograms coded via the symmetrization or orthogonal methods, double- phase encoding does not produce a central spot in the diffraction orders of the reconstructed image because the hologram is recorded in the phase medium without an amplitude bias. Fig.11-14 shows the pattern of diffraction orders in the reconstructed image plane for this case with $\boldsymbol{u} = -\boldsymbol{N}_1/\mathbf{2}$ and $\boldsymbol{v} = -\boldsymbol{N}_2/\mathbf{2}$.

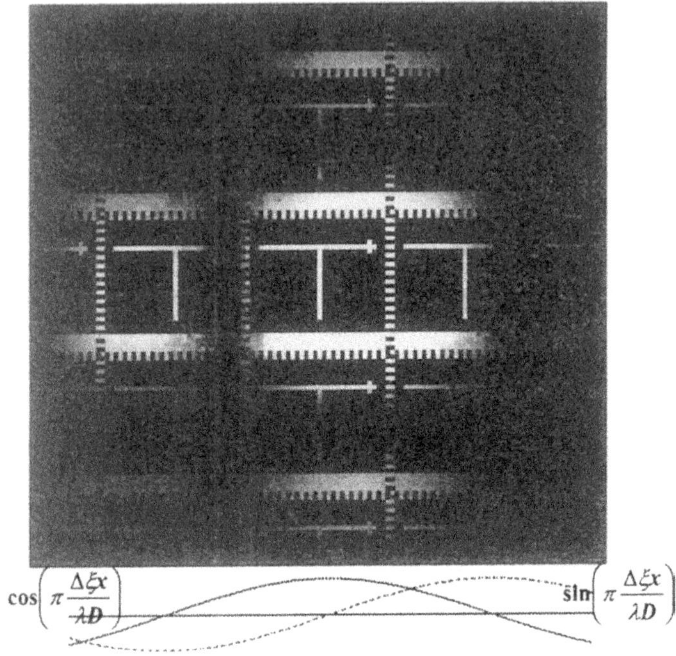

Figure 13-14. Hologram reconstruction for double phase recording method: a) -arrangement of diffraction orders (Solid and dashed arrows indicate the basic and aliasing images with diffraction order shown in boxes; spatial masking functions for both images are depicted at bottom correspondingly by solid and dashed lines); b) - reconstructed image for the test object of Fig. 13-12, c).

References

1. L. Yaroslavskii, N. Merzlyakov, Methods of Digital Holography, Consultance Bureau, N.Y., 1980
2. H. I. Bjelkhagen, Silver Halid Recording Materials, Springer Verlag, Heidelberg, 1993
3. R.S. Nesbitt, S. L. Smith, R.A. Molnar, S. A. Benton, Holographic recording using a digital micromirror device, Conf. on Practical Holography, Proc. SPIE, 3637, 12-20, 1999
4. T. Kreis, P. Aswendt, R. Hoeffling, Holographic reconstruction using a digital micromirror device, Optical Engineering, v. 40 (6), pp. 926-933, June 2001
5. W.H. Lee, Computer generated holograms: techniques and applications, In: Progress in Optics, E. Wolf, ed., v. XVI, North Holland, 1978, pp. 121-231
6. W. I. Dallas, Computer generated holograms, in: *The Computer in Optical Research*, Methods and Applications, B. R. Friede, Ed., Topics in Applied Physics, v. 41, Springer, Berlin, 1980, pp. 297-367
7. O. Bryngdahl, F. Wyrowski, Digital holography - computer-generated holograms, in: Progress in Optics, E. Wolf., ed.,v. XXVIII,Elsevier Science Publishers B.V., 1990, pp. 3-87
8. L. B. Lesem, P. M. Hirsch, J. A. Jordan, Kinoform, IBM Journ. Res. Dev., 13, 150, 1969
9. L. B. Lesem, P. M. Hirsch, J. A. Jordan, Computer synthesis of holograms for 3-D display, Commun. Of Associat. For Computing Machinery, v. 11, 1968, pp. 661-674
10. B. R. Brown, A. Lohmann, Complex spatial filtering woth binary masks, *Appl. Optics*, **5**, No. 6, 967-969 (1966)
11. B. R. Brown, A. Lohmann, Computer generated binary holograms, IMB J. Res. Dev., v. 13, No. 2, 160-168 (1969)
12. H. P. Herzig, Ed., Micro-optics - Elements, Systems, and Applications, Taylor&Francis, London, 1997
13. C. B. Burckhardt, A simplification of Lee's method of generating holograms by computer, Appl. Opt., v. 9, No.8, p. 1949 (1970)
14. P. Chavel, J. P. Hugonin, High quality computer generated holograms: the problem of phase representation, JOSA, v. 66, p. 989 (1976)
15. J. W. Goodman, D. C. Chu, I. R. Fienup, Recent developments in computer holograms, Proc. SPIE, v. 41, pp. 155-159, 1974
16. C. K. Hsueh, A. A. Sawchuk, Computer generated double phase holograms,, Appl. Opt., v. 17, No. 24, pp. 3874-3883, (1978)
17. J. J. Burch, Algorithms for computer synthesis of spatial filters, Proc. IEEE, v. 55, 999,1967
18. T. S. Huang, E. Prasada, Considerations on the generation and processing of holograms by digital computers, MIT/RLE Quat. Progr. Rept., 1966, v. 81
19. J. P. Kirk, A. L. Jones, Phase-only complex valued spatial filter, JOSA, v. 61, No. 8, pp. 1023-1028, 1971

INDEX

GPSR Compliance
The European Union's (EU) General Product Safety Regulation (GPSR) is a set of rules that requires consumer products to be safe and our obligations to ensure this.

If you have any concerns about our products, you can contact us on

ProductSafety@springernature.com

In case Publisher is established outside the EU, the EU authorized representative is:

Springer Nature Customer Service Center GmbH
Europaplatz 3
69115 Heidelberg, Germany

www.ingramcontent.com/pod-product-compliance
Ingram Content Group UK Ltd.
Pitfield, Milton Keynes, MK11 3LW, UK
UKHW022319190726
13856UKWH00001B/92

* 9 7 8 1 4 7 5 7 4 9 8 9 2 *